OLFACTION, TASTE, AND COGNITION

The human organs of perception are continually being bombarded with chemicals from the environment. Our bodies have in turn developed complex processing systems that manifest themselves in our emotions, memory, and language. Yet the available data on the high-order cognitive implications of taste and smell are scattered among journals in many fields, with no single source synthesizing the large body of knowledge, much of which has appeared in the past decade.

This book presents the first multidisciplinary synthesis of the literature in olfactory and gustatory cognition. The book is conveniently divided into sections, including linguistic representations, emotion, memory, neural bases, and individual variation. Leading experts have written chapters on many facets of taste and smell, including odor memory, cortical representations, psychophysics and functional imaging studies, genetic variation in taste, and the hedonistic dimensions of odors. The approach is integrative, combining perspectives from neuroscience, psychology, anthropology, philosophy, and linguistics, and is appropriate for students and researchers in all these areas who seek the authoritative reference on olfaction, taste, and cognition.

Catherine Rouby is an associate professor of neuroscience at Université Claude Bernard.

Benoist Schaal is a research director at the CNRS, Centre Européen des Sciences du Goût.

Danièle Dubois is a research director at the CNRS, Institut National de la Langue Française.

Rémi Gervais is a research director at the CNRS, Institut des Sciences Cognitives, Lyon.

A. Holley is a professor of neuroscience at Université Claude Bernard and director of the Centre Européen des Sciences du Goût.

OLFACTION, TASTE, AND COGNITION

Edited by

CATHERINE ROUBY
Université Claude Bernard, Lyon

BENOIST SCHAAL
Centre National de la Recherche Scientifique

DANIÈLE DUBOIS
Centre National de la Recherche Scientifique

RÉMI GERVAIS
Centre National de la Recherche Scientifique

A. HOLLEY
Centre Européen des Sciences du Goût

CAMBRIDGE
UNIVERSITY PRESS

CAMBRIDGE UNIVERSITY PRESS
Cambridge, New York, Melbourne, Madrid, Cape Town, Singapore, São Paulo

Cambridge University Press
The Edinburgh Building, Cambridge CB2 2RU, UK

Published in the United States of America by Cambridge University Press, New York

www.cambridge.org
Information on this title: www.cambridge.org/9780521790581

© Cambridge University Press 2002

First published 2002
This digitally printed first paperback version 2005

A catalogue record for this publication is available from the British Library

Library of Congress Cataloguing in Publication data

Olfaction, taste, and cognition / edited by Catherine Rouby ... [et al.]
    p. ; cm.
Includes bibliographical references and index.
ISBN 0-521-79058-1 (hardback)
1. Taste.   2. Smell.   3. Cognition.   I. Rouby, Catherine, 1944–
[DNLM: 1. Smell.   2. Cognition.   3. Taste. WV 301 O454 2002]
QP456 .O445   2002
612.8´6–dc21   2001043456

ISBN-13 978-0-521-79058-1 hardback
ISBN-10 0-521-79058-1 hardback

ISBN-13 978-0-521-02097-8 paperback
ISBN-10 0-521-02097-2 paperback

Contents

Contributors

Baeyens, Frank, Department of Psychology, University of Leuven, Tiensestraat 102, 3000 Leuven, Belgium, Tel: +32-(0)16/.32.59.63, Fax: +32-(0)16/.32.59.24, e-mail: Frank.Baeyens@psy.kuleuven.ac.be

Bartoshuk, Linda M., Department of Surgery (Otolaryngology), Yale University School of Medicine, 333 Cedar Street, New Haven, CT 06520-8041, USA, Tel: +1-203-785-2587, Fax: +1-203-785-3290, e-mail: linda_bartoshuk@quickmail.yale.edu

Bensafi, Moustafa, Laboratoire de Neurosciences et Systèmes Sensoriels, Université Claude Bernard, Lyon 1, 50 avenue Tony Garnier, 69366 Lyon, Cedex 07, France; CNRS, 69622 Villeurbanne, France, Tel: +33 4 37 28 74 97, Fax: +33 4 37 28 76 01, e-mail: bensafi@olfac.univ-lyon1.fr

Boireau, Nathalie, Laboratoire de Neurobiologie Sensorielle, École Pratique des Hautes Études, rue des Olympiades, Massy, France, and Laboratoire de Physiologie de la Manducation, Université Paris VII, Paris, France, Tel: +1 69 20 63 39, Fax: +1 60 11 61 94, e-mail: boireau@ccr.jussieu.fr

Cerf, Barbara, Laboratoire de Neurobiologie Sensorielle, École Pratique des Hautes Études, rue des Olympiades, Massy, France, and Laboratoire de Physiologie de la Manducation, Université Paris VII, Paris, France, Tel: +1 69 20 63 39, Fax: +1 60 11 61 94, e-mail: cerf@ccr.jussieu.fr

Chabaud, Pascal, Institut des Sciences Cognitives, CNRS/Université Claude Bernard, Lyon 1, 69675, Bron, France, Tel: +33 4 37 91 12 33, Fax: +33 4 37 91 12 10, e-mail: chabaud@isc.cnrs.fr

Chastrette, Maurice, Laboratoire de Neurosciences et Systèmes Sensoriels, Université Claude Bernard, Lyon 1, 50 avenue Tony Garnier, 69366 Lyon, Cedex

07, France; CNRS, 69622 Villeurbanne, France, Tel: +33 4 37 28 74 91, Fax: +33 4 37 28 76 01, e-mail: chastret@olfac.univ-lyon1.fr

David, Sophie, Modèles, Dynamiques, Corpus, CNRS, Université Paris X – Nanterre, Bâtiment L, 200 avenue de la République, 92001 Nanterre Cedex, Tel: 01 40 97 47 33, e-mail: sophie.david@u-paris10.fr

Degel, Joachim, Allmendring 12, 75203 Königsbach-Stein, Germany, Tel: 0171-31 98 275, Fax: 0171-31 42 968, e-mail: jde@itm-research.de

Distel, Hans, Institut für Medizinische Psychologie, Ludwig-Maximilians-Universität München, Goethestrasse 31, D-80336 München, Germany, Tel: +49-89-5996-240, Fax: +49-89-5996-615, e-mail: hdistel@imp.med.uni-muenchen.de

Dubois, Danièle, Laboratoire Languages, Cognitions, Pratiques, Ergonomie, CNRS, Institut National de la Langue Française, 44, rue de l'Amiral Mouchez, 75014 Paris, France, Tel: +33 1 43 13 56 50, Fax: +33 1 43 13 56 60, e-mail: danièle.dubois@inalf.cnrs.fr

Duffy, Valerie B., Department of Surgery (Otolaryngology), Yale University School of Medicine, 333 Cedar Street, New Haven, CT 06520-8041, USA, and School of Allied Health Sciences, University of Connecticut, Tel: +1-860-486-1997, Fax: +1-203-785-3290, e-mail: vduffy@uconnvm.uconn.edu

Fast, Katharine, Department of Surgery (Otolaryngology), Yale University School of Medicine, 333 Cedar Street, New Haven, CT 06520-8041, USA, Tel: +1-510-845-3668, Fax: +1-203-785-3290, e-mail: kf37@pantheon.yale.edu

Faurion, Annick, Laboratoire de Neurobiologie Sensorielle, École Pratique des Hautes Études, rue des Olympiades, Massy, France, and Laboratoire de Physiologie de la Manducation, Université Paris VII, Paris, France, Tel: +1 69 20 63 39, Fax: +1 60 11 61 94, e-mail: faurion@ccr.jussieu.fr

Faxbrink, Maria, Department of Psychology, Uppsala University, Box 1225, 751 45 Uppsala, Sweden, Tel: +46 18 18 21 50, Fax: +46 18 471 2123, e-mail: mariafaxbrink@hotmail.com

Gervais, Rémi, Institut des Sciences Cognitives, CNRS/Université Claude Bernard, Lyon 1, 69675, Bron, France, Tel: +33 4 37 91 12 33, Fax: +33 4 37 91 12 10, e-mail: gervais@isc.cnrs.fr

Hayreh, Davinder J. S., Department of Psychology, University of Chicago, 5730 Woodlawn Avenue, Chicago, IL 60637, USA, Tel: +1(773) 702-6016, Fax: +1(773) 702-0320, e-mail: djhayreh@midway.uchicago.edu

Heilmann, Stefan, Department of Otorhinolaryngology, University of Dresden Medical School, Fetscherstrasse 74, 01307 Dresden, Germany, e-mail: stefan-heilmann@hotmail.com

Heining, Maike, Ph.D. student, Department of Psychology, Institute of Psychiatry, DeCrespigny Park, London, SE5 8AF, U.K., Tel: +44 020 7848 0365, Fax: +44 020 7848 0379, e-mail: m.heining@iop.kcl.ac.uk

Hermans, Dirk, Department of Psychology, University of Leuven, Tiensestraat 102, 3000 Leuven, Belgium, Tel: +32-(0)16/.32.59.63, Fax: +32-(0)16/.32.59.24, e-mail: Dirk.Hermans@psy.kuleuven.ac.be

Herz, Rachel S., Department of Psychology, Box 1853, Brown University, Providence, RI 02912, USA, Tel: +1-401-863-9576, Fax: +401-863-1300, e-mail: Rachel_Herz@Brown.edu

Holley, André, Laboratoire de Neurosciences et Systèmes Sensoriels, Université Claude Bernard, Lyon 1/CNRS, 69366 Lyon, France; Centre Européen des Sciences du Goût, CNRS, and Université de Bourgogne, F-21000 Dijon, France, Tel: +33 3 80 68 16 20, Fax: +33 3 80 68 16 21, e-mail: holley@cesg. cnrs.fr

Howes, David, Department of Sociology and Anthropology, Concordia University, 1455 de Maisonneuve Ouest, Montréal, Québec, Canada H3G 1M8, Tel: +1(514) 848-2148, Fax: +1(514) 848-4539, e-mail: howesd@vax2.concordia.ca

Hudson, Robyn, Instituto de Investigaciones Biomédicas, Universidad Nacional Autónoma de México, Apartado Postal 70228, Ciudad Universitaria, 04510 México D.F., Mexico, Tel: +52-5-622-3828, Fax: +52-5-550-0048, e-mail: rhudson@servidor.unam.mx

Hummel, Thomas, Department of Otorhinolaryngology, University of Dresden Medical School, Fetscherstrasse 74, 01307 Dresden, Germany, Tel: +49-351-458-4189, Fax: +49-351-458-4326, e-mail: thummel@rcs.urz.tu-dresden.de

Issanchou, Sylvie, Laboratoire de Recherches sur les Arômes, Institut National de la Recherche Agrnomiques, 17 rue Sully, BP 86510, 21065 Dijon, France, Tel: +33 3 80 69 30 76, Fax: +33 3 80 63 32 27, e-mail: issan@arome.dijon.inra.fr

Jacob, Suma, Department of Psychology, University of Chicago, 5730 Woodlawn Avenue, Chicago, IL 60637, USA, Tel: +1(773) 702-6015, Fax: +1(773) 702-0320, e-mail: sj11@midway.uchicago.edu

Jönsson, Fredrik U., Department of Psychology, Uppsala University, Box 1225, 751 45 Uppsala, Sweden, Tel: +46 18 18 21 50, Fax: +46 18 471 2123, e-mail: fredrik.jonsson@psyk.uu.se

Köster, Egon Peter, ASAP Gesellschaft für Sensorische Analyse und Produktentwicklung, Drachenseestrasse 1, 81373 München, Germany, and The Royal Veterinary and Agricultural University, Rølighedsvej 30, DK1958 Frederiksberg C, Denmark, Tel: +31 30 2510387, Fax: +31 30 2546071, e-mail: ep.koster@wxs.nl

Larsson, Maria, Department of Psychology, Stockholm University, S-106 91 Stockholm, Sweden, Tel: +46 816 39 37, Fax: +46 8 15 93 42, e-mail: maria.larsson@psychology.su.se

Le Guérer, Annick, 1, chemin Es Pots, Agey, 21410 Pont-de-Pany, France, Tel/Fax: +33 3 80 23 61 88, e-mail: annick.le.guerer@wanadoo.fr

Lehrner, Johannes, Neurologische Universitätsklinik, Allgemeines Krankenhaus, Universität Wien, Währingergürtel 18-20, A-1097 Wien, Austria, Tel: +43-1-40400-3443 or 3433, Fax: +43-1-40400-3141, e-mail: Hannes. Lehrner@AKH-WIEN.AC.AT

Marlier, Luc, Centre Européen des Sciences du Goût, CNRS, Université de Bourgogne, 15 rue Hugues Picardet, 21000 Dijon, France, Tel: +33 3.80.68.16.10, Fax: +33 3.80.68.16.26, e-mail: marlier@cesg.cnrs.fr

McClintock, Martha K., Department of Psychology, University of Chicago, 5730 Woodlawn Avenue, Chicago, IL 60637, USA, Tel: +1(773) 702-2579, Fax: +1(773) 702-0320, e-mail: mkm1@midway.uchicago.edu

Mouly, Anne-Marie, Institut des Sciences Cognitives, CNRS/Université Claude Bernard, Lyon 1, 69675, Bron, France, Tel: +33 4 37 91 12 41, Fax: +33 4 37 91 12 10, e-mail: mouly@isc.cnrs.fr

Murphy, Claire, University of California, San Diego, School of Medicine, and Department of Psychology, San Diego State University, San Diego, CA 92129, USA, Tel: +1-619-594-4559, Fax: +1-619-594-3773, e-mail: cmurphy@sunstroke.sdsu.edu

Nordin, Steven, Department of Psychology, Umeå University, Sweden, and Department of Psychology, San Diego State University, San Diego, CA 92129, USA, Tel: +46-90-7866006, Fax: +46-90-7866695, e-mail: steven.nordin@psy.umu.se

Olsson, Mats J., Department of Psychology, Uppsala University, Box 1225, 751 45 Uppsala, Sweden, Tel: +46 18 18 21 50, Fax: +46 18 471 2123, e-mail: mats.olsson@psyk.uu.se

Pause, Bettina M., Institute of Psychology, Christian-Albrechts-Universität zu Kiel, Olshausenstrasse 62, 24098 Kiel, Germany, Tel: +49-431-880-3675, Fax: +49-431-880-1559, e-mail: bmpause@psychologie.uni-kiel.de

Phillips, Mary L., Department of Psychological Medicine, Institute of Psychiatry, 103, Denmark Hill, London SE5 8AZ, England, Tel: +44-171-740-5096/5089, Fax: +44-171-740-5129, e-mail: spmamlp@iop.kcl.ac.uk

Pillias, Anne-Marie, Laboratoire de Neurobiologie Sensorielle, École Pratique des Hautes Études, rue des Olympiades, Massy, France, and Laboratoire de Physiologie de la Manducation, Université Paris VII, Paris, France, Tel: +1 69 20 63 39, Fax: +1 60 11 61 94, e-mail: pillias@ccr. jussieu.fr

Ravel, Nadine, Institut des Sciences Cognitives, CNRS/Université Claude Bernard, Lyon 1, 69675, Bron, France, Tel: +33 4 37 91 12 42, Fax: +33 4 37 91 12 10, e-mail: ravel@isc.cnrs.fr

Rolls, Edmund T., Department of Experimental Psychology, University of Oxford, South Parks Road, Oxford OX1 3UD, England, Tel: +44-1865-271348, Fax: +44-1865-310447, e-mail: Edmund.Rolls@psy.ox.ac.uk

Rouby, Catherine, Laboratoire de Neurosciences et Systèmes Sensoriels, Université Claude Bernard, Lyon 1, 69 366 Lyon, France, Tel: +33 4 37 28 74 97, Fax: +33 4 37 28 76 01, e-mail: rouby@olfac.univ-lyon1.fr

Schaal, Benoist, Centre Européen des Sciences du Goût, CNRS, Université de Bourgogne, 15 rue Hugues Picardet, 21000 Dijon, France, Tel: +33 3.80.68.16.10, Fax: +33 3.80.68.16.26, e-mail: schaal@cesg.cnrs.fr

Sicard, Gilles, Laboratoire de Neurosciences et Systèmes Sensoriels, Université Claude Bernard, Lyon 1/CNRS, 69366 Lyon, France, Tel: +33 4 37 28 74 91, Fax: +33 4 37 28 76 01, e-mail: sicard@olfac.univ-lyon1.fr

Soussignan, Robert, Laboratoire Personnalité et Conduites Adaptatives, CNRS, and CHU Pitié-Sapêtrière, Pavillon Clérambault, 47 boulevard de l'Hôpital, Paris, France, Tel: +33 1 42 16 16 51, Fax: +33 1 53 79 07 70, e-mail: soussignan@wanadoo.fr

Sulmont, Claire, Laboratoire de Recherches sur les Arômes, Institut National de la Recherche Agrnomiques, 17 rue Sully, BP 86510, 21065 Dijon, France, Tel: +33 3 80 69 30 76, Fax: +33 3 80 63 32 27, e-mail: sulmont@arome.dijon.inra.fr

Valentin, Dominique, École Nationale Supérieure de Biologie Appliquée à la Nutrition et à l'Alimentation, Université de Bourgogne, 1 esplanade Erasme, 21000 Dijon, France, Tel: +33 3 80 39 66 43, Fax: +33 3 80 39 66 11, e-mail: valentin@u-bourgogne.fr

Walla, Peter, Neurologische Universitätsklinik, Allgemeines Krankenhaus, Universität Wien, Währingergürtel 18-20, A-1097 Wien, Austria, Tel: +43-1-40400-3443 or 3433, Fax: +43-1-40400-3141, e-mail: Peter.Walla@akh-wien.ac.at

Zatorre, Robert J., Montréal Neurological Institute, McGill University, 3801 University Street, Montréal, QC Canada H3A 2B4, Tel: +1-514-398-8903, Fax: +1-514-398-1338, e-mail: robert.zatorre@mcgill.ca

Zelano, Bethanne, Department of Psychology, University of Chicago, 5730 Woodlawn Avenue, Chicago, IL 60637, USA, Tel: +1(773) 702-6015, Fax: +1(773) 702-0320, e-mail: bzelano@dura.spc.uchicago.edu

Preface

This book arises from an acknowledgment: the lack, as far as we know, of a book dedicated to the cognition of chemical senses.

Although recent discoveries in the field of molecular biology raise the hope of a future understanding of the transduction and peripheral coding of odors and tastes, it seems to us that they imply a risk: to make us forget that in the other extreme of knowledge, that of maximal complexity, the evolution of cognitive sciences allows an epistemologically fruitful reformulation of information-processing problems.

Unlike the other senses, olfaction and taste do not have a learned discourse dealing with elementary aspects, that is, sensory processing, as well as the most abstract aspects, that is, symbolic processing. The purpose of cognitive science is to orient these processings into a continuity, and particularly to try to find out to what extent higher-order processes interact with the sensory level in order to produce sufficiently reliable representations of the world. We are still quite unaware of the nature of gustatory and olfactory representations, as compared with what we know about vision and audition, for example.

Faced with this relative ignorance, our prejudice was the following: If odors and tastes are ill-identified cognitive objects, then none of the available potential resources should be neglected: Expert and naive people, as well as "savage" and "civilized" ones, conscious knowledge and emotions, biology and social sciences – all of those can contribute first to an assessment of our knowledge, and then to confrontation of its inadequacies. This inter-disciplinary point of view first gave rise to a meeting held in Lyon, in June 1999: the European Symposium on Olfaction and Cognition, which tried to coordinate knowledge from several scientific disciplines with that from perfumery professionals. The other aim of that meeting was to publish a book, conceived not as a handbook gathering all the validated knowledge but as a book reflecting the questions and the divergences running through this field of knowledge, whose complexity biology and chemistry still cannot explore thoroughly.

This book is meant for all those studying taste and olfaction. We hope that it will foster other debates and other novel collaborations between neuroscience and social science.

Acknowledgments

We acknowledge the support of a number of people without whom this book would have gone no further than the stage of the outline. We are grateful to all the authors for taking time and putting effort into contributing to this collection. Apart from the editors, a number of people kindly refereed the chapters herein, and we are very appreciative of their help. These are the following doctors: Christiane Ayer-Lelièvre (Limoges), Olivier Bertrand (Lyon), Claude Bonnet (Strasbourg), Driss Boussaoud (Lyon), Pierre Cadiot (Paris), Paul Castle (Southampton), Bernard Croisille (Lyon), Philippe Descola (Paris), Hans-Jürgen Distel (Munich), Marie-Hélène Giard (Lyon), Avery N. Gilbert (Montclair, CA), James Hampton (London), Dirk Hermans (Leuven), Joseph Hossenlopp (Clermont-Ferrand), David Howes (Montréal), Robyn Hudson (Mexico City), Thomas Hummel (Dresden), Sylvie Issanchou (Dijon), Marilyn Jones-Gotman (Montréal), Egon P. Köster (Utrecht), Jan Kroeze (Utrecht), Serge Larochelle (Montréal), Annick Le Guérer (Agey), Johannes Lehrner (Vienna), Jean Marie Marandin (Paris), Julie Mennella (Philadelphia), Mats J. Olsson (Uppsala), Pierre Perruchet (Dijon), Paul Rozin (Philadelphia), Margret Schleidt (Andechs), Gilles Sicard (Lyon), Dana Small (Chicago), Guy Tiberghien (Grenoble), Rémy Versace (Lyon), Daniel Widlocher (Paris), and Don A. Wilson (Oklahoma City).

From the beginning, Michel Vigouroux was in charge of the different stages of the book preparation, especially the task of repeatedly formatting texts and editing illustrative material. Agnès Magron contributed to clear references and to the elaboration of the index.

Finally, Michael Penn and Ellen Carlin, our editors at Cambridge University Press, advised and supported us through time. To all of these persons, our wholehearted appreciation.

C. Rouby, B. Schaal, D. Dubois, R. Gervais, A. Holley

A Tribute to Edmond Roudnitska

This book is dedicated to Edmond Roudnitska, who was an artist, a creator of perfumes, and a writer. The organizers of the symposium from which this book originated feel that his work is one of the best illustrations of the relationship between olfaction and cognition.

Through his olfactory creations, which constitute landmarks in the relatively short history of the perfume industry, he was among the handful of people who helped elevate olfactory composition to the rank of an art. Beyond sheer sensorial pleasure, his great perfumes[1] convey to enlightened perfume lovers purely aesthetic emotions fueled by reference to the history of perfumes, leaving connoisseurs marveling at the balanced combinations of olfactory elements similar to a bold architectural construction, and finally bringing to the emotions this cognitive component without which no art is possible.

Another aspect of Roudnitska's work is of particular interest to scientists researching the cognitive component of olfactory perception: It consists in writings,[2] in which he expresses his views regarding perfume creation. Perfume creators, in accordance with a tradition of secrecy, rarely communicate about their work. Roudnitska, however, communicated about his art with remarkable freedom and pertinence. Throughout his books and articles, which clearly reveal pedagogical intent, he presents the results of a rigorous thought process going well beyond practical applications and leading to perceptive examination of the components of creation. One does not need to extrapolate or translate his thoughts to recognize many of the themes of modern cognitive research. The acuity of his analysis, combined with well-thought-out practice, finds its direct reflection in the fields of perception, learning and memory, attention, mental imagery, and the relationship between emotion and cognition.

This tribute to Edmond Roudnitska is an acknowledgment of our indebtedness. This profoundly rational spirit, who was confident that scientific research could enlighten any endeavor and someday facilitate the art of perfume creation, was also a generous man. Through creation of the Fondation Roudnitska he enabled many young researchers to progress in their knowledge of olfaction, and we thank him for this.

André Holley

Notes

1. Among E. Roudnitska's creations, let us mention "Femme" for Rochas (1944), "Diorama" for Dior (1948), "L'Eau" for Hermès (1951/1987), "Diorissimo" (1956) for Dior, "Eau sauvage" (1966) for Dior, and "Diorella" (1972) for Dior.
2. *L'Esthétique en question*, Presses Universitaires de France, Paris, 1977; *Le parfum*, Presses Universitaires de France, Paris, 1990; *Une vie au service du parfum*, Thérèse Vian, Paris, 1991.

OLFACTION, TASTE, AND COGNITION

Section One

A Specific Type of Cognition

Olfactory experience is difficult to define: From ineffable to unmentionable, it seems to remain in the limbo of cognition. More than any artist's work, the competence of the perfumer is a challenge for explication. The few artists who are able to communicate in writing about their creative processes are mainly plasticians (painters and sculptors), musicians, and, of course, writers. As regards chemical senses, the writings are extremely rare, and the very status of "artist" is not easily conferred. As an example, Edmond Roudnitska faced a difficult task in his effort to have olfaction accepted into the realm of aesthetics.

In Chapter 1, Annick Le Guérer proposes an explanation for that misappreciation that has to do with the history of Western philosophy: Our philosophical heritage denies any nobility to olfaction and taste, as compared with the other senses, and depreciates them almost systematically. Psychoanalysis has cited that fact as evidence that civilization can be built only if there is repression of smell. However, as Le Guérer points out, the history of psychoanalysis itself is marked by fantastic representations of the nose and its functions within the relationship linking Freud and Fliess – their unconscious *montre son nez* in the learned conception of smell.

Moving away from the neurophysiology of smell, André Holley, in Chapter 2, looks into the perfumer's knowledge, which remains largely secret and intuitive. What is the difference between expert and novice, artist and amateur in the cognitive treatment of odors? How does plain emotion become an aesthetic feeling, and which departure from the common meaning of "odors" does this imply? These questions require an understanding of how the composer perceives, recalls, and imagines odors and whether abstract representations and rules are superimposed on ordinary perception or replace it. Whereas audition research has revealed differences between musicians and non-musicians, similar research on expertise for smell has yet to be done: Is there an absolute sense of smell, as there is an absolute sense of pitch?

Egon Peter Köster has had long practical experience in research on preferences and olfactory memory in various industrial applications of the chemical senses. His research has been characterized by this pragmatic point of view, orienting him toward the implicit, emotional, and infra-attentional functioning of this modality. In Chapter 3 he examines here the differences between the chemical senses and the "far senses" and questions the relevance of many marketing surveys: "People will eat what they like, but do not know what they like, and certainly not why they like it."

1

Olfaction and Cognition: A Philosophical and Psychoanalytic View[1]

Annick Le Guérer

"Are you not ashamed to believe that the nose is a means to find God?" wrote Saint Augustine (1873) in his refutation of the Manichaeans in 389 C.E. Fifteen centuries later, Sigmund Freud, addressing the members of the Vienna Psychoanalytical Society, stated that "the organic sublimation of the sense of smell is a factor of civilisation" (Freud, 1978, p. 318).

These peremptory opinions expressed by the Christian philosopher and by the founder of psychoanalysis offer little support for any investigation into the sense of smell as a tool for knowledge. Furthermore, they are typical of an attitude widely held by both philosophers (Le Guérer, 1987) and psychoanalysts (Le Guérer, 1996). There are many possible explanations for the mistrust – indeed, the rejection – with which both groups have treated this sense, but all of them converge on its animal nature.

Philosophers have often slighted and underestimated the sense of smell. Plato (1961) and Aristotle (1959) maintained that the pleasures it provided were less pure, less noble, than those offered by sight and hearing. Aristotle also found it lacking in finesse and discernment. Descartes (1953) regarded it as vulgar, and Kant (1978) thought it a coarse sense and one best left undeveloped, leading as it did to more unpleasant experiences than pleasant ones. Schopenhauer (1966) considered it an inferior sense; Hegel (1979) eliminated it from his aesthetics. And at the beginning of the twentieth century, the German philosopher Georg Simmel (1912) went so far as to term it the antisocial sense par excellence. Animal, primitive, instinctive, lust-provoking, erotic, selfish, irrelevant, asocial, dictatorial, imposing upon us willy-nilly the most painful of sensations, making us unable to escape its subjective solipsism, incapable of abstraction or of leading to any artistic ends, not to mention thought – the philosophical reasons for denigrating the sense of smell are numerous (Le Guérer, 1988, pp. 225–93).

Why Such Negativity?

If we approach the question analytically, we find two basic approaches. The first derives from the hierarchy often established between the distanced, noble "intellectual" senses (sight and hearing), which require some external medium (e.g., the air), and the proximity senses, which are more physical, animal, and sensual (taste and touch) and which have as their medium the body itself, the flesh. Smell exists at the junction between these two groups and is regarded as an intermediate sensorial faculty. We find this basic belief in the works of many thinkers, from Aristotle to Jean Jaurès (1891), including Saint Thomas Aquinas, Hegel, and Cournot, who variously described the sense of smell as ambiguous, bastard, vague, and nonautonomous.

The second approach concerns the paucity of the olfactory vocabulary. In his Timaeus, Plato dealt with the difficulty of arriving at any precise description of odors. In his view, the only distinctions that can be made with any validity are affective, that is, based on the pleasure or displeasure to which they give rise. The list of subsequent philosophers making similar distinctions is a lengthy one. The sense of smell was regarded as a dead end, incapable of abstraction. Henning (1916, p. 66) noted that "olfactory abstraction is impossible. We can easily abstract the common, shared color – i.e., white – of jasmine, lily-of-the-valley, camphor and milk, but no man can similarly abstract a common odor by attending to what they have in common and setting aside their differences." Simmel (1912, p. 3) noted: "The difficulty of translating smell impressions into words is far different from that of translating the impressions of sight and hearing. They cannot be projected on an abstract level." Dan Sperber (1974, p. 128) wrote that the only possible way to classify smells was in relation to their sources (the smell of roses, or coffee with milk). In saying that, he was in agreement with many neurophysicists (Holley and MacLeod, 1977, p. 729): "Only the source of an odor is truly apprehended as an entity, to the point that we are unable to give a name to the latter save via the former. We lack the rigor of language required for more precise description and are forced to fall back on metaphors."[2]

This perpetual problem of abstraction has meant that smell has come to be viewed as a primitive, archaic, *needed* sense, one more important to sensory pleasure than to knowledge. Lacking in objectivity, and making us aware not of true distinctions but rather of elements that are "complementary," "congenerous," and "convenient" to us – making us aware, in a way, of ourselves – the sense of smell therefore fails to afford us, in the words of the philosopher Maurice Pradines (1958, p. 513), "any true knowledge, either of the world or of ourselves." Cournot (1851, p. 121) asserted that the sense of smell is too subjective to provide any precise information about external objects: "A smell, like a taste, is an affect of the feeling subject that affords no representation, that in and of itself neither implies nor determines any knowledge of the thing sensed . . . . What is incontrovertible is that the sense of smell alone can provide no notion of the outer world and that, in a normal man, it adds no theoretical or scientific knowledge of the outside world . . . scientific advances would in no way be hindered were this sense to be done away with completely."

Further, the sense of smell has been regarded as a hindrance to scientific progress. Smells, because of their direct and intimate nature, their insinuating, penetrating strength, and their permeability, are treated as active realities, as the surest messengers from the real world. This view is buttressed by the belief that smell contains the principle, the virtue, of any substance, a belief that the philosopher Gaston Bachelard (1938, p. 115) has termed "substantialist" and one that has often led scholars astray. Thus, for a long time, ozone failed to be identified as a gas because it was believed to be the odor emitted by electricity.

Unlike sight and hearing, which are viewed as higher senses and are endowed with a rich and specific vocabulary, the sense of smell has been regarded as incapable of serving as the basis for any art form. That view can be attributed to a long philosophical tradition wrought by many great thinkers – Plato, Kant, Hegel, Schopenhauer, Bergson – whose opinions have weighed heavily on all aestheticians.

However, there have been philosophers, though in fact very few, who have taken exception to the categorization of olfaction as somehow disreputable. The movement toward rehabilitation was begun by the eighteenth-century Sensualists, who, in opposition to the intellectualizing Philosophes of the seventeenth century, vaunted the importance of "feeling" as a part of knowledge and maintained that the sense of smell had been unjustly disparaged in the past. According to La Mettrie (1745), for example, all ideas derive from the senses, and Helvétius (1774, p. 135) went so far as to state that "to judge is to feel."

The famous example of the statue imagined by the Abbé de Condillac symbolizes this program of reevaluation. To explain how perceptions are assimilated to produce understanding, he imagined endowing a statue with each sense separately, and he began with the sense of smell because it supposedly was the one that made the least contribution to knowledge and understanding. Diderot (1955) went even further and stated (without demonstration) that the sense of smell was capable of abstraction. Rousseau (1969) considered it to be the sense of the imagination and of love; pity the man, he wrote, so insensitive as to be unmoved by his mistress's odor.

In the nineteenth century, the sense of smell found defenders among philosophers who set out to rehabilitate the role of the body in the search for knowledge. The German philosopher Ludwig Feuerbach (1960) was an advocate for the importance of the sense of smell. He declared that it was as capable as the sense of sight of rising above animal needs and that it could make spiritual and scientific contributions to both knowledge and art. Feuerbach condemned those thinkers who found it necessary to reject the importance of the sense of smell in order to improve their thinking: Without a "nose," their theories were condemned to emptiness.

It was Nietzsche, however, who spoke out most strongly against those who denigrated the sense of smell. Whereas Feuerbach attempted to revalorize the sense of smell by making it more spiritual, more intellectual, Nietzsche, on the contrary, lauded its animal nature: The refusal of the majority to acknowledge the sense of smell as a means for attaining knowledge was rooted in an absurd rejection of man's animal nature, in addition to revealing an exaggerated esteem for logic and reason. The utility of the sense of smell resides in this animality, obviating the need to employ language in order to advance thought and under-standing. It is, in fact, the most delicate instrument available to us. Nietzsche's apologia took the shape of a metaphor. "Flair," for him, was a real tool for psy-chological and moral investigation. Its links to instinct, judgment, and mental perception would make it a tool for the psychologist, who is guided by intuition and whose art consists not in reasoning but in "scenting out."

An accomplished psychologist, Nietzsche (1971, p. 333) claimed to have an especially remarkable flair that enabled him to read people's hearts and souls and to sniff out falsity and illusion. "I am the first to have discovered the truth by virtue of the fact that I am the first to have sensed, to have had the flair to scent out, falsehood as falsehood." As the sense most attuned to truth, smell, in its search for veracity, overturns the cold logic that is the product of the struggle against the instinctual and draws on the sure sources of the animal instincts that endow the body with such great wisdom. Above and beyond its basic function, therefore, the sense of smell assumes the function of a "sixth sense," the sense of intuitive awareness. All of the cognitive importance with which Nietzsche (1971, p. 333) endows this unjustly despised sense is expressed in this statement: "All my genius is in my nostrils."

However, Nietzsche's defense of the sense of smell did not have any lasting effect, and the few subsequent attempts to plead on its behalf have not been note-worthy. The most effective damper on such efforts came from psychoanalysis, which has played an important role in the cognitive devaluation of the sense of smell.

One might have expected the psychoanalysts, given their awareness of their patients' sexuality, to have undertaken a closer examination of this sense, which is so intimately bound up with a person's sensual side. Such, however, has not been the case, and the situation may perhaps be explained as follows: The story of the sense of smell with regard to psychoanalysis itself bears the mark of repression. Freud's olfactory investigations, which were undertaken at a time when hygiene and an odor-free environment were becoming increasingly important and during a period in which the scholarly discourse on the sense of smell tended to devalue it, were conducted within the framework of his transferential relationship with Fliess, in which the nose (frequently purulent) and the repression of concerns about certain olfactory events each played a considerable role (Le Guérer, 1996).

Freud had met the Berlin otorhinolaryngologist Wilhelm Fliess in 1887 at Vienna and had become fascinated by the wide-ranging scientific views expressed by that brilliant and original thinker, views in which the nose played a major role (Freud, 1985). Indeed, Fliess (1893, 1977) believed that he had discovered the existence of a neurosis linked to the nasal passages. He also believed that edemas of the nasal mucous membrane and infections of the sinuses and nasal turbinate cartilage were at the root of a whole group of symptoms: migraine headaches, neuralgias, and functional disorders of the heart and of the respiratory, digestive, and sexual systems. According to Fliess, all of these ills, notwithstanding the marked differences between them, shared one common characteristic: They could momentarily disappear when the appropriate nasal areas were anesthetized with cocaine. Both men suffered from serious rhinological problems, which further increased their interest in that organ. Fliess managed to persuade Freud that Freud's nasal problems were the root causes of his heart condition. In an attempt to correct that, Freud allowed Fliess to operate on him on several occasions. Indeed, they were joined by the nose, so to speak, and that connection was strengthened by their use of cocaine. Present throughout the course of their intellectual exchanges and a part of their daily lives, their fixation on the nose was to become even more pronounced as the result of an unfortunate surgical procedure.

In February 1895, Freud asked Fliess to perform an operation on one of his patients, a hysterical young widow named Emma Eckstein. Fliess came from Berlin to Vienna and performed the procedure. During the same visit, he cauterized Freud's nasal turbinate bones (Freud was suffering from acute rhinitis). After Fliess's departure, however, the young woman's nose remained extremely painful and began to give off a foul odor. Freud summoned another specialist to Emma's bedside, and the new doctor saw what appeared to be a string in the young woman's nasal cavity. He pulled on it and drew out what – to everyone's surprise – turned out to be a piece of surgical gauze nearly 50 cm long (~19.5 inches). The girl, streaming blood, promptly lost consciousness. The hemorrhage was finally stemmed, but Freud was so shocked at Fliess's professional negligence and his patient's alarming state that he became sick and was forced to leave the room and revive himself with a shot of brandy.

That failed operation was to play a fundamental role in psychoanalytic theory, for it was the source for the well-known dream known as "Irma's injection," in which the sense of smell is of prime importance, and it became the paradigm for Freud's theory of the dream as wish fulfillment. At the time, Freud greatly needed Fliess's friendship and support, and he did everything in his power to retain his admiration for Fliess and repress his doubts. That process of repression is especially clear in the dream, in which the Fliess character appears to be esteemed and guiltless. It is also much concerned with the noses of the three protagonists involved in the minor tragedy that had occurred at Vienna a few months earlier. In my view, it was within this transferential relationship with Fliess – an otorhinolaryngologist, the author of a work on the relationship between the nose and sexuality, and a man deeply involved in an incident featuring an evil-smelling nose – that Freud came to conceive his theory of smell, which is centered around the notion of repression (Le Guérer, 2001, p. 460). Indeed, on 14 November 1897 Freud wrote to Fliess as follows: "Memory now comes up with the same stench as an actual object. Just as we turn away in disgust from evil-smelling objects, so too do the preconscious and our conscious awareness turn away from the memory. This is what we call repression."

At the turn of the twentieth century, the olfactory observations cited in medical and scientific discourse tended, for the most part, to have a negative tone. Indeed, the great naturalist Charles Darwin (1971) considered that the sense of smell, so important to most mammals, was, in humans, merely a weakened and rudimentary function, a holdover from the distant ancestors for whom it had been extremely useful. Subsequently, the majority of contemporary psychiatrists and sexologists would connect the sense of smell not only to animality but also to psychopathologic conditions. The famous Austrian psychiatrist Richard von Krafft-Ebing (1931) and the German sexologist Albert Moll, as well as Charles Féré (1974) in France, all felt that the sense of smell had little to do with any normal sexuality. On the other hand, certain mental illnesses seemed to reveal a close connection between the sense of smell and sexuality.

Freud's theories were influenced by all of those ideas, but he was to go deeper into the question by explicitly positing an indissoluble link between repression and the almost universal contempt for the olfactory sense.

According to the founder of psychoanalysis, both our sense of smell and our sexuality were altered profoundly when our distant ancestors abandoned walking on all fours and adopted an upright position. When the nose was elevated higher above ground level, the once-dominant sense of smell weakened, the sense of sight moved to the fore, and hitherto-exciting olfactory sensations gradually came to be considered disgusting. That virtual obliteration of the sense of smell led to a broad-based repression of sexuality, though it facilitated the founding of families and the development of civilization. In short, when humans broke away from their animal nature, that entailed a dual abandonment: withdrawal from the sense of smell and from sexuality. The virtual disappearance of the sense of smell during phylogeny is a model of what occurs during ontogeny (Freud, 1961).

As a result, acute olfactory sensitivity came to be viewed as an archaic and even pernicious trait, revealing a fixation on anal sexuality. According to Freud, it persisted only in animals, in savages, and in very young children, who were not disgusted by excremental odors: Education teaches the child to shun such smells. The result held repression of the sense of smell and of anal sexuality. To have a keen sense of smell was seen as a symptom of latent animality, a failure of the socializing process.

Freud, in proposing a direct relationship between the development of civilization and the virtual eradication of the sense of smell, was implicitly suggesting that the olfactory sense was a hindrance to knowledge and aesthetics. Following in his footsteps, many psychoanalysts came to view it as an archaic, animal faculty that would have to be repressed if people were to function in society. Such was the view of Lacan (1973, pp. 61–2) when he noted that repression of the sense of smell in humans "has a great deal to do with [their] access to the dimension of the Other." It was also the view of Françoise Dolto (1980, p. 342), who agreed with the philosophical tradition that deemed the sense of smell inferior because it was incapable of abstraction. "Culture," she stated, "is speech and obviously not smell."

Nevertheless, some psychoanalysts, like Nietzsche before them, have attempted to rehabilitate the sense of smell by demonstrating that it is capable of leading to intuitive knowledge. For example, the Hungarian psychoanalyst Sandor Ferenczi (1974) sought for hidden links between the sense of smell and thought processes. Like animals, some humans may have a gift for "sniffing out" repressed feelings and tendencies. An extraordinarily subtle sense of smell and a powerful olfactory imagination may underlie the performances of spirit mediums, who may be sensitive to the odors emanating from certain people. Ferenczi (1985, p. 142) went so far as to state that "a large part of what has hitherto been regarded as an outré, occult or metaphysical performance may have some psycho-physiological explanation."

Owing to its ability to enable people to "sniff things out," the sense of smell plays a large role in transferral and countertransferral. Instances of sick persons who gave off extremely unpleasant odors when experiencing attacks of repressed rage made Ferenczi aware of the semiotic importance of odors: They could replace speech and reveal affects that might be hidden in social communication. The psychoanalyst went on to discuss the case of a patient who, when in a state of repressed anger, emitted highly unpleasant odors, "as though, lacking other weapons, she was attempting, like certain animals, to keep people away from her body by frightening them with these emanations of hatred" (Ferenczi, 1985, p. 141).

More recently, Didier Anzieu had an experience with the cognitive value of olfactory messages in psychoanalysis. His patient, whom he called Gethsemane (in reference to the Mount of Olives, where Jesus sweated blood), suffered from excessive and very malodorous sweating. The psychoanalyst, repelled and almost paralyzed by the stench, which was made even more unpleasant by the eau de cologne his patient liberally applied, probably to mask it, was unable to cope with this particularly intense sensory manifestation. He began by setting up an early countertransferential resistance to the manifestation, not verbalized but yet "the most pervasive factor in the session," which he considered devoid of any "apparent communicative validity" and as falling outside the purview of psychoanalysis. Anzieu (1985, p. 182) thus despaired of interpreting the condition, the treatment languished, and boredom set in. Gethsemane, continuing to exteriorize his anxiety and conflicted feelings in this unique manner, went on smelling worse and worse.

One day a patient protested about the fetid atmosphere in Anzieu's clinic. Anzieu shook off his lethargy and realized that he had almost reached a point at which he himself was no longer able to "sense" Gethsemane, "with all the meanings that that word entails." "Might it not be a transferral neurosis simultaneously concealing and expressing itself through these bad-smelling – and, in my case, slyly aggressive – emanations?" he asked, now alerted. But how to discuss such nauseating emanations without seeming inconsiderate or hurtful? Because he could find no psychoanalytic theory to guide him in answering that question, the analyst ventured to make "a compromise and fairly general interpretation" centered on the senses that, after several sessions, finally uncovered a memory from the past connected with smell.

Gethsemane had had a difficult birth, during the course of which his flesh had been lacerated and there had been considerable bleeding. He had been kept alive by the care of a slatternly wet nurse, who had taken him into her own bed. At the same time, his mother, an extremely well-groomed woman, was liberal in her use of eau de cologne. "Thus," concluded Anzieu (1985, p. 184), "the two contradictory odors with which he filled my consulting room represented a fantasy attempt to recreate in his own flesh the flesh of his wet-nurse and his mother. Did this mean that he had none of his own?" By becoming aware of his conflicted feelings, rather than sublimating them in the form of perspiration, Gethsemane, unconsciously and without suffering, began to show marked progress, and his bad odor gradually became less noticeable. The importance of the sense of smell in that case and the actual odors noted by Anzieu during that treatment led him, in a footnote, to venture a hypothesis rich in potential and in keeping with certain of Ferenczi's intuitions: "It may be that the analyst's intuition and empathy both rest on an olfactory basis that is difficult to study."

Gisèle Harrus-Révidi (1987, p. 62) deplored the total negation of the importance of the sense of smell in analysis, for that sense plays a most basic role in human communication. Citing the English psychoanalyst Donald Winnicott, she viewed any smell that entered into the "intermediate area of illusion" of the analytical space as a "transitional object" that called for interpretation: "When making resumés of sessions, little attention is paid [to] the stomach rumblings, hiccoughs or physical noises made by the patient's body (and we piously omit to mention those of the analyst as well!), all those things by which the patient makes us aware that he has a body. The patient may be attempting to fill the transitional space with his own odors in order to make it his own. 'Regression' may lead him to try to revive the odorous aura, basically of feces, surrounding his relationship with his mother." And the effluvia of the analyst are as meaningful as those of the analysand. His odors, like those of his apartment, the smells of the food prepared therein, the flowers that he has bought, can all play a part in the process of transferral. The sense of smell, which was studied by Freud's earliest heirs, essentially with regard to anal sexuality, has gradually been shown to have relationships to every libidinal stage, but it remains tainted with negative references. The notion of it as a decadent sense, but especially the linkage of the sense of smell to animality and anal eroticism, has contributed to this situation. The few analysts who have concerned themselves with this sensorial mode have often commented on the paucity of psychoanalytic documentation in this field.

Conclusion

The rare attempts by philosophers and psychoanalysts to provide a cognitive reevaluation of the sense of smell have led to the idea of a "non-rational" intelligence, a "flair" that cannot be expressed in words. Must we, then, with Aristotle, conclude that the sense of smell is not of intellectual benefit? Frequently denigrated as a tool for rational knowledge because of its resistance to abstraction, and because of its close links to sexuality, the sense of smell is nevertheless indispensable in grasping some extremely subtle, pre-rational factors, the indefinable "something" that emanates from a person, an object, a place, a situation. As the sense most closely linked with affect and contact, a sense that helps to establish a fusional relationship with the world, revelatory not only of substances but also of ambiences, climates, and even existential states, the sense of smell is a subtle tool for knowledge that allows for an intuitive and prelinguistic understanding.

Its links to respiration enable us to have a profound relationship with our environment and give the sense of smell a very special vocation, one that Tellenbach (1968) describes as "atmospheric flair." Thus, in the past it played an important role in medical diagnosis and had an essential function in communication.

Indeed, at a very early date, medical men knew that smells were one way of identifying diseases. Hippocrates recommended that a doctor be a "man with an open nose," capable of recognizing diseases by their smells. Doctors have long paid close attention to smells, and in the eighteenth and nineteenth centuries there were specialists, known as osphresiologists, who drew up highly detailed olfactory classifications for every sort of disease. Thus the patient suffering from smallpox would smell of onions, a person with ringworm would smell of cat, and a madman would emit the odor of a wild animal or of mouse urine. Monin (1885, p. 16), an osphresiologist, stated that "smell is the subtle soul of the clinical apparatus. Its language rings a faint bell in the practitioner's mind, recalling primal diagnostic ideas. It is a semiotic method in use from the earliest times." Recent surveys in medical circles have shown that, notwithstanding the decline in the teaching of olfactory diagnosis, the sense of smell continues to play a part in the acquisition of professional competence. Health-care personnel are expected to identify certain pathologic conditions by their odors and to know, solely on the basis of smell, whether or not they are serious (Candau, 1999, p. 185). Present-day research using an "electronic nose" to analyze morbid effluvia is in line with that age-old tradition.

Psychiatric phenomenology (Tellenbach, 1983) has demonstrated the importance of olfactory experiences in understanding the changes in one's relationships to one's self and to the world. Thus, when patients suffering from olfactory delirium complain that they are emitting obnoxious odors, their lives must be dominated by feelings of guilt and corruption. The vile thing conveyed by the evil smell can totally upset one's relationship with reality. Such patients are constantly humiliated at having to impose their supposed stench on those around them. In a world in which there is a growing tendency toward deodorization, those suffering from olfactory paranoia, trapped in their own stench, are continually fearful of being caught in the act of creating an unpleasant smell. Such a person, experiencing himself as an object of disgust, comes to think that people are laughing at his smell and even holding their noses when he goes by. Here, smell is the symbol of the judgment such sufferers believe the world is leveling against them: arrogant rejection, contempt, discrimination.

In *The Brothers Karamazov*, Dostoyevsky analyzes the critical changes that an odor can trigger. When, instead of the expected odor of sanctity, the corpse of Father Zosima exudes a stench of putrefaction, those in attendance interpret it as the odor of doubt and hatred. The atmosphere of purity and freshness, trust and idealism, that emanated from the monk when he was alive gives way to a pestilence that confuses, dismays, and causes profound spiritual unease among his followers.

Odors, more sheltered from intellectual analysis than the other sense impressions, are the tools for intuitive and emotional knowledge of the world, as is reflected in everyday language and in many colloquial expressions (we "sniff out" wrongdoing, a person is said to "have a nose for" news or something else, we say that we "smell a rat" or that someone is "in bad odor," or that something just plain "smells" – sometimes "to high heaven").

It is also from this special locus "between experience and representation, affect and percept" (Holley, 1999, p. 243), that odors derive their ability to evoke the past in all its freshness, a power that so often benefits writers of all kinds. Like individual, private signals, odors are the keys to hidden memories that often we no longer know we have retained. We thus see that smell, as the philosopher Gaston Bachelard (1960, p. 123) has written, "in a childhood, in a life, is . . . a vast detail."

Notes

1. Translated from the French by Richard Miller.
2. Research is under way into the hypothesis that odors, like colors, may have some "universal qualities" (Schaal et al., 1998, p. 146). See also the investigations into African languages in an attempt to discover terms employed exclusively to describe odors (Mouélé, 1997, pp. 209–22).

References

Anzieu D (1985). *Le Moi-peau*. Paris: Dunod.

Aristotle (1959). *On the Soul. On the Senses. Nicomachean Ethics*. In: *Collected Works*, trans. K Foster & S Humphries. New Haven, CT: Yale University Press.

Augustine (1873). *Des moeurs des Manichéens*. In: *Oeuvres complètes*. Paris: L. Vivès.

Bachelard G (1960). *La Formation de l'esprit scientifique, contribution à une psychanalyse de la connaissance objective*. Paris: Presses Universitaires de France. (Originally published 1938.)

Candau J (1999). Mémoire des odeurs et savoir-faire professionnels. In: *Odeurs et Parfums*, ed. C. Fabre-Vassas et al. Paris: Editions du Comité des Travaux Historiques et Scientifiques.

Condillac E B de (1947). *Traité des sensations*. In: *Oeuvres philosophiques*. Paris: Presses Universitaires de France. (Originally published 1754.)

Cournot A A (1851). *Essai sur les fondements de nos connaissances et sur les caractères de la critique*. Paris: Vrin.

Darwin C (1971). *The Descent of Man and Selection in Relation to Sex*. New York: Appleton. (Originally published 1871.)

Descartes R (1953). *Oeuvres et lettres*, ed. A. Bridoux. Paris: Gallimard.

Diderot D (1955). Lettre à Mademoiselle de la Chaux. In: *Correspondance* (1713–1757). Paris: éditions de Minuit.

Dolto F (1980). *Les Cahiers du nouveau-né. 3: D'Amour et de lait.* Paris: Stock.

Féré C (1974). *L'Instinct sexuel, évolution et dissolution.* Paris: Payot.

Ferenczi S (1974). *Oeuvres complètes (1913–1919).* Paris: Payot.

Ferenczi S (1985). *Journal clinique (January–October 1932).* Paris: Payot.

Feuerbach L (1960). *Principes de la philosophie de l'avenir.* In: *Manifestes philosophiques,* trans. L. Althusser. Paris: Presses Universitaires de France. (Originally published 1843.)

Fliess W (1977). *Les relations entre le nez et les organes génitaux féminins au point de vue biologique.* Paris: Le Seuil. (Originally published 1897.)

Freud S (1961). *Civilization and Its Discontents,* trans. J. Strachey. New York: Norton. (Originally published 1929.)

Freud S (1978). *Minutes de la Société psychanalytique de Vienne* (17-11-1909). Paris: Gallimard.

Freud S (1985). *The Complete Letters of Sigmund Freud to Wilhelm Fliess, 1887–1904.* Cambridge, MA: Harvard University Press (Belknap Press).

Harrus-Révidi G (1987). *La Vague et la Digue, du Sensoriel au Sensuel en Psychanalyse.* Paris: Payot.

Hegel G W F (1979). *Esthétique,* trans. S. Jankélévitch. Paris: Flammarion. (Originally published 1832.)

Helvétius C A (1774). *De l'homme, de ses facultés intellectuelles et de son éducation.* Liège.

Henning (1916). *Der Geruch.* Leipzig: J. A. Barth. (Originally published 1824.)

Holley A (1999). *Eloge de l'odorat.* Paris: Odile Jacob.

Holley A & MacLeod P (1977). Transduction et codage des informations olfactives. *Journal de Physiologie (Paris)* 73:725–828.

Jaurès J (1891). *De la réalité du monde sensible.* Paris: F. Alcan.

Kant E (1978). *Anthropology from a Pragmatic Point of View,* trans. V L Dowdell. Carbondale, IL: Southern Illinois University Press. (Originally published 1798.)

Krafft-Ebing R von (1931). *Psychopathia Sexualis. Eine Klinisch-Forensische Studie.* Paris: Payot. (Originally published 1886.)

Lacan J (1973). *L'identification, Séminaire (1961–1962).* Paris: Le Seuil.

La Mettrie J O de (1745). *Histoire naturelle de l'âme.* Oxford: aux dépens de l'auteur.

Le Guérer A (1987). Les philosophes ont-ils un nez? *Autrement (Special Issue "Odeurs")* 92:49–55.

Le Guérer A (1998). *Les Pouvoirs de L'odeur.* Paris: Odile Jacob. (Originally published 1988.)

Le Guérer A (1996). Le nez d'Emma. Histoire de l'odorat dans la psychanalyse. *Revue Internationale de Psychopathologie* 22:339–87.

Le Guérer A (2001). The psychoanalyst's nose. *Psychoanalytic Review* (New York: Guilford) 88(3):451–503.

Monin E (1885). *Les Odeurs du corps humain.* Paris: Doin.

Mouélé M (1997). L'apprentissage des odeurs chez les Waanzi. In: *L'odorat chez l'Enfant,* ed. B Schaal. *Enfance* 1:209–22.

Nietzsche F (1971). *Ecce Homo.* In: *Oeuvres philosophiques complètes,* ed. G Colli & M Montinari, trans. C Heim, I Hildenbrand, & J Gratien. Paris: Gallimard. (Originally published 1888.)

Plato (1961). *The Collected Dialogues,* ed. E Hamilton & H Cairns. Princeton University Press.

Pradines M (1958). *Traité de psychologie générale (1943–1950)*. Paris: Presses Universitaires de France.

Rousseau J J (1969). *Emile, ou de l'éducation*. In: *Oeuvres complètes*. Paris: Gallimard. (Originally published 1762.)

Schaal B, Rouby C, Maplier L, Soussignan R, Kontar F, & Tremblay R (1998). Variabilité et universaux au sein de l'espace perçu des odeurs: approches inter-culturelles de l'hédonisme olfactif. In: *Géographie des Odeurs*, ed. R Dulau & J R Pitte, pp. 143–65. Paris: L'Harmattan.

Schopenhauer A (1966). *The World as Will and Representation*, trans. E F J Payne. New York: Dover. (Originally published 1818.)

Simmel G (1912). *Mélanges de Philosophie Relativiste*. Paris: Alcan.

Sperber D (1974). *Le Symbolisme en Général*. Paris: Hermann.

Tellenbach H (1983). *Goût et Atmosphère*. Paris: Presses Universitaires de France. (Originally published 1968.)

Winnicott D W (1975). *Objets transitionnels et phénomènes transitionnels*. In: *Jeu et réalité*. Paris: Gallimard. (Originally published 1971.)

2

Cognitive Aspects of Olfaction in Perfumer Practice

André Holley

1. Introduction

For their survival, human beings are less dependent than many other mammals on the use of their olfactory systems. One of the reasons for this decreased biological role for smell in humans is that information about the world needed for everyday life is available to humans through a wide variety of channels, including language as a vehicle for scientific and technical knowledge. The cognitive content of olfactory cues, which is of vital importance to many vertebrate species, enabling them to behave efficiently in their physical and biological environments, is comparatively of very modest importance for humans, who have come to rely on more sophisticated and more accurate sources of information. This is especially manifest in advanced societies.

In contrast to the relative regression in the ways humans use the informative-cognitive content of odors, odors still exert a powerful affective dimension. Odors are pleasant or unpleasant. This emotional component presumably originated in an evolutionary strategy associating sensory pleasure with beneficial consequences of approaching various odor sources, and displeasure with dangerous consequences of such an approach. Although the link between the actual emotional effect of an odor and its potential behavioral meaning was not clear in most cases, there evolved a general attitude that consisted in attempts to avoid unpleasant odors and increase one's exposure to pleasant ones.

The art of concocting perfumes can be seen as a specifically human activity that leads a biological mechanism out of its natural domain of expression in order to make it serve a cultural purpose. In ecological conditions of olfactory perception, odors diffusing from objects (fruits, flowers, congeners, potential prey) indicate the presence of these objects, even though they may not be immediately perceptible through other sensory modalities. By contrast, in the realm of perfumes, the odors of the synthetic and natural compounds present in these complex mixtures are not attributable to identifiable objects. Odors are in some way separated from their primary referents and therefore lose their informative-cognitive potential. They become pure sensory qualities and positive affect inducers.

The reduction in the informative content of odors to the benefit of their affective content in perfume compositions might suggest that composing pleasant mixtures of odors is an activity that deals essentially with the emotional dimension of olfactory perception. However, it will be argued here that the reality is more complex: The art of perfumes actually is a privileged domain for investigating the involvement of cognitive processes in olfaction and, in addition, is a promising area in which to explore the links between emotion and cognition.

Perception, categorization, attention, memory, learning, mental imagery, language, innovation, and creation – all these traditional fields of cognitive studies are represented in the complex processes leading to the creation and appreciation of perfumes. The purpose of this essay is to identify specific lines of investigation that might lead to a better understanding of the process of perfume creation in particular, and thereby to a better appreciation of the cognitive involvements of olfaction in general. There will be several references to the writings of the French perfume composer Edmond Roudnitska, to whom the European Symposium on Olfaction and Cognition was dedicated.

2. Perception

Regarding perception, one may first ask if perfumers adopt behavioral or mental procedures different from those used by naive subjects when seeking to achieve maximum efficacy from their olfactory systems in assessing complex mixtures. Edmond Roudnitska instructed his students to use precise smelling procedures that were intended primarily to minimize sensory adaptation and increase attention and mental concentration. One can easily see that a full exploration of olfactory perception requires optimization of odor inhalation. It is less clear, however, whether or not trained practitioners develop procedures significantly different from those spontaneously adopted by naive subjects. Studies by Laing (1982, 1983) can be interpreted as giving a negative reply. Spontaneously, human subjects seem to adopt a sniffing technique that optimizes odor perception, and it is difficult to improve on the efficiency of this technique. A new finding could be relevant to this issue: Recently, Sobel et al. (1999) reported that alternate fluctuations in the resistance in the nostrils to the inhaled airflow resulted in concomitant variations in the analytical capacity of the nostrils for detecting the physical properties of various odorants. It could be interesting to investigate whether or not perfumers take advantage of this apparent cyclic functional differentiation of the nostrils in order to improve their perception.

It is currently recognized that odor attributes can be grouped in three main classes: quality, intensity, and hedonic valence. However, perfumers are inclined to avoid any reference to hedonism and instead insist on an additional dimension called "duration." For example, Roudnitska (unpublished notes) said that "for the perfumer, odors are classified with respect to quality, to intensity and duration. The latter must be distinguished from intensity, even though duration is expressed as intensity along time." Regarding this surprising lack of reference to the affective valence, one must understand that perfumers are more interested in the "behavior" of a compound introduced in a composition than in its intrinsic value in terms of pleasantness. An odor is classified primarily with regard to its potential utilization in compositions. Professionals need not consider the pleasant–unpleasant polarity as a criterion in designing perfumes; in general, they use a subset of odorants that all lie close to one pole of the hedonic dimension, the pleasant one. In addition, those odorants in their "palette" that have negative hedonic value when smelled individually can reveal positive properties when added to an odorant mixture in minute quantities.

Regarding the attribute called "duration," this term refers to purely physical characteristics of perfumes and, more precisely, to complex interactions taking place between the perfume and its support vehicle across time. In other cases, however, physical terms are used to designate perfume properties that can be said to include a sensorial, qualitative component. For example, "volatility" is used to describe a property that gives one the feeling that the odorous compound is propelled out from its support. At first examination, "volatility" would seem to designate physical properties having to do with the evaporation rates of the components. Nevertheless, one may wonder if that impression is really related to those physical properties or alternatively could be viewed as an aspect of the olfactory quality. Such was suggested by Roudnitska (unpublished notes): "I think that for a limited number of products, there is a particular phenomenon [volatility] distinct from evaporation." Along the same line, the term "volume," sometimes used to qualify a perfume, could refer more to the impression of harmony and plenitude of sensation induced by the perfume than to physical parameters such as evaporation.

3. Perceptual Analysis and Description of Perfumes

When attempting to describe the quality of sensations evoked by a fragrance, perfumers use notions such as "notes," "faces," and "sub-tones," terms whose meanings are not always immediately clear. One might think that these terms refer to odor qualities imparted to the mixture by individual components or subsets of components. However, it must be noted that the same kinds of descriptors are used to describe pure chemicals, as, for example, in the handbook published by Arctander (1986). It is no longer possible to identify these "notes" as being some primary odors that would be combined to produce the particular odor of any product, according to the principle popularized by Amoore (1967). A more reasonable interpretation would be that notes represent perceived similarities between a given odor (the odor of a pure compound or of a complex mixture) and several other odor qualities taken as references.

It should be noted that describing odor qualities as independent entities is not an ecologically founded activity. Odors are essentially attributes of objects and substances whose natural function is to reveal the presence of those objects and substances in the environment. As a consequence, odor naming turns out to be odor-source naming (see Chapters 4 and 6 in this volume for recent discussions). In some cultural practices, such as those exemplified by perfumery, odors become dissociated from their sources insofar as their qualities rather than their referents are brought into focus. However, the linguistic tools available to a speaker remain those relevant to source naming, and because a vocabulary to describe pure qualities is almost nonexistent (at least in French and English), qualities must therefore be designated by the names of their most representative sources.

Several factors can enter into the choice of which odor source will be seen as representative of a quality perceived in a complex fragrance. Obviously, one of these factors is the sensory similarity between the detected note and the quality of the referent. In addition, the referent must be readily available and be familiar to those who engage in linguistic exchange about odors. Meeting these requirements is not an easy task. Long sessions of "linguistic negotiation" among panels of professionals are needed to reach a consensus on the best term to describe a note. Still much more difficult would be selection of a closed set of descriptors intended to describe all fragrances. Because there is no theoretical reason to think that there is a finite number of odor qualities, attempting to reduce the list of reference terms to a few dozen items, as has sometimes been attempted, would be a project that necessarily would involve considerable arbitrariness, even though its usefulness would be indisputable.

In a successful composition whose odor components are adequately balanced, these components lose part of their perceptual individuality and fade away, to the benefit of the whole fragrance. Some professionals designate such a composition an "olfactory form." In this context a "form" is a Gestalt, a complex perceptual structure undergoing a spatiotemporal development that yields more than the sum of its components, just as a melody is not reducible to a sum of notes and can keep its individuality through several transpositions. Roudnitska (1983b) wrote: "A British specialist in perfumes asked me to describe as simply as possible, for a magazine, what I understand by the expression olfactory form. Here is my reply: When you hear a very well-known musical tune, even if you have never learned simple scales or an instrument, even if you are completely uninitiated in music, you nevertheless recognize immediately the tune in question, even without hearing a single word that goes with this tune. How is it that you recognize this familiar tune? Quite simply by its musical form."

If one accepts this musical metaphor, clearly borrowed from Gestalt theory, it follows that a given olfactory form may keep its personality even though there may be substitutions for several of the individual odors that compose it. There is a practical consequence of adopting this view. For example, in terms of patent rights, the perfumer may want to protect his creation from copying or imitation, basing his ownership claim not on a list of the products included in his perfume but on its "olfactory form." However, whereas there are objective means to compare two melodies, there is no objective criterion for determining if two fragrances have similar forms.

4. Learning and Memory

Learning and memory are essential components of the perfumer's expertise. It is well known that becoming a perfumer requires very long training. In the absence of any possible generalizations regarding the sensory behaviors of odorants entering into the composition of a complex mixture, perfumers must memorize large amounts of empirical information acquired from experience. Predicting the intensifying and qualitative effects of mixing several compounds is practically impossible if one has no previous experience with these products. What are the best conditions in which to encode odors? Can we draw interesting indications from the training practices of perfumers?

Roudnitska (1983a) described an exercise he used for his students: "Exercise 1 consists in a very simple mixture of six products, all of them constituents of the essence of Neroli, but intermingled in such a way as not to make it too easy. The student will then be asked to reconstruct the composition, i.e., to identify – by smell alone – the constituents and their proportions." That initial part of the training program was aimed at developing the analytical capacity of students by providing them conditions that would favor their familiarization with odorants perceived individually and in different combinations. It is interesting to note the teacher's concern "not to make it too easy," emphasizing that the identifiability of individual constituents would depend on their relative concentrations. Incidentally, the exercise provides additional insight into the perfumer's attitude when he creates a composition (i.e., when he works on the synthetic side of his art). Commenting on his exercise, Roudnitska wrote: "When preparing the Neroli exercise, for example, what do I really do? I try to schematize the arrangement worked out by Nature. . . . What then is my problem? It is precisely to try and schematize, i.e., obtain by means of six products only the Neroli effect, which Nature achieved with several hundred of constituents. And since it is more difficult to conceal the identity of six products than that of several hundreds well intertwined, this requires a great effort in conciseness and rigorous harmonization."

Another question related to odor and perfume memory is whether or not we can reach a consensus regarding the role of odor naming in odor encoding and recognition. As noted earlier, professionals wanting to exchange views about fragrances use descriptors. Does the fact of knowing the name of an odor and its semantic description have a positive influence on one's long-term recognition of this odor? There is an ongoing debate about the role of semantic processing in the encoding of odors (Engen, Kuisma, and Eimas, 1973; Engen and Ross, 1973; Lawless and Cain, 1975; Rabin and Cain, 1984; Walk and Johns, 1984; Lyman and McDaniel, 1986; Jehl, Royet, and Holley, 1997). Whatever the conclusion of this issue, there are conditions in which the perfumer's performance necessarily calls for the use of semantic knowledge. For example, one might well be skeptical if a perfumer claimed that he could recognize all the compounds in a complex fragrance made of 10 or 20 products. There have been controlled tests indicating that humans have great difficulty in discriminating and identifying more than three – exceptionally four – components in olfactory mixtures (Laing and Francis, 1989; Livermore and Laing, 1998). What is more, training and experience seem to have little influence on discriminative performance (Livermore and Laing, 1996). However, it must be supposed that the discriminative sensory capacity of the expert will be strongly aided by the semantic dimension of his expertise, insofar as the mixture to be analyzed will have been composed by another perfumer. Professionals have learned, for example, that a neroli "accord" can be made by combining six particular constituents. Recognizing a neroli note in a perfume makes it possible to hypothesize that the mixture contains the main constituents of neroli, thus somewhat orienting the sensory analysis. Because perfume composition is an art based partly on traditional rules, knowledge of these rules – the semantic knowledge – is synergistic with nonsemantic, purely sensory expertise. By contrast, when confronted with a mixture prepared by experimenters ignoring professional habits and traditions, perfumers must rely solely on their detection and recognition capacities, and their performances are more like those of naive subjects.

5. Attention

Olfactory perception is closely linked to attention. Whereas many studies have explored the roles of different kinds of attentional processes in vision, very few have investigated olfactory attention. When one's attention is not directed toward olfactory stimuli, and especially when it is directed toward other sensory modalities or non-olfactory tasks, stimuli of low and middle intensities, even though ordinarily they would be readily detected, fail to be detected. In one study, subjects presented with a fragrance while engaged in a demanding cognitive task (playing an electronic game) were not aware of the olfactory stimulation, whereas they readily detected the fragrance at the same concentration in a subsequent test when they were instructed to expect an odor (Bensafi et al., 1998). There is no doubt that perfumers make use of such modality-specific attention when exploring the quality of a fragrance.

Along the same line, it should be useful to further investigate the relationship between attention and learning. One study found that subjects exposed to odors without being aware of such exposure subsequently exhibited behaviors that were influenced by their exposure to those odors (Degel and Köster, 1999). The concept of "unconscious odor conditioning" was introduced in such a context by Kirk-Smith, Van Toller, and Dodd (1983). In a review of the acquisition of odor hedonics, Hermans and Baeyens (Chapter 8, this volume) discuss several examples of such acquisition taking place even in subjects who were not aware that they had been presented with an odor linked with a positive or negative event.

Another open question is whether or not attention can be focused on sub-modal features. Perfumers seem to be able to focus their attention on different "notes" or "sub-tones" of a complex fragrance. For example, one might concentrate on a given note during a first sniff, then on another note during a second inhalation. How can olfactory attention be directed toward particular components of a fragrance? Are attentional processes related to sniffing in humans, and if so, how are they related? It is tempting to speculate that experts who want to analyze a complex odor begin their analysis by assuming the presence of a certain note on the basis of a quick first impression, then evoke a mental representation of that note from memory, and confirm or abandon that initial assumption by comparing the mental image with actual perceptions during subsequent inhalations. Could this be systematically investigated? Laing and Glemarec (1993) reported that subjects asked to identify the components of mixtures consisting of up to six components performed at about the same level when they attempted to identify all the components present as when they attempted to identify only one component. In trials conducted under the latter condition, attempts at selective attention did not significantly influence the identification process.

6. Olfactory Mental Imagery

Another interesting topic is conscious mental representation of odors. There are divergent opinions on the possibility of achieving mental imagery in the olfactory domain. However, perfumers claim that they can elaborate a schematic for a perfume in the mind, sometimes for long periods, before formulating it. Roudnitska (1983a) wrote that "a perfume, like a piece of music, can be composed on paper without the help of any other sensorial references than those to be found in the mind or the memory, the senses being used only afterwards as a mean of checking up." Would it be possible for a perfume composer to conceive an "olfactory form" without representing it mentally in some way? One might argue that professionals possess a vast store of semantic knowledge of their materials. They know that a certain "accord" can be achieved by combining such and such odors in adequate proportions. They have learned a number of rules that must be followed in order to achieve a well-balanced composition. It is hardly conceivable, however, that one would be able to create original combinations of odors without being able to evoke a mental image of the Gestalt one wants to elaborate.

When questioned about their ability to imagine odors, perfume composers agree that they do have this capacity. Roudnitska (1983b) explicitly states that olfactory mental images are postulated by perfumers: "If you have been in love with a woman who used Arpège, and if, several years later, someone mentions in your presence the name Arpège, won't your mind call forth the particular form of this perfume just as quickly as if you had the bottle right under your nose?"

Demonstrating experimentally that humans are able to evoke mental olfactory images is difficult. Interesting attempts were those by Lyman (1988) and by Carrasco and Ridout (1993), who compared the patterns of odor similarity ratings for a set of odorants presented to subjects and the corresponding patterns of similarity ratings for the same odorants simply evoked by their names. The presented odors and the linguistically evoked odors showed quite similar grouping patterns. That was interpreted as indicating that subjects were able to evoke from memory mental representations of odors and compare those representations. However, it has been objected that odor names could evoke a list of semantic properties and that similarity evaluations could be based on previously acquired knowledge, rather than on iconic mental evocation of odors. An experimental strategy that should not incur that criticism has recently been suggested (Sulmont, 2000). It was inspired by some research on visual mental imagery conducted by Kosslyn and associates, who compared the imagined properties of visual stimuli with their physical properties. According to Sulmont, similarity judgments should be made in pairs between familiar odors evoked by their names and unfamiliar odors physically presented to subjects. It is assumed that subjects will evaluate the degress of separation between images of familiar odors and perceptions of unfamiliar odors. This condition should reduce the possibility that similarity judgments would be founded on purely conceptual properties, as unfamiliar odors would have minimal semantic properties. Finally, additional evidence in favor of the possibility of mental olfactory imagery can be found in functional magnetic-resonance imaging (MRI) studies of the human brain (Levy et al., 1999): Subjects asked to imagine familiar odors show brain activation, and the activation occurs in regions similar to those activated by real odors.

7. Creation

Artistic creation remains a very complex and mysterious process, whatever its domain of expression, whether music, painting, or olfactory composition. Creative abilities are no more easily explained in olfaction than in any other artistic domain. However, the art of composing in perfumery encounters an unusual problem that arises from possible confusion between sensory emotion and aesthetic emotion. When they want to practice art with odors, perfumers are challenged by the apparent ease of their task. Most odors that are the raw materials of their compositions are pleasant fragrances (even if perfumers claim that they do not take pleasantness into consideration). Intrinsically, odors are endowed with hedonic attributes, most often positive hedonic attributes, so that mixtures of such odors can rather easily evoke pleasant feelings provided that they are combined with at least a minimum of skillfulness. Nevertheless, aesthetic emotions, which are expected to be generated in the art lover's mind by fine pieces of art, must be distinguished from purely sensory emotions.

A fine perfume is much more than a good-smelling mixture. In perfumery, the innovative composition that is artistic creation involves cognitive components similar to those involved in other varieties of art. Like any other artwork, a fine perfume takes its place in our short artistic-cultural history made up of streams of tradition. Some famous perfumes have played leading roles in that tradition. A fine perfume achieves a delicate balance between tradition and novelty. It makes reference to earlier works, often in addition to introducing innovative departures in its construction, incorporating new notes or startling combinations of more familiar notes. Connoisseurs obviously are not immune to the sensory pleasure of a fine perfume. However, their sensory emotion is amplified and converted into aesthetic emotion when they feel that they have access to the cognitive dimension of the artwork.

8. Conclusion

The modern era of perfumery began more than a century ago with the discovery of new synthetic compounds and their introduction into perfumes. Traditionally, the relationship between perfumery and scientific knowledge has featured a privileged connection with chemistry, whose syntheses have enriched the palette of perfumers with original notes, stimulating the creative imagination. This brief survey shows that perfume design is open to another kind of scientific investigation. In the realm of perfumery the cognitive approach to olfactory processes should find a rich collection of scientific problems, and, reciprocally, cognitive investigations should provide perfumers with a better understanding of their art.

References

Amoore J E (1967). Specific Anosmia: A Clue to the Olfactory Code. *Nature* 214:1095–8.

Arctander S (1986). *Perfume and Flavor Chemicals.* Montclair, NJ: Arctander.

Bensafi M, Rouby C, Farget V, Vigouroux M, & Holley A (1998). Effet du contexte olfactif sur une tâche de prise de décision. In: *Proceedings of the VII ème Colloque de l'Association pour la Recherche Cognitive*, ed. D Kayser, A Nguyen-Xuan, & A Holley, pp. 250–4. Paris: Universités de Paris 8 et Paris 13.

Carrasco M & Ridout J (1993). Olfactory Perception and Olfactory Imagery: A Multidimensional Analysis. *Journal of Experimental Psychology: Human Perception and Performance* 19:287–301.

Degel J & Köster E P (1999). Odors: Implicit Memory and Performance Effects. *Chemical Senses* 24:317–25.

Engen T, Kuisma J E, & Eimas P (1973). Short-Term Memory of Odors. *Journal of Experimental Psychology* 99:222–5.

Engen T & Ross B (1973). Long-Term Memory of Odors with and without Verbal Description. *Journal of Experimental Psychology* 100:221–7.

Jehl C, Royet J P, & Holley A (1997). Role of Verbal Encoding in Short- and Long-Term Odor Recognition. *Perception and Psychophysics* 59:100–10.

Kirk-Smith M, Van Toller C, & Dodd G (1983). Unconscious Odor Conditioning in Human Subjects. *Biological Psychology* 17:221–31.

Laing D G (1982). Characterization of Human Behaviour during Odour Perception. *Perception* 11:221–30

Laing D G (1983). Natural Sniffing Gives Optimum Perception for Humans. *Perception* 12:99–117.

Laing D G & Francis G W (1989). The capacity of humans to identify odors in mixtures. *Physiology and Behavior* 46:809–14.

Laing D G & Glemarec A (1993). Selective Attention and the Perceptual Analysis of Odor Mixtures. *Physiology and Behavior* 52:1047–53.

Lawless H T & Cain W S (1975). Recognition Memory for Odors. *Chemical Senses* 1:331–7.

Levy L M, Henkin R I, Lin C S, Hutter A, & Schellinger D (1999). Odor Memory Induces Brain Activation as Measured by Functional MRI. *Journal of Computer Assisted Tomography* 23:487–98.

Livermore A & Laing D G (1996). Influence of Training and Experience on the Perception of Multicomponent Odor Mixtures. *Journal of Experimental Psychology: Human Perception and Performance* 22:267–77.

Livermore A & Laing D G (1998). The Influence of Odor Type on the Discrimination and Identification of Odorants in Multicomponent Odor Mixtures. *Physiology and Behavior* 65:311–20.

Lyman B J (1988). A Mind's Nose Makes Scents: Evidence for the Existence of Olfactory Imagery. Dissertation Abstracts International 48, 2807B. Quoted by M J Intons-Peterson & M A McDaniel (1991). In: *Imagery and Cognition*, ed. C Cornoldi & M A McDaniel, pp. 47–76. Berlin: Springer-Verlag.

Lyman B J & McDaniel M A (1986). Effects of Encoding Strategy on Long-Term Memory for Odors. *Journal of Experimental Psychology: Learning, Memory and Cognition* 16:656–64.

Rabin M D & Cain W S (1984). Odor Recognition: Familiarity, Identifiability, and Encoding Consistency. *Journal of Experimental Psychology: Learning, Memory and Cognition* 10:316–25.

Roudnitska E (1983a). The Physiologist and the Perfumer. *Perfumer and Flavorist* 8:1–7.

Roudnitska E (1983b). The Investigator and the Perfumer. *Perfumer and Flavorist* 8:8–18.

Sobel N, Khan R M, Saltman A, Sullivan E V, & Gabrieli J D E (1999). The World Smells Different to Each Nostril. *Nature* 402:35.

Sulmont C (2000). Impact de la mémoire des odeurs sur la réponse hédonique au cours d'une expérience répétée. Doctoral dissertation, Université de Dijon (France).

Walk H A & Johns E E (1984). Interference and Facilitation Effects in Short-Term Memory for Odors. *Perception and Psychophysics* 36:508–14.

3

The Specific Characteristics of the Sense of Smell

Egon Peter Köster

1. Introduction

In the nineteenth century, a distinction was commonly made between the "higher" senses, vision and audition, and the "lower" senses, touch, taste, and smell. In an age in which, at least in the Western world, faith in science and technological progress was almost absolute and bodily pleasures were viewed with suspicion, the senses of the intellect seemed to score a moral triumph over the senses of the body. Or was there more to the distinction than that? Are the two types of senses indeed different? And can they even today be grouped as they were then, but on more objective grounds?

Vision and hearing are involved in such vital human activities as spatial orientation (distance and depth perception, direction perception for sound sources, equilibrium) and communication (hearing, speaking and reading language, perception of body language, imitation of expressions and gestures). Furthermore, vision plays a very important role in form perception and in gross and fine manipulation of objects.

Finally, vision and hearing are the vehicles of the arts (painting, sculpture, architecture, dance, music, theatre, cinema, and photography). Compared with that, touch, kinesthesia, taste, and olfaction can show only lesser glories (perfumery and cooking, and, to a certain extent and only in combination with vision, pottery, sculpture, dance, and pantomime). They also seem rather subjective and less universal – more related to feelings and emotions than to thoughts and decisions. What, then, are the advantages of having these senses? And what are their characteristics? They seem to be involved mainly in contributing to security, well-being, and pleasure. In short, they tend to make us feel at home in our world. In this chapter, some fundamental differences between one of the lower senses, olfaction, and one of the higher senses, vision, will be discussed, and the methodological consequences of these differences for cognitive olfactory research and for sensory evaluation of foods and beverages will be shown.

2. Characteristics of the Sense of Smell

The sense of smell differs from vision and also from hearing in terms of a number of factors. Most of these differences are not absolute, but, at least in humans, the differences in those factors are spaced at rather large distances on a continuum. In certain animals, some of these distances will be much smaller, or even nonexistent. The main characteristics are discussed in the following sections.

2.1. The Sense of Smell Is a "Nominal" Sense

Unlike vision and hearing, in which non-conscious assessments of relative intensities, sizes, directions, hues, and tones play a major roles in informing us about the structure of the world around us, the sense of smell primarily provides us simple "nominal" data about the presence of qualitatively different odors in our surroundings. In order to provide such nominal information, olfaction combines good absolute sensitivity (for many odors, the nose is still the most sensitive instrument we have) with excellent quality discrimination (Holley, 1999). The latter feature enables us to distinguish almost any two odorous compounds presented to us. Even enantiomers like R-(−)-carvone (spearmint) and S-(+)-carvone (caraway) or S-(−)-limonene and R-(+)-limonene [the R-(+) being more orange-like than the other] can be distinguished (Laska and Teubner, 1999). The word "nominal" should not be taken too literally, however, for although it is easy to distinguish many different odors, it is difficult to identify them by name (Cain, 1979). In fact, it is much easier to detect that two successively presented odors are not the same than to detect that they are identical (Köster et al., 1997). This indicates that in olfaction, quality discrimination is much more important than identification. In vision, the reverse is true, and matches are always detected faster than non-matches in visual priming experiments (Posner, 1986).

At the same time, olfaction has rather poor intensity discrimination. When starting from a given odor concentration, the concentration of that odor has to be raised by about 20% in order for a difference to be perceived. In vision, an increase of only 2% in brightness suffices for detection of a difference. Obviously detection, discrimination, and recognition of odors as familiar or unfamiliar are more important than assessments of intensity gradients or verbal identification of the odor. All of this is in accordance with the function of olfaction as a warning system against dangers in the immediate environment or in food taken into the mouth. Changes in quality must be detected immediately, because we cannot stop breathing, and we must not swallow toxic food.

2.2. The Sense of Smell Is a "Near" Sense

In vision and audition, information about distant objects is transmitted to the eye and the ear by the mediation of light and sound waves, and the structural properties of a distant object are reconstructed mentally with the help of inborn rules. In olfaction and taste, all information about the quality of an object is contained in the molecules that make direct contact with the receptors. The same is true, to a certain degree, for the sense of touch when the textural properties of food in the mouth are monitored, but during active manipulation of objects in the mouth or outside of the body, touch and kinesthesia cooperate in a form of object reconstruction over time. Clearly, in regard to touch, even with the same degree of receptor stimulation, there is a distinction between the experience of active intentional touching and the experience of the passive stimulation of being touched (Gibson, 1968). In fact, such distinction is seen with all the senses, but active intentional searching is a much more prominent feature in vision than in any of the other senses. Olfaction and taste tend to be rather passive until a stimulus arrives, and even then we rely mostly on vision to find its source. When no head movements are made, an odor can offer no clue as to the direction in which its source can be found, but merely serves as a cue for the visual search. Kobal, Van Toller, and Hummel (1989) found that only when olfaction was accompanied by pain sensations transmitted by the trigeminal nerve did it give some directional information. However, recent findings by Boireau et al. (2000) using vanillin as a non-trigeminal odor have raised some doubt about that conclusion. Vision dominates, however. A well-hidden coffee cup will be immediately detected in a visual display when the odor of coffee is presented to a subject (Degel and Köster, 1998, 1999).

The fact that "far" senses like vision and hearing have such prominent roles in our perceptions of the structural properties of the world, such as spatial orientation, distance and form perception, whereas olfaction is truly a "near" sense, suggests the basis for another major distinction between the two types of senses. In regard to spatial orientation, inter-subjectivity (all subjects perceiving the world in exactly the same way) has great survival value for most animals that live in groups, including humans. Imagine having to deal with automobile traffic without being sure that the other participants have the same perceptions of distance, form, and size. Nobody would dare to cross a street. Also, we rely on the precision of other people's directional hearing many times each day. They simply know where we are, even if they do not see us. It is therefore convenient that the mechanisms for these perceptions are inborn, even if they still have to be somewhat further developed in the first years of life, as in the case of size constancy. This guarantees that we see the same things (forms, distances), even though our interpretations of what we see may be different. A violin will look the same (form, size) to us and to a man who has never seen one before, but he may not interpret it as a musical instrument and thus may devote less attention to the differences between the strings. In olfaction, which is not involved in spatial orientation (but see Howes, Chapter 5, this volume), there is no necessity for such strict inborn inter-subjectivity. Perhaps as a result, the only inborn feature of olfaction and taste seems to be aversion to things that are decaying (the odor of mushrooms, cadavers) or bitter, and even some of those smells and tastes we can learn to appreciate (Limburger cheese, durian fruit, Campari). In fact, the absence of inborn mechanisms in olfaction is an evolutionary advantage in an omnivore (see Section 2.4). At the same time, however, that means that olfactory variability among people is much greater than their variability for most other senses. Threshold sensitivities for certain odorants (amyl acetate, pyridine) can easily vary by a factor of 1,000 between individuals (Koelega and Köster, 1974), and people who are very sensitive to one substance may be quite insensitive to another (Punter, 1983). Specific anosmia, a strongly reduced sensitivity to one particular odor in a person who shows normal sensitivity to other substances, is the rule rather than the exception, although it seems to occur more often with certain odorous substances than with others (Amoore, 1977). A large number of such specific anosmias have been described, and they can be found in different combinations. Even in its deficiencies, vision follows more general rules: The number of receptor types is small, and as a consequence the number of forms of color blindness is small. It is clear that people's wide-ranging differences in specific olfactory sensitivities also influence their perceptions of the very complex odor mixtures to which they are exposed in everyday life. Thus it can be concluded that people differ much more in the way in which they perceive odors than in the way they perceive visual objects. These large inter-individual differences in olfaction have their origins not only in differences in interpretation, as in vision, but also in the sensory basis of the perception itself. That such differences do not harm us is related to the fact that we do not use odor information in orientation and movement and that we *learn* to attach (emotional) meaning to odors. Even though a rose, with its mixture of flowery and fecal odors, may smell different to each of us, we all have learned to like the smell and to connect it with love and tenderness (and thorns).

2.3. The Sense of Smell Is a "Hidden" Sense

Unlike vision, olfaction is seldom the focus of attention. In vision, much information is processed without explicit awareness (one can drive a car for many miles while thinking about other things and without paying close attention to the surrounding traffic), but in olfaction, awareness of odors is the exception rather than the rule. We inhale all day, and smells can influence our moods (Ehrlichman and Halpern, 1988; Knasko, Gilbert, and Sabini, 1990; Bastone and Ehrlichman, 1991; Kirk-Smith and Booth, 1992; Knasko, 1993; Gilbert, Knasko, and Sabini, 1997), the time we spend in various locations (Teerling, Nixdorf, and Köster, 1992), and our perceptions of other people (Teerling and Köster, 1988). In some cases, odors can influence performance on vigilance tasks (Warm, Dember, and Parasuraman, 1990) and mathematical tasks (Baron, 1990), even when the subjects are unaware of the presence of the odor (Degel and Köster, 1999). The latter research suggests that the presence of unnoticed and unidentifiable odors may have a stronger effect than that of clearly perceptible odors or odors that can be identified. In fact, the sense of smell seems to be trying to hide its presence. Adaptation in olfaction, which is loss of sensitivity as a result of prolonged stimulation, is very strong and often complete (no remaining awareness of the stimulus). The same holds for habituation, which is a reduction of attention and responsiveness to monotonous stimuli (see Dalton, 1999, who erroneously calls it adaptation). Furthermore, suppression of odor intensities when odors are mixed in uneven ratios is much more frequently seen than is mutual enhancement or synergism (Köster, 1969; Köster and MacLeod, 1975). But even when olfactory stimuli can no longer be consciously perceived or are no longer attended to, they continue to exert influences on behavior and mood. It is in this way that malodorous substances of animal origin – civet, for instance – are completely hidden in perfumes that without them would not seem nearly so attractive.

The fact that humans are seldom aware of the odors that influence them is one of the reasons for their difficulty in talking about and describing odors. We have only a few abstract words to describe odors, like "fresh" or "musty," and even these words do not have the same meaning for everyone (see Howes, Chapter 5, this volume). By far the majority of odors are described by the names of their sources (strawberry, coffee, etc.). It is well known from sensory analysis and from perfumery practice that lengthy training is required before people can reliably describe olfactory experiences in detail. Furthermore, it has been demonstrated (Degel and Köster, 1998, 1999; Degel, Piper, and Köster, 2001) that odors can be implicitly remembered without any awareness of the learning event, and such memories can be disturbed by explicit knowledge about the odor (see Section 2.5 and Issanchou et al., Chapter 13, this volume).

If olfaction is seen as a "hidden" sense, then in normal everyday life perhaps odors are best not talked about, but simply experienced, with all the surprisingly emotional reactions and memories they entail.

2.4. The Sense of Smell Is an "Associative" and "Emotional" Sense

As indicated earlier, in olfaction there seem to be few inborn features or mechanisms: the aversive reactions to smells of decay and dead bodies, along with the strong tendency for adaptation, mixture suppression, and habituation. In contrast to the situation with taste, where, apart from our aversion to bitter tastes, at least our preference for sweet seems to be inborn, there are no inborn preferences in olfaction, and all that we like, we probably have learned to like. Because humans, like rats and pigs, are omnivores, not having inborn preference restrictions in our food choices has been advantageous and has permitted humans to live almost everywhere in the world. Very early in life, and perhaps even prenatally (Schaal, 1988; Schaal, Marlier, and Soussignan, 1998, 2000), we learn to distinguish between the edible and the nonedible. In this distinction, olfaction and taste play the major role. Many of our first food perceptions are related to motherly warmth and care and may remain with us for a good part of our lives (Haller et al., 1999). In general, odors are easily linked to emotions by association (Kirk-Smith and Booth, 1987; Robin et al., 1999), and olfactory memories have been found to be more emotional than visual memories (Herz and Cupchik, 1992, 1995). Many, if not most, of these links are created without any intention to learn and result simply from contingent circumstances in which the subject may be unaware of the presence of the odor. Thus the emotions evoked by a given odor can be very different from one individual to another, depending on the different emotional encounters with that odor in each person's personal history.

Dalton and colleagues (1999) have demonstrated that perceptual responses to odors can easily be influenced by emotional suggestions. They exposed subjects to clearly perceptible odors and asked them to rate the odor intensities regularly during periods as long as 20 min. Before going into the odor room, different groups of subjects were given the impression that the odors were "hazardous to health," "neutral," or "relaxing." As would be expected, that created rather strong response biases that were expressed in differences in response habituation (not "adaptation" as they stated). Those authors seemed to think that was typical for olfactory perception, but they never compared their results with those from similar experiments in other senses. If one went to do the same with a tone of a certain frequency and were to suggest to the groups that prolonged exposure to it might lead to "epileptic seizures" or to "healthier sleep," one probably would obtain similar results. If it could be shown that under circumstances less unnatural than those used in these experiments olfaction was indeed less resistant to suggestion than other senses, it would emphasize that odors are excellent elicitors of emotions.

2.5. The Sense of Smell Has a "Special Memory"

From the foregoing it should be clear that learning and memory play important roles in olfaction. Is this the same type of memory as the memory for words and word associations that has been studied in psychology for about a century, ever since Ebbinghaus wrote his book on memory in 1885? In a critical review, Herz and Eich (1995) presented a number of arguments to support the view that olfactory memory is different from verbal memory: Olfactory memory is episodic and not semantic in nature. We can remember where and when we encountered an odor before, but in most cases we cannot call up the name of the odor, and if we do, often it is by deduction: "When I was small, I smelled this odor in the attic of my grandparents' house. They kept apples there during winter. It is dried-apple odor!" Other differences between verbal memory and olfactory memory have been reviewed by Herz and Engen (1996). Thus, forgetting seems to follow a different time course for olfaction then for verbal memory. Most odors that are presented to people in test sessions are more rapidly forgotten than are words presented under the same conditions, but those few odors that are remembered seem to linger on in memory almost forever, whereas words are gradually forgotten completely. Furthermore, it is clear that olfactory memory, as measured by nonverbal methods such as recognition, is better in women than in men. Such differences are not found with verbal or visual memory. Because of the episodic nature of olfactory memory and its longevity, odors play an important role in making us feel at home in the world. Without any explicit awareness on our part, we link emotionally meaningful situations to them and carry them with us in an implicit memory that renders them up (usually involuntarily) when those odors are encountered again.

Listening to people who have recently lost their olfactory sensitivity is perhaps the best way to learn about the real meaning of the sense of smell and olfactory memory in human life. Not only have such people lost the pleasure of eating and drinking, but also they do not seem to come home by the same routers, and they miss that sudden appearance of their most vivid memories. Generally, they complain that their emotional lives have become less eventful and grayer in tone. Some of them suffer to the point of depression (Van Toller, 2000), though others clearly are less affected by the loss. It may well be that those in the latter group have the capacity to imagine odors, a gift that seems to be found in only part of the population (Köster et al., 1997). They probably can compensate for the perceptual loss by imagining the odors they can no longer perceive. Visual imagery is much more common than olfactory imagery.

In conclusion, it is clear that vision and olfaction differ in terms of a large number of factors. Table 3.1 summarizes the differences and sketches a number

Table 3.1. *Differences between vision and olfaction and their consequences for future research*

Vision	Olfaction	Consequences for olfactory research and sensory analysis
Relative perception: *structural* (intensity, direction, depth, size, distance, movement, form, etc.); good intensity discrimination	Absolute perception: *nominal* (qualities and discrimination); poor intensity discrimination	More research on the processing of olfactory quality is needed
Strict inter-subjectivity ("far") (inborn perception mechanisms); individual variations only in interpretation	No strict inter-subjectivity ("near") (almost nothing inborn); individual variations in both sensitivity and interpretation	Large panels necessary; analysis of smaller subsets of perceiver (and consumer) populations
Usually at the center of attention ("overt" sense); also subconscious information processing	Almost never at the center of attention ("hidden" sense); subconscious processing the rule rather than the exception	Explicit (conscious) methods may lead to artifacts; implicit methods (reaction time, authenticity) needed
Prone to forming more objective associations	Prone to forming more subjective and emotional associations	More indirect methods of investigation needed
Semantic memory prominent; more *explicit* learning and memory	Episodic memory prominent; more *implicit* learning and memory	Situational analysis needed; implicit memory methods needed; questions about actual behavior rather than about feelings and opinions

of the consequences those differences will have for the development of methods and techniques for research in the fields of olfaction and sensory analysis. These consequences will be discussed in the remainder of this chapter.

3. Methodological Consequences

Although many authors have pointed out differences between olfaction and other senses, the methodological consequences of those differences have seldom been drawn. Thus, in the area of fundamental research there is a vast literature on olfactory intensity, but very little on odor quality and odor discrimination. Since the odor classifications of Linnaeus (1764), Zwaardemaker (1895), and Henning (1916), which were based on similarity judgments, there have been few attempts to study qualitative relationships in the world of odors (Harper, Bate-Smith, and

Land, 1968). Besides the odor theories of Wright (1982) and Amoore (1962, 1963), who used certain examples of odor similarities to try to illustrate the validity of their models, and subsequently were shown other examples of similarities that tended to discredit the generality of their theories, there have been some systematic efforts to classify pure odors using more quantitative methods. Amoore (1967, 1977) based his effort on studies of specific anosmias, considering the odorous stimuli for which specific anosmias occurred as primary odors and measuring the relative hyposmias for other, but related, odors in specific anosmics in order to assess the part played by the "primary receptor" in the perception of these related odors. Woskow (1968) used similarity scaling and multidimensional scaling techniques to assess qualitative distances among a group of 25 odors. He found that the most important dimension in their three-dimensional solution was strongly related to independent measures of odor liking, a variable that obviously is highly culture-dependent. In an attempt to circumvent such culture dependence, Köster (1971) tried to classify odors on the basis of their cross-adaptational relationships. Although he was able to show a pecking order and some general patterns in nonreciprocal cross-adaptations between substances, which indicated that some odorous substances had access to more extensive parts of the receptive system than did others, the method was too complicated and time-consuming to lead to a realistic odor classification.

Since then, odor quality has seldom been investigated in fundamental research. Even problems of odor mixing, in an area that would seem predestined for qualitative research, have been studied almost exclusively with intensity measures (Köster, 1969; Köster and MacLeod, 1975; Laffort, 1989; Berglund and Olsson, 1993; Olsson, 1994; Cain et al., 1995). Laffort and Olsson attempted to study qualitative aspects, but only in binary mixtures. Meanwhile, in applied odor research and in sensory analysis of foods and drinks, methods for description of the odorous qualities of the much more complex mixtures encountered in real life have been developed. A large vocabulary of descriptive terms has been developed, and data banks of odor descriptors for special products have been compiled. Forced by the need to produce data that would have external validity, those working in applied research went ahead and attacked the problems from which fundamental research had shied away. And even though some of the methods they used can be criticized, they have contributed much more to our knowledge of odor quality than have the sterile and intensity-dominated methods of the psychophysicists. Furthermore, the attention paid to odor quality in applied research made it clear that strict inter-subjectivity was not to be found in olfaction: Different observers will have somewhat different perceptions of a given stimulus. That led not only to the use of larger testing panels but also to the development of new statistical methods allowing for analyses of various

subgroupings of observers and reanalysis of the data based on differences between them (Schlich, 1996; Dijksterhuis, 1997). Unfortunately, such methods
remain largely ignored in fundamental research, where averaging over subjects
(based on the false assumption that all observers are equivalent) is still the predominant way of dealing with data. Thus it becomes clear that what may be a
valid approach in the higher, "far" senses like vision and hearing, with their strict
inter-subjectivity, may not be appropriate for a lower, "near" sense like olfaction,
even if only simple sensory perception is involved. Such differences appear even
more pronounced when one considers the roles that attention plays in vision and
in olfaction. Whereas with an "overt" sense like vision the objects usually are at
the center of attention, that is not the case for odors, which most of the time are
not observed at all.

Nevertheless, odors influence our moods and our emotions, and they can
brighten our world with memories of places where we encountered them before,
without being aware of perceiving them. In the literature, it is often said that we
find it so difficult to name odors because, unlike the situation with colors, our
parents never taught us to name them. It is true that parents take great delight in
teaching children to name colors, but the proponents of that theory forget that
such learning usually is limited to about 10 words and that such a number is
ridiculously low if one considers the vast numbers of completely different odors
that surround us. There exist greenish blues and yellowish greens, but are there
tea-ish coffees or fruitish sweats and so on ad infinitum? To try to learn the names
of odors would be nonsensical for everyday life. But that it can be done is proved
daily by perfumers, flavorists, and people trained to work with descriptive panels.
But why should an ordinary person learn more than the handful that cannot be
avoided because they are encountered every day? And even if we did not know
the names of these, would that in any way change our lives? Usually we would
not notice that they were there, and if we did, they might just make us hungry
or thirsty or interested in the other sex. No identification is needed for that.
Gastronomy is an interesting pursuit, but it is highly questionable that analyzing
food and determining its components can add to the pleasure of eating it. Great
cooks need words to buy the materials they use, but their senses of smell and
taste to make the dish.

Degel and Köster (1999) and Degel, Piper, and Köster (2001) have shown
that knowing the name of an odor can block the encoding (or contribute to the
forgetting) of a new implicit episodic memory for it. Furthermore, it is well
known that after thorough training in odor identification, experts find odors
more interesting, but they no longer experience the natural emotional impact of
odors (Holley, Chapter 2, this volume). That is precisely why experts in sensory evaluation cannot be used to predict consumer reactions to products and

why consumers should not be asked to make descriptive statements. In the latter case, one would force consumers to do something that they normally would not do, thus changing their natural way of dealing with a product and defeating one's goal: insight into normal consumer behavior. The fact that olfaction is a "hidden" sense, dealing mostly with non-conscious emotional associations, has consequences for the types of questions that can be put to consumers. Unfortunately, many investigators, both in fundamental research and in applied research, grossly overestimate the capacity of people to answer their questions. One of the most serious problems for psychology is that whenever one asks a question, one almost always will get some kind of answer. Worse yet, the more unanswerable the question is, the more uniform will be the responses of a group of respondents, and the more rational the answers will seem.

In 1969, Weber and Bach instructed their experimental subjects to imagine the letters of the alphabet; then they asked the subjects to indicate where in the head the imagination had taken place. Without exception, all subjects pointed to areas of their heads, and the vast majority pointed to the forehead. That surprised the two psychologists, who were disappointed that the subjects did not point to the back of the head, where visual signals are processed in the area striata. They ascribed the uniform behavior of their subjects to "implicit cultural stereotypes" and never realized that they had asked an absurd question. If such a nonsensical and unanswerable question can be put by psychologists working in the field of an overt sense like vision, it is not surprising that one finds many more similar unanswerable questions in olfactory and flavor research. Here is the strangest and most complex one that I have encountered: "How many just noticeable differences [JNDs] is the whiteness of this coffee away from your ideal whiteness?" All subjects had answered the question and given numbers. Moreover, for all subjects those numbers became larger when the coffee whiteness was nearer to the extremes (completely black or almost completely white). The spurious conclusion was that people could use such a scale based on JNDs very well to express their feelings about coffee whiteness. In fact, the subjects, faced with this impossible question, must have resorted to a rationalization and chosen numbers that were more extreme when the coffee seemed more extreme to them. They did not use the intended scale units at all.

Whereas with such complex questions the absurdity of the questions is easy to understand, many people find it more difficult to comprehend that a simple question like "Why do you like this food?" is just as absurd. If one were looking for the real sensory reasons for food choices, that would be the best question to guarantee erroneous responses. People usually do not know what sensory properties in a given food or drink really attract or repel them. The food may be linked to implicit and non-conscious memories, or there may be sensory sensations and feelings attached to it that are extremely difficult to describe. Nevertheless, people feel that they have to give answers in order not to appear stupid, and thus they come up with all sorts of notions as to what the attraction is. Most of these notions are derived from advertisements for the type of food or drink involved. In Germany, all coffee is advertised as being "mild." Therefore, if consumers like a particular coffee, they will say that it is mild. Unfortunately, all research efforts to find out which sensory properties make a coffee mild have been in vain. Certainly a very sour or bitter coffee is less mild than a less sour or bitter one, but of two equally sour or bitter coffees, one may be judged mild and the other not mild at all. Other notions often used by consumers to answer "why" questions are rooted in their attitudes and their convictions about healthfulness, vitamins, fat or fiber content, and the naturalness of foods. In Europe, the importance of these convictions will vary from country to country, but usually it is easy to show that such convictions lead to answers that are contradicted by actual consumer behavior (Köster, Beckers, and Houben, 1987).

Unfortunately, in market research, "why" questions and questions requiring descriptive answers are often asked, and belief in the answers that consumers give to such questions probably is the reason for the many flops (8 out of 10 new products disappear from the market within a year) in the food industry. Questions about actual consumer behavior and simple likes and dislikes are much better predictors: "How many glasses of water do you drink from the tap each day?" (Zoeteman et al., 1978; Köster et al., 1981) and "Do you serve tap water when you have close friends over for dinner?" prove to be far better questions to assess water quality than are evaluative questions like this: "How do you like the drinking water from your tap?" For measurement of the implicit and non-conscious feelings that are connected with odors and flavors, more indirect methods have been developed. Thus, to estimate the liking of the odor of a laundry powder, it is better to ask about the whiteness and softness of the clothes washed with it. For foods also, special methods for evaluating hidden preferences and changes in such preferences during repeated exposures have been developed (Köster, 1998). Indirect ways of obtaining information often are even better than questions about preferences for different specimens of the same type of product. If people are told that there is a new way of making a traditional product, beer, for instance, but that it is questionable that this new product has the right taste, and these people are then asked to select the genuine old-fashioned beers in tests in which four existing beers are presented a number of times, they discriminate better between those beers than they do when they are asked which beer they like best in a paired comparison experiment. This method has also been used to find out how far one can reduce the alcohol content in beer before the beer is judged to be no longer a "real" beer (Mojet and Köster, 1986).

The special memory for odors, with its strongly associative and episodic characteristics, can lead to problems in both fundamental research and applied research. Many people have taken home a few bottles of the inexpensive wine they found so pleasing while on a holiday trip or a vacation, only to be disappointed when they tasted the wine at home. Why is that? And why is that not the case with the car they rented and also liked very much in that same country? Obviously, appreciation of food and drink is highly dependent on situational factors, and a given product can have quite different effects depending on the circumstances in which it is consumed. If that is the case, it is strange that even in the best research in sensory analysis all efforts are made to exclude situational factors: People taste small portions of pasta at nine in the morning in a sterile cubicle under "normalized" artificial daylight, or, even worse, we conduct so-called in-home-use tests, in which a product is given to people and is to be tasted under "normal" circumstances at home, with the result that instead of eating or drinking it normally, the family members begin "analyzing" and discussing the product in the most uncontrolled way, paying attention to aspects of the product that had never before entered into their normal patterns of choosing. No wonder that 8 out of 10 new food products are flops. Situational analysis, in which people are asked to imagine and helped to imagine the use of products in different eating situations, has been proposed (Köster, 1996) and can be used to test foods that will be marketed in special circumstances, such as snacks in service stations, foods to be eaten while watching television, and special foods for parties (fondues, etc.). The best predictive results are obtained with indirect and implicit methods in which the subjects have no idea that their opinions about the food are what the investigation is all about. Observation of their behaviors (choices, quantities consumed, eating speed and its variation over time during the meal) would seem to provide much better indications of their likes and dislikes than could the direct questions and scaling methods employed in market research and in traditional sensory analysis. Situational analysis should be based on such methods and on data about the frequencies of occurrence of specific situations in consumers' daily lives. Such data could provide much better information for predictions of real consumer choices and behaviors than can answers to questionnaires about attitudes, convictions, and "values."

4. Conclusion

In this chapter we have examined the differences between olfaction, as a representative of the lower senses, and vision and audition, as the higher senses, and a number of the methodological consequences of those differences for both fundamental research and applied research have been mentioned. In many of the examples given, flavor and food were used to illustrate the functioning of olfaction. Although olfaction plays an important part in the appreciation of foods, it is by no means the only factor determining food choices and eating behaviors. Other factors, such as one's internal physiologic state, the accompanying intestinal feelings, the memory of such feelings, and well-studied phenomena like sensory-specific satiety (Rolls, 1999, and Chapter 23, this volume), probably are more important in determining food choices and eating behaviors. In fact, there are indications that some of these factors modulate the electrical activity of the olfactory bulb and probably the reactivity to food flavors (Pager, 1974). This chapter has stressed the differences between olfaction and the other senses. Obviously there are also many similarities in the ways the various senses are used to interact with the outside world. Nevertheless, it is argued here that failure to properly take into account the obvious sensory differences has led many a researcher to a dead end in the maze of scientific research and has hampered progress in targeted marketing of food products. People *eat* what they like, but do not *know* what they like, and certainly *not why* they like it.

References

Amoore J E (1962). The Stereochemical Theory of Olfaction. *Proceedings of the Scientific Section of the Toilet Goods Association* 37(suppl.):1–13.

Amoore J E (1963). Stereochemical Theory of Olfaction. *Nature* 198:271–2.

Amoore J E (1964). Current Status of the Steric Theory of Odor. *Annals of the New York Academy of Sciences* 116:457–76.

Amoore J E (1967). Specific Anosmia: A Clue to the Olfactory Code. *Nature* 214:1095–8.

Amoore J E (1977). Specific Anosmia and the Concept of Primary Odors. *Chemical Senses and Flavor* 2:267–81.

Baron R A (1990). Environmentally Induced Positive Effects: Its Impact on Self-efficacy, Task Performance, Negotiation and Conflict. *Journal of Applied Social Psychology* 20:368–84.

Bastone L M & Ehrlichman H (1991). Odor Experience as an Affective State: Effects of Odor Pleasantness on Cognition. Poster presented at the ninety-ninth meeting of the American Psychological Association, San Francisco, 16–20 August 1991.

Berglund B & Olsson M J (1993). Odor-Intensity Interaction in Binary and Ternary Mixtures. *Perception and Psychophysics* 53:475–82.

Boireau N, Pinault G, Kirsche L, & MacLeod P (2000). Lateralization of Vanillin Odor Perception by Birhinal Stimulation – Differential Olfactometer Method. In: Abstracts, ISOT/ECRO 2000 Congress, July 20–24, 2000, Brighton, U.K.

Cain W S (1979). To Know with the Nose: Keys to Odor Identification. *Science* 203:467–70.

Cain W S, Schiet F T, Olsson M J, & De Wijk R A (1995). Comparison of Models of Odor Interaction. *Chemical Senses* 20:625–37.

Dalton P (1999). Cognitive Influences on Health Symptoms from Acute Chemical Exposure. *Health Psychology* 18:579–90.

Degel J & Köster E P (1998). Implicit Memory for Odors: A Possible Method for Observation. *Perceptual and Motor Skills* 86:943–52.

Degel J & Köster E P (1999). Odors: Implicit Memory and Performance Effects. *Chemical Senses* 24:317–25.

Degel J, Piper D, & Köster E P (2001). Implicit Learning and Implicit Memory for Odors: The Influence of Odor Identification and Retention Time. *Chemical Senses* 26:267–80.

Dijksterhuis G B (1997). *Multivariate Data Analysis in Sensory and Consumer Science*. Trumbull, CT: Food and Nutrition Press.

Ebbinghaus H (1885). *Über das Gedächtniss*. Leipzig: Duncker & Humbolt.

Ehrlichman H & Halpern J N (1988). Affect and Memory: Effects of Pleasant and Unpleasant Odors on Retrieval of Happy and Unhappy Memories. *Journal of Personality and Social Psychology* 55:769–79.

Gibson J J (1968). *The Senses Considered as Perceptual Systems*. London: Allen & Unwin.

Gilbert A N, Knasko S C, & Sabini J (1997). Sex Differences in Task Performance Associated with Attention to Ambient Odor. *Archives of Environmental Health* 52:195–9.

Haller R, Rummel C, Henneberg S, Pollmer U, & Köster E P (1999). The Influence of Early Experience with Vanillin on Food Preference Later in Life. *Chemical Senses* 24:465–7.

Harper R, Bate-Smith E C, & Land D (1968). *Odour Description and Odour Classification*. London: J & A Churchill.

Henning G (1916). *Der Geruch*. Leipzig: Barth.

Herz R S & Cupchik G C (1992). An Experimental Characterization of Odor-evoked Memories in Humans. *Chemical Senses* 17:519–28.

Herz R S & Cupchik G C (1995). The Emotional Distinctiveness of Odor-evoked Memories. *Chemical Senses* 20:517–28.

Herz R S & Eich E (1995). Commentary and Envoi. In: *Memory for Odors*, ed. F R Schab & R G Crowder, pp. 159–75. Mahwah, NJ: Lawrence Erlbaum.

Herz R S & Engen T (1996). Odor Memory: Review and Analysis. *Psychonomic Bulletin and Review* 3:300–13.

Holley A (1999). *Eloge de l'odorat*. Paris: Odile Jacob.

Kirk-Smith M D & Booth D A (1987). Chemoreception in Human Behaviour: Experimental Analysis of the Social Effects of Fragrances. *Chemical Senses* 12:159–66.

Kirk-Smith M D & Booth D A (1992). Effects of Natural and Synthetic Odorants on Mood and Perception of Other People. *Chemical Senses* 17:849–50.

Knasko S C (1993). Performance, Mood, and Health during Exposure to Intermittent Odors. *Archives of Environmental Health* 48:305–8.

Knasko S C, Gilbert A N, & Sabini J (1990). Emotional State, Physical Well-being, and Performance in the Presence of Feigned Ambient Odor. *Journal of Applied Social Psychology* 20:1345–57.

Kobal G, Van Toller S, & Hummel T (1989). Is there Directional Smelling? *Experientia* 45:130–2.

Koelega H S & Köster E P (1974). Some Experiments on Sex Differences in Odor Perception. *Annals of the New York Academy of Sciences* 237:234–46.

Köster E P (1969). Intensity in Mixtures of Odorous Substances. In: *Olfaction and Taste III*, ed. C Pfaffmann. New York: Rockefeller University Press.

Köster E P (1971). Adaptation and Cross-adaptation in Olfaction. Doctoral thesis, Utrecht University, Rotterdam, Bronder Offset N.V.

Köster E P (1996). The Consumer? The Quality? In: *Production industrielle & qualité sensorielle*, ed. B Colas, pp. 11–19. Paris: Lavoisier Tec & Doc.

Köster E P (1998). Épreuves Hédoniques. In: *Évaluation sensorielle: Manuel Méthodologique (2ème édition)*, ed. F Depledt & F Strigler, pp. 182–207. Paris: Lavoisier Tec & Doc.

Köster E P, Beckers A W J M, & Houben J H (1987). The Influence of Health Information on the Acceptance of a Snack in a Canteen Test. In: *Flavour Science and Technology*, ed. M Martens, G A Dalen, & H Russwurm, pp. 391–8. New York: Wiley.

Köster E P & MacLeod P (1975). Psychophysical and Electrophysiological Experiments with Binary Mixtures of Acetophenone and Eugenol. In: *Methods in Olfactory Research*, ed. D G Moulton, A Turk, & J W Johnston, Jr, pp. 431–44. New York: Academic Press.

Köster E P, Van der Stelt O, Nixdorff R R, & Linschoten M R I (1997). Olfactory Imaging: A Priming Experiment. *Chemical Senses* 22:201–2.

Köster E P, Zoeteman B C J, Piet G P, Greef E de, Oers H van, Heijden B G van der, & Veer A J van der (1981). Sensory Evaluation of Drinking Water by Consumer Panels. *The Science of the Total Environment* 18:155–6.

Laffort P (1989). Models for Describing Intensity Interactions in Odor Mixtures: A Reappraisal. In: *Perception of Complex Smells and Tastes*, ed. D G Laing, W S Cain, R L McBride, & B W Ache, pp. 173–88. New York: Academic Press.

Laska M & Teubner P (1999). Olfactory Discrimination Ability of Human Subjects for Ten Pairs of Enantiomers. *Chemical Senses* 24:161–70.

Linnaeus (1764). *Odores medicamentorum. Amoenitates Academiae*, vol. 3, pp. 183–201. Stockholm: Lars Salvius.

Mojet J & Köster E P (1986). Investigation into the Appreciation of Three Low-Alcohol Beers (in Dutch). Report, Psychological Laboratory, Utrecht University (confidential).

Olsson M J (1994). An Interaction Model for Odor Quality and Intensity. *Perception and Psychophysics* 55:363–72.

Pager J (1974). A Selective Modulation of the Olfactory Bulb Electrical Activity in Relation to the Learning of Palatability in Hungry and Satiated Rats. *Physiology and Behavior* 12:189–96.

Posner M I (1986). *Chronometric Explorations of Mind*. Oxford University Press.

Punter P H (1983). Measurement of Human Olfactory Thresholds for Several Groups of Structurally Related Compounds. *Chemical Senses* 7:215–35.

Robin O, Alaoui-Ismaïli O, Dittmar A, & Vernet-Maury E (1999). Basic Emotions Evoked by Eugenol Odor Differ according to the Dental Experience. A Neurovegetative Analysis. *Chemical Senses* 24:327–35.

Rolls E T (1999). *The Brain and Emotion*. Oxford University Press.

Schaal B (1988). Olfaction in Infants and Children: Developmental and Functional Perspectives. *Chemical Senses* 13:145–90.

Schaal B, Marlier L, & Soussignan R (1998). Olfactory Function in the Human Fetus: Evidence from Selective Responsiveness to the Odor of Amniotic Fluid. *Behavioral Neurosciences* 112:1–12.

Schaal B, Marlier L, & Soussignan R (2000). Earlier-than-Early Odour Learning: Human Newborns Remember Odours from Their Pregnant Mother's Diet. In: Abstracts, ISOT/ECRO 2000 Congress, July 20–24, 2000, Brighton, U.K.

Schlich P (1996). Defining and Validating Assessor Compromises about Product Distances and Attribute Correlations. In: *Multivariate Analysis of Data in Sensory Science*, ed. T Naes & E Risvik, pp. 259–306. Amsterdam: Elsevier.

Teerling A & Köster E P (1988). Fragrance and Mood. A Study on the Effect of Fragrances on the Human Mood. Report, Psychological Laboratory, University of Utrecht (confidential).

Teerling A, Nixdorf R R, & Köster E P (1992). The Effect of Ambient Odours on Shopping Behavior. *Chemical Senses* 17:886 (abstract).

Van Toller S (2000). Assessing the Impact of Anosmia: Review of a Questionnaire's Findings. *Chemical Senses* 24:705–12.

Warm J S, Dember W N, & Parasuraman R (1990). Effects of Fragrance on Vigilance, Performance and Stress. *Perfumer and Flavorist* 15:15–18.

Weber R J & Bach M (1969). Visual and Speech Imagery. *British Journal of Psychology* 60:199–202.

Woskow M H (1968). Multidimensional Scaling of Odors. In: *Theories of Odor and Odor Measurement*, ed. N Tanyolaç, pp. 147–88. Bebek Istanbul: Robert College.

Wright R H (1982). The Sense of Smell. Boca Raton: CRC Press.

Zoeteman B C J, De Greef E, Van Oers H, Köster E P, & Rook J J (1978). Monitoring Drinkwater Taste by Consumers (in Dutch). H_2O 25:660–8, 673–4.

Zwaardemaker H (1895). *Die Physiologie des Geruchs*. Leipzig: Engelmann.

Section Two

Knowledge and Languages

The chapters in this section will clearly illustrate the multidisciplinary approach of this volume, interweaving issues from different domains – chemistry, anthropology, psychology, and linguistics – to examine the complex relationships between language for odors and knowledge of odors.

In Chapter 4, from a psycholinguistic point of view, Danièle Dubois and Catherine Rouby's analysis of verbal answers in laboratory identification tasks for odors shows that scoring is not solely a technical issue, but also raises numerous theoretical questions about cognitive representations and the naming of odors. They first demonstrate that the "veridical label" is the name of the odorant source rather than the name of an olfactory property. Therefore, because subjects lack adequate labels for the olfactory properties of objects, they have to resort to the use of a large diversity of linguistic devices to account for their olfactory perceptions, and those must be interpreted by researchers as cues to subjects' knowledge of odors.

In Chapter 5, David Howes examines a large body of data borrowed from the history of Western culture and from distant cultures. Referring to Classen's "archeology of sense words" for intellect in the English language, he shows that the sense of smell has been differently valued and interpreted in various cultures and that, for example, intelligence was more "tactile" before the Enlightenment than it is now. A cross-cultural study of sensory vocabularies from different languages and a precise description of the Ongee people of Little Andaman Island demonstrate that olfaction can serve as a medium for construction of an elaborate cosmology and epistemology, according only a minor role to vision.

In Chapter 6, Sophie David presents a precise analysis of the diversity of linguistic resources in the French language to account for olfactory experiences. Her specific methodology (called lexical grammar) allows her to show that most French olfactory terms do not refer to odors independently of their sources. Furthermore, the contrasts between acceptable and unacceptable assertions regarding odor naming allow her to argue that odors cannot be considered as "objects," as entities located in time and space (as in visually grounded experience), but rather as properties of objects experienced by subjects.

In Chapter 7, Maurice Chastrette presents the point of view of a chemist in his overview of the various systems for classifying smells, ranging from empirical classifications grounded in our taxonomic tradition for natural objects to more recent systems based on semantic descriptions, odor profiles, similarity data, and statistical methods. One conclusion is that even in expertise domains such as perfumery, odor structures do not readily fit into the classical taxonomic patterns used in the natural sciences. The complexity of the picture of knowledge of odors raises interesting questions about relationships between chemical structures and cognitive structures. A second finding is that the classification systems are strongly associated with the descriptive systems that account for odors as substances or as cognitive representations. Therefore it becomes important to pay closer attention to the relationships between knowledge of odors and language for odors.

A common theme in all the chapters in this section is the contrast between the language and body of knowledge that emerges from the olfactory sense and the more widely used language and knowledge elaborated from the visual modality. Because the visual realm has always been taken as the proper frame of reference for interpreting olfactory knowledge and representation, odors have not been conceptualized as effects experienced by a subject, but rather as entities in the world. Thus their relationships to language and cognition need to be reconsidered, emphasizing the contributions of linguistic devices available in various languages to the construction of both individual and collective cognitive representations.

4

Names and Categories for Odors: The Veridical Label

Danièle Dubois and Catherine Rouby

The research presented here is aimed at clarifying the relationships between cognitive categories and the naming of odors, our method being a close analysis of the verbal answers given to odorants presented in various experiments.[1] This is part of a work in progress within a cognitive science research program on categorization. By definition, such a program is multidisciplinary, involving chemistry, neurophysiology, psychology, and, in this case, an emphasis on psycholinguistics and linguistics. The concerns of these latter two disciplines led us to pay special attention to the conceptualizations and wording chosen for odors. We shall focus on the verbal answers given by subjects: On the one hand, they are considered as direct and obvious, like any other outputs or responses from the cognitive system in any experimental setting, like neural or behavioral responses; on the other hand, they may be scored at the same time as "erroneous." The questions raised by this ambiguity when scoring verbal answers in olfactory research led to examination of some theoretical issues regarding the conceptual representation of "an odor" in human memory, as compared with the descriptions of odorants in the natural sciences.

1. Odors and Their Names

When scoring the verbal performances of subjects in identification tasks and in diagnostic tests, words such as "orange," "apple," and "vanilla" are considered the "correct odor names" or the "veridical labels" (e.g., Engen, 1987; Doty, 1992). However, little consideration has been given to what, precisely, a "word" is and what it means to researchers and subjects. Even in cross-cultural studies, great attention is paid to the choices of odorant samples to be used as stimuli, but little or nothing is said about the choices of distractors in multiple-choice questions (Doty et al., 1985). Thus, "words that may not be culture-free for the response alternatives" are " substituted" when translating tests from one culture to another: "garlic" is changed for "pumpkin pie," "dog" for "skunk," and so forth (Doty, Marcus, and Lee, 1996). Such scoring relies on the commonsense evidence that the words simply refer to the things to be named. That may work quite well with vision (Dubois, Resche-Rigon, and Tenin, 1997), but its applicability is not so clear for olfaction.

Some authors have noted peculiarities of verbal behavior in olfactory research. As a first contrast with the visual realm, where objects and their names are so closely linked that they seem cognitively inseparable, "the linking of names and odors is usually weak" (Engen, 1987, p. 498). A second contrast is that between the vividness of odors in subjects' memories and the poor performances in naming those odors. For example, "we can recognize odors virtually forever, though we have difficulty conjuring up a particular smell and remembering its name," and "people are not good at naming even familiar odors" (Engen, 1987, p. 497).

Engen, like many other researchers, has suggested that such "poor" performance may depend on some property of the stimulus: for example, the "real thing" being a better stimulus than the "scratch-and-sniff" odorants, and some odors simply being more easily identified than others. He has also suggested that poor performance may be due to linguistic factors, such as "the nature of the verbal response," and psycholinguistic factors, such as "the subjects' verbal abilities." For example, the better performances of women relative to men in naming odors could be due to their greater verbal ability. In short, the theoretical grounding for the relationship between odors and their names remains obscure and appears to involve mainly problems in linguistics and psycholinguistics. An understanding probably will also require input from psychologists involved in basic empirical research into memory: Why should one score a verbal answer as "correct" or "veridical"? For example, as asked by Engen (1987, p. 500), "What is a correct answer for 'musk'?"

We propose here that such difficulties in grasping the relationship between odors and names (and therefore in scoring answers) are due mainly to an implicit theory of lexical semantics rooted in the study of vision that fosters the illusion that a veridical label should exist for anything in the world. If such a theory holds for visual objects, why not for odors? Conversely, if such a theory of naming is questionable for odors, it should lead us to reconsider the following implicit assumptions:

(1) the linguistic conception of naming that exclusively determines word meanings with reference to the "entities of the world," a commonsense conception inherited from a philosophical tradition, usually discussed in lexical semantics (e.g., Rastier, 1991; Wierzbicka, 1992; Lucy, 1992; Eco, 1997);

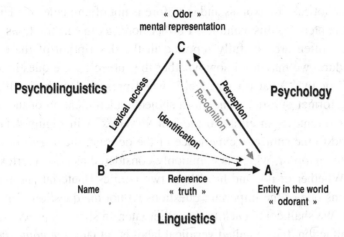

Figure 4.1. The semantic conception underlying the use of a veridical label.

(2) the corresponding psychological conception of "lexical access," which supposes that subjects can automatically access a mental lexicon that will provide the correct label for the previously stored odor representation;

(3) the logical entailment of (1) and (2) implying that odors are cognitive representations of entities in the world.

We shall therefore examine the conceptual frameworks that attempt to account for the relationships among odorants, odors, and their names (Figure 4.1)

2. The Veridical Label: An Odor Name?

The notion of "veridical label" or "correct response," when used in olfactory research, implies that there are names that adequately (truthfully) refer to odors, just as color terms adequately refer to colors. However, experimenters themselves are not always sure how to name what is presented to the subjects, whether using the names of the chemical substances (amyl acetate, phenylethyl alcohol) or the names of the sources, that is, the names of the odorant objects (banana and rose, respectively). To clarify the ambiguity, some explanations are needed: Strictly speaking, and in contrast to the situation with colors, odors have no specific names, at least not in English or in French (Dubois, Rouby, and Sicard, 1977). The correct linguistic form for a veridical label would be of the form "the odor of *X*," *X* being the name of a source (e.g., the odor of rose), with the word "rose" referring to the name of the odorant object (see David et al., 1997, for a description of the linguistic ways of naming odors in French).

Such a lexical gap means that the distinction between the name of the odorant (the substance) and the name of the odor (a subject's verbal answers to a stimulation) is not easy to express and therefore is not often made (see Hudson and Distel, Chapter 25, this volume). Furthermore, as the instructions given to the subjects often are not fully reported in the description of the experimental procedure, we may not know whether the subjects were questioned by "What is it?" or by "What do you feel?" when presented with an olfactory stimulus. The former question concerns a subject's identification of the objectivity of the odorant, as an entity in the world (vertex A in Figure 4.1). The latter is related to the subject's experience of the odorant, the *cognitive representation of the odorant*, which is commonly considered as *odor* (vertex C in Figure 4.1). Whether or not and how these two entities (material and mental) overlap remains one of the important questions (if not *the* question) for olfactory research. We shall come back to this point later. In short, in psychological research on olfaction, the so-called veridical label is actually *the name that the experimenter expects*, as a relevant and obvious answer to the question within the experimental setting, in other words, the noun for the commonly encountered object that produces an odor quite similar to the one produced by the presented odorant.

The complexity of this statement, as contrasted with the simple "veridical" label, points to the need for further examination of the realities of the three entities (odorant, odor, and name) and the mapping relationships among them. We shall therefore reconsider what is represented at each vertex of the "semiotic triangle" (Ogden and Richards, 1923; Eco, 1988; Rastier, 1991) as presented in Figure 4.1.

We shall also reexamine the criteria for determining that a verbal response to an odorant is correct or incorrect, as well as why that question arises only for verbal responses. Do we, in neurophysiology, talk about the "correct response" of a receptor when it is firing, or of any type of central response (fMRI, PET, or electrophysiological)? Why should verbal responses to experimental questions be susceptible to being declared "erroneous," whereas electrophysiological responses are always true?

2.1. The Stimuli

Olfactory research does not specify how similar a presented stimulus is to the "real thing." In other words, we do not know if the odorant presented as an experimental stimulus is a good material (chemical) representation of the odorant regularly and previously experienced by subjects. Consequently, we do not know whether subjects are responding to "something that is considered similar enough to a previously experienced odorant" or just to the "presented stimulus," something never experienced before, but to which they are asked to give some "appropriate" name. As mentioned by Engen (1987), there are significant

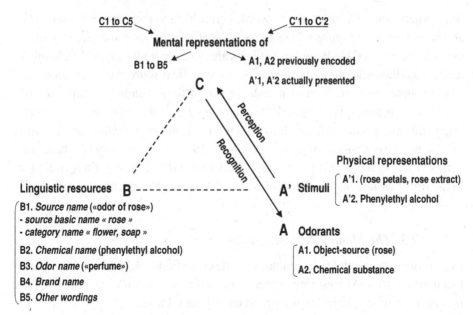

Figure 4.2. Relationships among verbal (left side) and nonverbal (right side) representations, linguistic resources, and percepts.

differences in odor-naming performances depending on the mode of presentation or representation: The scratch-and-sniff odorants lead to poorer performances than do the brand names (e.g., Crayola crayons, Johnson's baby powder in opaque bottles). In short, at vertex A, two classes of entities have to be explicitly distinguished: the commonly encountered odorants of the real world (A) and their physical representations (A′) as stimuli within experimental settings (Figure 4.2).

Therefore, when subjects are asked to name an odorant, at least two processes should be identified:

> First, a psychological process of recognition of the stimulus as something identical with, close to, or different from something already experienced. (For an experimental exploration of just how representational visual stimuli are vis-à-vis both the "real thing" and the subject's previous experience, as well as the influence of that variable, see Tenin, 1991.)
>
> Second, a psycholinguistic process of giving a name to "something" quite unspecified or at least implicit, the subject playing along with the experimenter as if the presented stimulus were some "thing" they agree on.

2.2. The Names

The problem now is to decide which name (vertex B in Figure 4.1) is the genuine veridical label: the chemical name or the name of the object that is the source (the "object-source"), or a term or wording referring to the substance (vertex A, and now A or A′ in Figure 4.2) or to the subject's experience and memory for A (a name for vertex C)? Of course, subjects are not expected to know the chemical names, and therefore they have to rely on the specific resources of their language. The resources in French and English for use in the olfactory domain are still largely unexplored (see David, 1997, and Chapter 6, this volume), and recent linguistic investigations have shown large variations in lexicality and naming of smells across various languages (Classen, 1993; Boisson, 1997). Therefore, vertex B in Figure 4.1 also has to be split into the different wordings available in the different languages. We shall focus on this point in the next section.

2.3. The Mental Representations

Like a name, the mental representation (vertex C in Figure 4.1) cannot be reduced to a unique entity. At least three types of mental representations should be taken into account when subjects respond on experimental tests:

1. The sensory representations of the presented stimuli ($C'1$ or $C'2$ as a representation of $A'1$ or $A'2$ in Figure 4.2).
2. The representations of previously encountered odorant objects, stored in memory, that are similar enough to the actual stimuli to be activated; this type of representation (C1 or C2 as a representation of A1 or A2 in Figure 4.2) is investigated in recognition-memory experiments.
3. The representations of linguistic resources referring to olfactory experiences in the subject's linguistic community: names in the "mental lexicon" if there are any, or any other device, including complex phrasing such as "the odor of an object-source" commonly used in English or French, or any other idiosyncratic wording available in the subject's memory (C1 to C5 as mental representations of B1 to B5 in Figure 4.2).

Besides the idea that naming as a psychological process has to do with the perceived "thing to be named" (an idea that protects us from absolute relativism), and besides the ambiguity of what this "thing" is in olfaction (discussed later), we would emphasize that naming also relies on an *implicit contract of communication* between the experimenters and the subjects (Grossen, 1989). The mapping of words onto things can therefore be considered as an illusion grounded in the consensus between subjects and experimenters (the partners in the transaction). However, this illusion can collapse, depending on the stimulus qualities and on the subjects' previous knowledge, expertise, familiarity with the stimuli, and ability to "play the game" and to come up to the experimenters' expectations. In tests of visual stimuli, Clark and Wilkes-Gibbs (1986) showed that when presented with photographs of New York City buildings, New Yorkers named them by proper names, referring to the precise buildings (e.g., the Empire State Building), whereas non–New Yorkers described the picture itself, therefore referring to the representation of a specific type of building (a skyscraper). Thus naming is not unique and depends not only on individual perceptual processes but also on what the two partners in the transaction will accept as the "correct response" within this specific communication setting. Inasmuch as French and English lexicons do not have specific terms for odors, subjects do "what they can with what is available" to communicate and give overt accounts of their subjective, covert experiences in order to satisfy the experimenter's expectations (Dubois, 1996). Instead of a "true" reference (as represented in Figure 4.1), a scoring of a "correct response" by the experimenter can therefore be viewed as a *référence heureuse* (lucky reference) (Eco, 1997), the partners agreeing on a name that is known to both. In the case of odors, we can come back to the formulation proposed earlier: The correct and expected response is *the name of a commonly encountered source (odorant object) that produces a sensation, an odor, quite similar to the one produced by the presented odorant stimulus.*

Considering the naming process within this frame of reference therefore leads to reconsideration of the scoring of the answers. Instead of simply sorting out correct versus incorrect responses, we shall reassess the verbal answers as one way of getting at the knowledge that subjects want (or try) to communicate to the experimenter about their olfactory experience, using the resources available in their language itself (B1 to B5 in Figure 4.2) and in their individual linguistic representations (C1 to C5 in Figure 4.2). This will lead us to reconsider the validity of the triangle in Figure 4.1 and introduce new entities represented in Figure 4.2.

3. Odors and Their Wording in Identification Tests

In sum, scoring verbal answers to odorants is not simply a technical question, but involves the elaboration of theoretical knowledge in three domains:

1. an adequate semantic theory of lexical meaning (domain of linguistics), describing the entities at vertex B (Figure 4.2)
2. a theory of cognitive representations in memory (domain of psychology), representing entities at vertex C, specific to humans, as humans are the only primates who can give linguistic answers to olfactory stimulations
3. a theory of the relationship between language and cognition (domains of linguistics, psycholinguistics, and also philosophy)

The analysis here is a first attempt to employ this rationale in close coordination with work done in linguistics (point 1). It will involve explicit hypotheses about the structural properties of categories in human memory (point 2), still largely unknown for olfaction, and explored through other nonverbal paradigms, such as pair comparisons and free-sorting tasks with odorants (see Rouby and Sicard, 1997, for a review; Rouby et al., 1997). The psychological approach differs from the linguistic approach, even though it may process the same data. Whereas linguistics considers words only as parts of the linguistic system per se and reasons about their structural properties (formal and/or semantic) independently of any psychological functioning, psychology accounts for mental processes that interact with the linguistic constraints involved in producing and understanding language. Finally, the relationship between language and cognition (point 3) will be considered here only within psycholinguistics, by studying the individual processes involved in verbal productions. Empirically, that leads to a reanalysis of the "kinds of words or descriptions used by subjects misidentifying odors" (Engen, 1987, p. 500). However, we shall no longer consider that failure to produce a "correct response" is a misidentification of an odor. Notwithstanding that there is no a priori veridical label available (neither in language, nor consequently in subjects' memories), we shall consider all answers given by subjects as attempts by them to communicate to the experimenter "something" about their memories and identifications of the presented odors.

This analysis has been carried out using the answers of 40 subjects in a spontaneous identification task involving 16 familiar odorants (Rouby and Dubois, 1995). The answers to only 3 of the 16 odorants (hereafter, LEMON, ORANGE, and APPLE) will be presented here as examples of our rationale. We also have relied on other sources of knowledge and psychological measurements of memory for odors that were collected in rating tasks and in free-sorting tasks.

Before proceeding, it must be emphasized that this analysis is partial and is restricted to isolated lexical items, overlooking the context, the discourse, within which the subjects produced these words. It has been shown elsewhere (Mondada and Dubois, 1995; Mondada, 1997) that the discursive context contributes to the construction of categories and also provides relevant information on the personal involvement in cognitive representations in subjects' memories.

In the cases of complex phrasing, we had to make choices in classifying the answers.[2] Our choices relied on psychological hypotheses about categorization, not on linguistic motivations.

The resulting list of verbal answers is given in Table 4.1 and includes the produced French words, English translations, and their frequencies of occurrence (*N*).

The verbal answers are presented along with explicit hypotheses about the relationships between linguistic descriptions and cognitive structures. These latter refer to conceptualizations of the categorical structures in memory for odors: Seven major semantic types of answers (T.1–T.7) were identified, clustering 17 finer-grained categories (simply numbered 1–17). Table 4.1 shows the answers categorized into these seven semantic types. Table 4.2 summarizes and quantifies the occurrences of these different semantic types for the three odorants considered here.

Type 1: The expected source

This first group of answers includes any answer (simple word, constructed, or polylexical forms) that has some wording to describe the expected source (any verbal form that includes the veridical label). This group includes the following subtypes:

1. the simple name of the expected source (commonly called the veridical label): LEMON, ORANGE, or APPLE. This canonical scoring was in accordance with the olfactory literature: The frequency of this type of response ranged from 10% (APPLE) to 24% (ORANGE) up to 36% (LEMON), with 22% as a mean value. [Such a result is consistent with (even if lower than) the data of Engen (1987), who found a mean value of 44% "correct" answers. The range was from 27% for ROSE to 75% for BAZOOKA BUBBLE GUM in the set of "odorants including the brand names." The set of scratch-and-sniff odorants ranged from 0% for MUSK, ("never correctly identified," according to such a strict scoring) to 83% for LICORICE.]

The robust conclusion is that some odors are regularly better identified than others. However, it cannot yet be determined which factors most affected the performances: That may have depended on the properties of the odorant, on the type of odor, or on the link between the odor and some available name for some potential source. The only provisional conclusions that can be drawn from tests of these three odorants concern the level of categorization: Thus, "citrus fruits" (clustering LEMON and ORANGE) would be at the same level of categorization as "other fruits" (this latter including APPLE), with the better-identified LEMON as a prototype candidate within the first category (Rosch, 1978).

2. the name of the expected source plus an adjective that refers to a more specific level of categorization than the veridical label (*citron sucré*, sweet lemon; *pomme verte*, green apple). This type of answer indicates that subjects are able to discriminate odors below the preceding level, or that they can specify a

Table 4.1. *Spontaneous responses to LEMON, ORANGE, and APPLE*

Citron (Lemon)			Orange (Orange)			Pomme (Apple)		
French words	English words	N	French words	English words	N	French words	English words	N
citron	lemon	28	orange	orange	20	pomme	apple	10
citron sucré	sweet lemon	1	peau d'orange	orange peel	1	pomme verte	green apple	3
citron acidulé	sour lemon	1	zeste d'o.	orange zest	1	shampoing p.	a. shampoo	1
c. très doux	very sweet l.	1	sirop d'o.	orange syrup	1		**T.1**	**14**
citronnée	lemon-like	2	sirop X d'o.	o. brand syrup	1	fruit	fruit	1
bonbon c.	lemon candy	3	bonbon o.	orange candy	4	fruité	fruity	1
glace au c.	l. ice cream	1	fleur d'oranger	o. tree flower	1	fruit rouge	red fruit	1
tarte au c.	lemon pie	1	orangée	orange-like	1	fruit exotique	exotic fruit	1
gélatine c.	lemon jelly	1		**T.1**	**30**	fleur	flower	3
liq. vaisselle c.	l. detergent	1	agrume	citrus	1	bonbon	candy	4
citronnelle	citronella	1	fruit	fruit	3	bon à manger	good to eat	1
	T.1	**41**	fruit sucré	sweet fruit	1	déodorant	deodorant	1
fruitée	fruity	1	fleur	flower	1	chimique	chemical	1
fleur	flower	1	bonbon	candy	1		**T.2**	**14**
nourriture	food	1	agent nettoyant	cleaning agent	1	orange	orange	3
mangeable	edible	1		**T.2**	**8**	mélange fruits	mixed fruits	1
bonbon	candy	2	citron	lemon	18	fruit passion	passion fruit	2
condiment	spice	1	citron vert	lime	2	fraise	strawberry	20
	T.2	**7**	citronnée	lemon-like	2	fraise Tagada	Tagada strawb.	3
fleur d'oranger	o. tree flower	4	mandarine	tangerine	2	pêche	peach	5
orange	orange	4	fruit passion	passion fruit	1	abricot	apricot	4
framboise	raspberry	1		**T.3**	**25**	banane	banana	1
fraise	strawberry	1	citronnelle	citronella	1	mûre	blackberry	1
	T.3	**10**	tilleul	lime tree	1	framboise	strawberry	1
bonbon sucré	sweet candy	1	bonbon citron	lemon drop	2		**T.3**	**41**
sucrée	sweet	2	sucrée	sweet	1	chewing gum f.	fruit ch. gum.	1

Left block:

French	English	n
caramel	caramel	1
chamallow	chamallow	1
gâteau kiwi	kiwi cake	1
biscuit espagn.	spanish cake	1
dans gâteau	"in cake"	1
	T.4	**8**
agréable	pleasant	2
sent bon	smells good	2
	T.5.	**4**
indéterminée	undertermined	1
connue	known	1
inconnue	unknown	1
	T.6	**3**
parfum	perfume	1
odeur	odor	4
	T.7	**5**

Middle block:

French	English	n
chewing gum c	lemon c. g.	1
sucette	lollipop	1
Pétrole Hahn	Pétrole Hahn	1
pansement	bandage	1
limonene	limonene	1
	T.4	**10**
fraîche	fresh	1
agréable	pleasant	1
bon	good	1
	T.5	**3**
indéterminé	undertermined	1
familière	familiar	1
oubliée	forgotten	1
	T.6	**3**
senteur	scent	4
odeur	odor	1
	T.7	**5**

Right block:

French	English	n
chg. fraise	strawberry c. g.	2
chg. menthe	mint c. g.	1
yaourt aux f	fruit yogurt	1
bonbon Mentos	Mentos candy	1
bonbon fruité	fruit candy	1
bonbon acidulé	sour candy	1
b. chimique	chemical candy	1
b. arome abricot	apricot flavor c.	1
bonbon fraise	strawberry c.	1
sur les bonbons	"on candies"	1
sucrée	sweet	1
acide	sour	1
vanille	vanilla	1
grenadine	grenadine	1
rouge à lèvre	lipstick	1
Tahiti douche	T. shower gel	1
Tahiti douche fr.	T. s. fruit gel	1
shampoing	shampoo	1
sh. parfumé	perfumed sh.	1
sh. pêche	peach sh.	1
	T.4	**22**
agréable	pleasant	1
	T.5	**1**
connue	known	1
	T.6	**1**
odeur	odor	3
parfum	perfume	1
senteur	scent	1
	T.7	**5**

property of the "smell of *X*." It is consistent with the hypothesis that odor categories could be structured along typicality like other categories of objects (Rosch, 1978; Dubois, 1991).

3. an adjectival form constructed on the source noun (*citronné*, lemony; *orangé*, orange-like): Instead of giving a noun that would refer to an object, subjects might suggest that odors are some property or quality of the object. Such wording can be highly influenced by the linguistic resources of each language, and that will require further investigation.[3]

4. the name of another source or artifact (e.g., candies) syntactically constructed with the name of the expected source (as a compound noun: *bonbon citron*, lemon drops; *sirop d'orange*, orange syrup; *shampooing pomme*, apple shampoo): This type of answer mainly occurs with food or drugs and cosmetics now flavored with a wide variety of aromas. These answers can be considered as relevant to a "genuine" source, inasmuch as they belong to subjects' experiences, following the same rationale that avocados and tomatoes are exemplars of the category of vegetables (whereas they are "true" fruits), or that a whale is a fish, as currently analyzed within the "ecological" approach for "natural" categories (Neisser, 1987; Dubois and Resche-Rigon, 1995). This type occurs for the three odors discussed here.

To sum up, the answers in this group reflect primarily the different levels of categorization that structure odors in subjects' memories (vertex C in Figure 4.2), worded in terms of the basic or derived names for known sources of the odors (vertex B). It is still an open question whether that reflects a genuine basic level for odors or a basic level for object-sources, an issue that cannot be settled because the same word is used for the odor and for the object. The cumulative values for these four forms of "correct answers" (Type 1 in Table 4.2), referring to the possible experienced sources of odors, range from 14% for APPLE to 36% for ORANGE and up to 53% for LEMON, with an average of 33% for the three odorants. These findings support data collected from other psychological indicators: Memory for the odor of LEMON seems to be most accurate and can be considered as a candidate prototype within the category of "odors of citrus fruits," the odor of APPLE being also, but less frequently, categorized as a fruit odor. Therefore the findings are consistent with the hypothesis that the different odors of fruits are not structured at just a single level, isomorphic to the categorical organization of the corresponding objects (sources), and therefore the categorization of odors cannot be adequately represented by a taxonomic hierarchy like those used for "natural" objects (cf. Chastrette, Chapter 7, this volume).

Type 2: Generic terms for sources

The second group of answers includes the generic terms for the sources specified in the preceding group. Here we count as "correct answers" the generic terms that are acceptable for the preceding specific Type-1 sources in Table 4.2:

Table 4.2. *Semantic typology of the denominations of three odors in an identification task*

	Stimulus						Total		
	LEMON		ORANGE		APPLE				
Types of answers	N	%	N	%	N	%	N	%	cumulated%
1. expected basic-level source (e.g., veridical label)	28	36	20	24	10	10	58	22	22
2. expected source + adj. (specified)	3		0		3		6		
3. adjectival form /expected source	2		1		0		3		
4. other source + expected source	8		9		1		18		
T.1 = expected source	41	52.6	30	35.7	14.	14.3	85	32.7	32.7
5. fruit (or constructed on "fruit")	1		5		4		10		
6. *fleur* (flower)	1		1		3		5		
7. *aliment* (food, including candy)	5		1		5		11		
8. drugs and cleaning	0		1		2		3		
T.2 = generic sources	7	9	8	9.5	14.	14.3	29	11.2	43.9
9. *citron* (lemon + derivation)	0		22		0		22		
10. *orange* + der. derivation	4		0		3		7		
11. *pomme* (apple)	0		0		0		0		
12. other fruits	6		3		38		47		
T.3 = co-hyponyms of fruits	10	12.8	25	29.8	41	41.8	76	29.2	73.1
13. co-hyponyms of foods	8		7		16		31		
14. co-hyponyms of drugs & chemicals	0		3		6		9		
T.4 = other co-hyponyms	8	10.3	10	11.9	22	22.4	40	15.4	88.5
15. -**T.5 = hedonic judgment**	4	5.1	3	3.6	1	1	8	3	91.5
16. **T.6 = knowledge/familiarity**	3	3.8	3	3.6	1	1	7	2.7	94.2
17. **T.7 = generic olfactory term**	5	6.4	5	5.9	5	5.2	15	5.8	100
Total	78	100	84	100	98	100	260	100	100

5. *fruit*, fruit; *fruité*, fruity [considered as a "related odor" (fruity) in Engen's "liberal scoring"], more frequent for ORANGE and APPLE than for LEMON
6. "flower," mainly given for APPLE
7. generic names for foods (or any different wordings for different types of food, e.g., *bonbon*)
8. drugs, cosmetics, and cleaning products, which are more and more frequently being perfumed or flavored with "natural scents," also given as Type-1 responses using more specific wording.

These words also provide some relevant information about the categorical organization of odors, in the present case their superordinate level of knowledge organization: ORANGE is frequently considered as fruit, LEMON is considered as food (mainly candy), and APPLE is identified as fruit, flower, food, or cleaning product. Such diversity of attributions of membership in multiple superordinate categories is also encountered in the naming of "natural objects." As previously mentioned, the tendency to think that there is just one veridical category for each exemplar relies on a logical conception of categories and concepts that has been discussed in recent contributions to ecological cognition, the latter being viewed as rather experiential than veridical (Lakoff, 1987; Varela, Thompson, and Rosch, 1991). Within this framework, the diversity of attributions of membership in superordinate categories simply reflects the diversity of objects that can now be flavored because of developments in food technology. Therefore, we consider the scoring in terms of veridical labels for odors to reflect a belief in an implicit theory of knowledge rather than an empirical result, because it fails to account for a large proportion of the subjects' answers. As far as the organization of odor categories is concerned, it cannot be determined whether such responses reflected the organization of the categories of odors per se or the categories of objects whose representations would necessarily be "activated" through the mediation of the source names ("lexical access" from vertex C to vertex B in Figure 4.1), as was required by the task. The cumulative scorings (Type 2) for these answers added to the previous ones (Type 1) now range from 29% (APPLE) to 45% (ORANGE) up to 62% (LEMON), with an average score of 44%.

Type 3: Co-hyponyms of the acceptable source names

Psychological hypotheses about categorization allow us to account for quite an important set of answers that linguistically can be considered as co-hyponyms, that is, as nouns referring to objects that belong to the same superordinate categories as the expected (acceptable) sources – for example, any specific name for a fruit, except the expected one (lines 9–12, summarized under Type 3 in Table 4.2). Such answers can be interpreted as reflecting some categorical knowledge that subjects want to communicate about the odor. In the current case, if we take into account that bit of knowledge, we find that ORANGE was frequently given as LEMON (but not the reverse) and that APPLE was frequently named by various fruit names (frequently as "strawberry"). Once more, we cannot determine solely from the current data whether that result depended on the quality of the odorants (A′ at vertex A) and their representativeness of the "real odor of *X*" (A at vertex A), on the accuracy of memory for the odors (vertex C), or on the structural properties of the subject's mental lexicon for fruits (vertex B). However, these findings correlate with other nonverbal measures (such as similarity distances in free-categorization tasks) and support the interpretation that the better score for LEMON reflects the fact that it is a more typical representation of the category "citrus fruits" than is ORANGE, whereas APPLE is an atypical exemplar of the category "fruit" and was attributed to a diversity of categories.

Taking those answers into account in scoring leads to cumulative scores (Type 1 + Type 2 + Type 3) of 74%, 75%, and 70% for LEMON, ORANGE, and APPLE, respectively, with an average score of 73%. That is, whereas there were qualitative differences related to their categorical structuring in memory, the three odors prompted quite similar quantitative scores as "extended acceptable answers."

Type 4: Other co-hyponyms

The same rationale can be applied to the other categories listed in Table 4.2:

13. co-hyponyms for foods
14. co-hyponyms for chemicals, which could actually be flavored or odorized "real-world objects" experienced by the subjects as smelling like the stimuli: As such, these answers can also be taken into consideration in scoring. Such scoring mainly concerns types of food and candies for LEMON, ORANGE, and APPLE, but also soaps and cosmetics for ORANGE and APPLE.

Scoring that information, the acceptable answers were 85%, 87%, and 93% for LEMON, ORANGE, and APPLE, respectively (average: 88%). It will be noted that APPLE got a higher score than LEMON, a reversal of the order seen when calculations were for the veridical label. That might be due to the fact that apple is used to flavor a larger variety of objects.

Types 5–7

Finally, the last three types refer to the following:

15. hedonic judgments, which can be considered as relevant factors in knowledge of odors, often being considered the main feature in olfactory classification. The scoring of answers such as *"agréable*, pleasant" increased the acceptable answers by 1% (APPLE) up to 5% (LEMON).

16. elusive sensations: When subjects were able to state that they felt "something" that was "known", "unknown," or "familiar," being in the state of "tip of the nose" (Engen, 1987), or "in lack of word," as aphasic patients are, we counted such responses as acceptable and gained 1% up to 4% in acceptable answers (summed in T.6 of Table 4.1 and Table 4.2).

17. unidentifiable sensations: When subjects reported that they smelled "something," instead of nothing, either using plain generic olfactory terms (*odeur*, odor) or generic terms referring to the subclass of pleasant olfactory sensations (*senteur*, smell; *parfum*, perfume), we gained answers to reach, in each case, the 100% of answers that represent the bits of knowledge we have taken into account.

In short, when verbal scoring incorporates hypotheses concerning both the possible lexicalizations for odors and their structural properties as cognitive categories, we can count all the answers given by the subjects as acceptable. That is, even in the absence of attested and negotiated names for odors, subjects managed to communicate their (even partial) knowledge about their cognitive representations of odors.

4. Conclusion

The absence of specific names for odors forced us to confront fundamental issues regarding the relationship between language and cognition and their consequences for data scoring. Our analysis led us to question the validity of the triangle in Figure 4.1 and to subdivide each vertex to a degree that has the disadvantage of preventing any simple geometric representation (Figure 4.2), but also the advantage of helping to cope with the full range of verbal outputs obtained in human olfactory research. Whereas numerous studies have postulated a straightforward relationship between objects and words (mainly nouns), in our study we faced the problem of how to consider that relationship in the absence of "basic terms" for the categories of smells we were using. Our analysis is therefore an attempt to base the scoring of the verbal answers on an explicit theory of the psychological processes involved in naming. In other words, the empirical issue of scoring a "correct" verbal response depends on progress in understanding "whether these (as yet) unnamed categories share the same status as labeled categories" (Waxman, 1999, p. 274).

The inherently weak link between odors and names should not divert us from considering the contributions of the social sciences. Anthropological studies (Howes, Chapter 5, this volume) and numerous linguistic and psycholinguistic studies (Gentner, 1982; Markman, 1989; Taylor, 1995; Wierzbicka, 1992; Imai and Gentner, 1993; Foley, 1997; Engen and Engen, 1997; Waxman, 1999) have begun to indicate that the diversity of linguistic forms may constrain our ontological view of entities and lead to different distances between the "subject" and the "objects" of the world, such diversity ranging from complex phrasing expressing the effects of the world on the subject to simple "basic" names, which suggests that things are "crying out to be named" (Berlin, 1973) and that words, as labels, can be mapped onto things. Further experimental investigations will have to account for the fact that our culture has diversely lexicalized the different senses, and other cultures have also diversely lexicalized the different senses in different ways than our culture, increasing diversity (Classen, 1993). If we always perceive "something" through the diversity of senses, language diversely objectifies and "stabilizes" our cognitive representations of the world into a large variety of linguistic forms. Concerning olfaction, the diversity of "entities" we conceptualized at each vertex of the triangle leads us to reconsider the "mapping hypothesis" that allows a simple scoring according to a "veridical" criterion. Scoring has to take into account the different entities presented in Figure 4.2.

Ultimately there arises the question What is an odor? Is an odor (American) an odour (British), as well as *une odeur* (French)? The issue is not trivial, because dictionaries for these languages reveal interesting differences:

1. from the *American Heritage Dictionary*: "odor: the property of a thing perceived by the sense of smell."
2. from the *Concise English Dictionary*: "odour: any scent or smell, whether pleasant or offensive." Also: "scent: that which issuing from a body affects the olfactory nerves of an animal."
3. From the *Robert* (French common) *Dictionary*: *"odeur: Emanation volatile, caractéristique de certains corps et susceptible de provoquer chez l'Homme ou chez un animal des sensations dues à l'excitation d'organes spécialisés"* (volatile emanation specific of some bodies and able to elicit in humans or in animals some feelings provoked by the stimulation of specialized organs).

Odors or odorants?

These definitions indeed reflect diverse conceptualizations of odors. If an odor is always defined as "something" (David, 2000), as stated in the first part of such definitions, the expansions diverge from an objectivist description, ranging from (in 1) where the word "odor" refers to some thing in the world (therefore almost synonymous with "odorant") to a more physiological response (as stated in 2) or psychological object (as in 3), this latter being defined as a sensation related to something from the world, an emergent result of different processes of eliciting, provoking. Besides revealing different philosophical (ontological) assumptions, this issue entails consequences for the design of experimental research and, as discussed here, for scoring verbal answers. In contrast to experiments in vision and audition, where the stimuli and the correlated psychological objects have the same "name" (color, sound), olfactory research has to account for the fact that the description of the stimulus is concerned with the odorant (the substance), not with the odor it emits. Awareness that such questions are still open in olfactory research (Dubois and Rouby, 1997; Hudson, 1999) should spare us general and unproductive debates about past cognitive research, contrasting, for example, universalist (biological) versus relativist (cultural) conceptions of cognition. The challenge is rather to avoid opposing natural and therefore "veridical" sciences to social sciences, and instead develop interdisciplinary research that specifies how language articulates to categorization at every level of description within the cognitive sciences.

Acknowledgments

This study was part of a research project funded by the cognitive science program of the French Ministry of Education and by the CNRS ("Cognisciences"). We are also indebted to recent linguistic work with S. David.

Notes

1. Such an analysis is required to understand how the different lexical forms and cognitive representations can constrain one another. A first review of this general research program is available (Dubois, 2000).
2. We had, for example, to decide whether to classify a polylexical form composed of two source names, such as *bonbon à l'orange,* orange candy, as a "correct answer" or as an "associated object" (Engen, 1987), or, within our scheme, to classify it with the "expected source name" or with some "other possible source name."
3. It is worth noting, once more, that all odors are not "equal" in their relationships to word constructions and lexicalizations as in the current example (N + suffix *é*): Thus *citronné* and *orangé*, as well as *fruité*, are possible in French, but **pommé* is not acceptable, and there is no strict correspondence in English: One can have "fruity" or "lemony," but not "*orangy". The consequences of such differences in lexicalization for cognitive structures remain to be explored.

References

Berlin B (1973). Folk Systematics in Relation to Biological Classification and Nomenclature. *Annual Review of Ecology and Systematics* 4:259–71.
Boisson C (1997). Quelques généralités sur la dénomination des odeurs: variations et régularités linguistiques. *Intellectica* 24:29–49.
Clark H H & Wilkes-Gibbs D (1986). Referring as a Collaborative Process. *Cognition* 22:1–39.

Classen C (1993). *Worlds of Senses: Exploring the Senses in History and across Cultures*. London: Routledge.

David S (1997). Représentation sensorielles et marques de la personne: contrastes entre olfaction et audition. In: *Catégorisation et cognition: de la perception au discours*, ed. D Dubois, pp. 211–42. Paris: Kimé.

David S (2000). Certitudes et incertitudes dans les domaines olfactif, gustatif et auditif. *Cahiers du LCPE (Langages, cognitions, pratiques, ergonomie)* 4:77–107.

David S, Dubois D, Rouby C, & Schaal B (1997). L'expression des odeurs en français: analyse lexicale et représentation cognitive. *Intellectica* 24:51–83.

Doty R L (1992). Diagnostic Tests and Assessment. *Journal of Head Trauma Rehabilitation* 7:47–65.

Doty R L, Applebaum S, Zusho H, & Settle G (1985). Sex Differences in Odor Identification Ability: A Cross Cultural Analysis. *Neuropsychologia* 23:667–72.

Doty R L, Marcus A, & Lee W (1996). Development of the 12-Item Cross-Cultural Smell Identification Test (CC-SIT). *Laryngoscope* 106:353–6.

Dubois D (1991). Les catégories sémantiques naturelles: prototype et typicalité. In: *Sémantique et cognition: Catégories, concepts et typicalité*. ed. D Dubois, pp. 15–27. Paris: Éditions du CNRS.

Dubois D (1996). Matériels et consignes: un type de questionnaire social dans la recherche experimentale en psycholinguistique. In: Le questionnement social, ed. J Richard-Zappella, pp. 89–98. *Cahiers de Linguistique Sociale (numéro spécial)*.

Dubois D (2000). Categories as Acts of Meaning: The Case of Categories in Olfaction and Audition. *Cognitive Science Quartely* 1:35–68. http://www.iig.unifreiburg.de/cognition/csq

Dubois D & Resche-Rigon P (1995). De la "naturalité" des catégories sémantiques: des catégories d'objets naturels aux catégories lexicales. *Intellectica* 20:217–45.

Dubois D, Resche-Rigon P, & Tenin A (1997). Des couleurs et des formes: catégories perceptives ou constructions cognitives. In: *Catégorisation et cognition*, ed. D Dubois, pp. 7–40. Paris: Kimé.

Dubois D & Rouby C (1997). Olfaction et cognition: du linguistique au neuronal. *Intellectica* 24:9–20.

Dubois D, Rouby C, & Sicard G (1997). Catégories sémantiques et sensorialités: de l'espace visuel à l'espace olfactif. *Enfance* 1:141–51.

Eco U (1988). *Le signe*. Paris: Labor.

Eco U (1997). *Kant e l'ornotorinco*. Milan: Bompiani. (French translation, 1999, *Kant et l'ornithorynque*. Paris: Grasset.)

Engen T (1987). Remembering Odors and Their Names. *American Scientist* 75:497–503.

Engen T & Engen E (1997). Relationship between Development of Odor Perception and Language. *Enfance* 1:125–40.

Foley W A (1997). *Anthopological Linguistics*. London: Blackwell.

Gentner D (1982). Why Nouns Are Learned before Verbs: Linguistic Relativity versus Natural Partitioning. In: *Language Development: Language, Thought, and Culture,* ed. S Kuczaj, pp. 301–34. Hillsdale, NJ: Lawrence Erlbaum.

Grossen M (1989). Le contrat implicite entre l'expérimentateur et l'enfant en situation de test. *Revue Suisse de Psychologie* 48:179–89.

Hudson R (1999). From Molecule to Mind: The Role of Experience in Shaping Olfactory Function. *Journal of Comparative Physiology* 185:297–304.

Imai M & Gentner D (1993). What We Think, What We Mean, and How We Say It. *Chicago Linguistic Society* 29:171–86.

Lakoff G (1987). *Women, Fire, and Dangerous Things*. University of Chicago Press.

Lucy J (1992). *Language Diversity and Thought.* Cambridge University Press.

Markman E (1989). *Categorization and Naming in Children*. Cambridge, MA: MIT Press.

Mondada L (1997). Processus de catégorisation et construction discursive des catégories. In: *Catégorisation et cognition*, ed. D Dubois, pp. 291–313. Paris: Kimé.

Mondada L & Dubois D (1995). Construction des objets de discours et catégorisation: une approche des processus de référenciation. *TRANEL* 23:273–302.

Neisser U (1987). *Concepts and Conceptual Development: Ecological and Intellectual Factors in Categorization*. Cambridge University Press.

Ogden C K & Richards I A (1923). *The Meaning of Meaning*. London: Routledge & Kegan Paul.

Rastier F (1991). *Sémantique et recherches cognitives.* Paris: Presses Universitaires de France.

Rosch E (1978). Principles of Categorization. In: *Categorization and Cognition*, ed. E Rosch & B Lloyd, pp. 28–47. Hillsdale, NJ: Lawrence Erlbaum.

Rouby C, Chevalier G, Gautier B, & Dubois D (1997). Connaissance et reconnaissance d'une série olfactive chez l'enfant préscolaire. *Enfance* 1:152–71.

Rouby C & Dubois D (1995). Odor Discrimination, Recognition and Semantic Categorization. *Chemical Senses* 20:78–9.

Rouby C & Sicard G (1997). Des catégories d'odeurs? In: *Catégorisation et cognition*, ed. D Dubois, pp. 59–81. Paris: Kimé.

Taylor J (1995). *Linguistic Categorization*. Oxford University Press.

Tenin A (1991). Catégorisation et cognition: 10 ans après. In: *Sémantique et cognition: Catégories, concepts et typicalité*, ed. D Dubois, pp. 7–25. Paris: Éditions du CNRS.

Varela F J, Thompson E, & Rosch E (1991*). The Embodied Mind: Cognitive Science and Human Experience*. Cambridge, MA: MIT Press.

Waxman S (1999). The Dubbing Ceremony Revisited: Objects Naming and Categorization in Infancy and Early Childhood. In: *Folkbiology*, ed. D Medin & S Atran, pp. 233–84. Cambridge, MA: MIT Press.

Wierzbicka A (1992). The Meaning of Color Terms: Semantics, Culture and Cognition. *Cognitive Linguistics* 1:99–150.

5

Nose-wise: Olfactory Metaphors in Mind

David Howes

In *Visual Thinking*, the psychologist Rudolf Arnheim argued that there is no division between seeing and thinking: "Visual perception is visual thinking" (Arnheim, 1969, p. 14). As for smell and taste, "one can indulge in smells and tastes, but one can hardly think in them" (Arnheim, 1969, p. 18). That dismissive judgment as regards the cognitive potential of olfaction by such a prominent psychologist does not bode well for a book that is dedicated to the theme of olfaction and cognition.

Denigration of olfaction is a common theme among Western thinkers (cf. Le Guérer, Chapter 1, this volume). The eighteenth-century philosopher Condillac, for example, remarked that "of all the senses smell is the one that seems to contribute the least to the operations of the human mind" (Condillac, 1930, p. xxxi). His contemporary, Immanuel Kant, similarly wrote: "To which organic sense do we owe the least and which seems to be the most dispensable? The sense of smell. It does not pay us to cultivate it or to refine it in order to gain enjoyment; this sense can pick up more objects of aversion than of pleasure (especially in crowded places) and, besides, the pleasure coming from the sense of smell cannot be other than fleeting and transitory" (Kant, 1978, p. 46).

The Kantian line of thought continues to inform the way in which aesthetic experience is conceptualized in the West. Aesthetic experience is defined as the disinterested contemplation of some object in which beauty inheres. Such experience is to be had in the context of the art museum, where art objects are placed on display, and visitors are required to keep their distance. This definition encodes an essentially visual relation to the art object, as Jim Drobnick (1998) has observed, which is enforced by the basic sensual sterility of the art gallery or "white cube." The Kantian line of thought is also manifest in the cognitive sciences. For example, in the literature of cognitive anthropology there are numerous studies having to do with the categorization of sense data. However, the vast majority of those studies pertain to the classification of visual stimuli – most notably color, following the lead of Berlin and Kay (1969) – and only a few concern the categorization of olfactory stimuli (Classen, Howes, and Synnott, 1994). Berlin and Kay drew a series of grand conclusions about the evolution of language and the universals of human cognition on the basis of their comparative study of color lexicons. Yet they never bothered to inquire whether or not the same conclusions would follow from a comparative study of olfactory or gustatory terminologies. That major oversight has gone largely unchallenged in the literature, as if there were indeed nothing to be learned from study of the language of gustation or olfaction, as if Condillac were right and smell and taste do have little to contribute to the operations of the human mind.

This chapter poses a series of questions: Why the devaluation of olfaction by cognitive psychologists such as Arnheim and philosophers such as Condillac and Kant? What insights into the nature of cognition might be gleaned from taking the so-called lower senses of olfaction, gustation, and touch, in place of vision, as metaphors for cognition? What can anthropology teach us about cognition in the context of societies that are more nose-minded than our own? And, finally, what can a comparative study of olfactory terminologies tell us that is different from what we have learned in cross-cultural study of color terminologies?

1. The Rise of Visualism

In Western culture, sight is commonly held to be the most important of the senses and the sense most closely allied with reason. This bias in favor of sight was already present in ancient philosophy (cf. Le Guérer, Chapter 1, this volume). Aristotle, for example, considered sight to be the most informative of the senses. Although vision had been ranked first among the senses since antiquity, it had remained "first among equals" and did not distance itself significantly from the other senses in terms of cultural importance until the period of the Enlightenment. Prior to the Enlightenment, the exalted position enjoyed by sight was tempered by the conviction that whereas the senses existed in a hierarchy, all were "perfectly formed in their own right," as evidenced by the fact that "colours, sounds, smells, tastes and touch can each be received by one and only one organ" (Sears, 1993, pp. 33–5). Sight, therefore, had its limits. Full knowledge could be acquired only if *all* the senses worked together.

The groundwork for the separation of sight from the other senses was laid in the eighteenth century when, as Constance Classen informs us, "vision became associated with the burgeoning field of science. The enquiring and penetrating gaze of the scientist became the metaphor for the acquisition of knowledge at this time. [Then, in the nineteenth century:] Evolutionary theories propounded by prominent figures such as Charles Darwin [and, later, Sigmund Freud] supported the elevation of sight by decreeing vision to be the sense of civilization. The 'lower,' 'animal' senses of smell, touch and taste, by contrast supposedly lost importance as 'man' climbed up the evolutionary ladder. In the late nineteenth and twentieth centuries, the role of sight in Western society was further enlarged by the development of such highly influential visual technologies as photography and cinema" (Classen, 1997, p. 402).

The cumulative effect of the developments described by Classen was that vision came to usurp many of the roles previously played by the other senses as means of cognizing reality. This process is apparent in the domain of medical science, for example. Whereas in premodern medicine, touch was used to ascertain the patient's pulse and temperature, and taste and smell were used to test the patient's urine and other bodily emanations, in modern medicine there are instruments and laboratory tests to perform these functions and generate graphic results, which are destined for the physician's eyes alone (Howes, 1995).

The Enlightenment thus marked the beginning of a fantastic mutation in the Western hierarchy of sensing. The balance or ratio of the senses was tipped in favor of vision, and the champions of sight have never looked back. The lower senses (especially the sense of smell) were pushed beyond the pale of culture and cognition and continue to languish in relative obscurity to this day. This process was aided by the invention of numerous technologies for the extension of sight (and hearing), from telescopes to television, which have no olfactory channel, thus effectively erasing olfactory stimuli from the modern consciousness. However, the traces of a different consciousness remain embedded in the very language we use to talk about intelligence, as will be shown in the next section.

2. Worlds of Sense

In *Worlds of Sense*, Constance Classen (1993) presents an in-depth study of the sensory etymology of many current and some moribund English words for cognitive operations. Her research shows that the senses that today are commonly considered "lower" or "animalistic" senses were once associated with the intellect. For example, "nose-wise," a word now obsolete, could mean either "clever" or "keen-scented." In Latin, both the sense of taste and the sense of smell were linked with wisdom, and that association is continued in some Latin words that are retained in English. The English words "sagacious" and "sage," both referring to intelligence, are based on Latin words meaning to have a good sense of smell. Similarly, the word "sapient," meaning wise, is based on the Latin word for taste; hence the term *Homo sapiens* means "tasting man" as well as "knowing man." In addition to the olfactory and gustatory links to cognition discussed earlier,

Classen's archeology of the sensory subconscious of the English language reveals that many English terms for "thought" are, in fact, tactile or kinesthetic in origin. These include "apprehend," "brood," "cogitate," "comprehend," "conceive," "grasp," "mull," "perceive," "ponder," "ruminate," and "understand." The predominance of tactile imagery in words dealing with intellectual functions indicates that thought is, or was, experienced primarily in terms of touch. Thinking was therefore less like looking than like weighing or grinding, and knowing was less like seeing than like holding. What difference does this make? "The use of tactile and kinaesthetic terms for thought expresses a more active involvement with the subject matter than visual terms do. To understand is to stand under or among, to be part of the picture, whereas to see is to view the picture from without. In light of this it seems possible that an emphasis on visual metaphors for intellectual functions, such as one finds in scholarly writing, for example, has to do in part with a desire to have or convey a certain detachment from the subject under consideration: to be 'objective' [see Jonas, 1970]. At the same time, visual metaphors for thought convey an accessibility of meaning, which tactile metaphors do not. There is a great deal more tension involved in grasping or weighing a subject than in looking it over" (Classen, 1993, p. 58).

Classen's examination of terms for intelligence further revealed that touch-based words such as "acumen," "acute," "keen," "sharp," "smart," "clever," and "penetrating" consistently outnumbered sight-based words such as "wise," "bright," "brilliant," and "lucid." It appears that it was only during the Enlightenment that words with strongly visual associations began to be used to refer to intelligence, perhaps as a consequence of the general rise of visualism at that time. Prior to that period, intelligence "was conceived of not so much as sight or light, as of sharpness. A knowledgeable person does not simply illuminate a subject, but cuts into it. Likewise, difficult subject matter would be characterized as 'hard' or 'complicated' (twisted together), resisting being cut or penetrated. Intelligence itself, along with intellect, is a tactile-visual metaphor, as its basic meaning is 'to pick between'" (Classen, 1993, p. 58).

With Classen's account of the sensory underpinnings of thought and intelligence in mind, it becomes apparent that Rudolf Arnheim's theorization of an intrinsic link between vision and cognition in *Visual Thinking* represented but a tiny fraction of the actual historical linkages between sensation and cognition. Moreover, his denigration of the cognitive potential of olfaction and gustation perpetrated a serious elision of the role these senses have played historically in the construction of intelligence. It is understandable, given the hypervisualism of contemporary Western culture, that Arnheim would choose to privilege the connection between vision and cognition and overlook the contributions of the other senses, but that does not make his position any more sound.

Of course, not every treatment of olfaction in the annals of Western thought is marred by the visualist bias manifest in the work of psychologists such as Arnheim and philosophers such as Kant. Think of the perfumer-philosopher Edmond Roudnitska's brilliant refutation of Kant and rehabilitation of the aesthetic power of smell in *L'esthétique en question: Introduction à une esthétique de l'odorat* (Roudnitska, 1977). Or, in a more cognitivist vein, consider the following remarks of the naturalist Lewis Thomas: "The act of smelling something, anything, is remarkably like the act of thinking itself. Immediately, at the very moment of perception, you can feel the mind going to work, sending the odor around from place to place, setting off complex repertoires throughout the brain, polling one center after another for signs of recognition, old memories, connections. This is as it should be, I suppose, since the cells that do the smelling are themselves proper brain cells" (Thomas, 1983, p. 42).

Earlier in the same essay, Thomas proclaims that "we might fairly gauge the future of biological science, centuries ahead, by estimating the time it will take to reach a complete, comprehensive understanding of odor" (Thomas, 1983, p. 41). These are bracing words. However, it seems doubtful that biological science will be able to accomplish this task on its own, unaided by the sorts of examples that historical or anthropological research on the senses can yield by way of models, for biological science remains preoccupied with issues of odor detection and identification, if Thomas's essay is any indication. Those issues are important, but for us to arrive at a comprehensive understanding of odor it will be necessary to explore how the medium of smell can, in some cases, provide the conceptual framework for construction of an understanding of the entire universe. Ongee cosmology offers a paradigm case of such "olfactory thinking."

3. A Nose-minded Society

Among the Ongee, a hunting and gathering people of Little Andaman Island in the Bay of Bengal, smell is the primary sensory medium through which the categories of time, space, and the person are conceptualized. Odor, according to the Ongee, is the vital force that animates all living, organic beings. A newborn is said to possess little scent. On growing up, people increase their olfactory strength. The odor that a person scatters about during the day is said to be gathered up during sleep by an inner spirit and returned to the body, making continued life possible. Death occurs when one loses one's odor – through illness or an accident or because it is absorbed by an odor-hunting spirit. Once dead, a person becomes an odorless and inorganic spirit, devoted to seeking out the odors of the living in order to be reborn. Thus the life cycle is conceptualized in terms of an olfactory progression. Indeed, the Ongee word for growth, *genekula*, means "a process of smell" (Pandya, 1987, pp. xii, 19, 108, 111–12, 312).

Life for the Ongee is a constant game of olfactory hide-and-seek. They seek out animals in order to kill them by releasing their odors, and at the same time they try to hide their own odors, both from the animals they hunt and from the spirits who hunt them. "To hunt" in the Ongee language is expressed by *gitekwabe*, meaning "to release smell causing a flow of death." The word for hunter, in turn, is *gayekwabe*, "one who has his smell tied tightly" (Pandya, 1987, p. 102).

The Ongee employ different techniques to keep their smell "tied tightly." Living in community is believed to unite the odors of individuals and lessen their chances of being smelled out by hungry spirits. When moving as a group from place to place, for example, the Ongee are careful to step in the tracks of the person in front, as this is thought to confuse personal odors and make it difficult for a spirit to track down an individual. The Ongee also screen their odor through the use of smoke. When traveling in single file, the person at the head of the group carries burning wood so that the trail of smoke will cover the odors of all those walking behind. For the same reason, the Ongee keep fires burning at all times in their villages and have smoke-filled, unventilated homes. Indeed, for the Ongee, a true fire is characterized not by the heat or light it produces, but by its smoke. Even the sun has an olfactory identity, as it is believed to produce an invisible smoke (Pandya, 1987).

The Ongee also limit their smell emission by painting themselves with clay. Clay paints are believed to help bind smells to the body, while the different designs used in painting alter the ways in which the body releases smell. An Ongee whose body has been painted will declare: "The clay paint has been good! I feel that my smell is going slowly and in a zig-zag manner like the snake on the ground!" (Pandya, 1987, p. 137). The Ongee also dab clay paint on their skin after eating meat in order to prevent the smell of the consumed meat from warning living animals that one of their kind has been killed and eaten (Pandya, 1987).

Space is conceived of by the Ongee not as a static area within which things happen but as a dynamic environmental flow. The olfactory space of an Ongee village fluctuates: It can be more expansive or less, depending on the presence of strong-smelling substances, such as pig's meat, the strength of the wind, and other factors. Given that smell is believed to guide both harmful and helpful spirits to human sites, it is the olfactory ambience of the village that is of the greatest concern to the Ongee, not its extension in physical space. The Ongee "smellscape," then, is not a fixed structure, but rather a highly fluid pattern that can shift and change according to differing atmospheric and seasonal conditions.

This fluidity finds expression in the use of the same word, *kwayaye*, by the Ongee for both the emission of odors and the ebb and flow of tides (Pandya, 1993).

A distinct succession of scents wafts through the jungle of the Andaman Islands as one after another of the indigenous trees and climbing plants come into flower throughout the year. The Ongee have constructed their calendar on the basis of this cycle, naming the different periods of their year after the fragrant flowers that are in bloom at different times. The Ongee year is a cycle of odors, their calendar a "calendar of scents" (Radcliffe-Brown, 1964).

The Ongee seasonal cycle is further based on the winds that blow in from different directions throughout the year, dispersing odors and bringing scent-hungry spirits. The Ongee conduct their own migrations from the coast to the forest according to this cycle: During the seasons when the spirits are believed to be hunting at sea, the Ongee hunt in the forest, and during the seasons when the spirits are believed to be hunting in the forest, the Ongee hunt along the coast. The information the shaman brings from the spirit world helps the Ongee plan their movements so as to continue to succeed in their game of olfactory hide-and-seek with the spirits (Pandya, 1987).

Both time and space therefore acquire meaning for the Ongee in terms of the movements of winds and spirits, humans and animals, and their odors. When asked to draw a map, an Ongee will depict a line of movement from one place to another, rather than the locations of the places themselves. The ethnographer Vishvajit Pandya found that during his stay among the Ongee his official map of the island was of little help in making sense of the routes his Ongee guides took him along. When he complained to an Ongee friend that his experience of the island did not coincide with the map of its geographical layout, the man replied: "Why do you hope to see the same space while moving? One only hopes to reach the place in the end. All the places in space are constantly changing. The creek is never the same; it grows larger and smaller because the mangrove forest keeps growing and changing the creek. The rise and fall of the tidewater changes the coast and the creeks.... You cannot remember a place by what it looks like. Your map tells lies. Places change. Does your map say that? Does your map say when the stream is dry and gone or when it comes and overflows? We remember how to come and go back, not the places which are on the way of going and coming" (Pandya, 1991, pp. 792–3).

Here a static visual layout of space, such as is contained in a map, is opposed to a lived experience of movement through space, and of the movement *of* space – swelling streams, shifting coastlines, expanding forests. Space is thus as imprecise and changing as the odors that animate the world for the Ongee. Time, in turn, is a cycle of olfactory production, loss and gain. It is impossible to adequately represent such an olfactory cosmology by means of a map or a calendar: It can only be scented.

The Ongee may appear to constitute an exceptional or unusual case in terms of the extent to which they use smell to order their world. However, from the standpoint of the "anthropology of the senses" (Howes, 1991), the olfactory cosmology of the Ongee is no more unusual than the thermal cosmology of the Tzotzil of Mexico (Classen, 1993) or the gustatory cosmos of Hindu India (Pinard, 1991) or the visual universe of the modern West. Different cultures attach different meanings and uses to the different senses, and this affects the way in which they imagine or represent the world. Each culture must be approached on its own sensory terms if its perceptual world is to be apprehended and described accurately by the anthropologist. We are fortunate that Vishvajit Pandya, the principal ethnographer of the Ongee, had the presence of mind to follow his nose in studying the Ongee life world.

Other students of cross-cultural psychology interested in issues of culture and perception have not been so sensitive to the possible implications of variations in the cultural construction of the sensorium. They often have opted instead to repeat a little experiment involving Munsell color chips that was pioneered by Brent Berlin and Paul Kay in the 1960s, not hesitating to draw grand conclusions about "universals" of human cognition on the basis of the responses they received (Brown, 1991). In the next section, a review and critique of the basic-color-terms paradigm is presented, followed by a discussion of how to go about studying olfactory classifications in their cultural contexts.

4. Sensory Vocabularies

In *Basic Color Terms*, Brent Berlin and Paul Kay make two claims, one of which is empirical, and the other evolutionary. The empirical claim holds that there exist certain universal "focal colors" that can be identified by test subjects when presented to them in the form of Munsell color chips, independently of how few or how many color terms their mother tongues possess. The evolutionary claim holds that there is a progressive order to the sequence in which basic color terms enter languages and that it is possible (in view of that order) to plot the various languages of the world on a single evolutionary scale. In Berlin and Kay's own words, "the overall temporal order [to the encoding of perceptual categories into basic color terms] is properly considered an evolutionary one; color lexicons with few terms tend to occur in association with relatively simple cultures and simple technologies, while color lexicons with many terms tend to occur in association with complex technologies" (Berlin and Kay, 1969, p. 101).

In recent years, researchers interested in the study of nonvisual sensory vocabularies have begun to raise serious questions about the universality and applicability of Berlin and Kay's conclusions. For example, as regards Berlin and Kay's empirical claim, Dubois and Rouby, in numerous publications (e.g., Dubois, 1997; Dubois and Rouby, 1997; Dubois, Rouby, and Sicard, 1997), have pointed to some of the experimental difficulties involved in simply extending the basic-color-terms paradigm to the study of odor terms. They note that there is no determinate spectrum of smells in the way that there is for colors; moreover, odors have no names in most Indo-European languages. Joel Kuipers (1991) expressed similar reservations about the applicability of Berlin and Kay's methods to the study of taste in his research on gustation among the Weyéwa of Sumba.

Dubois and Rouby, as well as Kuipers, have nevertheless gone on to study odor and taste categorization. What is most interesting about the sorts of considerations they advance in discussing their research findings is the importance they attach to taking contexts – the diversity of human activities and practices – into account because of the orienting effect contexts have on categorization. This emphasis on customary activities and practices seriously contradicts the abstractness of Berlin and Kay's research design and results.

As regards Berlin and Kay's evolutionary claim, even the briefest comparative study of smell and taste lexicons reveals that there is no one-to-one correspondence between the complexity of a given culture's technology and the number of terms in its taste or smell vocabulary. For example, the English language contains four taste terms, and the Japanese language has five (Doty, 1986), but in the language of the Weyéwa of Sumba there are seven taste terms (Kuipers, 1991), even though Weyéwa culture and technology are ostensibly much "simpler" than the culture and technology of the English- or Japanese-speaking world. The language of the Sereer Ndut of Senegal contains only three taste terms; however, the Sereer Ndut recognize five odor categories (Dupire, 1987). That is considerably more than English (which contains two odor terms at best), but not as comprehensive as the language of the Bororo of Brazil, which counts eight odor terms (Crocker, 1985), or the language of the Kapsiki of Cameroon, which boasts 14 odor categories (Classen et al., 1994)!

This brief survey shows that technological complexity is not a significant predictor of the encoding of smell or taste categories into language, however strongly it may be associated with the encoding of so-called basic color terms. Berlin and Kay were remiss not to repeat their study of color perception and categorization for each of the other fields of sense before advancing their grand evolutionary scheme. The apparent superiority of English-speaking cultures in their scheme turns out to be an illusion based on the arbitrary selection of color terms and technological sophistication as "universal" criteria for the ordering of cultures.

What would a research design for the study of odor categorization that would be sensitive to cultural context entail? Let me draw on my own experience in studying the sensory order of the Kwoma of northwestern Papua New Guinea by way of example. The Kwoma inhabit the Washkuk Hills, which are situated about halfway up the Sepik River. They subsist on a diet of sago and yams, with the cultivation of the latter being the focus of their annual ritual cycle.

During my stay in the Washkuk Hills, each time I visited a new Kwoma village I would bring out my kit of odor samples. This kit contained over 30 plastic cards, each one impregnated with a different scent and corresponding color: rose is red, coconut is white, cinnamon is brown. The odor samples did not constitute an exhaustive set, but rather a haphazard one, for they were supplied courtesy of a major international flavor and fragrance manufacturer and consisted of the sorts of scents found in typical North American bath, cleaning, cosmetic, and confectionary products.

I used the odor samples as conversation-starters, rather than conversation-stoppers or label elicitors (the way Berlin and Kay used their Munsell color chips). That is, I was more interested in the associations the samples provoked than in their identification, and the Kwoma did not disappoint. It often happened that adults would send children off into the forest to bring back some tree bark or other substance that would match the odor sample. Every one of the substances they brought back had some practical, medicinal, magical, or gustatory virtue that would be explained to me. My wintergreen sample, for instance, was matched by the bark of a hardwood tree used for houseposts, known in the local language as *mijica*. The cinnamon sample put many people in mind of their native ginger, and because that plant (in its many varieties) was used in magic, it got me into many discussion of magic and sorcery.

Not all of my samples were matched with natural products. For example, the coconut sample was not identified with the coconuts that grew on trees, but with the coconut-flavored cookies from the trade store; the lemon sample was classified not as a fruit, but as soap, in keeping with the lemon-scented detergent also available from the trade store. Thus, there were definite limits to the extent to which the Kwoma were prepared to propose matches or analogies from the natural world for my synthetic scents.

I was also interested in exploring Kwoma olfactory preferences. Interestingly, at first my Kwoma friends consistently picked the rose sample as their favorite. Could it be that humans are predisposed to prefer rose over all other odors, the same way we are predisposed to prefer sugar over all other savors? As it turned out, the redness of the rose sample must have been influencing responses – red being the color associated with the pleasures of chewing betel, as well as sexual arousal. Later, after I had learned to enclose the odor samples in envelopes so that their color could not be seen, there was no longer the same consistency in selecting rose as the favorite scent. The question whether or not there exist any universal human olfactory preferences will therefore have to remain open.

Eliminating color as a variable might seem like a good procedure when investigating smell, but I was actually more interested in studying how the senses interact than in separating them out, for it is only by analyzing the interplay of the senses that one can arrive at a proper understanding of a culture's sensory order (Classen and Howes, 1991). The Kwoma language is interesting to study from this perspective, particularly with the aid of Ross Bowden's fine recent dictionary (Bowden, 1997). For example, it appears that the Kwoma conceive of visual attraction on the same model as olfactory attraction. This is evidenced by the Kwoma word *hirika*, which means "smoke," "steam," "aroma" – as of an object that has been rubbed with a strong-smelling magical substance – and "aura," or the aesthetic quality of people and sculptures when they have been painted and decorated with shells and feathers in preparation for a ceremony that will make them visually attractive.

Sight is definitely the dominant sense in the Kwoma sensory order, as suggested by the importance the Kwoma attach to making visual representations (paintings, sculptures) of the spirits that control their universe, the way they privilege visual knowledge over aural knowledge in the context of male initiation rituals, and other indicia (Howes, 1992). Smell complements sight, but it is also frequently subversive or destructive of the visual order of society and the cosmos. For example, in one Kowma myth, an old man who magically sloughs off his aged skin and cavorts with two women in the forest is eventually found out when his dog is brought along by the women and sniffs him out. This myth suggests that olfactory identity transcends visual appearance.

In another myth explaining the origin of Ambon Gate, a particularly treacherous narrows in the Sepik River, it is told that there used to be a land bridge at that point, but that a spirit destroyed it because a menstruating woman broke the taboo on remaining isolated during her menses and her smell offended the spirit. The Kwoma in fact have a special category to refer to the allegedly unpleasant smell of a woman menstruating or one who has just given birth, *maba gwonya*. It is the opposite of the category of *mukuske gwonya*, which refers to the smell from the armpits and genitals of a pubescent girl, which Kwoma men claim to find sexually attractive.

There is a significant gender dimension to the Kwoma sensory order. Men are associated with the controlling power of sight, whereas women are associated with the alternately seductive or debilitating and destructive power of smell. Of course, men are also recognized to possess body odor among the Kwoma, and like the women are very concerned to control it by washing and decorating themselves with leaves. But the body odor of males is unmarked compared with the marked significance of female body odors, which stand for the opposed poles of attraction and repulsion. The hierarchy of the senses, and within that framework the hierarchy of smells, is thus a powerful means by which the hierarchy of the sexes is conceptualized and enforced among the Kwoma. Olfactory categorizations are rarely neutral discourses for describing the world.

5. Sensuous Intelligence

This chapter has shown that the hypervaluation of the visual sense in Western science, beginning in the period of the Enlightenment, had the effect of dissociating the so-called lower senses of olfaction, gustation, and taste from the intellect. However, the idea that the lower senses are devoid of cognitive potential can be contradicted on the basis of Classen's archeology of sense words for intellect in the English language and a consideration of the nose-wise culture of the Ongee of Little Andaman Island. The Ongee case, in particular, demonstrates that it is possible for smell to serve as the medium for construction of a highly elaborate cosmology and epistemology, with only a residual role played by vision. This point is further confirmed by the cross-cultural study of sensory vocabularies, which shows that cultures differ markedly in the intensity with which they attend to each of the five senses and exploit them for classificatory purposes.

The preceding discussion illustrates the importance of conceptualizing intelligence in a more sensuous manner than has been common in cognitive science (Gardner, 1983; Johnson, 1987). Indeed, bearing the senses in mind appears fundamental to the future of cognitive science. In this respect, a highly stimulating and suggestive sense-based model of cognitive (or, more accurately, brain) functioning is provided by the Desana Indians of Colombia. Desana ideas about the brain derive partly from their experience with hallucinogenic drugs. Such drugs are used by shamans to manipulate brain functions. Desana ideas about the brain are otherwise grounded in the knowledge they have gained from butchering game and observing victims of head injury: "The Desana imagine the brain to be divided into many small compartments. In one image the brain is compared to a huge rock crystal subdivided into many smaller hexagonal prisms, each containing a sparkling element of color energy. . . . In another image a brain consists of layers of innumerable hexagonal honeycombs; the entire brain is one huge humming beehive. . . . Each tiny hexagonal container holds honey of a different color, flavor, odor or texture, or it houses a different stage of insect larval development" (Reichel-Dolmatoff, 1981, pp. 82–3).

Pointing to different compartments in a drawing he had made of the brain, one Desana man explained: "Here it is prohibited to eat fish; here it is allowed;

here one learns to dance; here one has to show respectful behavior," and so on, demonstrating that each compartment is linked to social behavior (Reichel-Dolmatoff, 1981).

The two cerebral hemispheres are said by the Desana to have complementary functions. The left hemisphere is dominant and male and represents moral authority, shamanic wisdom, and divine law. It is the seat of music, dreams, and hallucinatory visions. The right hemisphere is female and subservient. It is concerned with practical affairs, customary rules and rituals, physical nature, illness and death. The left hemisphere presents the ideal, the right the potential to put it into practice. The central role of the Desana shaman is to interpret all sensory phenomena – from the song of a bird to the perfume of a flower – in terms of the abstract social and cosmological ideals they represent and direct people to put these ideals into practice.

Can the brain usefully be compared to a huge, humming beehive as the Desana suggest? Needless to say, no Western-trained neuroscientist has ever asked this question. True, the Desana understanding of the complementary functions of the two cerebral hemispheres may not quite square with what Western neuroscience teaches, but perhaps that is only because neuroscientists are more inclined to conceive of the brain as a computer and have yet to ask the right questions. It is time to start experimenting with the alternative models of perception and cognition presented by the world's indigenous cultures. For example, conceiving of the belly – or, more precisely, the intestines – as the seat of the emotions and the intellect the way the Kwoma (and many other indigenous peoples) do could open up new possibilities for investigation. It is also instructive to ponder the homology between hearing something and smelling something that the Kwoma language posits by virtue of the fact that both of these perceptual processes are designated by the same term, *meeji*. It appears that there are many possible ways of combining the senses, which in turn create different patterns of consciousness. The Ongee, with their notion of time as dispersion and their fluid conception of space, represent a prime example of such a differently constituted consciousness.

In addition to mapping cross-cultural variations in the tone and shape of consciousness in accordance with variations in cultural constructions of the sensorium, there is a need for further study of the sociological implications of divergent sensory orientations. As was shown in the case of the Kwoma, sensory hierarchies and gender hierarchies reinforce each other, and there is also a hierarchy in the ordination of smells that has important implications for relations between the sexes. Olfactory categories may be contested by those who find themselves stigmatized by virtue of their representation in the sensory and social order (Classen et al., 1994). There exists, in other words, a cultural politics of olfaction that needs to be studied in concert with the more cognitive dimensions of olfaction.

Acknowledgments

Parts of the research on which this chapter is based were made possible by grants from the Fonds pour la Formation de Chercheurs et l'Aide à la Recherche and from the Olfactory Research Fund. I also wish to thank Catherine Rouby, Danièle Dubois, and Benoist Schaal for their invitation to participate in the conference.

References

Arnheim R (1969). *Visual Thinking*. Berkeley: University of California Press.
Berlin B & Kay P (1969). *Basic Color Terms: Their Universality and Evolution*. Los Angeles: University of California Press.
Bowden R (1997). *A Dictionary of Kwoma*. Canberra: School of Pacific Linguistics.
Brown D (1991). *Human Universals*. New York: McGraw-Hill.
Classen C (1993). *Worlds of Sense: Exploring the Senses in History and across Cultures*. London: Routledge.
Classen C (1997). Foundations for an Anthropology of the Senses. *International Social Science Journal* 153:401–12.
Classen C & Howes D (1991). Sounding Sensory Profiles. In: *The Varieties of Sensory Experience: A Sourcebook in the Anthropology of the Senses*, ed. D Howes. University of Toronto Press.
Classen C, Howes D, & Synnott A (1994). *Aroma: The Cultural History of Smell*. London: Routledge.
Condillac E B de (1930). *Treatise on the Sensations*, trans. G Carr. Los Angeles: University of Southern California Press.
Crocker J C (1985). *Vital Souls: Bororo Cosmology, Natural Symbolism and Shamanism*. Tucson: University of Arizona Press.
Doty R (1986). Cross-cultural Studies of Taste and Smell Perception. In: *Chemical Signals in Vertebrates*, ed. D Duvall, D Müller-Schwarze, & R Silverstein, pp. 673–84. New York: Plenum Press.
Drobnick J (1998). Reveries, Assaults and Evaporating Presences: Olfactory Dimensions in Contemporary Art. *Parachute* 89:10–19.
Dubois D (1997). Cultural Beliefs as Non-trivial Constraints on Categorization: Evidence from Colors and Odors. *Behavioral and Brain Sciences* 20:188.
Dubois D & Rouby C (1997). Une approche de l'olfaction: du linguistique au neuronal. In: *Olfaction: du linguistique au neurone*, ed. D Dubois & A Holley. *Intellectica* 24:9–20.
Dubois D, Rouby C, & Sicard G (1997). Catégories sémantiques et sensorialités: de l'espace visuel à l'espace olfactif. *Enfance* 1:141–50.
Dupire M (1987). Des goûts et des odeurs. *L'Homme* 27:5–25.
Gardner H (1983). *Frames of Mind: The Theory of Multiple Intelligences*. New York: Basic Books.
Howes D (1991). Sensorial Anthropology. In: *The Varieties of Sensory Experience: A Sourcebook in the Anthropology of the Senses*, ed. D Howes, pp. 167–91. University of Toronto Press.
Howes D (1992). The Bounds of Sense. Unpublished Ph.D. dissertation, Université de Montréal.
Howes D (1995). The Senses in Medicine. *Culture, Medicine and Psychiatry* 19:125–33.

Johnson M (1987). *The Body in the Mind: The Bodily Basis of Meaning, Imagination and Reasoning.* University of Chicago Press.

Jonas H (1970). The Nobility of Sight: A Study in the Phenomenology of the Senses. In: *The Philosophy of the Body*, ed. S Spicker, pp. 133–52. Chicago: Quadrangle Books.

Kant E (1978). *Anthropology from a Pragmatic Point of View*, trans.V L Dowdell. Carbondale: Southern Illinois University Press.

Kuipers J (1991). Matters of Taste in Weyéwa. In: *The Varieties of Sensory Experience: A Sourcebook in the Anthropology of the Senses*, ed. D Howes, pp. 111–27. University of Toronto Press.

Pandya V (1987). Above the Forest: A Study of Andamanese Ethnoamenology, Cosmology and the Power of Ritual. Ph.D. dissertation, University of Chicago.

Pandya V (1991). Movement and Space: Andamanese Cartography. *American Ethnologist* 17:775–97.

Pandya V (1993). *Above the Forest: A Study of Andamanese Ethnoamenology, Cosmology and the Power of Ritual.* Oxford University Press.

Pinard S (1991). A Taste of India. In: *The Varieties of Sensory Experience: A Sourcebook in the Anthropology of the Senses*, ed. D Howes, pp. 221–38. University of Toronto Press.

Radcliffe-Brown, A R (1964). *The Andaman Islanders.* New York: Free Press.

Reichel-Dolmatoff G (1981). Brain and Mind in Desana Shamanism. *Journal of Latin American Lore* 5:73–98.

Roudnitska E (1977). *L'esthétique en question: Introduction à une esthétique de l'odorat.* Paris: Presses Universitaires de France.

Sears E (1993). Sensory Perception and Its Metaphors in the Time of Richard of Fournival. In: *Medicine and the Five Senses*, ed. W F Bynum & R Porter, pp. 17–39. Cambridge University Press.

Thomas L (1983). *Late Night Thoughts on Listening to Mahler's Ninth Symphony.* New York: Viking Press.

6

Linguistic Expressions for Odors in French

Sophie David

> Cette première pièce exhale une odeur
> sans nom dans la langue, et qu'il
> faudrait appeler l'odeur de pension.
> – Balzac (1843, p. 53)

In this chapter I shall describe the grammar of a subset of French substantives associated with olfaction, with two aims: (1) to identify, from a linguistic point of view, the properties of these words and to examine the extent to which they resemble or differ from others, especially those that are linked to other sensory modes such as vision (colors) and hearing; (2) to evaluate some consequences of the hypothesis that linguistic expressions contribute to the structuring of representations in memory (Dubois, 1991, 1997, 2000). Indeed, the results of such an analysis (i.e., an examination of the various properties of words) can have consequences for the methods and analyses employed in experimental approaches to olfactory cognition.

It is important to note that the French lexicon has few terms to refer to olfactory entities. That paucity of terms is shared by many Indo-European languages, but such is not the case in many other language groups.[1] Basically, French speakers are restricted to terms such as these:[2]

(1) | *odeur* | "odor" | *effluve* | "emanation" | *remugle* | "fustiness" |
parfum	"perfume"	*exhalaison*	"exhalation"	*miasme*	"miasma"
senteur	"scent"	*puanteur*	"stench"	*fraîchin*[3]	
fragrance	"fragrance"	*bouquet*	"bouquet"		

Excepting the words *remugle* ("fustiness"), *miasme* ("miasma"), and *fraîchin*, which name specific smells, the terms allow us to refer to olfactory entities only generically. On a primary level, they appear to be "classified" according to the criteria of pleasant and unpleasant.

Any sort of linguistic analysis necessarily involves, first, an ordered collection of empirical data. This is why, in the context of this multidisciplinary volume, I shall begin with a brief recapitulation of the aims of linguistics and of the manner in which its facts are constructed, followed by a brief description of the theoretical framework of lexical semantics I have adopted. Then I shall examine some of the properties of the terms listed earlier in (1) and go on to draw some conclusions about the various methodological consequences that could be of use for empirical cognitive sciences.

1. Lexical Grammar: A Theoretical Framework

Linguistics is an empirical science that derives its data from descriptions and analyses of real or potential verbal productions. Such data are given in the form of examples (Milner, 1989; Kerleroux and Marandin, 1994). They are combined by pairs to create a contrast: Two examples that differ by a single element are set out in parallel, and we can then form a judgment concerning them. That judgment is binary in nature, "well formed" versus "not well formed," and hinges on grammatical properties. Thus, we do not judge examples according to aesthetic, moral, or cultural criteria,[4] nor by the possible interpretations to which they may give rise (Godard and Jayez, 1993). For example:

(2) (a) The dog is running
     (b) *Dog the is running
     (c) He fell asleep during the show
     (d) *He fell asleep during the pavement

Examples (2b) and (2d) are not well formed; we shall henceforth identify such a statement with an asterisk. The contrasts (2a) versus (2b) and (2c) versus (2d) are considered as *facts*. The difference between (2a) and (2b) involves the position of the determiner in English. The difference between (2c) and (2d) involves the lexical relationship between the preposition "during" and the nouns "show" and "pavement."

My work involves lexical semantics, a particular sphere of linguistics that is concerned with attempting to describe what is involved in the meaning of words. I cannot here give a detailed account of the variety of earlier theoretical approaches [i.e., the componential approach (Katz, 1972), the category and prototype approaches (Lakoff, 1987; Kleiber, 1990), the index approach (Cadiot and Nemo, 1997), the object-classes approach (Gross, 1994), etc.]. I work within a lexical semantics framework initiated by Marandin (1984a,b) and recently formalized by Godard and Jayez (1993, 1996) and Jayez and Godard (1995); but see also Grimshaw (1990) and Pustejovsky (1995). My work focuses on the combinatorial properties of lexical units, which are of two kinds: those based on argumental properties and those based on sortal properties, which are the grammaticalized correlates of referential properties (i.e., the properties of an "object in the world").

Two types of properties will be examined: (I) syntactic/semantic properties and (II) semantic properties:

> (I) Syntactic/semantic properties can be observed from the complementation system, in this instance a complement applied to a noun (the nature of the complement and its interpretation) (Grimshaw, 1990). For example:

(3)   The destruction of the city

In example (3), the complement phrase "of the city" is interpreted as the object of the noun phrase "the destruction"; in terms of thematic relations it is the theme of "destruction." These properties imply a judgment based on the structural (syntactic) well-formedness of the expressions (preposition type, noun type, the compulsory presence of a complement, the semantic interpretation of the complement, etc.).

> (II) Semantic properties can be observed in "selectional-restriction-based" contrasts. We use the term "selectional restriction" when the elements explaining a particular possibility or impossibility are lexical in nature, as in the following example:

(4)   (a) *The boy may frighten sincerity (Chomsky, 1965, p. 109)
       (b)   The boy may frighten his sister
       (c)   Sincerity may frighten the boy

"Sincerity" cannot be the complement of "frighten," because that verb requires an object denoting an animate entity. Such properties imply a judgment based on the semantic well-formedness of the expressions.

From these considerations, it follows that we separate the properties of meaning from the properties of things. Although, from a conceptual viewpoint, some words can share some properties, they may not share the same linguistic properties. They can belong to quite different linguistic classes. For example, *plage* ("beach") and *désert* ("desert") can refer to entities that share properties on the conceptual level (i.e., "place," "sandy," etc.), but we have the following contrast:

(5)   (a) *Je suis dans le désert* (Berthonneau, 1998)
            "I am in the desert"
       (b) **Je suis dans la plage*
            "I am in the beach"
       (c) *Je suis dans la rue*
            "I am in the street"

In these examples, from a linguistic point of view, *désert* ("desert") and *rue* ("street") have more in common than *désert* and *plage* do.

Thus, this approach does not a priori describe all of the semantic considerations that can be associated with a word. Only the meaning properties that are based on the analysis of contrast are taken into account, which ensures that the properties underpinning word meaning are linguistic in nature.

2. Odor Lexicon: Syntactic and Semantic Properties

In this section, some of the syntactic and semantic properties of terms appearing in list (1) will be presented. However, this review is neither exhaustive nor systematic. We shall focus primarily on *odeur* ("odor") and *parfum* ("perfume"), indicating some properties for the terms *senteur* ("scent"), *fragrance* ("fragrance"), *effluve* ("emanation"), *puanteur* ("stench"), and *exhalaison* ("exhalation").

A first focus will be on the linguistic heterogeneity of such terms. Then we shall examine the nature of the relationship between a smell and its "source" (as defined later) and consider whether or not the referent consists of parts, especially if the "source" can be considered as a "part of" a smell. Lastly, we shall consider whether or not these terms can have temporal properties. These issues are related to questions concerning the typology of lexical units (here, nouns). We adopt the hypothesis of Godard and Jayez (1996) to distinguish the following types: material and informational objects, strong and weak events, property, and activity, as shown in Figure 6.1.

Sections 2.1, 2.2, and 2.3 are concerned with different kinds of material objects, or with the differences between material objects and properties. Section 2.4 deals with temporal properties (a category that gathers events and activities).

For the sake of clarity, we shall henceforth employ the word "term" or "expression" to denote a language unit (a word, a sentence); the word "referent" stands for an entity in the world to which a given phrase points. For convenience, we employ the locution "olfactory terms" in referring to terms in list (1).

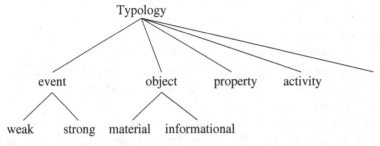

Figure 6.1. A portion of the typology proposed by Godard and Jayez (1996).

2.1. Linguistic Heterogeneity of Olfactory Terms

Olfactory terms do not appear to constitute a homogeneous class. They do not all have the same types of properties:

(I) We note simple names and also deverbal constructions such as *puanteur* and *exhalaison* (i.e., constructions that are morphologically derived from verbs).

(II) Some of these terms can be linked to different fields of knowledge. Such is the case with *miasme* in its more prevalent usage (naming a gas), or *parfum*, which is also frequently associated with the sphere of taste (see note 3). In the remainder of this presentation, we restrict our interest to the olfactory sphere alone.

(III) Constructions using *odeur, parfum, senteur, fragrance, effluve*, and *exhalaison*, followed by the preposition *de* and a noun, are frequently employed in French when a particular smell is being considered. Examples in which the second noun (N2) names an odorous entity, corresponding to a "source" [from either a psychological or a grammatical point of view (thematic role)], are our concern here:[5] For example: *vanille* ("vanilla") or *poisson* ("fish") in the phrases *odeur de vanille* ("odor of vanilla") [vanilla smell] and *odeur de poisson* ("odor of fish") [fish smell].

Odeur and *parfum* can both be in the N1 position: *Il a senti une odeur de rose* ("He smelled an odor of rose") [a rose smell]. *Il a senti un parfum de rose* ("He smelled a perfume of rose") [a rose perfume]. *Parfum* can also name a source, as shown by the following contrast (6), where *parfum* is well formed in the N2 position (6a), whereas *odeur* is not (6b):

(6) (a) *J'aime l'odeur de parfum*
          "I like the odor of perfume"
          [I like the perfume odor]
     (b) **J'aime le parfum d'odeur*
          "I like the perfume of odor"

This is confirmed by the following examples:[6]

(7) (a) *Ça (sent + pue) (le parfum + le poisson)*
          "It (smells + stinks) (the perfume + the fish)"
          [It smells of (perfume + fish), it stinks of (perfume + fish)]
     (b) **Ça sent (un parfum + un poisson)*
          "It smells like (a perfume + a fish)"
     (c) *Ça sent un parfum . . . je ne sais plus lequel . . . peut-être Chanel N°5*
          "It smells a perfume . . ." [I cannot remember which one . . . maybe Chanel No. 5]
     (d) *Ça sent un poisson . . . je ne sais plus lequel . . . peut-être le saumon*
          "It smells a fish . . ." [I cannot remember which one . . . maybe salmon]

In the context of *ça sent/ça pue* ("it smells/it stinks"), *parfum* behaves exactly like any noun signifying an odorous source, examples (7a) and (7b). Example (7c) is well formed, as is (7d), if the noun phrase *un parfum* ("a perfume") refers to an entity considered as "sort of," as clarified by the sentences that follow. Once again, *odeur* behaves differently: **Ça sent l'odeur* ("It smells the odor") and **Ça sent une odeur* ("It smells of an odor"). In the same way, *parfum* passes the test of material-object type (8a) (Godard and Jayez, 1996), whereas the other terms do not (8b):

(8) (a) *Le parfum se trouve sur l'étagère*
       "The perfume is on the shelf"
   (d) **(L'odeur + la senteur + la fragrance + l'effluve) se trouve sur l'étagère*
       "(The odor + the scent + the fragrance + the emanation) is on the shelf"

  (IV) *Parfum* is the only term that allows for mass determination [(9a) versus (9b) and (9c)]:

(9) (a) *Il s'est mis*
       (du + un peu de + peu de + trop de + beaucoup de) parfum
       "He has put
       (some + a little bit of + little + too much + a lot of) perfume"
   (b) **Dans cette pièce, il y a*
       (de l' + un peu d' + peu d' + trop d' + beaucoup d')
       (odeur + effluve)
       "In this room, there is
       (some + a little bit of + little + too much + a lot of)
       (odor + emanation)"
   (c) **Dans ce flacon, il y a*
       (de la + un peu de + peu de + trop de + beaucoup de)
       (senteur + fragrance)
       "In this bottle, there is
       (some + a little bit of + little + too much + a lot of)
       (scent + fragrance)"

In the following examples, *parfum* behaves as a "substance name" (according to Van de Velde, 1995), like *eau* ("water"), *fer* ("iron"), *or* ("gold"), and so forth:

(10) (a) *(Un flacon + une bouteille) de parfum*
       "(A flask + a bottle) of perfume"
       [A perfume bottle, a perfume flask]

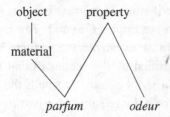

Figure 6.2. Linguistic types for *parfum* and *odeur.*

(b)   *(Deux litres + une goutte) de parfum*
    "(Two liters + a drop) of perfume"
(c)   **Le parfum de deux litres*
    "The perfume of two liters"
(d)   *Il a acheté du parfum en bouteille*
    "He has bought some perfume in bottle"

These terms present distinct properties having to do with their (possible) mor-phological constructions, the knowledge field to which they can be related, the (possible) various entities to which they allow reference, and the question of their determination. From these various tests, we can conclude that *odeur* has only one type, the property type, whereas *parfum* has two types: the property type and the material-object type, as shown in Figure 6.2.

2.2. Relationships between Olfactory Terms and Their Source Names

In the construction "*odeur* + preposition + (det) + N2," where "det" represents a "determiner" and N2 is the name of a source, *odeur* is a characteristic, a property, of the source, like *couleur* ("color"):

(11)   (a) *J'aime l'odeur de cette rose*
      "I like the odor of this rose"
    (b) *J'aime la couleur de cette chemise*
      "I like the color of this shirt"

And it is interesting to note that the preposition *de* is the only one possible and that it is compulsory (David et al., 1997):

(12)   (a) **Je déteste l'odeur gaz, les odeurs fumier*
      "I hate the odor gas, the odors manure"

(b) **Les odeurs fleur sont celles que je préfère*
   "The odors flower are those that I prefer"
(c) **La pièce sent une de ces odeurs produit d'entretien!*
   "The room smells one of these odors" [cleaning product!]

These examples show that this characteristic is not constructed independently of the source [the compulsory presence of the preposition *de*; examples (12a), (12b), and (12c) are not well formed]. However, one can at times note (exceptionally) the absence of *de*, as in the following extract:[7]

(13)   *La méthylionone . . . permet de mettre en relief successivement ces facteurs psychologiques. Ce produit, présentant simultanément ou successivement des* **odeurs d'iris, de réglisse, de cuir, de bois, de vétyver, de violette**, *etc. . . . , il est clair que plusieurs individus n'observeront pas forcément en même temps la même face ou la même phase. Chacun suit son inclination naturelle et s'accroche à un* repère *qui l'a spécialement frappé,* **l'odeur bois** *par exemple. Si nous attirons son attention sur un autre* repère, **l'odeur cuir**, *il finira par la remarquer à son tour. . . . De proche en proche . . . , nous lui ferons faire un* inventaire *complet de l'* odorant. . . . (Roudnitska, 1980, p. 14)

   [Methylionone . . . enables us to pinpoint these psychological factors successively. Since this product displays simultaneously or in succession **odors of iris, of liquorice, of leather, of wood, of vetiver, of violet** and so on . . . , it is clear that different people will not perceive the same aspect or the same phase at the same time. Each person will follow his natural intuition and will be responsive to the particular marker that has especially struck him, for example **the odor wood**. If we draw his attention to another marker such as **the odor leather**, he will then go on to notice it. . . . Little by little, therefore, we shall bring him to make a complete inventory of the odorant.]

This is a quotation from an expert who deals with perfume synthesis, as shown in particular by the use of the terms *méthylionone* ("methylionone") (a perfume component), *odorant* ("odorant") (an odorous mixture), *inventaire* ("inventory") (a whole made up of parts), and *repère* ("marker") (one of the components); in this context, all of these terms add up to a material approach to odor. However, this context is not one in current usage, and even so, we can employ the preposition, as in *odeurs d'iris, de réglisse* ("odors of iris, of licorice") [iris smells, licorice smells].

In French, things are different in the realm of color. *Couleur* ("color") can be followed by a substantive, whether preceded by *de* or not (14a and 14b). The ensuing nominal form is then interpreted like a color, even if it is a "discourse color" (i.e., an expression of the speaker's creativity that may not be lexicalized):

(14)   (a) *Un tapis couleur de feuille morte*
          "A carpet color of leaf dead"
          [A dead-leaf carpet color]
       (b) *Un tapis couleur feuille morte*
          "A carpet color leaf dead"
          [A dead-leaf carpet color]

These remarks are in line with the analyses of Corbin and Temple (1994, pp. 24–5): "Neither eucalyptus nor musk nor thyme alone, for example, can serve to name categories of objects [entities] having the smell characteristics of eucalyptus, musk or thyme," unlike such terms as *poire* ("pear"), which can name a fruit and any pear-shaped entity (e.g., an "enema bag" in French). To corroborate this, Corbin and Temple (1994) point out that no adjective associated with a smell can be converted[8] into a noun to name an entity having that smell as a characteristic, unlike adjectives naming colors, as shown by the following examples:

(15)   (a)   *fétide* ("fetid")
             *un fétide ("a fetid") (Corbin and Temple, 1994)
       (b)   *bleu* ("blue")
             un bleu de travail ("a blue of work") [overalls]
             une ecchymose ("a bruise") (Corbin and Temple, 1994, p.13)
       (c)   *noir* ("black")
             une noire ("a black") [a quarter note in music]

Parfum behaves like *odeur* (in the olfactory realm), but it can support a direct construction in specific contexts, such as in advertisements, where *parfum vanille* ("perfume vanilla") [vanilla perfume] can be associated with a deodorant, or on room-deodorant packaging, where *parfum pin* ("perfume pine") [pine perfume] may appear. However, we must emphasize the noncanonical character of this usage (plays on words, diversions of the semantic properties of words, etc). For *senteur*, I found only one example in Frantext, and the same for *effluve*, but none for *fragrance*:[9]

(16)   (a) *Les odeurs lui reviennent: poussière de charbon, fumées, fond permanent,*
          mégot refroidi, fameux café-chicorée, gnole, pisse de chat, d'enfant, ... vin
          *rouge, absinthe des comptoirs et sur chaque seuil l'ardente **senteur savon***
          ***noir** de la propreté ouvrière, de la pauvreté nickel.* (Chabrol, 1977, p. 196)

          [The smells return to his memory: coal-dust, smoke, permanent background,
          stale cigarette butt, a coffee-chicory mix, cheap booze, cat's and children's piss,
          red wine and absinthe, and on every doorstep the pungent **scent soap black**
          [smell of strong black soap], the smell of shining working-class poverty.]

(b) [*J*]*e regretterai sans doute la 208 ... 480, la 512 ... toutes les cellules de mes prisons et leurs punaises, les **effluves gogues**, les gaffes et peut-être même le mitard.* (Boudard, 1963, p. 363)

[I'll probably miss the 208 ... the 480, the 512 ... all my prison cells and their bugs, the **emanations shithole** [shithole smells], the guards and maybe even the solitary confinement.]

With the exception of these rare examples, which we always find in specific contexts (the speech of experts, advertisements, slang, etc.), French terms associated with the olfactory field, and the term *odeur* in particular, refer to entities that cannot be separated from their sources (Strawson, 1973). Moreover, the syntactic contrast between color and olfactory terms (14) is congruent with the contrast (15): A common noun denoting an object can be used as denoting the prototypical color of this object, whereas that is impossible with olfactory terms. In other words, in French, the source name can never be the name of an odor.

2.3. Olfactory Terms and "Part of"

Using the tests designed by Van de Velde (1995) with regard to "substance names" (which do not consist of identifiable parts), we first note that we cannot use the formulation *partie de* ("part of") before the term *odeur*:

(17) (a) **Il a senti une partie de l'odeur*
         "He has smelled a part of the odor"
     (b) *Il a lu une partie du livre*
         "He has read a part of the book"
     (c) **La partie de l'odeur que je préfère est la rose*
         "The part of the odor that I prefer is the rose"

Moreover, if the term under consideration consists of parts, it is possible to create expressions like the following: "N1 + *de* + numeral + N2." Thus, *une maison de trois pièces* ("a house of three rooms") [a three-room house]. Or "N1 + *à* + N." Thus, *un verre à pied* ("a glass with foot") [a stem glass] (Cadiot, 1992; Van de Velde, 1995), with N1 naming the entity containing the parts, and N2 naming one of those parts. For the term *odeur*, there are no such names:

(18) (a) **Il a senti une odeur de trois ??*
         "He has smelled an odor of three ??"
     (b) **Il a senti une odeur à ??*
         "He has smelled an odor with ??"

Nor can the source name be introduced by the preposition *à*, whether the name is determined or not:

(19)   (a) *Une odeur à vanille*
                "An odor with vanilla"
        (b) *Une odeur à la vanille*
                "An odor with the vanilla"

Again, *parfum* behaves differently, for it can be followed by "*à* + det + noun":

(20)   (a) *Un parfum à vanille*
                "A perfume with vanilla"
        (b) *Un parfum à la vanille*
                "A perfume with the vanilla"
                [A vanilla perfume]
        (c) *Il se promet d'acheter du parfum à l'iris* (Bienne, 1990, p. 71).
                "He promises himself to buy some 'perfume with the iris' [iris perfume]"

For Van de Velde (1995, p. 112), the sequence *à la* ("with the") is possible if N1 denotes a man-made object and the complement expresses a means; for Cadiot (1992, p. 203), N2 can be interpreted as an ingredient of N1. Thus, in French, the "means" or "ingredient" relations are different from the "part of" relation. Whatever the semantic relation may be, *parfum* can name an odorous mixture, an artifact, the source of which cannot be interpreted as a part; compare Section 2.1, where (20a) is not well formed; moreover, test (17) is negative for *parfum*: *Il a senti une partie du parfum* ("He has smelled a part of the perfume"). *Effluve*, *fragrance*, and *senteur* behave in the same way as *odeur*, although current usage may be changing in the case of *senteur*, as in the use of *phyto-senteur* ("phyto-scent") on a cosmetics package (see note 3).

(21)   (a) *Il a senti (une fragrance + un effluve) à la vanille*
                "He has smelled (a fragrance + an emanation) with the vanilla"
        (b) *?Une senteur à la vanille*
                "A scent with the vanilla"
                [A vanilla scent]

Finally, we note that we cannot have a phrase like (22a), whereas (22b) and (22c) are well formed:

(22)   (a) *La rose de l'odeur*
                "The rose of the odor"
        (b) *La plume du stylo*
                "The quill of the pen"
        (c) *Le guidon du vélo*
                "The handlebars of the bicycle"

Every phrase built as "N1 + *de* + det + N2" does not systematically mean "N1 = part of N2" (Fradin, 1984a,b). However, if N1 is categorematic (i.e., N1 denotes an "object in the world" and has a stable and non-processive meaning) (Fradin, 1984b), and if the two terms have a "part of" relationship, then it is possible to create the types of phrases illustrated by (22b) and (22c). That is never the case for the term *odeur* (22a). We do, however, find a few instances of *pois de senteur* ("pea of scent") [sweet peas], *eau de parfum* ("water of perfume") [perfume], *foin d'odeur* ("hay of odor") (which names a plant).[10] These are lexicalizations that allow for reference to natural or man-made objects in which the second term is taken as a distinct classificatory characteristic. The proof is that the second term cannot have a determiner. Thus, we cannot have the following:

(23) (a) **Les pois de la senteur*
           "The peas of the scent"
      (b) **L'eau du parfum*
           "The water of the perfume"
      (c) **Le foin de l'odeur*
           "The hay of the odor"

The source associated with the term *odeur* is not a part or an ingredient of the smell. *Odeur* cannot mean an odorous mixture, unlike *parfum*.

2.4. Olfactory Terms and Temporal Properties

In this section, I adopt the following tests or requirements devised by Godard and Jayez (1996); both test (a) and test (b) must be passed for the term to have temporal properties, and test (e) must be passed for the term to be a strong event:

(24) (a) **Il a dormi*
           (avant + après + pendant + durant + au cours de + lors de + depuis + tout au long de + au moment de) l'odeur
           "He has slept
           (before + after + during/for + during + in the course of + at the time of + since + throughout + at the time of) the odor"
      (b) **Il a senti deux heures d'odeur*
           "He has smelled two hours of odor"
           *Il a senti une odeur de deux heures
           "He has smelled an odor of two hours"
      (c) **L'odeur (a commencé à 8 h + a fini deux heures après + est interrompue + s'est interrompue quelques minutes après + s'est achevée + est suspendue)*
           "The odor (began at 8 o'clock + finished two hours later + is interrupted + broke off a few minutes later + ended + is suspended)"
      (d) *?L'odeur a duré quelques minutes*
           "The odor lasted a few minutes"

(e) **Une odeur a eu lieu. Une odeur s'est produite*
"An odor happened"

Example (24d) seems dubious, and with regard to the other verbs proposed by Godard and Jayez (1996) – *se prolonger pendant/jusqu'à* ("to go on during/until"), *se continuer* ("to continue"), *s'étaler sur* ("to spread over"), *traîner en longueur* ("to drag on"), *n'en plus finir* ("to take ages"), *s'éterniser* ("to drag on interminably") – the sentences are not well formed. Thus *odeur* has no temporal properties, nor do *parfum*, *effluve*, *fragrance*, and *senteur*.[11]

These results stand in contrast to those in the auditory field, in which terms like *bruit* ("noise"), *vacarme* ("din"), and *brouhaha* ("hubbub"), as well as many deverbal constructions such as *grondement* ("roaring"), *aboiement* ("barking"), and so forth, pass several of these tests. Let us consider here the results for some of them:

(25)   (a) *Il est parti (après + avant) ce (bruit + brouhaha + vacarme) assourdissant*[12]
             "He has left (after + before) this (noise + hubbub + din) deafening"
        (b1) *Ils ont fait quelques minutes de bruit*
             "They have made a few minutes of noise"
             Il y a eu un bruit de quelques minutes
             "A noise of a few minutes happened"
        (b2) *Il a dû supporter deux heures (de brouhaha + de vacarme)*
             "He has had to suffer two hours (of hubbub + of din)"
             ?Il a dû supporter (un brouhaha + un vacarme) de deux heures
             "He has had to suffer (a hubbub + a din) of two hours"
        (c) *Le (bruit + brouhaha + vacarme) (a commencé + a fini) à 8 h*
             "The (noise + hubbub + din) (began + finished) at 8 o'clock"
        (d) *Le (bruit + brouhaha + vacarme) a duré trois heures*
             "The (noise + hubbub + din) lasted three hours"
        (e1) *Un (bruit + clac) s'est produit*
             "A (noise + clac) happened"
        (e2) **Un (bruit + clac) a eu lieu*
             "A (noise + clac) happened"

These terms function correctly in tests (a), (c), and (d). Test (e), using *se produire* ("to happen"), is acceptable if it is "a fortuitous event" (Gross and Kiefer, 1995), which is not the case, for example, with *brouhaha* ("hubbub"). Complementation (test b) shows different acceptabilities. Beyond these difficulties (which call for further analysis of the terms associated with the auditory field), the main point is to show that results for smells, expressed in the most current olfactory terms, are clearly negative, and thus that these terms have no temporal properties.

3. Conclusion

The various contrasts demonstrated here show that the terms linked with the olfactory field do not make up a homogeneous class, that most of them do not allow independent reference to their sources, nor do they allow for reference to entities with identifiable parts, and, finally, that the most common among them have no temporal properties. In linguistic terms, this means that the linguistic type for words currently associated with the olfactory realm is one of property and that *parfum* has two different types, property and material object. These results can have important consequences for other fields, especially psychology. The instructions issued for psychology experiments, as well as analyses of subjects' verbal replies, must take into account the fact that French is ill-suited for testing the characteristics and properties of smells. The problems are of several types: Instructions for identifying the various components of an odorous mixture are difficult to formulate using the term *odeur*. Similarly, instructions dealing with smell as a perceptive event are particularly difficult to formulate, for most of the terms are not events. For example, it is difficult to express in words notions about the origin or the cause of a smell, its location, the duration of the perception, and so forth (Gross and Kiefer, 1995). Lastly, the analysis of results derived from the various tasks of categorization must always take into account the fact that smells, which have no autonomy vis-à-vis their sources, tend to induce classifications based on resemblances with the characteristics of the source objects (Dubois, 2000). In other words, analysis of the meanings of most olfactory terms, and especially *odeur*, shows that the referent of a phrase built on that term is not considered as an "object" independent of its source and of the person who perceives it, which is clearly obvious through the widespread use of such deverbal adjectives as *agréable* ("pleasant"), *piquant* ("pungent"), and so forth, to identify types of smells (David et al., 1997). Thus, from a cognitive point of view, the entities named by *odeur* do not have the concrete objectivity of visual or sound entities (David, 1997) and must be considered as effects (i.e., as manifestations of a sensation that cannot be dissociated from the subject sensing it).

Acknowledgments

I would like to thank Françoise Kerleroux and Rachel Panckhurst for their helpful remarks and criticism.

Notes

1. That situation has been studied by Engen (1987), and some of its consequences have been examined by Dubois and Rouby (Chapter 4, this volume) and David (2000). For additional details about other lexicologic surveys, see the following: Boisson (1997), who has studied the olfactory vocabularies of 60 languages from nine linguistic families by analyzing various dictionaries; Mouélé (1997) and Traill (1994), who have listed, respectively, the olfactory terms of Li Waanzi (a language of Gabon) and of !xo@o) (a language of Botswana); and Classen (1993, pp. 60ff.). In a different way, Camargo (1996) and Mennecier and Robbe (1996) described the properties of evidentials operating, respectively, in Caxinaua (Pano family) and in Tunumiisut (one of the Inuit languages).
2. French examples are given in italics, and they are followed by their literal translations in English. If the literal translation in English is incorrect, it is followed by a translation in square brackets reflecting the meaning of the French expression. To simplify reading, the initial translation of a word into English will be maintained throughout the text.
3. *Miasme* can name the particular smell associated with certain diseases, and *fraîchin* names a fishy sea smell. It is a regional term in the west of France that does not seem to have any equivalent in English. Some of these terms are not currently in use. The Frantext data-base search I made concerned all literary genres, covered the nineteenth and twentieth centuries, and referred to 2,012 different works. It produced the following results (occurrences of the singular or plural forms): *odeur*, 11,047; *parfum*, 5,664; *bouquet*, 3,999; *senteur*, 918; *effluve*, 389; *puanteur*, 355; *exhalaison*, 223; *miasme*, 213; *remugle*, 63; *fragrance*, 30; *fraîchin*, 0. We must note that, in French, *parfum* can refer to both a perfume and a flavor, and *bouquet* can refer to both a bunch and a smell, so there are numerous examples that have nothing to do with smells (e.g., *une glace au parfum vanille*, "an ice cream with the flavor vanilla" [a vanilla-flavored ice cream], or *il avait apporté un bouquet de fleurs*, "he brought a bunch of flowers"). *Senteur* and *fragrance* are quite rarely used, although the former seems to have been rejuvenated recently through the development of new odor-related technologies (for masking smells but also for the eradication of unpleasant smells).
4. For example:

    (1) *Il a mangé les grenouilles, les cafards*
        "He has eaten the frogs, the cockroaches"

    From a linguistic viewpoint, this example is not ungrammatical, because the only reason for rejecting it would be based on cultural considerations.
5. This is not always the case, for the second term can name a time entity, as in *odeur d'autrefois* ("odor of the past"), a state, as in *odeur de pourri, de renfermé* ("odor of rotten, of fusty") [rotten smell, fusty smell], a place, as in *odeur d'hôpital* ("odor of hospital") [hospital smell] or *l'odeur de chez le coiffeur* ("the odor of the hair-dresser's") (Romains, 1932, p. 294).
6. By convention, a plus sign means the various possibilities of reading:

    (1) *(Une odeur + un parfum) de rose*
        "(An odor + a perfume) of rose"

    Example (1) gathers two different examples, (2a) and (2b):

(2) (a) *Une odeur de rose*
"An odor of rose"
[A rose odor]
(b) *Un parfum de rose*
"A perfume of rose"
[A rose perfume]

7. In the longer quotations, only the words in boldface have been translated literally.
8. A "conversion" is a morphological operation of word construction (Corbin, 1987; Kerleroux, 1996). We note a semantic and categorical change without any material change (e.g., prefix or suffix) – for example, *bleu* ("blue") (noun) constructed from *bleu* ("blue") (adjective).
9. The bibliographic references in literary extracts are those of Frantext.
10. I do not include expressions such as *flacon de parfum* ("flask/bottle of perfume") [perfume bottle], *bouteille de parfum* ("bottle of perfume") [perfume bottle], and *bouffée de parfum* ("whiff of perfume") [perfume whiff], which are not denominations.
11. That is not the case for *puanteur* and *exhalaison*, which are deverbal constructions (Gross and Kiefer 1995):

(1) *Alors, commençaient les puanteurs* (Zola, 1873, p. 827).
"Then, began the stenches" [The stenches then began]

12. Considering the linguistic type of terms denoting noise, a complement is compulsory.

References

Balzac H de (1843). *Le Père Goriot.* (Reprinted 1976 in *La Comedie humaine*, vol. 3, Paris: Gallimard.)
Berthonneau A M (1998). *Je me suis butée dedans. A propos de dedans et de ses relations avec dans. Traitements actuels des prépositions du français.* Paris: Journée Conscila.
Bienne G (1990). *Les jouets de la nuit.* Paris: Gallimard.
Boisson C (1997). La dénomination des odeurs: variations et régularités linguistiques. *Intellectica* 24:29–49.
Boudard A (1963). *La cerise.* Paris: La table roude.
Cadiot P (1992). A entre deux noms: vers la composition nominale. *Lexique* 11: 193–240.
Cadiot P & Nemo F (1997). Propriétés extrinsèques en sémantique lexicale. *Journal of French Language Studies* 7:127–46.
Camargo E (1996). Valeurs médiatives en Caxinaua. In: *L'énonciation médiatisée*, ed. Z Guentchéva, pp. 271–83. Paris: Peeters.
Chabrol J P (1977). *La folie des miens.* Paris: Gallimard.
Chomsky N (1965). *Aspects de la théorie syntaxique.* Paris: Le Seuil.
Classen C (1993). *Worlds of Sense: Exploring the Senses in History and across Cultures.* London: Routledge.
Corbin D (1987). *Morphologie dérivationnelle et structuration du lexique.* Tübingen: Max Niemeyer.

Corbin D & Temple M (1994). Le monde des mots et des sens construits. *Cahiers de Lexicologie* 65:5–28.

David S (1997). Représentations sensorielles et marques de la personne: contrastes entre olfaction et audition. In: *Catégorisation et cognition: de la perception au discours*, ed. D Dubois, pp. 211–42. Paris: Kimé.

David S (2000). Certitudes et incertitudes dans les domaines olfactif, gustatif et auditif. *Cahiers du LCPE (Langages, cognitions, pratiques, ergonomie)* 4:77–108.

David S, Dubois D, Rouby C, & Schaal B (1997). L'expression des odeurs en français: analyse lexicale et représentation cognitive. *Intellectica* 24:51–83.

Dubois D (1991). *Sémantique et cognition*. Paris: Éditions du CNRS.

Dubois D, ed. (1997). *Catégorisation et cognition: de la perception au discours*. Paris: Kimé.

Dubois D (2000). Categorization as Acts of Meaning: The Case of Categories in Olfaction and Audition. *Cognitive Science Quaterly* 1:35–68.

Engen T (1987). Remembering Odors and Their Names. *American Scientist* 75:497–503.

Fradin B (1984a). Anaphorisation et stéréotypes nominaux. *Lingua* 64:325–69.

Fradin B (1984b). Hypothèses sur la forme de la représentation sémantique des noms. *Cahiers de Lexicologie* 44:63–83.

Godard D & Jayez J (1993). Le traitement lexical de la coercion. *Cahiers de Linguistique Française* 14:123–49.

Godard D & Jayez J (1996). Types nominaux et anaphore. *Chronos* 1:41–58.

Grimshaw J (1990). *Argument Structure*. Cambridge, MA: MIT Press.

Gross G (1994). Classes d'objets et description des verbes. *Langages* 115:15–30.

Gross G & Kiefer F (1995). Structure événementielle des substantifs. *Folia Linguistica* 24:43–65.

Jayez J & Godard D (1995). Principles as Lexical Methods. In: *Proceedings of AAAI Spring Symposium on the Representation and Acquisition of Lexical Knowledge: Polysemy, Ambiguity and Generativity*, pp. 10–16. Stanford University Press.

Katz J J (1972). *Semantic Theory*. New York: Harper & Row.

Kerleroux F (1996). *La coupure invisible. Etude de syntaxe et de morphologie*. Villeneuve d'Ascq: Presses Universitaires du Septentrion.

Kerleroux F & Marandin J M (1994). La linguistique théorique est une linguistique de terrain. Presented at Colloque terrain et théorie en linguistique, Paris, September 26–28.

Kleiber G (1990). *Sémantique du prototype. Catégories et sens lexical*. Paris: Presses Universitaires du France.

Lakoff G (1987). *Women, Fire and Dangerous Things. What Categories Reveal about the Mind*. University of Chicago Press.

Marandin J M (1984a). Miniatures sentimentales. Syntaxe et discours dans une description lexicale. *LINX* 10:75–95.

Marandin J M (1984b). Distribution et contexte dans une description lexicale. *Cahiers de Lexicologie* 44:137–49.

Mennecier P & Robbe B (1996). La médiation dans le discours des Inuits. In: *L'énonciation médiatisée*, ed. Z Guentchéva, pp. 233–47. Paris: Peeters.

Milner J C (1989). *Introduction à une science du langage*. Paris: Le Seuil.

Mouélé M (1997). L'apprentissage des odeurs chez les Waanzi: note de recherche. *Enfance* 1:209–22.

Pustejovsky J (1995). The Generative Lexicon. Cambridge, MA: MIT Press.

Romain J (1932). *Les hommes de bonne volonté*. Paris: Flammarion.

Roudnitska E (1980). *Le parfum*. Paris: Presses Universitaires du France (Coll. "Que sais-je," 3ème édition n° 1888).

Strawson P F (1973). *Les individus*. Paris: Le Seuil.

Traill A (1994). *A !xo@o) Dictionary*. Köln: Rüdiger köppe Verlag.

Van de Velde D (1995). *Le spectre nominal*. Paris: Peeters.

Zola E (1873). *Le ventre de Paris*. In: *Les Rougon-Macquart*, ed. A Laroux & M Mitterand, vol. 1. Paris: Gallimard.

7

Classification of Odors and Structure–Odor Relationships

Maurice Chastrette

In spite of considerable recent progress in olfaction studies, the sense of smell remains poorly understood, and the description of perceived odors is still a difficult task. When our understanding of olfaction is compared with that for other senses, such as vision, there are striking differences: In the field of vision, the absorption spectra of electromagnetic radiation are easily measured and provide information to predict the perceived colors of substances. As the molecular features responsible for the characteristics of light absorption are well known, the colors of substances, such as dyes, can be predicted with reasonable accuracy. The situation is much more obscure for the olfactory properties of substances, as in olfaction there is no equivalent for absorption spectra, and descriptions of odors, which are absolutely necessary to establish structure–odor relationships, have been far from rigorous.

To describe an odor, we generally search through our past experience to try to make comparisons with similar chemical compounds encountered earlier, attempting to find appropriate words. However, even if we decide that compound A smells somewhat like compound B, but does not smell exactly the same, we nevertheless feel justified in using the same words to describe both. Moreover, unless we are carefully trained experts, there will be emotional, personal, and cultural factors that will intervene in our perceptions and strongly influence our choices of descriptors. Concern about such factors was long ago expressed by the Dutch physiologist Zwaardemaker (1899), who called for a quantitative psychology of olfaction. In the first part of this chapter, the point of view of a chemist regarding classifications of odors will be presented. As classifications are strongly associated with description systems, these two aspects will be considered successively. In the second part, problems associated with the determination of structure–odor relationships will be briefly discussed.

1. Description of Odors

Usually, the description of an odor supposes that both its quality and intensity characteristics can be precisely defined. Several description systems, based on a thesaurus of words referring to pure substances or to complex mixtures, have proved their utility in the various domains where olfactory descriptions are necessary. Unfortunately, no general agreement has ever been reached, and any compilation of a large data base would be difficult because of the many differences between the various description systems.

1.1. Qualities of Odors

The qualities of odors usually are specified using specific words (descriptors), or by detailing similarities between odors, or by using olfactory profiles. Whatever the description system used, the distinction between olfactory and trigeminal perceptions, as well as the factor of an odor's concentration, must be considered. For instance, the odors of thiols are known to be strongly dependent on their concentrations: Thus p-menth-1-en-8-thiol has a grapefruit "note" at very low concentrations, whereas the smell of the pure liquid is extremely powerful and causes nausea (Demole, Enggist, and Ohloff, 1982). Contextual effects on odor quality, frequently encountered in human perception, such as different sorts of perceptual contrast, have been discussed by Lawless and associates. (Lawless, 1991; Lawless, Glatter, and Hohn, 1991)

1.2. Descriptions Using Word Descriptors

When naive subjects are used in testing odors, their descriptions of odors are strongly influenced by emotional and subjective factors (Weyerstahl, 1994). Olfactory descriptions are most often compiled, by panels of subjects trained to use a given terminology in a more uniform and less personal way. However, even perfumers sometimes use subjective descriptions, as shown by Ellena (1987), who compiled a list of 122 descriptors used by perfumers that related to taste, tactile sensations, affective responses, and so forth. Clearly, the various description systems are not equivalent, and a given word used in the various description systems will not always have the same status, as shown by the results of linguistic research (e.g., David, 1997).

Arctander (1969), in his reference work, cited about 260 different descriptors and 88 classes of odors. A much shorter list of descriptors (32) has been used by Swiss perfumers, and the international standard for training in the recognition of odors features only 24 reference substances (ISO, 1992). Jaubert, Tapiero, and Doré (1995) have described a system with 42 odor descriptors referring only to pure substances.

It has been suggested (Dravnieks and Bock, 1978; Dravnieks, 1982) that semantic descriptions are reproducible for many odorants. One would hope, and expect, to find something close to a consensus among different experts sampling a given odorant. However, when Brud (1986) asked 120 perfumers to associate with each of 20 odor descriptions the name of only one substance, a total of 507 different substances were cited, nearly 400 of which were cited by only one person. The consensus was good for odors like "musk" and very poor for odors like "floral" and "fatty". Harper (1975) asked seven experts to determine which pure chemical substances were the best representatives for 44 particular odor qualities and concluded that "the views of at least ten persons (i.e. their qualitative descriptions) should normally be combined together to be reasonably sure that all relevant qualities or notes have been identified." It therefore seems important in classifications to use homogeneous data sets originating from a single source (e.g. a restricted group of experts using the same description system or at least similar description systems).

1.3. Descriptions Using Profiles

Crocker and Henderson (1927) suggested that four odor "characters" ("fragrant," "acid," "burnt," and "caprylic") "are the principal and perhaps the only units which make up all the odors we perceive" and could perhaps explain all smells: Any substance could be rated on the four character scales, with values from 0 to 8, to generate a profile. That was the first description system based on profiles. The simplicity of their approach seems attractive, although practicing perfumers no doubt found their four dimensional description system too crude. A more complex system was used by Wright and Michels (1964). Brud (1986) also used odor profiles based on 10 descriptors ("green," "fruity," "floral," "fatty-aldehydic," "herbal," "animal-musky," "amber-woody," "spicy-balsamic," "earthy-fungoid," and "chemical-unpleasant"), with values from 0 to 4. Dravnieks (1985) asked a group of 120–140 panelists to rate, on a scale from 0 to 5, the applicability of each of 146 descriptors or odor characteristics for 160 products, including 145 pure substances and 15 mixtures. Statistical measures were derived from such data and used in establishing profiles. In a study by Moskowitz and Gerbers (1974), observers rated each of 15 odors on 17 attributes (14 descriptors of particular qualities, such as "minty" or "burnt," and three descriptors related to intensity, pleasantness, and familiarity) using a method of magnitude estimation.

To obtain odor profiles without using such scales, Boelens and Haring (1981) asked a panel of experienced perfumers to sample 309 compounds and assess their similarities to 30 reference materials selected to represent different odor aspects. The profiles derived by Boelens and Haring, on one hand, and Dravnieks, on the other, are not always in agreement, and there are striking differences for an odorant so frequently encountered as eugenol. A primary source of the divergence between their systems may have been the very different numbers of reference materials (30 compounds, against 146 descriptors) used in the two systems, a second source being differences in the nature of their panels of testers.

1.4. Descriptions Using Similarities

Useful information that avoids language and semantic problems can be obtained by measuring odor similarities. Polak (1983) has discussed the problems associated with descriptions of odor qualities and measurements of odor similarities and concluded that "odor quality as such is not yet definable in scientific terms. . . . The result is at best quantifiable as an order of similarities on an ordinal scale." The principal description systems based on similarities will be discussed later in Section 2.3.

2. Classification of Odors

The general problems of classification of odors have been discussed in depth by Kastner (1973). A brief survey of attempts to produce general classifications of odors, based on somewhat arbitrary distinctions among empirical, semiempirical, and statistically based classifications, will be given. It must be remembered that all these classifications were established with different aims in mind, and they differ also in terms of their theoretical foundations, which often are implicit and sometimes absent. Therefore, comparisons should be made cautiously.

2.1. Empirical Classifications

The first attempt to classify odors in a general system seems to have been that of Aristotle, who divided odors into six classes, a number that remained stable until Linnaeus (1756) proposed seven classes in his *Odores medicamentorum*. He described two classes (*fragrantes* and *aromatici*) as pleasant, two classes (*tetri* and *nauseosi*) as fetid, and two classes (*ambrosiaci* and *hircini*) as pleasant for some persons and unpleasant for others. The pleasantness of the last class (*alliacei*) was not considered. That classification was then used to describe the medicinal properties of substances according to class. Fourcroy (1798) proposed a classification of odors in five classes (or genera), based on their olfactory characteristics and as well as on their extraction techniques and chemical stability. Rouby and Sicard (1997) have discussed old and recent research on categorization and reported numerous examples of classification.

Classifications by and for perfumers have been much more numerous and useful: Following the development of a fragrance industry at the end of the nineteenth century, large numbers of empirical classifications were proposed. A few have been based on physical or psychophysical properties such as volatility, intensity, and odor threshold, and they are more or less convergent. Classifications based on qualities or similarities of odors are strongly dependent on the description system used, which sometimes constitutes in itself the core of the classification. They are more numerous and are in only partial agreement, the number and nature of the classes remaining highly variable and far from being stabilized. The major examples of those classifications were briefly discussed by Wells and Billot (1988). Several important classifications, mostly taken from their work, will now be considered.

Rimmel (1895) proposed an 18-group classification in which only three classes referred to pure substances, and Zwaardemaker (1895) designed a classification based on nine groups ("fragrant," "ethereal," "aromatic," "ambrosiac," "alliaceous," "empyreumatic," "hircine," "foul," and "nauseous") with a total of 30 subclasses. Some of the reference substances were pure compounds possessing strong primary notes and only weak secondary notes.

Another interesting classification was suggested by Billot (1948), who proposed a system of nine classes, each containing 2 to 10 subclasses. Brud (1986) described an odor ring with 12 sectors ("floral," "aldehydic-ozone," "woody," "conifer/evergreen," "mossy," "green-herbal," "spicy," "citrus," "balsamic," "fruity," "animal," and "lavender like") that attempted to serve both classification and description purposes. Cerbelaud (1951) designed a more detailed system with 45 classes and noted that each group could share some characteristics with other groups (e.g., the group of "rose" odors has features common with eight other groups). Arctander (1969) decided on 88 classes and also assumed some cross references among classes.

In a classic paper on the art of perfumery, the famous designer of perfumes Edmond Roudnitska (1991) offered some general recommendations for beginners, based on a system of 15 classes covering the most important series of odors. He wrote that "an odorant may have several 'faces' or several nuances. The observer must remain objective and not linger on any one pleasant or unpleasant characteristic as this could completely prevent him from discerning the other faces."

The latter three authors clearly recognized that their classes and subclasses were not mutually exclusive, that a substance could belong, to different degrees, in more than one category, thus suggesting that appropriate data processing (e.g., fuzzy logic and neural networks) could be used productively in classifications and studies of structure–odor relationships. The numbers of subclasses found in several recent systems are between 40 and 45, a seemingly acceptable compromise between the desired generality and ease of use.

2.2. Classifications Based on Primary Odors

These classifications are based on a small number of reference odors and on a more or less strong belief in some sort of primary odors. Henning (1915) proposed his famous olfactory prism of six reference odors ("spicy," "resinous," "burnt," "floral," "fruity," and "foul"). Crocker and Henderson (1927) used a system in which four characters ("fragrant," "acid," "burnt," and "caprylic") constitute, in a way, primary odors. The idea of primary odors was developed on the basis of studies of specific anosmia by Amoore (1967, 1977, 1982), who postulated, at least during the first stages of development of his theory, the existence of seven primary odors, a number that later was greatly increased. Because the idea of primary odor is no longer as popular as it once was, the importance of those classifications is declining.

2.3. Classifications Based on Statistical Methods

Several researchers have tried to avoid *a priori* decisions regarding the number and nature of classes by using multidimensional statistical methods applied to large sets of data. Essentially, three types of data have been used: classic semantic descriptions, descriptions emphasizing similarities, and odor profiles featuring estimation of the intensity of each feature.

Classifications Based on Semantic Descriptions. No doubt it has long been tempting to use the large numbers of high-quality olfactory descriptions accumulated by perfumers (although they were not compiled using a single description system) to produce odor classifications. Descriptors (often called "notes") can be considered as variables that describe chemical compounds, and they should be, at least to some extent, independent and discriminating. Correlatively, notes can be considered as "described" by the list of chemical compounds that possess them.

In a large data set consisting of olfactory descriptions of pure compounds, odor classification depends on correct estimation of the distances between notes, seen as points in an unknown n-dimensional olfactory space. When no intensity indication is given, distances between notes can be estimated on the basis of their occurrences and co-occurrences in the descriptions of all the compounds in the sample, and the usual multidimensional statistics can be applied, followed by more or less complex statistical treatment of the data.

Jaubert, Gordon, and Doré (1987a,b) compiled, from books by Arctander (1969) and Fenaroli (1971) and several other sources, a set of 1,396 pure substances with detailed olfactory descriptions. Culling of descriptors led to a list of 64 descriptors, then to 41 "archetypes" represented by 41 pure substances. Finally, the olfactory space was described using 45 descriptors structured into six principal poles ("amine," "citrus," "terpene," "sulfur," "pyrogenated," and "sweet"). A very similar description system is currently used by Jaubert in the training of future perfumers.

Chastrette, Elmouaffek, and Zakarya (1986) compiled a large data bank from the book of Arctander (1969) only. They used a set of 2,467 pure substances, the odors of which were described using 270 words, reduced to 233 after elimination of descriptors such as "strong," "cool," "warm," and so forth that did not provide purely olfactory information. There was an average number of 2.8 notes per compound, and the 34 most frequent notes accounted for 80% of the total number of occurrences. In a preliminary study, similarity coefficients were calculated from occurrences and co-occurrences of the 24 most frequently encountered notes, and multidimensional methods were used to represent the olfactory space. It appeared that those notes were largely independent and that the olfactory space was not hierarchically organized. Some notes like "amber" and "anise" were rather isolated, and others such as "minty/camphoraceous" constituted pairs of notes.

In a more detailed study of the same data set, Chastrette, Elmouaffek, and Sauvegrain (1988) selected the 74 most frequently encountered notes and calculated the similarities between pairs of notes. They found generally low similarities (no similarity for about 63% of the pairs) and strong associations for a few pairs only, such as "pineapple" and "banana," "camphor" and "minty," "apricot" and "peachy." A cluster analysis resulted in a distribution of 60 notes into 27 groups, and 14 other notes were not parts of groups. It was concluded that no primary odors could be identified and that the olfactory space, as defined by a set of notes, should be seen as a continuum, as proposed by Holley and MacLeod (1977). The so-called isolated notes were considered as good candidates for future studies of structure–odor relationships, in contrast to ill-defined notes such as "fruity" and "floral."

Chastrette, de Saint Laumer, and Sauvegrain (1991) analyzed a system of description used by a group of perfumers at Firmenich S.A. in which olfactory descriptions contained a maximum of four descriptors, chosen by consensus on smelling a given compound, from a list of 32 words. A data matrix obtained for 628 pure compounds was analyzed by means of four multidimensional statistical methods that gave results both convergent and in agreement with the practice of perfumers.

Abe et al. (1990) used a data set of 1,573 compounds, also taken from the book by Arctander, and selected 126 odor descriptors. A cluster analysis based on occurrences and co-occurrences of descriptors resulted in 19 "obvious" clusters, in good agreement with those found in previous studies. The four most frequent descriptors ("fruity," "floral," "herbaceous," and "green") were considered as fundamental odor descriptors. It seems that those words would be used in the first step of a description being drafted by an expert, who, for example, would recognize first a fruity odor and later would specify which fruit. The descriptors characterizing the clusters were compared to 38 odor descriptors chosen by Jennings-White (1984) to represent the primary odors. In a cautious conclusion, the authors considered that "semantic description is the only practical method of representing odor quality, but it tends to be too emotional and subjective."

In earlier studies, unfortunately not based on semantic descriptions, but also using multidimensional statistics, Abe et al. (1987, 1988) sought to determine if purely physicochemical results obtained from sensor data could be used to classify and predict odors. They used cluster analysis on data obtained for 30 substances using eight gas-sensing semiconductor elements. A few obvious clusters were observed and were found to correspond to "ethereal," "ethereal-minty," "minty," "ethereal-pungent," and "pungent" substances (Abe et al., 1987). Those authors suggest that their "semiconductor sensor system, though its sensing mechanism is quite different from that of biosystems, makes it possible to identify the odors of chemical substances." Their underlying hypothesis that "because odor is an inherent property of substances, identification of substances should correspond to identification of odors" might be seen with some suspicion by physiologists and psychologists working in olfaction. Abe et al. (1988) extended their studies to a set of data produced in response to vapors from 47 pure substances and applied three pattern-recognition techniques.

Classifications Based on Odor Profiles. Kastner (1973) has traced the origins of the profile methods. The classification by Crocker and Henderson (1927) was based on very simple odor profiles. As profile data are rather difficult to obtain, only a few good data sets have been used in several studies using different statistical treatments.

A first data set was produced by Wright and Michels (1964), who used nine standards and asked members of a panel to rate the similarities of 45 compounds or mixtures to each of those nine standards. From that set, Schiffman

(1974), using a nonlinear multidimensional scaling technique, obtained a two-dimensional map of the nine-dimensional space defined by those similarities. She found two large clusters (one more pleasant than the other, in agreement with several studies showing the role of hedonic factors) and several subsets and related them to physicochemical properties. She concluded that "there are no clear psychological groups or classes of stimuli, merely trends."

McGill and Kowalski (1977) used 43 physical properties to see how well they could explain that nine-dimensional olfactory space. A factor analysis led them to the representation of a two-dimensional continuum retaining 74% of the total variance, with only two identifiable axes, reflecting electron distribution and symmetry on the first axis, and the degree of electron delocalization on the second axis. Coxon, Gregson, and Paddick (1978) used odor profiles for 18 compounds and 5 reference compounds, each rated on how it exemplified each of the nine selected descriptors. Four- and five-dimensional classifications were obtained (the first dimension, again, being related to hedonic aspects).

Yoshida (1975) selected 32 standard odors from several sources, including the paper by Wright and Michels (1964), and asked 20 students to rate the similarities of 40 essential oils to each of those standard stimuli, on a nine-point scale. Statistical analyses (principal-component analysis, PCA, and multidimensional scaling, MDS) were performed on the similarity-data matrix. PCA yielded seven factors, and on factorial maps, stimuli were again clearly separated. The first axis was linked to the kind of stimulus, the second axis discriminated unpleasant and pleasant groups, and the third reflected the opposition between "resinous-spicy" and "sweet." The meaning of the other dimensions was unclear.

Another set was compiled by Boelens and Haring (1981), who defined profiles for 309 compounds using 30 reference materials. A multivariate statistical treatment led to the definition of 14 groups, without an attempt to obtain a graphic representation. Ennis et al. (1982) analyzed the same set of data using several methods of multivariate analysis and obtained 27 groups, some containing only a few compounds.

In the descriptions proposed by Dravnieks (1985), each substance can be seen as defined in an olfactory space with 146 dimensions. Correlation coefficients calculated in our laboratory for all possible pairs of descriptors show some expected correlations, positive (e.g., "lemon," "grapefruit," and "orange") and negative ("light" and "heavy"), but a surprising absence of a negative link between the two opposite descriptors "cool" and "warm."

Both linear and nonlinear multidimensional analyses have confirmed that no clear hierarchical structure can be detected in the olfactory space defined by descriptors, and they have shown the dominance of hedonic characteristics even in experiments using carefully planned protocols. This brief survey shows that profile ratings provide interesting data for classification studies. However, it should not be forgotten that they are also expensive and time-consuming.

Classifications Based on Similarity Data. Similarity judgments for all pairs of stimuli in a group of nine pyridyl ketones, provided by a panel of nine subjects, were used by Southwick and Schiffman (1980) and Schiffman (1983) to represent those compounds in a three-dimensional space and to investigate relationships between odor quality and physicochemical parameters.

Jaubert et al. (1987b) selected 50 molecules to span the whole olfactory space, as defined in a previous work (Jaubert et al., 1987a), and asked a panel of experts to rate the similarities of all possible pairs. The 50×50 similarity matrix was submitted to multidimensional statistical analyses, which resulted in an olfactory space in which nine zones could be distinguished. Moskowitz and Gerbers (1974) used similarity ratings among 15 odors and 17 descriptors, evaluated across four days, to generate a two-dimensional representation of odorants and descriptors. In a pilot study preceding some work mentioned earlier, Coxon et al. (1978) asked subjects to estimate the similarities among nine pure substances. The study showed consistency in responses between groups of subjects and between replications a week apart, but was not otherwise exploited.

Lawless (1989) examined a more limited section of the olfactory space and used simple sorting procedures to obtain similarity estimates for 18 chemicals or essential oils that fell into three a priori categories: "woody," "citrus," and "partially woody and citrus." A multidimensional analysis (MDS) was performed on the similarity estimates. The implications of that research and of several other studies for the structure of the olfactory space were discussed, and the conclusion was that "odor space may resemble an *n*-dimensional head of cauliflower more than a simple Euclidean map."

It seems fair to conclude from this survey of a large number of published classifications of odors that all studies indicate a weak structure of the olfactory space. The dimensionality of olfactory space, as defined by the data and methods used to establish it, appears to be rather high, and the nature and significance of these dimensions remain unclear. A hierarchical structure was never observed, and, moreover, all multidimensional studies confirmed that classes of odors are not sharply delineated.

2.4. Classification of Perfumes

Classification of perfumes is a difficult task, as they are complex mixtures that can smell quite differently to customers and to perfumers. The most important classification was proposed by the Société Française des Parfumeurs in 1988, followed a decade later by a new version (SFP, 1998). Perfumes, without distinction as to masculine or feminine, are classified in seven families, with a total of 45 subfamilies. Several other interesting classifications, such as the fragrance octagons of Dragoco that have been proposed by some perfume companies, will not be discussed here. Jellinek (1992) presented a critical study of existing classifications showing that perfumers and consumers have different perceptions, and proposed "a classification method that is more in line with consumer perceptions and needs than the ones currently in use."

3. Structure–Odor Relationships

Chemical structure is highly determinative of the quality and also the intensity of an odor. It might seem at first sight that studies of quality would be almost the sole interest of the perfume industry, as they offer ways to design new compounds possessing interesting olfactory properties. In fact, intensity is also taken into account, and compounds with high intensities are avidly sought. Nevertheless, the vast majority of structure–odor relationships are concerned mainly with quality.

3.1. Intensity Data

Intensity parameters of interest in various domains of olfaction are olfactory thresholds and dose–response relationships linking the perceived intensity and the stimulus concentration. Dose–response relationships are always difficult to establish because of the often poor quality of intensity data. This discussion will be restricted to olfactory thresholds, which have been more frequently studied.

Odor thresholds have been measured for more than a century (Cain, 1978), and hundreds of compilations of threshold data have been published. Olfactory thresholds show large person-to-person variability. They can differ by four and even five orders of magnitude, and variations by a factor of a thousand are quite common (Brown, McLean, and Robinette, 1968). As a result, even the published thresholds, which are means obtained from panels of several persons, involve enormous variability. The relatively large body of data available is very difficult to use because of the intrinsic variability in the data and the variety of methods used to obtain them. Devos et al. (1990) tried to circumvent those difficulties in publishing a useful compilation of olfactory thresholds.

In a series of papers, Schnabel, Belitz, and Von Ranson (1988), Von Ranson and Belitz (1992a,b), and Von Ranson, Schnabel, and Belitz 1992) reported odor profiles and olfactory thresholds for a large number of oxygenated compounds. Threshold values were measured in aqueous solutions and correlated with structural parameters. In the first paper, Schnabel et al. (1988) reported those values for a set of 281 compounds, including a subset of 99 aliphatic alcohols (from C_1 to C_{12}) that later was used by several groups.

Anker, Jurs, and Edwards (1990) generated a set of 112 descriptors for 53 of those alcohols. From that set, four descriptors were found significant, and attempts were made to understand their physical meaning. Zakarya (1992) also used 53 alcohols from the Schnabel set and obtained good correlations using different sets of variables, including autocorrelation vectors. Edwards, Anker, and Jurs (1991), using a data set of 53 of those alcohols, obtained a good four-parameter regression equation, after removal of four outliers. With a data set including 60 pyrazines (Mihara and Masuda, 1988) and 14 other pyrazine derivatives, they obtained a five-parameter regression equation, those five parameters having been selected from a set of over 100 variables including various sorts of descriptors as well as physical properties.

Chastrette, Crétin, and El Aidi (1996), thinking that the quality of odors might be important even in a study of thresholds, selected a subset of 45 alcohols described as "camphoraceous." By means of a three-layer back-propagation neural network, with input variables describing the sizes of substituents on a common skeleton, they were able to correctly calculate 91% of the logarithms of concentrations at the threshold in the learning phase, but only 78% in the prediction phase, probably as a result of a structural description that was too crude.

3.2. Quality Data

Lucretius probably was the first to suggest a theoretical basis for olfactory properties by relating them to the shape of atoms. Guillot (1948) discussed the problem of structure–odor relationships (SORs) in relation to specific anosmia. More ambitious SORs have been based on the large variety of so-called theories of olfaction, which actually are theories of the molecule–receptor interaction. An impressive number of SORs for a large variety of odors, generally of interest in the perfume industry, have recently been published. For detailed descriptions of many successful SORs, see the recent reviews by Rossiter (1996) and Chastrette (1997). Only a brief description of the principal strategies used to establish SORs will be presented here.

The first step in establishing an SOR is compilation of a data bank including olfactory descriptions for a large number of molecules. Such descriptions must be obtained from perfumers and flavorists because of the difficulty of such a task for untrained persons. They should also be homogeneous (i.e., provided by the same perfumers for all the substances in the sample). The substances in the sample must be clearly distinguished chemical compounds, as pure as possible (minute impurities can significantly modify odors). Isomers and even enantiomers must be considered as distinct compounds, and consequently racemic mixtures should be avoided. Molecules with low degrees of flexibility, for example, cyclic or polycyclic molecules, are preferred to those that can assume many different conformations.

The second step consists in the selection of variables to define the chemical structure. This is not as simple as it might appear, for such a selection relies on a more or less implicit theory of the interaction between molecules of odorant and receptors. Recent progress in understanding the structures of receptors has led to wider application of methods classically used in pharmacologic studies, and researchers are continually searching for some aspect or part of a molecule that reasonably could be expected to interact with proteins. They will be looking for structural elements possibly involved in hydrogen bonding and dispersion interactions, for example.

The third step is to establish a relationship between a set of variables describing the chemical structure and a set of variables describing the odor. Researchers in this field have used a vast array of statistical methods, including linear and nonlinear procedures.

4. Conclusion

The description of an odor supposes that both its quality and intensity characteristics can be precisely defined. Many different description systems, based variously on profiles or on similarities or on a thesaurus of words referring not only to pure substances but also to complex mixtures, have proved useful in the domains where olfactory descriptions are necessary. Unfortunately, as no general agreement has ever been reached, comparisons between any two systems are not easy, and compiling large data bases remains a difficult task. Moreover, olfactory perceptions are always more or less mixed with psychological aspects, and consequently physiological data are not so pure as one would wish. When intensity data are collected, huge inter-individual variations and the use of different measurement techniques introduce enormous variability into published data. However, some useful compilations have been achieved.

Notwithstanding the often poor quality of olfactory data, large numbers of meaningful structure–odor relationships have been established in recent years, in academic laboratories as well as in industrial research, by means of a vast array of multidimensional statistical methods.

Numerous classifications based on different description systems and taxonomic procedures have been published. These classifications can be quite different, as they reflect the different needs of their authors. As stated by Harper

(1966), "classifications can be carried out for a variety of purposes. . . . In the widest possible context the aim is to develop a universal system, within which all existing and all future odours could be placed." We are still far from reaching that goal, but researches generally agree that the olfactory space is a sort of continuum, where the elements are not hierarchically organized.

In fact, classifications reflect complex cognitive processes and depend on the needs and motivations of their authors, as well as on their choices of description systems and statistical treatments.

Acknowledgments

I am grateful to the Fondation Roudnitska, which generously supported our research on odor classification and structure–odor relationships.

References

Abe H, Kanaya S, Komukai T, Takahashi Y, & Sasaki S (1990). Systematization of Semantic Descriptions of Odors. *Analytica Chimica Acta* 239:73–85.

Abe H, Kanaya S, Takahashi Y, & Sasaki S (1988). Extended Studies of the Automated Odor-sensing System Based on Plural Semiconductor Gas Sensors with Computerized Pattern Recognition Techniques. *Analytica Chimica Acta* 215:155–68.

Abe H, Yoshimura T, Kanaya S, Takahashi Y, Miyashita Y, & Sasaki S (1987). Automated Odor-sensing System Based on Plural Semiconductor Gas Sensors and Computerized Pattern Recognition Techniques. *Analytica Chimica Acta* 194:1–9.

Amoore J E (1967). Specific Anosmia: A Clue to the Olfactory Code. *Nature* 214:1095–8.

Amoore J E (1977). Specific Anosmia and the Concept of Primary Odors. *Chemical Senses and Flavor* 2:267–81.

Amoore J E (1982). Odor Theory and Odor Classification. In: *Fragrance Chemistry. The Science of the Sense of Smell*, ed. E T Theimer, pp. 27–76. New York: Academic Press.

Anker L S, Jurs P C, & Edwards P A (1990). Quantitative Structure-Retention Relationship Studies of Odor-active Aliphatic Compounds with Oxygen-containing Functional Groups. *Analytical Chemistry* 62:2676–84.

Arctander S (1969). *Perfume and Flavor Chemicals*. Montclair, NJ: Arctander.

Billot M (1948). Classification des odeurs. *Industrie de la Parfumerie* 3:87–92.

Boelens H & Haring H G (1981). *Molecular Structure and Olfactive Quality*. Internal report, Naarden International, Bussum, The Netherlands.

Brown K S, MacLean C M, & Robinette R R (1968). Sensitivity to Chemical Odors. *Human Biology* 40:456–72.

Brud W S (1986). Words versus Odours: How Perfumers Communicate. *Perfumer and Flavorist* 11:27–44.

Cain W S (1978). History of Research on Smell. In: *Handbook of Perception. Vol. 6A: Tasting and Smelling*, ed. E C Carterette & M P Friedman, pp. 197–229. New York: Academic Press.

Cerbelaud R (1951). *Formulaire de parfumerie*, 3 vols. Paris: Opera.

Chastrette M (1997). Trends in Structure–Odor Relationships. *SAR & QSAR in Environmental Research* 6:215–54.

Chastrette M, Crétin D, & El Aidi C (1996). Structure–Odor Relationships Using Neural Networks in the Estimation of Camphroraceous or Fruity Odors and Olfactory Thresholds of Aliphatic Alcohols. *Journal of Chemical Information and Computer Sciences* 36:108–13.

Chastrette M, de Saint Laumer J Y, & Sauvegrain P (1991). Analysis of a System of Odors by Means of Four Different Multivariate Statistical Methods. *Chemical Senses* 16:81–93.

Chastrette M, Elmouaffek A, & Sauvegrain P (1988). A Multidimensional Statistical Study of Similarities between 74 Notes Used in Perfumery. *Chemical Senses* 13:295–305.

Chastrette M, Elmouaffek A, & Zakarya D (1986). Etude statistique multi-dimensionnelle des similarités entre 24 notes utilisées en parfumerie. *Comptes Rendus de l'Académie des Sciences, Paris, sér. II* 303:1209–14.

Coxon J M, Gregson R A M, & Paddick R G (1978). Multidimensional Scaling of Perceived Odour of Bicyclo[2.2.1]heptane, 1,7,7-Trimethylbicyclo[2.2.1]heptane and Cyclohexane Derivatives. *Chemical Senses and Flavor* 3:431–41.

Crocker E C & Henderson F L (1927). Analysis and Classification of Odors: An Effort to Develop a Workable Method. *American Perfumer and Essential Oil Review* 22:325–56.

David S (1997). Représentations sensorielles et marques de la personne: contrastes entre olfaction et audition. In: *Catégorisation et cognition: de la perception au discours*, ed. D Dubois, pp. 211–42. Paris: Kimé.

Demole E, Enggist P, & Ohloff G (1982). 1-*p*-Menthene-8-thiol: A Powerful Flavor Impact Constituent of Grapefruit Juice (*Citrus paradisi MacFayden*). *Helvetica Chimica Acta* 65:1785–91.

Devos M, Patte M, Rouault F, Laffort J P, & Van Gemert L J (1990). *Standardized Human Olfactory Thresholds*. Oxford: IRL Press.

Dravnieks A (1982). Odor Quality: Semantically Generated Multidimensional Profiles Are Stable. *Science* 218:799–801.

Dravnieks A (1985). *Atlas of Odor Character Profiles. ASTM Data Series DS 61*. Baltimore: ASTM.

Dravnieks A & Bock F C (1978). Comparison of Odors Directly and through Profiling. *Chemical Senses and Flavor* 3:191–9.

Edwards P A, Anker L S, & Jurs P C (1991). Quantitative Structure–Property Relationship Studies of the Odor Threshold of Odor-active Compounds. *Chemical Senses* 16:447–65.

Ellena J C (1987). Des odeurs et des mots. *Parfum, Cosmétiques, Arômes* 76:63–4.

Ennis D M, Boelens H, Haring H, & Bowman P (1982). Multivariate Analysis in Sensory Evaluation. *Food Technology* 11:83–90.

Fenaroli G (1971). *Handbook of Flavor Ingredients*, vol. 2 (3rd ed. 1990), ed. G A Burdock. Boca Raton: CRC Press.

Fourcroy A F (1798). Sur l'esprit recteur de Boerhave, l'Arome des Chimistes français, ou le principe de l'odeur des végétaux. *Annales de Chimie* 26:232–50.

Guillot M (1948). Sur la relation entre l'odeur et la structure moléculaire. *Comptes Rendus de l'Académie des Sciences, Paris* 226:1472–4.

Harper R (1966). On Odour Classification. *Journal of Food Technology* 1:167–76.

Harper R (1975). Some Chemicals Representing Particular Odor Qualities. *Chemical Senses and Flavor* 1:353–7.

Henning H (1915). Der Geruch, I. *Zeitschrift für Psychologie* 73:161–257.

Holley A & MacLeod P (1977). Transduction et codage des informations olfactives chez les vertébrés. *Journal de Physiologie* 73:725–828.

ISO (1992). Sensory Analysis: Methodology: Initiation and Training of Assessors in the Detection and Recognition of Odors. ISO 5496. International Standards Organization.

Jaubert J N, Gordon G, & Doré J C (1987a). Une organisation du champ des odeurs. *Parfum, Cosmétiques, Arômes* 77:53–6.

Jaubert J N, Gordon G, & Doré J C (1987b). Une organisation du champ des odeurs. II. Modèle descriptif de l'organisation de l'espace odorant. *Parfum, Cosmétiques, Arômes* 78:71–82.

Jaubert J N, Tapiero C, & Doré J C (1995). The Field of Odors: Towards a Universal Language for Odor Relationships. *Perfumer and Flavorist* 20:1–16.

Jellinek J S (1992). Perfume Classification: A New Approach. In: *Fragrance. The Psychology and Biology of Perfume*, ed. S Van Toller & G H Dodd, pp. 229–42. London: Elsevier.

Jennings-White C (1984). Human Primary Odors. *Perfumer and Flavorist* 9:46–7, 50–4, 56–8.

Kastner D (1973). Die Beschreibung und Klassifizierung von Gerüchen. *Parfüm und Kosmetik* 54:97–106.

Lawless H T (1989). Exploration of Fragrance Categories and Ambiguous Odors Using Multidimensional Scaling and Cluster Analysis. *Chemical Senses* 14:349–60.

Lawless H T (1991). A Sequential Contrast Effect in Odor Perception. *Bulletin of the Psychonomic Society* 29:317–19.

Lawless H T, Glatter S, & Hohn C (1991). Context-dependent Changes in the Perception of Odor Quality. *Chemical Senses* 16:349–60.

Linnaeus C (1756). *Odores medicamentorum. Amoenitates Academicae*, vol. 3, pp. 183–201. Stockholm: Lars Salvius.

McGill J R & Kowalski B R (1977). Intrinsic Dimensionality of Smell. *Analytical Chemistry* 49:596–602.

Mihara S & Masuda H (1988). Structure–Odor Relationships for Disubstitued Pyrazines. *Journal of Agricultural and Food Chemistry* 36:1242–7.

Moskowitz H R & Gerbers C L (1974). Dimensional Salience of Odors. *Annals of the New York Academy of Sciences* 237:1–16.

Polak E (1983). Is Odor Similarity Quantifiable? *Chemistry and Industry* 3:30–6.

Rimmel E (1895). *The Base of Perfumes*. London. (French translation 1990, *Le Livre des Parfums*. Paris: Les éditions 1900.)

Rossiter K J (1996). Structure–Odor Relationships. *Chemical Reviews* 96:3201–40.

Rouby C & Sicard G (1997). Des catégories d'odeurs? In: *Catégorisation et cognition: de la perception au discours*, ed. D Dubois, pp. 59–81. Paris: Kimé.

Roudnitska E (1991). The Art of Perfumery. In: *Perfumes. Art, Science and Technology*, ed. P M Müller & D Lamparsky, pp. 3–48. London: Elsevier.

Schiffman S S (1974). Physicochemical Correlates of Olfactory Quality. *Science* 185:112–17.

Schiffman S S (1983). Future Design of Flavour Molecules by Computer. *Chemistry and Industry* 3:39–42.

Schnabel K O, Belitz H D, & Von Ranson C (1988). Investigations on the Structure–Activity Relation of Odorous Substances. I. Detection Thresholds and Odor Qualities of Aliphatic and Alicyclic Compounds Containing Oxygen Functions. *Zeitschrift für Lebensmittel-Untersuchung und -Forschung* 187:215–23.

SFP (1998). *Classification des parfums et terminologie*. Paris: Société Française des Parfumeurs.

Southwick E & Schiffman S S (1980). Odor Quality of Pyridyl Ketones. *Chemical Senses* 5:343–57.

Von Ranson C & Belitz H D (1992a). Structure–Activity Relationships of Odorous Substances. Part 2. Detection and Recognition Thresholds and Odor Qualities of Saturated and Unsaturated Aliphatic Aldehydes. *Zeitschrift für Lebensmittel-Untersuchung und -Forschung* 195:515–22.

Von Ranson C & Belitz H D (1992b). Structure–Activity Relationships of Odorous Substances. Part 3. Detection and Recognition Thresholds and Odor Qualities of Alicyclic and Aromatic Aldehydes. *Zeitschrift für Lebensmittel-Untersuchung und -Forschung* 195:523–6.

Von Ranson C, Schnabel K O, & Belitz H D (1992). Structure–Activity Relationships of Odorous Substances. Part 4. Structure and Odor Quality of Aliphatic, Alicyclic and Aromatic Aldehydes. *Zeitschrift für Lebensmittel-Untersuchung und -Forschung* 195:527–35.

Wells F V & Billot M (1988). *Perfumery Technology. Art : Science : Industry*, pp. 160–73. New York: Wiley.

Weyerstahl P (1994). Odor and Structure. *Journal für Praktische Chemie/Chemiche Zeitung* 336:95–109.

Wright R H & Michels K M (1964). Evaluation of Far Infrared Relation to Odors by a Standard Similarity Method. *Annals of the New York Academy of Sciences* 116:535–51.

Yoshida M (1975). Psychometric Classification of Odors. *Chemical Senses and Flavor* 1:443–64.

Zakarya D (1992). Use of Autocorrelation Components and Wiener Index in the Evaluation of the Odor Threshold of Aliphatic Alcohols. *New Journal of Chemistry* 16:1039–42.

Zwaardemaker H (1895). *Die Physiologie des Geruchs*. Leipzig: Engelmann. (French translation 1925, *L'odorat*. Paris: Doin.)

Zwaardemaker H (1899). Les sensations olfactives, leurs combinaisons et leurs compensations. *L'Année Psychologique* 5:202–25.

Section Three

Emotion

Going against the Cartesian tradition, Darwin considered emotions as behavioral adaptations, "useful habits" inherited during phylogeny. Many authors think of olfaction in much the same way: Like emotion, olfactory perception challenges rational explanations of the world, and language itself. In the nineteenth century, efforts were made to find links between such perceptions and knowledge, which in turn implied a questioning of established knowledge. As psychologists, Hermans and Baeyens (Chapter 8) are specialists in a form of learning: evaluative conditioning, which supposes no awareness of the contingency between conditioned and unconditioned stimuli and is resistant to extinction. They have designed "ecologically valid" experiments, such as a bathroom study and a massage study, showing how emotional valuations of odors can be acquired and also changed during adult life without awareness. Such experiments, if they confirm that olfaction is an emotional sense, may also indicate that emotional processing of odors does not differ from emotional processing in other modalities. This is in contrast with memory processing, as will be discussed in Section 4.

In Chapter 9, Rouby and Bensafi question the concept of a hedonic dimension as a continuous variation between pleasantness and unpleasantness. They gather convergent cues from different disciplines indicating that hedonic judgment can distinguish two main types of odors: those that may have meanings and possibly attributes, and those that primarily have effects on the perceiver (see Section 2). Moreover, they propose the hypothesis that faster and more automatic neural networks subserve the processing of unpleasant odors.

What Herz, in Chapter 10, calls affective cognition is manifested as odor-induced modulations of moods, attitudes, memories, performances, and even perceptions of health: Odors can operate as conditioning stimuli, and later exposure to the same odors may influence behavior in a covert way. Beliefs and misinformation in turn are able to influence perceptual processes like adaptation. Challenging the claim that odors are better cues to memory than pictures and sounds, several experiments by Herz have shown that memories associated with odors are more emotional, but not more efficient, than memories associated with other sensory modalities.

The review by Jacob and colleagues (Chapter 11) enlarges our conception of odor effects beyond perception and explains how chemical substances that are not even smelled or do not even have any odor may affect us in different ways. Avoiding magical explanations for marketing techniques and phenomena, they discuss the topic of human pheromones, maintaining their necessary distance in a field deeply invested by passions and by economy, where desires can be mistaken for reality. Do we react to specific chemical signals in the same way as other animals? Their recent studies, if they do not cast doubt on the physiologic effects of some chemical signals, question the idea that chemicals can trigger complex behaviors in humans as in other mammals; rather, they show that chemical signals can influence mood in rather subtle ways.

Conceiving emotions as a set of basic, qualitatively different states, instead of points on a positive–negative continuum, Phillips and Heining, in Chapter 12, report on the use of one of the most recent tools of cognitive neuroscience, namely, functional magnetic-resonance imaging, to highlight the complex relationships involving taste, odor, the experience of disgust, and the brain. Disgust is especially representative of human emotion because it evolves from the unconditional, inborn responses of newborns to become elaborate social and symbolic representations that remain deeply rooted in the body experience.

8

Acquisition and Activation of Odor Hedonics in Everyday Situations: Conditioning and Priming Studies

Dirk Hermans and Frank Baeyens

1. Odor Hedonics: Acquired Evaluative Meaning

Odors and the reactions of liking and disliking are so intimately intertwined that it would be difficult to object to the statement (Richardson and Zucco, 1989) that "it is clearly the *hedonic meaning* of odor that dominates odor perception" (p. 353). The fact that the affective/emotional consequences of odor stimuli are so powerful makes it possible for an economically important industry – perfumery – to thrive on the production of substances whose only real function is to elicit highly positive reactions. Also, because the principal distinctive properties of food flavors are provided by olfaction rather than by taste cues (Rozin, 1982), it could reasonably be argued that the whole of culinary culture is based largely on the same strong connection between evaluative meaning and odor. Besides eliciting the reactions of mere liking or disliking, odors can have considerable emotional impact (Ehrlichman and Halpern, 1988; Miltner et al., 1994). This bond between odors and emotions has long been recognized, and hence it may not be that surprising that the subcortical limbic system, which is considered to be of critical importance in the generation of emotions, was originally known as the rhinencephalon or the "smell brain" (Van Toller, 1988).

Although the evaluative and emotional components and consequences of odors have been given a lot of thought, it is less well appreciated that most human evaluative reactions toward odor stimuli are not fixed and innate, but are largely the products of associative learning (Engen, 1988; Bartoshuk, 1994). Remarkably enough, the experimental evidence for evaluative learning about odors in humans is surprisingly meager. This is true for odors perceived retronasally as part of consumed odor/taste mixtures ("flavors"), and even more so for odors perceived orthonasally as sniffed substances ("smells"). More specifically, the literature on smell-preference learning in humans is limited to a few scattered demonstrations of the mere existence of the phenomenon (e.g., Hvastja and Zanuttnni, 1989), and the conditions of and the processes involved in human evaluative flavor-taste learning have just recently begun to attract the experimental attention they deserve (Baeyens et al., 1998), as discussed later. Hence, it remains a challenge and an important and interesting enterprise to investigate experimentally the (associative) processes by which odors acquire positive or negative valence, that is to say, their affective meanings.

It seems clear that nonassociative effects of mere exposure can contribute to changes in odor preferences (e.g., Balogh and Porter, 1988). Nevertheless, associative processes, and more specifically Pavlovian associative conditioning, have been suggested as major sources of evaluative odor learning. The theory of Pavlovian learning as it applies here is that when an odor that is perceived as neutral is paired with any liked or disliked event, that may be sufficient to change one's perception of the originally neutral odor into perception of a stimulus with positive or negative valence. A study by Hvastja and Zanuttini (1989) provides a good example of that phenomenon. They exposed groups of six- and eight-year-old children to simultaneous presentations of various odors and positive or negative images on photographic slides. One month after that learning phase, there were significantly more favorable ratings for odors that had previously been paired with the positive images. As the odors were presented equally often in both the positive and negative contexts, mere exposure cannot account for the acquired evaluative differentiation. Also, given the fact that testing took place one month after the original odor–slide pairings, a nonassociative mood effect can be ruled out as an alternative explanation. Although it was not conducted fully within the parameters of fundamental research on learning processes, that study provides an empirical demonstration of the idea that associative processes may play an important role in the acquisition of odor hedonics. It shows a laboratory analogue of an experience that most of us will have encountered before, namely, that a previously neutral odor acquired a positive or negative valence for us because of its contiguous presentation with a positive or negative event. For instance, the odor of a given perfume originally perceived as neutral may have taken on a positive valence through its association with an attractive and charming person who used to wear that fragrance, perhaps the person with whom you were in love at the time.

If we conceptualize the odor as a conditioned stimulus (CS), and the positive/negative stimulus or event as an unconditioned stimulus (US), both the perfume anecdote and isolated studies, such as the one by Hvastja and Zanuttini (1989), can be situated within the empirically rich research tradition on evaluative conditioning that has been studied extensively in our laboratory in Leuven. In the remainder of this chapter we shall explore the concept of evaluative conditioning and describe two experimental field studies of evaluative odor conditioning that were conducted at our laboratory. Next, we shall discuss how associative learning might contribute to the emotional influence that odors can have in our daily lives. A replication of an odor-conditioning study that was originally conducted by Kirk-Smith, Van Toller, and Dodd (1983) will be the basis for a more fundamental reflection on the automatic (unconscious) affective influences of odors. This discussion will be in the context of research on automatic affective processing, which has been a second major line of research at our laboratory.

2. Acquisition of Odor Hedonics: Evaluative Odor-Conditioning Studies

"Evaluative conditioning" refers to the situation in which simultaneous exposure to a neutral stimulus (conditioned stimulus, CS) and a positively or negatively valenced stimulus (unconditioned stimulus, US) can result in the originally neutral CS itself acquiring a valence congruent with the affective value of the US (Levey and Martin, 1987). For example, after a neutral picture of a human face has repeatedly been paired with a liked (or disliked) picture of another face, research participants typically demonstrate an acquired liking (or disliking) for the originally neutral face (e.g., Baeyens et al., 1992). That pattern has now been demonstrated using stimuli from different sensory modalities. Besides a wide range of studies using visual stimuli as CSs and USs, evaluative conditioning has also been demonstrated with olfactory and gustatory stimuli. Although in the bulk of this chapter we shall discuss some interesting studies from our laboratory in which evaluative learning with odors was demonstrated, we shall first focus on a series of flavor-conditioning studies to clarify some of the most important characteristics of evaluative conditioning in general.

The evaluative flavor–flavor learning paradigm is a conditioning framework that has proved to be a very efficient tool for study of human evaluative learning. In this differential-conditioning paradigm, which was first developed by Zellner et al. (1983) and extended at our laboratory, research participants were exposed to a series of flavored and colored water solutions containing the appropriate CS+/US and CS− contingencies. Several experiments showed that after an originally neutral fruit flavor A (CS+) had been presented several times mixed with the bad-tasting substance Tween 20 (Polysorbate 20) as the US, research participants developed a dislike for the pure flavor A relative to a similar fruit flavor B (CS−) that had been presented equally often in plain water (e.g., Baeyens et al., 1990b).

We are beginning to achieve a more detailed understanding of the conditions and characteristics of this type of evaluative learning with flavors. The available evidence indicates that this learning effect is quite robust, in that (1) it involves rapid learning, (2) it survives the passage of time, (3) it is not influenced by the color of the solutions, (4) it can be achieved with sequential rather than simultaneous presentations of CS and US flavors, albeit in a weaker form, and (5) it allows for an observational mode rather than a direct mode of presenting the crucial CS–US information (Baeyens et al., 1990b, 1995a, 1998).

Two properties of evaluative flavor–flavor learning bring it into line with the functional characteristics of other evaluative-conditioning designs and at the same time justify a distinction between evaluative learning and other forms of Pavlovian learning, such as expectancy learning (Baeyens, Eelen, and Crombez, 1995b). The first property concerns the role of explicit knowledge about the CS–US relation (i.e., "contingency awareness"). Expectancy learning, involving the acquisition of preparatory (defensive, appetitive, or orienting) conditioned responses, has repeatedly been shown to require awareness of the CS–US contingency (Dawson and Shell, 1982). In contrast, evidence is accumulating from several study designs that contingency awareness is not a necessary condition for evaluative conditioning (e.g., Baeyens, Eelen, and Van den Bergh, 1990a; De Houwer, Baeyens, and Eelen, 1994). In line with this, evaluative flavor–flavor learning has been demonstrated to proceed orthogonally to conscious identification of the CS–US flavor contingencies. For example, in two studies (Baeyens et al., 1990b, 1996a) a double dissociation between evaluative learning and conscious acquisition of knowledge was observed, in that with flavors functioning as the CS+/CS−, none of the participants demonstrated awareness of the stimulus relations, but they evidenced clear conditioning; on the other hand, when colors acted as the CS+/CS−, a substantial number of participants were contingency-aware, but no evaluative conditioning was observed (Baeyens et al., 1996a).

The second important functional property of evaluative learning for flavors pertains to the interrelated phenomena of "extinction" and "contingency sensitivity." In cases of expectancy learning, unreinforced CS presentations after the acquisition phase (i.e., an extinction procedure, during which the CS+ is no longer followed by the US) result in a decrease or complete disappearance of the conditioned preparatory responses. Similarly, presentation of unreinforced CSs during the learning phase, thus lowering the CS–US contingency, clearly attenuates the acquisition of conditioned responses. Descriptively, both phenomena can be thought of as the result of expectancy disconfirmation (the expected US does not arrive). The main reason to conceive of evaluative learning as qualitatively different from expectancy learning is that unreinforced CS presentations, either during or after acquisition, do not seem to have an effect on this type of conditioning: Evaluative conditioning is resistant to extinction (Baeyens et al., 1988,

1995b, 1989) and to CS–US contingency manipulations (Baeyens, Hermans, and Eelen, 1993). In line with this, in the evaluative flavor–flavor paradigm it was observed that post-acquisition unreinforced CS presentations (extinction procedure) did not attenuate the acquired flavor dislike (Baeyens et al., 1995a, 1996a) and that an acquisition schedule containing equal numbers of Tween-20-reinforced and nonreinforced flavor trials resulted in an acquired flavor dislike equally as strong as that produced by an acquisition schedule containing the same number of reinforced, but no unreinforced, learning trials (Baeyens et al., 1996a).

These findings justify considering evaluative learning as a prototypical example of a Pavlovian learning process mediated by what we have named the "referential system" (Baeyens et al., 1995a). According to this hypothesis, during acquisition an association is built up (consciously or unconsciously) between CS and US representations following a rudimentary learning algorithm, the parameters of which include only temporal contiguity and stimulus salience: Synchronous activation of CS and US representations results in an increase in associative strength, whereas activation of the representation of the CS alone (or of the US alone) is causally ineffective. Because of this associative link, activation of the representation of the CS flavor activates (consciously or unconsciously) a representation of the US (i.e., the bad-tasting substance Tween 20) that is behaviorally expressed in a dislike for the CS flavor. Thus, it is hypothesized that this whole process can (but will not necessarily) take place in the absence of any engendered expectation that the US is really going to arrive momentarily, and without triggering defensive preparatory responses.

Translated to the study of odors, all this information would seem to imply that an originally neutral odor could acquire a positive or negative valence through a contiguous occurrence of a positive or negative event, even without the person being aware of the crucial association between the odor (the CS) and the US with which it was associated. Nevertheless, it is assumed that the newly acquired odor valence is based on this CS–US association, in that the odor is linked, consciously or unconsciously, to the positive or negative US. From this perspective, the fragrance mentioned in the earlier perfume anecdote is assumed to be evaluated as pleasant and agreeable not because it has gained any new "intrinsic" valence but because it makes one think (unconsciously) of the person who wore it and with whom one was in love. Moreover, one would expect the odor to retain its acquired positive valence, even if the real-life contingency between the perfume and one's beloved no longer held. Although the evaluative differentiation reported in the study of Hvastja and Zanuttini (1989) likely was due to such a process of evaluative conditioning, the potential for extraneous effects of social desirability or other demand effects was not clearly eliminated in their study.

Therefore, in two experimental field studies, we attempted to further investigate the possibility of evaluative odor conditioning.

2.1. Bathroom Study

In a first study, Baeyens et al. (1996b) investigated whether or not a room's background odor (CS) that was quite "natural" for the specific environment could be perceived as acquiring the affective-evaluative tone of the situation and activities with which it was associated. In this study, we implemented a real-life contingency between an odor as the CS and toilet rooms or toilet-room activities as the conceptual US. We made use of the fact that each unit in our department has its own nearby bathroom, and the majority of people frequent only their own unit's bathroom. Also, at the time of this study, air-refreshers were not used in the toilet rooms. Thus it was possible to establish a multi-trial real-life conditioning schedule with simultaneous presentations of one particular odorant (CS) and the toilet-room and toilet-activity stimuli (US).

During the conditioning phase, two different odors were used on a between-subjects basis: lavender and pine. The odors were presented by means of stand-alone automatic odor dispensers that were placed inconspicuously above the ceiling slats of 10 toilet rooms in the faculty building. Five were located in toilet rooms on the back side of the building (location A), and five were located on two floors on the front side of the building (location B). Participants from location A in the building received several days of exposure to the lavender odor in their toilet rooms. For that group, pine was the control odor. Participants from location B in the building were exposed to the pine odor in their toilet rooms, and for them lavender functioned as the control odor.

After the conditioning phase, the odor dispensers were removed from the bathrooms, and after a delay of several days, employees whose offices were in the proximity of the toilet rooms treated with the odors were invited to participate in "a short pilot study on odor perception." During that part of the experiment, participants were asked to rate the two odors on 11 semantic rating scales covering Osgood's three dimensions of semantic meaning: "valence" (5 items), "potency" (3 items), and "activity" (3 items) (Osgood, Suci, and Tannenbaum, 1957). To assess the participants' awareness of the fact that one of the odors was the same as the one that had earlier been presented in their toilet rooms, they were asked if they had ever smelled either odor before, and if so, in what context or product. Finally, because it was assumed that the people probably differed in terms of the frequency with which they visited the toilet rooms (some might visit the toilet room as a relaxing break from work, whereas others might experience it as a necessary evil), they were asked to indicate whether or not they usually liked entering the toilet room, using a −10 ("necessary evil") to +10 ("agreeable break from work") rating scale.

When the group of participants was subdivided on the basis of the latter ratings, it was possible to demonstrate an evaluative odor discrimination in line with what would be expected on the basis of the evaluative-conditioning hypothesis. That is, the participants who rather liked going to the toilet room evaluated the toilet-paired conditioning odor more positively than the control odor, whereas, the reverse was true for people who rather disliked going to the toilet room. That effect was observed in a situation in which few of the participants (15%) recognized the CS odor as being the odor that had earlier been presented in the toilet rooms, making it very implausible that experimental demand could account for the observed differences. Moreover, when the data for the participants who were aware of the CS (odor)–US (toilet) contingency were excluded from the analysis, a similar result was obtained. Also, no conditioning effects could be demonstrated for the "activity" and "potency" ratings.

Consequently, these data provide clear support for the possibility of Pavlovian evaluative learning about odors in adult human participants. A second study (Baeyens et al., 1996b) was devised in an attempt to replicate those findings conceptually in the context of therapeutic massages performed by a professional physiotherapist.

2.2. Massage Study

In this second experimental field study, we obtained the cooperation of a physio-therapist who was instructed to use massage oil containing one of two CS odors during a series of therapeutic massages (US) on clients who frequented his prac-tice. On the basis of their medical diagnoses, half of the participants were treated with positive-relaxing massages, whereas the other half of the participants were treated with less positive, and sometimes even negative-painful, massages. Orthogonal to that manipulation, and also on a between-subjects basis, two differ-ent odors were used during the conditioning phase. Half of the patients were treat-ed with odor A, and half with odor B. Both odors were etheric oils and were selec-ted on the basis of a pilot study in which they were similarly rated for Osgood's semantic dimensions (valence, potency, and activity) (Osgood et al., 1957).

The main purpose of the study was to demonstrate in an adult population that a rather salient odor (the CS) that was quite natural in the given setting might acquire the affective-evaluative tone of the events with which it was associ-ated. That was achieved by establishing a contingency between an odor (the CS) and the bodily sensations resulting from either positive (relaxing) or less posi-tive (negative-painful) massage treatments (USs). Hence, it was predicted that participants in the "positive" massage group would rate the treatment odor more positively than the control odor, whereas the reverse was expected for the "negative" massage group. That was indeed confirmed. About one week after the final treatment session, the patients returned to the physiotherapist for the standard follow-up visit. At that time the physiotherapist explained that a pharmacological company had asked him to invite his patients to evaluate two massage oils that they were planning to bring onto the market, and they were interested primarily in the odors of the oils. Next the patients were asked to rate the two odors on a series of items in the "valence" and "dynamism" dimensions (where "dynamism" is a combination of Osgood's original "activity" and "potency" dimensions; see also Osgood et al., 1957). Patients were also asked to rate the massage treatment they had gone through during the preceding weeks.

As predicted, the crucial interaction involving conditioning (experimental odor versus control odor) and massage type (positive versus negative reactions to massage) was highly significant. Subsequent analyses demonstrated that for the "positive" massage group, the odor that was used in the massage oil was rated as more pleasant than the control odor. Also, the massage odor was experienced as less dynamic (active/potent) than the control odor. For the "negative" massage group, however, there was no significant impact of conditioning on the "valence" and "dynamism" ratings of the odors. However, that was not really surprising, given that we apparently failed to administer a negatively valenced US (massage), because the intended negative-painful massage treatments were experienced by the patients as rather neutral events.

Interesting from a theoretical point of view was that the conditioning effect in the "positive" massage group was not dependent on conscious recognition of the experimental odor. In fact, when participants were explicitly questioned whether or not they had ever smelled the experimental odor before, quite surprisingly only 13 of the 34 participants recognized the experimental odor as the odor to which they had been exposed during the massage treatments. When only those participants who were unaware of the crucial CS (odor)–US (massage) contingency were taken into account, the conditioning effects for odor "valence" and odor "dynamism" could still be demonstrated in the "positive" massage group. In other words, awareness of the fact that the odor had been encountered before during the therapeutic massage treatment was not a necessary precondition for the conditioning effect to occur.

In sum, the data of the massage study demonstrate that not only odor "valence" but also odor "dynamism" can be influenced by Pavlovian contingencies. As it has often been argued that "valence" and "dynamism" (or "activity/arousal") are the two basic dimensions underlying emotional meaning (Osgood et al., 1957;

Lang, Bradley, and Cuthbert, 1990), these data seem to indicate that Pavlovian conditioning may be able to shape or modify the core emotional meanings evoked by odors.

3. Acquisition and Activation of Odor Hedonics: Emotional Influences of Aversive Conditioning

In the two preceding studies it was demonstrated that an evaluative-conditioning process can take place even if the subject is unaware of the crucial CS–US contingency. In both studies, most of the participants were unaware of the CS–US contingency, for although the odors were used at clearly noticeable concentrations, they were used in situations in which people are accustomed to the presence of odor stimuli (scented massage oil, air-refreshers in toilets). Another strategy was used by Kirk-Smith et al.(1983), who published the article "Unconscious Odour Conditioning in Human Subjects" in *Biological Psychology*. There the CS–US contingency was masked by using an unfamiliar odor that was presented at a very low, hardly noticeable concentration. Their study is particularly interesting because it demonstrated that the associative learning that occurs in the relationship between an odor (CS) and an emotionally toned event (US) not only can change one's evaluation of that odor but also can have other affective/emotional influences when that odor is encountered later in another situation.

3.1. The Study by Kirk-Smith et al. (1983)

In that study, participants were subjected to the stressful task of completing block patterns at a rather demanding rate. On completion of the task, participants were told that their performances were below average. For half of the participants (experimental group), a low-intensity odor (trimethylundecylenic aldehyde, TUA) was present, but for the control group the unobtrusive odor was not presented. In what the participants were led to believe was an unrelated exercise, several days later, they were invited to participate in an experimental task on "moods and the assessment of people." They were instructed to fill in mood-rating sheets and to assess a series of photographs of people in terms of pairs of attributes. During that second phase, the odor was present for both the experimental group and the control group. It was hypothesized that participants who had experienced the association between the odor (CS) and the stressful task (US) would describe themselves and the photographs of people as more anxious than would the control participants, for whom the odor had not been associated with the stressful experience.

Although the expected group differences in these ratings were not significant, the data showed that women made more mistakes and were slower on the block-pattern task than men. Hence it was concluded that the task may have been more stressful for the women and that as a result the women would be more likely to be conditioned to the odor than would the men. Indeed, the mood scales indicated increased anxiety ratings for the female experimental group in that second phase, relative to the control group. In addition, women rated the photographs as showing more anxiety. A post-experimental test of awareness indicated that those participants were not aware of any association between the odor and the stressful experience. It was therefore concluded that the data provided support for the theory of unconscious emotional conditioning with odors. The one-trial contingency between the odor (CS) and the stressful situation (US) seems to have been sufficient to change the previously neutral reactions to the odor (CS) such that it would elicit affective reactions in those subjects when the odor was subsequently encountered in a situation unrelated to the original conditioning context.

Nevertheless, that study by Kirk-Smith and colleagues has been severely criticized on methodological as well as statistical grounds (Black and Smith, 1994); see Kirk-Smith (1994) for a reply to those comments. One of the comments by Black and Smith was that no evidence was presented to show that the block-pattern task was effective in inducing stress/anxiety as intended. Also, and more importantly, they argued that it could not be excluded that initial group differences in the extent to which the block-pattern task was experienced as stressful might have been responsible for the observed findings (that was not explicitly tested), rather than (group) differences due to being exposed to the CS (odor)–US (stressful-task) contingency.

In spite of those methodological and statistical flaws, we still believe that the procedure proposed by Kirk-Smith and colleagues provides an elegant way of investigating the behavioral and emotional impacts of aversive conditioning with odors in a laboratory analogue of typical natural situations. For that reason, we decided to replicate their experiment conceptually in a more controlled study (Hermans, Pauwels, and Baeyens, 1992).

3.2. Conceptual Replication of the Study by Kirk-Smith et al. (1983)

We collected data from a group of psychology students who were submitted to the stressful task of completing patterns in a Chinese tangram puzzle. They were told that they were participating in the validation of a new intelligence test and that the puzzle was a measure of their "general problem-solving ability and level of motor skill." A pilot study in our department had shown that the "tangram task" was rather stressful, with the mean total of correctly assembled patterns in that pilot study only about one pattern in 15 minutes. However, we told our students that related studies abroad had indicated that an average person should be capable of assembling one pattern every two minutes, adding that we did not expect any differences for the Flemish population, and certainly not for university students. During a period of 15 minutes the participants were to assemble as many patterns as possible. As in the original study by Kirk-Smith and associates, the TUA odor was present at an unobtrusive intensity for the participants in the "stress, odor" condition, whereas participants in the "stress, no odor" condition did not receive the odor during that first session. And in order to accommodate the Black and Smith (1994) critique of the original study by Kirk-Smith et al. (1983), at the end of that first phase a brief "test-stress evaluation questionnaire" was presented. It consisted of five bipolar 10-point visual-analogue scales to assess the extent to which the task had been experienced as stressful.

Six days after the first session, all participants received letters in which they were invited to participate in "a brief study on moods and the assessment of people." No reference was made to the first session, and the letter was signed by a second experimenter. On arrival, participants entered a TUA-odorless room and were informed that the purpose of the study was to investigate how certain moods may affect a person's perception and assessment of other people. Then they were asked to complete the "General Mood Ratings" for the first time. In that questionnaire, which is similar to the one used by Kirk-Smith and associates, participants have to rate themselves in terms of "current affective state" ("anxiety," "depression," and "hostility") on a series of 11-point visual-analogue scales (0 = not at all; 10 = very). Next, participants were invited into another room in which the odor was present. During that phase of the experiment, the odor was present for both the experimental group and the control group. As was the case during the first phase of the experiment (the stressor task), the concentration of the odor was such that it would not be noticed unless one's attention were drawn to it. In that room, the participants were asked to complete the "mood" scale for a second time, as well as both the "state" and "trait" parts of the Dutch version of the State Trait Anxiety Inventory (STAI) (Van der Ploeg, Defares, and Spielberger, 1980). In addition, just as in the procedure employed by Kirk-Smith and associates, participants were asked to judge a set of photographs of four male and female persons on a series of bipolar semantic differential scales. After completion, participants were asked to evaluate the room's odor on a 10-point visual-analogue scale (pleasant–unpleasant) and were asked where they had encountered that odor before. Subsequently they were asked (1) if they had noticed anything unusual about the conditions of the experiment, (2) what they thought the experiment was really about, and

(3) if they had noticed any similarity between the conditions of the most recent experiment and those of the tangram task they had carried out six days
earlier.

Initial analysis revealed that the "stress, odor" group and the "stress, no odor"
group did not differ in the way they had experienced the stressor task, as indicated
by their answers on the five questions of the "test-stress evaluation questionnaire"
and their numbers of completed puzzles. Although all participants experienced
the task as difficult and rather stressful, there were nonetheless strong inter-
individual differences in their ratings.

Because emotional effects from the odor (CS) would be expected for only
those participants who experienced the tangram task as rather unpleasant and
distressing, we subdivided the total sample of participants by means of a cluster
analysis for the data from the "test-stress evaluation questionnaire" and the numbers of completed puzzles. That analysis revealed two clusters of participants:
cluster 1 = highly stressed participants ($N = 18$); cluster 2 = slightly stressed
participants ($N = 15$). Second, a principal-components analysis was performed
on all "emotional/affective" ratings that were carried out in the presence of the
odor during the second part of the study, including the photograph ratings, the
different items of the General Mood Ratings, the "state" and "trait" scores on
the STAI, and the evaluative ratings of the odor. That analysis (varimax rotation) revealed three major factors (Mood 1, Mood 2, Mood 3) that explained the
majority of variance (34%, 15%, and 10%, respectively).

Next, a multivariate analysis of variance (MANOVA) for the participants' ratings on those mood factors (Mood 1, Mood 2, Mood 3), with "condition"
(odor group/no-odor group) and "stress" (low/high) as between-subject variables, showed that the crucial "condition" × "stress" interaction was marginally
significant [Pillais $F(3, 27) = 2.6$; $p = .072$]. Subsequent univariate analyses for
the three mood factors separately revealed a significant interaction for Mood 3
[$F(1, 29) = 7.6$; $p = .01$]. The three variables that loaded highest in that mood
factor were the "assertiveness," "hostility," and "sexiness" ratings for the pictures.
Relative to the control group, participants in the experimental condition (odor
group) rated the photograph images as more hostile, more assertive, and less
sexy when that odor had been associated with an experience of stress during the
first session. That was not the case, however, for those who described the stressor
test as not so stressful.

So although the data did not reveal a main effect of odor, we can nevertheless
conclude that for those participants who actually experienced the stress-inducing
task as difficult and stressful, encountering the odor again in an apparently unrelated context influenced their affective ratings for a series of pictures. The
fact that the effect was restricted to that group is not really surprising, given

that the stressor task could be regarded as a genuine negative US only for those participants.

4. Automatic Activation of Odor Hedonics: Affective Odor-Priming Studies

An important aspect of the Kirk-Smith paradigm is that the data suggest that after the subjects were exposed to the odor (the CS), they became nonconsciously susceptible to its influence in their affective/emotional behavior. The idea of automatic/unconscious/implicit emotional influences by affectively valenced stimuli is certainly not new. Whereas most older studies had investigated the emotional influence of subliminally presented stimuli, recent research on automatic affective influences has focused more on the automatic affective impact of supraliminal and clearly perceptible emotional stimuli. This approach is more attractive from the perspective of ecological validity. One possible method within this approach is to study the impact of perceptible stimuli that are not within the focus of attention. A second method is to employ clearly perceptible stimuli that can be within the central focus of attention, but to assess their emotional impact by means of a procedure that cannot be influenced by consciously controlled strategies: Emotional effects should hence be attributed to automatic (unconscious) processes. A good example of such a research paradigm is the affective-priming procedure, which was introduced by Fazio et al. (1986) and has been under further investigation in our laboratory (e.g., Hermans, De Houwer, and Eelen, 1994, 1996, 2001). Before describing some very recent data from affective-priming studies using odor stimuli, it seems appropriate to first introduce some basic findings on affective priming that were obtained using stimuli from a sensory modality other than olfaction: vision. In an extensive series of affective-priming experiments it has been demonstrated that response latencies to affectively valenced stimuli are mediated by the affective context against which they are presented. In a standard affective-priming study, a series of target words (e.g., "constructive," "jealous") is presented, to be evaluated as quickly as possible as either "positive" or "negative." Each target word is preceded by a priming stimulus, which can be positive, negative, or neutral (e.g., "music," "dentist," "circle") and is to be ignored by the subject. Of crucial importance in these priming studies is the affective relation between the valence of the prime and the valence of the target, which typically is manipulated over three levels. Prime–target pairs can be affectively congruent (e.g., "music-constructive;" "dentist-jealous"), affectively incongruent (e.g., "dentist-constructive;" "music-jealous"), or affectively unrelated (e.g., "circle-constructive;" "circle-jealous;" or control pairs). It has now repeatedly been demonstrated that this affective relation mediates the time needed to respond to the target stimulus. Response latencies are shortened for affectively congruent prime–target trials, as compared with control trials, and are relatively lengthened for affectively incongruent trials. This data pattern can be explained only if one assumes that the participants have evaluated the primes, even though they have been asked to ignore those stimuli. Moreover, the parameters that typically are used in affective-priming research preclude consciously controlled processes that could be responsible for these priming effects. For instance, in most affective-priming studies, the interval between the onset of the prime and the onset of the target – the stimulus-onset asynchrony (SOA) – has been only 300 msec [prime = 200 msec; inter-stimulus interval (ISI) = 100 msec], which is assumed to be too brief for participants to deploy controlled response strategies. Hence, affective-priming effects that are observed under these conditions should be attributable to automatic processes. Given the importance of the automatic character of the affective-priming effects, it has been thoroughly investigated (Hermans, Van den Broeck, and Eelen, 1998b); it is at-tributed to fast-acting, goal-independent, relatively efficient processes that can operate even without any awareness of the instigating stimulus. Taken together, these data indicate that humans are able to evaluate stimuli as "positive" or "negative" in a rather unconditional and automatic fashion, an idea that has been one of the central tenets of several modern cognitive-representational theories of emotion.

Following the original study by Fazio et al. (1986), wide generality of this affective-priming effect has now been demonstrated. Not only has it been pos-sible to demonstrate affective-priming effects using different types of response tasks (target evaluation, target pronunciation, lexical decision tasks), but also the evidence for automatic stimulus evaluation, derived from the affective-priming paradigm, has been generalized toward different types of stimulus materials, such as words (e.g., Fazio et al., 1986; Hermans et al., 1994), nonsense words for which affective meanings have only recently been learned (De Houwer, Hermans, and Eelen, 1998), simple line drawings (Giner-Sorolla, Garcia, and Bargh, 1994), and complex real-life color pictures (Hermans et al., 1994, 1996). Moreover, the effect has been obtained with a wide range of stimuli varying in content as well as in the accessibility of their affective valence in memory. But whereas the models of automatic affective processing have assumed that automatic stimulus evaluation (and hence affective priming) should apply to all sensory modalities, in all of the studies of the affective-priming effect that have been published, the stimulus material has been visual in nature. In that context (here we return to the main issue of this chapter) we decided to use odors as primes in an affective-priming procedure. Accordingly, to test whether or not the affective-priming effect could be generalized to nonvisual stimuli, we used a cross-modal affective-priming paradigm in which selected positive and negative odors provided the affective-processing context for word evaluation.

4.1. Affective Odor-Priming Study

In this study (Hermans, Baeyens, and Eelen, 1998a), each participant selected a most liked odor and a most disliked odor from a series of 10 odors. Ten positive and 10 negative target words were also selected on an individual basis. In the second phase of the experiment, the individually selected positive and negative odors and words served as primes and targets, respectively, in an affective-priming procedure. During the priming procedure, participants were asked to evaluate each target word as quickly as possible by saying, out loud, "Positive" or "Negative." Each word was immediately preceded by a 10-sec presentation of either the "positive" odor or the "negative" odor. Analysis of the evaluative-response latencies to the words demonstrated a significant main effect of "affective congruence." Target words were evaluated significantly faster when preceded by a similarly valenced prime (odor), as compared with trials in which prime and target were of opposite valence. Because of known gender differences in the way odors are processed, the gender of the participants was also entered into the analysis as a between-subjects variable. The analysis showed that the effect of "affective congruence" was moderated by "gender." For female participants, the data were clearly in line with the affective-priming hypothesis. For the male group, however, there was no difference between response latencies for affectively congruent and incongruent trials.

The fact that the affective-priming effect was limited to the female participants may simply be attributable to the fact that the female olfactory sense is superior in most respects to that of males (Doty et al., 1985; Doty, 1991). Thresholds for odor identification are much lower in women for several odorants, and odors are experienced as more intense by women; they have a better capacity to discriminate odors and are more pronounced in their judgments of whether a particular odor is pleasant or unpleasant (Vroon, Van Amerongen, and De Vries, 1994, p. 95). That would seem to imply that the absence of affective priming for males is not a consequence of their inability to automatically evaluate odors, but of their lower capacity to detect and/or discriminate odors. The fact that additional analyses showed that the means for the males were in the expected direction in the second experimental block supports this hypothesis. The olfactory system of a male participant probably needed more time to perceive the odor primes. All in all, given the presentation parameters that were employed, the data of this study provide support for the idea that the positive/negative valence of odor stimuli can be processed automatically (Hermans et al., 1998a).

4.2. Evaluative Odor Conditioning and Priming Combined

The first part of this chapter focused on the role of associative processes in the acquisition of odor valence. But given the predominance of like/dislike reactions in the perception of odors, and given the fact that affective-priming data suggest that these like/dislike reactions can be triggered automatically on perception of odors, the question can also be reversed: How do odors, once they have acquired a positive or negative valence for an individual, in turn influence evaluative reactions to non-odor stimuli, and, more specifically, what is the role of associative learning here? Again, this is a largely unexplored research issue, but the little evidence that is available seems promising. For example, Todrank et al. (1995) convincingly demonstrated that pairings of valenced odors with pictures of human faces resulted in reliable shifts in evaluation during subsequent viewing of the pictures presented without odors. Similar results have been described by Schneider et al. (1999).

In the final experiment to be discussed here, we wanted to take these observations one step further (Hermans, Baeyens, and Natens, 2000): (1) Would it be possible to effectively use positive and negative odors as USs in an evaluative-conditioning procedure, with neutral pictures of objects as CSs? (2) Would the newly acquired valences of the CSs (the pictures) in turn exert an automatic affective influence, as measured by an affective-priming procedure?

The CSs used in this study were four different color pictures of a bottle of liquid soap, each placed in a transparent plastic freezer bag together with a cotton ball. For one of the pictures (CSpositive), a pleasant odor (lemon; essential oil) was applied to the cotton ball, and for a second picture (CSnegative) a negative odor (civet; dissolved in ethyl alcohol) was applied to the cotton. For the last two pictures (CSneutral), which served as controls, there was no odor. As part of the cover story, an electrode was attached to the participant's right hand, and each was told that the experiment was designed to determine whether or not there was a relationship between preferences for complex sensory stimuli and physiological (skin-conductance) responses to those stimuli, and whether or not those skin-conductance responses depended on preferences for those visual and olfactory stimuli. On each trial, the participant was asked to take a look at the picture, to open the bag and to sniff it, to close the bag again, and finally to take a second look at the picture. For each CS–US pair there were eight acquisition trials. At the end of the acquisition phase, participants were asked to rate each of the four CSs on a number of dimensions. The first two dimensions were evaluative: to what extent they thought the product (the bottle of liquid soap) was attractive, and to what extent they thought the CS was pleasant/unpleasant. Also, it was asked to what extent this product would attract their attention in a supermarket, what they anticipated the quality of the product to be, and to what extent they would be inclined to buy this product on seeing it in a supermarket.

The results showed that there was a strong effect of evaluative conditioning. The CSpositive was rated as more attractive and pleasant than the CSnegative. Also, participants indicated that the CSpositive probably would attract their attention more in a supermarket and probably would be of better quality. Finally, participants indicated that they would be more inclined to buy the CSpositive than the CSnegative. We can thus conclude from these data that the mere pairing of relatively neutral pictorial CSs with a positive or negative odor (the US) can result in an evaluative shift for these visual CSs.

However, it cannot be excluded that those effects may have been due in part to demand effects. Most participants probably were aware of the specific CS–US contingencies, and it is possible that effects of social desirability or other demand effects may have influenced their ratings. Therefore, in the second phase of the experiment, the four CSs were entered as primes in an affective-priming procedure, which, as argued earlier, can be viewed as an indirect and unobtrusive measure of stimulus valence that is not likely to be influenced by demand effects (Hermans et al., 1999). During the priming phase, participants were asked to evaluate a series of positive and negative target words that were preceded by one of the four CSs. It was predicted that if the conditioning phase had indeed changed the valences of the CSs in the expected directions, presentation of the CSpositive would facilitate the processing of a positive target, but inhibit the processing of a negative target, whereas an opposite pattern would be expected for the CSnegative. The results showed that there was a significant effect of "affective congruence" for the male subgroup. Response latencies for affectively congruent prime–target pairs were significantly shorter than for affectively incongruent trials. For the female participants, an effect of "affective congruence" was not seen. We have no valid explanation for this gender difference in the affective-priming data, but the data from the male subgroup offer a clear indication that contingent pairing of a neutral visual CS with a positive or negative odor (US) can result in reliable evaluative shifts for these non-odor CSs. Moreover, this newly acquired valence can be automatically activated, as was shown by the results of the affective-priming procedure.

5. Conclusion

This chapter has focused on recent research into the acquisition and activation of odor valence. With respect to the acquisition of odor likes and dislikes, we can conclude that associative processes are justifiably assumed to play an important role. Even in natural situations such as toilets and therapeutic massages, processes of evaluative conditioning seem to provide a sufficient basis to modify odor preferences. An important observation from the reported studies is that awareness of the contingency between the CS and the US is not a necessary precondition for evaluative odor conditioning. In other words, the mere contingency between a relatively neutral odor and a positive or negative stimulus/event is sufficient to change one's evaluation of that odor, even if one is unaware of the fact that the odor was presented together with the positive or negative stimulus/event. Hence, these data provide a fine demonstration that the acquisition of odor hedonics can take place outside our conscious awareness. In addition, it is important to note that to our knowledge these data represent the first clear demonstration of plasticity in adults' odor hedonics due to a Pavlovian stimulus contingency. Hence, even though it still may be the case that the fundamentals of odor hedonics are acquired during childhood, there is certainly room left for additional evaluative plasticity during adulthood. Moreover, there is no reason to assume that such Pavlovian associative processes should not play an equally important role in the acquisition of odor likes and dislikes in early childhood. Hence, it is our conviction that the kind of studies discussed here, founded on theoretical and empirical knowledge regarding evaluative conditioning, provide a firm and attractive basis for studying the acquisition of odor hedonics over all age groups. Not only can associative processes alter the liking or disliking of an odor, but also the results of our conceptual replication of the study by Kirk-Smith et al. (1983) suggest that these associative processes can even be the basis for unconscious affective/emotional influences of odors on our behavior. This observation is commensurate with recent more general studies of the automatic affective/emotional impact of stimuli employing experimental paradigms borrowed from experimental cognitive psychology, as modified to study automatic affective influences. The affective-priming procedure discussed here is a good example of such a paradigm. On the basis of studies that have employed this methodology, we can now conclude that humans are capable of automatically evaluating external stimuli as positive or negative. The data of our affective odor study (Hermans et al., 1998a) show that this postulate not only holds for visual stimuli but also can be generalized to olfactory stimuli. Odors are capable of automatically activating positive or negative connotations. All in all, the data presented here lead to the conclusion that acquisition of odor hedonics can take place outside conscious awareness, and once acquired, odor valence can exert an automatic (unconscious) influence on our behavior. And according to the data from the liquid-soap study, odor valence not only can directly influence our perception and behavior but also can have an indirect impact, as a clearly valenced odor can by itself act as an evaluative US and alter the valence of stimuli that go together with that odor. Moreover, these latter stimuli can in turn be evaluated automatically, as was indicated by the data from the affective-priming procedure. Relating this to the perfume example from our introduction, this can be translated as follows: Not only can an attractive or beloved person change one's perception of the perfume that person is wearing, but once it has come to be seen as pleasant, that perfume can subsequently elicit positive feelings for an unknown person wearing the same perfume.

References

Baeyens F, Crombez G, De Houwer J, & Eelen P (1996a). No Evidence for Modulation of Evaluative Flavor–Flavor Associations in Humans. *Learning and Motivation* 27:200–41.

Baeyens F, Crombez G, Hendrickx H, & Eelen P (1995a). Parameters of Human Evaluative Flavor–Flavor Conditioning. *Learning and Motivation* 26:141–60.

Baeyens F, Crombez G, Van den Bergh O, & Eelen P (1988). Once in Contact Always in Contact: Evaluative Conditioning Is Resistant to Extinction. *Advances in Behaviour Research and Therapy* 10:179–99.

Baeyens F, Eelen P, & Crombez G (1995b). Pavlovian Associations Are Forever: On Classical Conditioning and Extinction. *Journal of Psychophysiology* 9:127–41.

Baeyens F, Eelen P, Crombez G, & Van den Bergh O (1992). Human Evaluative Conditioning: Acquisition Trials, Presentation Schedule, Evaluative Style, and Contingency Awareness. *Behaviour Research and Therapy* 30:133–42.

Baeyens F, Eelen P, & Van den Bergh O (1990a). Contingency Awareness in Evaluative Conditioning: A Case for Unaware Affective-Evaluative Learning. *Cognition and Emotion* 4:3–18.

Baeyens F, Eelen P, Van den Bergh O, & Crombez G (1990b). Flavor–Flavor and Color–Flavor Conditioning in Humans. *Learning and Motivation* 21:434–55.

Baeyens F, Eelen P, Van den Bergh O, & Crombez G (1989). Acquired Affective-Evaluative Value: Conservative but Not Unchangeable. *Behaviour Research and Therapy* 27:279–87.

Baeyens F, Hendrickx H, Crombez G, & Hermans D (1998). Neither Extended Sequential nor Simultaneous Feature Positive Training Results in Modulation of Evaluative Flavor Conditioning in Humans. *Appetite* 31:185–204.

Baeyens F, Hermans D, & Eelen P (1993). The Role of CS-US Contingency in Human Evaluative Conditioning. *Behaviour Research and Therapy* 31:731–7.

Baeyens F, Wrzesniewski A, De Houwer J, & Eelen P (1996b). Toilet Rooms, Body Massages, and Smells: Two Field Studies on Human Evaluative Odor Conditioning. *Current Psychology: Developmental, Learning, Personality* 15:77–96.

Balogh R D & Porter R H (1988). Olfactory Preferences Resulting from Mere Exposure in Human Neonates. *Infant Behavior and Development* 9:395–401.

Bartoshuk L M (1994). Chemical Senses. *Annual Review of Psychology* 45:419–49.

Black S L & Smith D G (1994). Has Odor Conditioning Been Demonstrated? A Critique of "Unconscious Odour Conditioning in Human Subjects." *Biological Psychology* 37:265–7.

Dawson M E & Shell A M (1982). Electrodermal Responses to Attended and Nonattended Significant Stimuli during Dichotic Listening. *Journal of Experimental Psychology: Human Perception and Performance* 2:315–24.

De Houwer J, Baeyens F, & Eelen P (1994). Verbal Evaluative Conditioning with Undetected Stimuli. *Behaviour Research and Therapy* 32:629–33.

De Houwer J, Hermans D, & Eelen P (1998). Affective and Identity Priming with Episodically Associated Stimuli. *Cognition and Emotion* 12:145–69.

Doty R L (1991). Psychophysical Measurement of Odor Perception in Humans. In: *The Human Sense of Smell,* ed. D G Laing, R L Doty, & W Breipohl, pp. 95–134. Berlin: Springer-Verlag.

Doty R L, Applebaum S, Zusho H, & Settle R G (1985). Sex Differences in Odor Identification Ability: A Cross-Cultural Analysis. *Neuropsychologia* 23:667–72.

Ehrlichman H & Halpern J N (1988). Affect and Memory: Effects of Pleasant and Unpleasant Odors on Retrieval of Happy and Unhappy Memories. *Journal of Personality and Social Psychology* 55:769–79.

Engen T (1988). The Acquisition of Odour Hedonics. In: *Perfumery: The Psychology and Biology of Fragrance*, ed. S Van Toller & G H Dodd, pp. 79–90. London: Chapman & Hall.

Fazio R H, Sanbonmatsu D M, Powell M C, & Kardes F R (1986). On the Automatic Activation of Attitudes. *Journal of Personality and Social Psychology* 50: 229–38.

Giner-Sorolla R, Garcia M T, & Bargh J A (1994). The Automatic Evaluation of Pictures. Unpublished manuscript, New York University.

Hermans D, Baeyens F, & Eelen P (1998a). Odours as Affective Processing Context for Word Evaluation: A Case of Cross-Modal Affective Priming. *Cognition and Emotion* 12:601–13.

Hermans D, Baeyens F, & Natens E (2000). Evaluative Conditioning of Brand CS with Odor USs: A Demonstration Using the Affective Priming Technique. Unpublished internal research report, University of Leuven.

Hermans D, Crombez G, Vansteenwegen D, Baeyens F, & Eelen P (1999). Expectancy-Learning and Evaluative Learning in Human Classical Conditioning: Differential Effects of Extinction. Unpublished internal research report, University of Leuven.

Hermans D, De Houwer J, & Eelen P (1994). The Affective Priming Effect: Automatic Activation of Evaluative Information in Memory. *Cognition and Emotion* 8:515–33.

Hermans D, De Houwer J, & Eelen P (1996). Evaluative Decision Latencies Mediated by Induced Affective States. *Behaviour Research and Therapy* 34:483–8.

Hermans D, De Houwer J, & Eelen P (2001). A Time Course Analysis of the Affective Priming Effect. *Cognition and Emotion* 15:143–5.

Hermans D, Pauwels P, & Baeyens F (1992). Nonconscious Odour Conditioning. Unpublished internal research report, University of Leuven.

Hermans D, Van den Broeck A, & Eelen P (1998b). Affective Priming Using a Colour-Naming Task: A Test of an Affective-Motivational Account of Affective Priming Effects. *Zeitschrift für Experimentelle Psychologie* 45:136–48.

Hvastja L & Zanuttini L (1989). Odour Memory and Odour Hedonics in Children. *Perception* 18:391–6.

Kirk-Smith M D (1994). Comments on "Has Odour Conditioning Been Demonstrated?" *Biological Psychology* 37:269–70.

Kirk-Smith M, Van Toller C, & Dodd G (1983). Unconscious Odour Conditioning in Human Subjects. *Biological Psychology* 17:221–31.

Lang P J, Bradley M M, & Cuthbert B N (1990). Emotion, Attention, and the Startle Reflex. *Psychological Review* 97:377–95.

Levey A B & Martin I (1987). Evaluative Conditioning: A Case for Hedonic Transfer. In: *Theoretical Foundations of Behaviour Therapy,* ed. H J Eysenck & I Martin, pp. 113–31. New York: Plenum Press.

Miltner W, Matjak M, Braun C, Diekmann H, & Brody S (1994). Emotional Qualities of Odors and Their Influence on the Startle Reflex in Humans. *Psychophysiology* 31:107–10.

Osgood C, Suci G, & Tannenbaum P (1957). *The Measurement of Meaning.* Urbana: University of Illinois Press.

Richardson J T E & Zucco G M (1989). Cognition and Olfaction. *Psychological Bulletin* 105:352–60.

Rozin P (1982). "Taste–Smell Confusions" and the Duality of the Olfactory Sense. *Perception and Psychophysics* 31:397–401.

Schneider F, Weiss U, Kessler C, Mueller G, Posse S, Salloum J, Grodd W, Himmelman F, Gaebel W, & Birbaumer N (1999). Subcortical Correlates of Differential Classical Conditioning of Aversive Emotional Reactions in Social Phobia. *Biological Psychiatry* 45:863–71.

Todrank J, Byrnes D, Wrzesniewski A, & Rozin P (1995). Odors Can Change Preferences for People in Photographs: A Cross-Modal Evaluative Conditioning Study with Olfactory USs and Visual CSs. *Learning and Motivation* 26:116–40.

Van der Ploeg H M, Defares, P B, & Spielberger C D (1980). *Handleiding bij de Zelf-Beoordelings Vragenlijst ZBV.* Lisse: Swets & Zeilinger.

Van Toller S (1988). Emotion and the Brain. In: *Perfumery: The Psychology and Biology of Fragrance,* ed. S Van Toller & G H Dodd, pp. 121–43. London: Chapman & Hall.

Vroon P, Van Amerongen A, & De Vries H (1994). *Verborgen verleider. Psychologie van de reuk* [Hidden Seducer. Psychology of Smell]. Baarn: AMBO.

Zellner D A, Rozin P, Aron M, & Kulish C (1983). Conditioned Enhancement of Human's Liking for Flavor by Pairing with Sweetness. *Learning and Motivation* 14:338–50.

9

Is There a Hedonic Dimension to Odors?

Catherine Rouby and Moustafa Bensafi

The goal of psychophysics is to understand the relationships between variations in the physical environment and variations in our mental states. During the past century, psychophysics has evolved from pure sensory and physicalist conceptions to more perceptive and ecological ones, allowing us to include decision, attention, and expectation in the processing of sensory stimuli (Tiberghien, 1984). In the olfactory domain, memory psychophysics has been compared to perception psychophysics (Algom and Cain, 1991), and multidimensional scaling methods have been used to compare perception and imagery (Carrasco and Ridout, 1993). Thus, psychophysics has come to include more and more cognitive issues. Perception is recognized as categorical, inferential, and generic, leading to meaning and knowledge.

Hedonic responses to odors are very salient in folk psychology (David, Dubois, and Rouby, 2000) as well as in scientific accounts (see Rouby and Sicard, 1997, for a review). But can we study hedonic responses to odors with the usual psychophysical tools? Empirical descriptions and measures of preferences are clearly possible (Moncrieff, 1966; Köster, 1975), allowing a kind of affective psychophysics. But beyond preferences, if one is interested in the links between emotional responses and cognitive processing of odors, methodological questions arise concerning what is measured and how. Three main questions will be examined in this chapter:

1. Are the degree of pleasantness and the intensity of an odor reflecting the same dimension?
2. Is there a hedonic axis? Is a representation of affect possible on a psychophysical continuum, or are there clear-cut odor categories?
3. Are unpleasant odors symmetric with pleasant odors in relation to "zero affect"? That is, do they carry the same weight with respect to a neutral reference?

140

To discuss these theoretically important questions, we shall gather contributions from different sciences, namely, psychology, anthropology, psycholinguistics, and neurophysiology. Insights from these disciplines will show that pleasantness and intensity can be considered as separable aspects of odor perception, relying on different neural networks, that a hedonic continuum is only partially adequate for an understanding of the affects accompanying odors, and that the negative pole of the hedonic axis is more relevant for humans.

1. Pleasantness and Intensity: A Single Dimension?

Strong odors are perceived as being unpleasant: A negative correlation is often found between intensity and pleasantness, leading some authors to consider them as reflective of a single dimension (Henion, 1971). Doty (1975), in comparing the pleasantness and intensity curves with concentrations for 10 odorants, did not support that view: For 3 of them (anethole, eugenol, geraniol), no correlation was found. He also found that the range of variation was smaller for pleasantness than for intensity, pleasantness being not described by a power function. Moreover, even when a strong correlation was found, the relationship of pleasantness and intensity to one another, as well as their relationship to concentration, varied strongly across odorants. Royet et al. (1999), using a very large sample of 185 odorants, found no correlation between intensity and pleasantness, a result they attributed to the wide range of odors they studied, from very pleasant to very aversive, which was not the case in the Doty study. However, they found significant correlations between intensity and familiarity and between pleasantness and familiarity, thus supporting earlier observations (Engen and Ross, 1973) of a link between familiarity and preferences that is addressed in food sciences as neophobia (Rozin and Vollmecke, 1986). Neophobia appears to be a key to hedonic responses, because only familiar foods lead to a confident approach, with unfamiliar ones entailing avoidance or minimal, cautious consumption. Transcultural studies have shown that these familiarity/novelty effects extend to non-food odors and that familiarity seems to influence not only hedonic evaluation but also intensity psychophysical evaluation, suggesting that perceived intensity itself does not depend solely on stimulus concentration, but also on cultural and ecological influences. These top-down influences are reflected in different biases characterizing different groups: Japanese, Mexican, German (Ayabe-Kanamura et al., 1998; Distel et al., 1999).

Thus, one attributing an intensity level to an odorant could be relying not only on a "pure" sensory coding but also on recognition of the intensity of a familiar object. This interpretation is evident when the familiar object is a food: For example, de Graaf, van Staveren, and Burema (1996) measured perceived

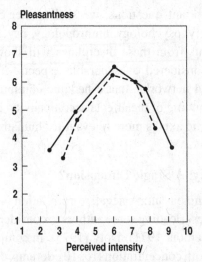

Figure 9.1. Mean responses on the pleasantness of bouillon as a function of perceived intensity, judged by elderly (dashed line) and young (solid line) subjects. The possible range in responses was from 1 (very weak) to 10 (very strong) for intensity, and from 1 (very unpleasant) to 10 (very pleasant) for pleasantness. (Adapted from de Graaf et al., 1996.)

intensity for foods like bouillon, tomato soup, chocolate custard, and orange lemonade, according to concentration (Figure 9.1). They found that different psychophysical curves characterized the variation of pleasantness with concentration. The psychohedonic function relates two psychological variables: perceived intensity and pleasantness. When perceived intensity rises, pleasantness goes up to a maximum, before going down. Our understanding of this phenomenon is that this optimum corresponds to a memory of what should be a good bouillon or chocolate custard, or what it was in the past: The apex of this curve will correspond to the intensity that best matches the memory the subject has of this particular food. The difference between the curves for older and yonger subjects in Figure 9.1 derives from the elevated thresholds of the elderly, but the optima are nevertheless very similar.

These diverse data show that the pleasantness/intensity relationship is far from simple, being linked to bottom-up sensory processes as well as to top-down mechanisms (Figure 9.2). This is in accordance with psychophysical models that take into account context effects and anchor effects in judgments (Parducci, 1995).

From a theoretical point of view, it is important to know if pleasantness can be dissociated from intensity and if these aspects of olfactory perception reflect different cognitive processes. But assessing the independence of these processes, from a practical point of view, has been difficult because of lack of a hedonic test. Thus we had to design one in which a set of odorants could be rated by subjects

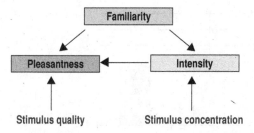

Figure 9.2. Schematic representation of the observed relationships among intensity, pleasantness, and familiarity. Arrows indicate top-down and bottom-up processes.

according to their intensity, on the one hand, and according to their pleasantness, on the other (Rouby, Jones-Gotman, and Zatorre, in press). Such a design was used to measure the response times for hedonic judgments.

According to Craik and Lockhart (1972), various sensory stimuli may be processed at different levels, from superficial processing (such as pitch or intensity in the case of audition) to deep semantic processing (such as identification of a sound, noise, word, or melody). The influence of the depth of processing on word-recognition performance has been shown by Craik and Tulving (1975). Response times are commonly used in cognitive psychology in order to evaluate the complexity of cognitive tasks: Short response times are most often seen when there is superficial encoding of stimuli, whereas semantic processing involves longer response times. Thus we used response times to determine the depth of processing for hedonic judgments, intensity judgments, and judgments of the dangerousness of odorants.

Our hypothesis was that response times would be shortest for the intensity judgments, and longest for the dangerousness judgments, the latter requiring semantic processing before a decision would be possible. If the hedonic judgments were dependent only on stimulus properties, we would expect that they would not differ from intensity judgments in terms of response times; alternatively, if the hedonic judgments required deeper processing, we would expect longer response times (Burnet, Dubois, and Rouby, 1997).

To measure response times, we used a device combining a respiratory sensor placed in one nostril, a timer, and a response box with two keys. The respiratory sensor was inserted in the left nostril. Pairs of odorant vials were presented to the right nostril. Each subject was instructed to sniff at each presentation of an odorant vial (only one sniff was allowed) and, after sniffing the second odor of the pair, to respond as quickly as possible by pressing one key on the response box; for the second odor, the sniff was detected by the sensor, which started the timer; the timer was stopped when the subject pressed one of the two

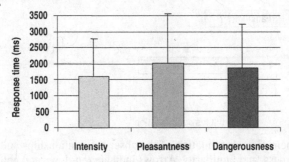

Figure 9.3. Mean response times for 31 subjects making pair comparisons during three judgments: intensity, pleasantness, and dangerousness.

keys to respond. Six odors [thymol, pyridine, isovaleric acid, isoamyl acetate, cyclodecanone, R-(+)-limonene] were presented by pairs (15 pairings, each pair presented twice). The order of the pairs was random and different for each task (intensity, pleasantness, dangerousness), as was the order of tasks across subjects.

After some training trials to synchronize the subject's sniff and the odor presentation by the experimenter, the test began and responses were recorded after presentation of the second odor of each pair: The subject had to press the right button if the second odor was less intense, or the left button if the second odor was more intense than the first one. The same procedure was used for pleasantness judgments and for dangerousness judgments, with subjects having to decide if the second odor was more or less pleasant (or dangerous) than the first one.

The results are shown in Figure 9.3. Dangerousness and pleasantness comparisons did not differ in response times, but both took significantly longer than the intensity judgments. The difference between each of them and the intensity judgments revealed a 345-msec mean effect ($F[1, 29] = 15$, $p < .001$). This is a large effect, if we consider that the apparatus allowed a temporal resolution of 10–20 msec for sniff detection, 0.1 msec for the timer, and 0.1 msec for the response box. Of course, the natural sniff action of the subject was the main source of measurement error: It is possible that some subjects may have detected an odor before it was 1 cm from the nostril, and thus some response times may have been shortened in an undetermined way; the large standard deviations in response times (intensity, 1,185 msec; dangerousness, 1,340 msec; pleasantness, 1,550 msec) may in part reflect such variation in stimulus input. However, that was not the sole explanation, because in sets of odorants that were judged for intensity, pleasantness, and dangerousness, only for the intensity judgments were both the mean and standard deviation smaller. Therefore, it is unlikely that variations in the onset of sniffing and in the presentation of the odorant under the subject's nostril could explain the robust difference of 345 msec between judgments.

These findings indicate that hedonic judgment is elaborated beyond the perceptive level and that perceived intensity does not require a deep level of processing. Although this does not mean that hedonic and intensity judgments cannot be correlated in psychophysical tasks, it lends support to the idea that they rely on different processes. Such processes could differ in two ways: Either they could be completely dissociated, or they could rely on the same early neural process, which might be sufficient for the intensity judgment, but not for the pleasantness and dangerousness judgments. Such an issue is too difficult to resolve using only response times.

Other support for the idea of dissociation between intensity and pleasantness can be found in neuropsychological observations: The well-known patient H.M., whose surgery produced bilateral medial temporal damage, was unable to identify odors or to describe them, but was completely normal in terms of odor-detection and odor-intensity tasks (Eichenbaum et al., 1983). To our knowledge, no example of the reverse pattern has been reported, that is, subjects discriminating and describing odors, but impaired in terms of intensity processing. Unfortunately, H. M. was not tested for his hedonic responses to odors. Nevertheless, we have tried to document in normal subjects a possible dissociation between intensity and pleasantness. In collaboration with neuropsychologists, we have sought to further test the specificity of hedonic judgment using brain-imaging techniques, namely, positron-emission tomography (PET), with the same kind of intensity/pleasantness test, in order to try to determine if intensity and pleasantness rely on different neural networks.

The tests showed right orbitofrontal activation that was common to both tasks, but robust activation of the hypothalamic area that was specifically attributable to the situation of judging pleasantness/unpleasantness (Zatorre, Jones-Gotman, and Rouby, 2000; Zatorre, Chapter 20, this volume). Thus, in the absence of evidence of double dissociation between pleasantness and intensity, which would argue in favor of completely distinct neural networks, the response times and the PET study tend to indicate that intensity and pleasantness ratings share common processes and activate common cortical structures, but nevertheless rely on different neural networks and require different computing durations.

2. Is There a Hedonic Axis? Or Discontinuous Hedonic Categories?

The second question concerns the ecological validity of the idea of a continuous hedonic axis to study the affective processing of odors. To tackle this problem, we shall consider different attempts that have been made to understand and describe the psychological odor space, as well as the linguistic odor space, across cultures.

A first problem transcends the odor space: the existence of an emotional continuum. Theorists of emotion do not agree on this point. To present things briefly, there are two main positions: One group argues that there is a single emotional dimension, characterizing and underlying approach/withdrawal behaviors (Davidson and Irwin, 1999; Rolls, 1999). The other group believes in a limited number of primary, basic emotions (sadness, happiness, fear, surprise, anger, disgust) with distinctive facial expressions that are associated with distinctive physiological events (Ekman, Levenson, and Friesen, 1983): Basic emotions rely on innate mechanisms, and during ontogeny they give rise to secondary emotions, evoked by a variety of stimuli, situations, and meanings, that shape individual and cultural differences by association with the primary, basic emotions (Damasio, 1994). For the historical grounds for those positions, see Phillips and Heining (Chapter 12, this volume).

Thus, in the olfactory domain, the idea of continuous hedonic axis presupposes that odors can be ranked at least on an ordinal scale, at best on an interval scale. If the assumption of a continuum does not hold, and if different odors evoke different psychological and physiological events, then their placement on a continuous scale will mask its purely nominal character and will not help us to understand the psychological and physiological mechanisms underlying subjects' responses.

The status of the hedonic "dimension" can be evaluated by reviewing the scientific attempts to organize the olfactory space. All psychophysical odor classifications and odor-similarity descriptions in the twentieth century have attributed great weight to the dimension "pleasantness" (Harper et al., 1966; Schiffman, 1974; Schiffman, Robinson, and Erickson, 1977). Berglund et al. (1973) considered that dimension to be the only one that was common among the olfactory spaces of individuals. But as we do not know much about the organization of this space (or spaces), the number of dimensions can only be approximated, and their physiological meanings are still undetermined (Chastrette, Chapter 7, this volume).

Scientific attempts at odor classification in the eighteenth century (Table 9.1), when the language of science was Latin, tried to impose some order with crude tools: categories. Albrecht von Haller (1763) erected three categories: good (literally, "sweet-smelling"), neutral, and bad (stinking/infectious odors). The well-known Linnaeus (1765) believed in seven groups, three of which referred to unpleasant odors: *odores fragrantes* (fragrant odors), *odores tetri* (repelling odors), and *odores nausei* (nauseating odors). A fourth category, *odores hircini* (male-goat smell), also would seem to include a hedonic characteristic, although its description did not explicitly refer to affect or evaluation; however, that odor is a prototypical malodor in the Mediterranean world. More recently, Zwaardemaker (1925) proposed a classification that kept and refined most of

Table 9.1. *Examples of European scientific and popular odor classifications*

Date	Authors		Odor classifications by scientists and laypeople
1763	von Haller		1. *odores suaveolentes* (musk, camphor, mint, apple, violet, rose) 2. *odores medii* (wine, vinegar, empyreumatic odors) 3. *odores foetores* (animal perspiration, corpses, corrupted organic compounds)
1765	Linnaeus		1. *odores aromatici* (laurel) 2. *odores fragrantes* (lime or jasmine blossom) 3. *odores ambrosiaci* (amber, musk) 4. *odores alliacei* (garlic) 5. *odores hircini* (male goat) 6. *odores tetri* (solanaceous plants) 7. *odores nausei* (foul-smelling plants)
1925	Zwaardemaker		1. ethereal odors 2. aromatic odors 3. floral and balsamic odors 4. ambrosiac odors 5. alliaceous odors 6. empyreumatic odors 7. caprylic odors (capronic acids and homologs) 8. repelling odors (pyridine, alkaloids) 9. nauseating odors (indole, scatole)
1997	David et al.	Descriptions referring to	% of responses
		effect	40.75
		source	27.75
		effect/source	13.5
		effect/memory	5
		effect/intensity	3.75
		effect/intensity/memory	2.25
		others	7

the seven Linnaean categories, including the same three malodorous groups (Table 9.1: group 7, caprylic odors; group 8, repelling odors; group 9, nauseating odors). Thus, all scientific attempts at odor classification have paid considerable attention to the hedonic and affective aspects of odor perception.

Up to this point, several terms have been used somewhat indifferently: "pleasantness," "hedonicity," "preference," "affect," "hedonic judgment," and "emotional value." From a psychological and linguistic point of view, it is interesting to consider to what extent these different expressions refer to the same reality. Can we speak of hedonic valence or hedonic connotation as describing some attribute of a stimulus – nothing more than a side effect of the perception of an odor? Can we speak of emotional reactions triggered by some odorants? Can we speak of emotional judgments attributable to an external stimulus?

Linguistic inquiries have led to a more thorough examination of the hedonic dimension (David et al., 1997). We asked 250 subjects to write their own answers to open questions such as "What is an odor for you?" and obtained a wide diversity of responses. But the linguistic analysis showed that the words used to speak about odors were constructed mainly on verbs referring to an effect of an odor on the perceiver (disgust*ing*, nause*ating*). Moreover, descriptions referring to the effect were far more frequent than descriptions referring to the source: Spontaneous expressions tend to replicate scientific classifications using essentially the same words (Table 9.1).

Thus, what linguistics tells us about odor perception is that it is subjective, which we already knew, but also that this hedonic valence is not referred to as external; it is a relationship between the odor and one's body. What laypeople put into words in a spontaneous, nonscientific way is not a description of objective, simply quantitative external "dimensions," symbolized by positions on a continuous line, but rather descriptions of effects on their bodies or their well-being. We must remember that when we use psychophysical hedonic "scales," the scientific classifications also take into account the hedonic valence, in almost the same words as laypeople (nause*ating*, repell*ing*), mainly on the negative side of the ledger. Thus the supposed continuum between good and bad could rather imply, in the case of olfaction, that there are two types of odors, **those that have qualities that can be processed semantically and described (mainly by their sources), and those that have effects:** For some aspects of the olfactory space that appear to be most important for human beings, an auto-centered reference is used (the effect); for some other aspects, an allo-centered reference is possible (a semantic processing referring to the source). This duality of representation has also been deployed in the cognitive representations of space (Berthoz and Viaud-Delmon, 1999).

3. Are Pleasant Odors Symmetric with Unpleasant Ones?

In an inquiry into odor descriptions in 60 languages from nine language families, Boisson (1997) confirmed that the hedonic dimension is dominant everywhere, an observation reported earlier by Buck (1949) for Indo-European languages:

Table 9.2. *Summary of transcultural studies of odor denomination and hedonic categorization*

Date	Authors	Odor classification or comments
1997	Boisson	Among 60 languages:
		35 have specific terms for sweat and body odor
		34 have specific terms for strong animal odors
		31 have specific terms for rotten-things odor
		31 have specific terms for burnt-things odor
		26 have specific terms for odor of molded things or places
		23 have specific terms for odor of fish, rotten (15) or fresh (8)
		13 have specific terms for urine odor
		12 have specific terms for fresh-meat odor
		11 have specific terms for rancid-food odor
		8 have specific terms for dampness odor
		6 have specific terms for breath malodor
		4 have specific terms for feces odor
1998	Schaal et al.	There is better agreement between and within cultures on the negative side of the hedonic dimension. Far more variability is found on the positive and neutral sides.
1997	Mouélé et al.	The negative pole is the one that is linguistically dominant.

"The only widespread popular distinction is that of pleasant and unpleasant smells – good and bad smells, to use the briefest terms – and this is linguistically more important than any similar distinction, that is, of good and bad, in the case of other senses." Thus the ecological value of the hedonic dimension has been fully confirmed using a very large sample from different language families. Some malodors are the most frequently named odors across languages, probably because they are more salient for perceptual reasons and/or for cultural reasons (Table 9.2). Boisson distinguishes four main groups: body odors, decaying organic substrates, animal foods, and live animals.

A comparative olfactory ethnography remains to be done, but this survey shows that negative terms are always more numerous in a given language than are positive terms: Even in Hawaii, where there are many different terms for pleasant odors, they are outnumbered by words for unpleasant odors.

Although many different odorant sources are employed in various cultures to represent the positive and negative poles of the hedonic dimension, the portion of the ecological environment that is common across cultures may have room for at least some olfactory universals: body odors or any of the four main groups described by Boisson (1997).

A review of studies on preferences by children and adults from different cultures (Schaal et al., 1998) revealed a strong consensus between and within cultures on the negative side of the hedonic dimension; that agreement was little influenced by age and sex variables. Far more variability was found on the positive and neutral sides, with strong influences from age and sex variables. Thus, malodors and human body odors may be olfactory universals in the sense that they are common to all cultures and salient for most of them. Moreover, perceived pleasantness and unpleasantness seem to be rather independent of each other during early development, and negative facial responses appear to be acquired earlier than positive responses in newborns (Schaal, Soussignan, and Marlier, Chapter 26, this volume).

Some ethnolinguistic studies have confirmed that view: In a short inquiry into five Bantu languages in Gabon, Hombert (1992) found that they had from 6 to 14 specific odor terms, each referring not to a source, as in Indo-European languages, but to the odor of that source. Those languages have names for odors for which the Indo-European lexicon is empty. Those names refer mainly to body odors or to strong odors: In Li Wanzi and in four other languages, the odor of the civet has its specific name. Comparisons with other populations of hunter-gatherers have yielded similar findings. Moreover, ethnolinguistic studies have confirmed the differential attention given to malodors: **The negative pole is the one that is linguistically marked** (Mouélé et al., 1997; Mouélé, 1997).

Our detour through linguistics and psychology led us to examine further the negative side of the supposed hedonic continuum and to ask whether or not unpleasant odors are symmetrical with pleasant odors. Linguistic studies indicate that such is not the case. What does neuroscience suggest? Psychophysiology and neurophysiology do not trust only in words; they consider behavior and its neural correlates. Behavioral responses to odors have been studied through facial expressions (Steiner, 1979; Gilbert, Fridlund, and Sabini, 1987; Soussignan et al., 1997; Schaal et al., Chapter 26, this volume) and psychophysical measures of preferences with different methodologies (Herz, Chapter 10, Hermans and Baeyens, Chapter 8, this volume). Here we shall focus only on the recording of physiological responses, using two main approaches: One relies on recordings of response times or responses of the autonomic nervous system (startle reflex, changes in skin conductance, heart rate). The other approach uses cerebral imaging: evoked potentials, functional magnetic-resonance imaging (fMRI), and positron-emission tomography (PET). Their main results are summarized in Table 9.3.

One of the first reports of differential treatment of unpleasant versus pleasant odors in the brain was that of Kobal, Hummel, and Van Toller (1992), who showed differences in the latency and amplitude of the P_2 wave, with a positive potential appearing around 500 msec after odorant stimulation: Latencies were shorter and amplitudes smaller after stimulation of the left nostril by an unpleasant odor

Table 9.3. *Main results of psychophysiological studies contrasting pleasant and unpleasant odors*

Date	Authors	Parameters studied	Pleasant odors	Unpleasant Odors
1997b	Alaoui-Ismaïli et al.	Skin conductance	Shorter duration	Longer duration
1997a	Alaoui-Ismaïli et al.	Skin ohmic perturbation	Shorter duration	Longer duration
1995	Brauchli et al.	Skin conductance	No difference with odorless air	Increase in conductance
1997b	Alaoui-Ismaïli et al.	Heart rate	Decrease	Increase
1997a	Alaoui-Ismaïli et al.	Heart rate	Decrease	Increase
1995	Brauchli et al.	Heart rate	Decrease	Increase
1997	Ehrlichman et al.	Heart rate	No effect	Increase
1994	Miltner et al.	Startle reflex	No effect	Increase
1995	Ehrlichman et al.	Startle reflex	No effect	Increase
1997	Ehrlichman et al.	Startle reflex (between-subjects design)	Decrease in amplitude	Increase in amplitude
1992	Kobal et al.	Odor-evoked potentials (P_2 wave, ~500 msec)	Short latencies and smaller amplitude after stimulation of the right nostril	Short latencies and smaller amplitude after stimulation of the left nostril
1997	Zald & Pardo	Cerebral blood flow	Amygdala: no difference with a no-odor condition. Orbito-frontal cortex (left and right): more activated than in a no-odor condition	Amygdala (left and right): more activated than in a no-odor condition. Left orbito-frontal cortex: more activated than in a no-odor condition
1998	Fulbright et al.	fMRI	Brodmann area 32, (left) BA 46/9 BA 6 (right). Insula BA 8	Brodmann area 32, BA 46/9 BA 6. Insula
2000	Bensafi et al.	Response times	Around 1,300 msec to perform a pleasantness judgment	Around 1,000 msec to perform an unpleasantness judgment

(hydrogen disulfide), whereas for a pleasant odor (vanillin) shorter latencies and smaller amplitudes were obtained after stimulation of the right nostril. Those authors interpreted their data as showing a predominance of the left hemisphere for positive emotions, and of the right hemisphere for negative ones. Zald and Pardo (1997) also found hemispheric differences in neuronal activation, with unpleasant odors activating the left orbito-frontal cortex (OFC) and the amygdala on both sides, whereas less unpleasant odors activated the OFC bilaterally and caused no difference in amygdala activation compared with a no-odor condition. The fMRI study of Fulbright et al. (1998) also revealed hemispheric differences: The left Brodmann area 8 was activated specifically by pleasant odors.

Autonomic activity evoked by olfactory stimuli presents two patterns: either opposite effects for pleasant and unpleasant odors, which is the case for skin conductance and heart rate (Alaoui-Ismaïli et al., 1997a,b; Brauchli et al., 1995; Ehrlichman et al., 1997), or effects due to unpleasant odors only, which is the case for the startle reflex (Miltner et al., 1994; Ehrlichman et al., 1995). These latter results favor the idea of automatic emotional arousal by malodors.

Recent psychophysical studies have shown that the right nostril dominates in discrimination of unfamiliar odors, but not of familiar odors (Savic and Gulyas, 2000), and familiar, moderately pleasant odors are rated as more pleasant when sniffed by the right nostril (Herz, McCall, and Cahill, 1999). Thus far, it is not possible to describe clear hemispheric patterns of activation for pleasant and unpleasant odors, but evidence seems to be accumulating to suggest differential involvements of brain hemispheres according to the pleasantness/unpleasantness of odors. Royet et al. (2000) have shown that such hemispheric differences are not limited to olfaction, but extend to reactions to emotional stimuli in the visual and auditory modalities. Regarding our questioning of the idea of a continuous hedonic dimension, it is important to emphasize that Zald and Pardo (1997) considered the odors they presented as "aversive" rather than "unpleasant": They may have elicited fear, anger, or disgust. Some evidence from the study of Alaoui-Ismaïli et al. (1997a) shows that unpleasant odorants (methyl methacrylate and propionic acid) activate primarily the autonomic responses of disgust and anger. As recent neuropsychological studies have shown differential brain activations for disgust (Phillips and Heining, Chapter 12, this volume) and have lent support to theories of distinct basic emotions, it seems likely that the situation is as follows:

1. Unpleasant odors do not form a continuum with pleasant odors.
2. Unpleasant odors do not form a homogeneous group with respect to the emotional experience.

Fear has been studied extensively by LeDoux and collaborators (Armony et al., 1997) with animal models, in the form of fear conditioning in the auditory modality. Their studies have revealed the critical role of the amygdala in fear conditioning, and they converge with human studies implicating the amygdala in emotional processing, especially of negative emotions (Aggleton, 1992). Fear conditioning with odors has also been studied (Otto, Cousens, and Herzog, 2000), and a crucial role for the amygdala in odor-mediated fear conditioning has been confirmed in animals. Although amygdala lesions in humans do not produce the same drastic emotional disturbances observed in animals, the amygdala's role in emotional processing is clear (Halgren, 1992): Aversions and other negative emotions like fear imply amygdala activation (Morris et al., 1996).

A study by Hermann et al. (2000) used odors to study aversive and appetitive conditioning in humans, that is, building conditioned approach / withdrawal responses based on unconditioned odor stimuli in the laboratory. In that electroencephalographic (EEG) study, autonomic correlates of the startle reflex and emotional activation were also recorded. Neutral images of faces were paired with either an aversive unconditioned stimulus (fermented yeast) or a pleasant unconditioned stimulus (vanilla); the experiment produced differential aversive conditioning with yeast, but failed to achieve any appetitive conditioning with the vanilla odorant. EEG and autonomic recordings showed no effect of the aversive conditioning on EEG tracings nor on the startle reflex, but there were effects on skin conductance and muscular nonvoluntary responses. The authors interpreted their data as indicating subcortical processing of aversive responses induced via olfaction.

Although one may be surprised by the certitude of the latter authors concerning the idea of olfactory unconditioned stimuli driving appetitive as well as aversive behaviors in humans (Chapters 8, 10, and 26, this volume), the important point for our purpose is that a negative affect (aversion) could be manipulated with odors, but not a positive affect. That is in agreement with the work of Zald and Pardo (1997), who found activation of the amygdala only for aversive odors.

Thus, several studies using different techniques to record activities in the central nervous system and autonomic nervous system argue for asymmetry between responses to pleasant odors and responses to unpleasant odors. For that reason, we decided to use response times to check for possible differences between hedonic judgment and other judgments, as well as differences between the processing of pleasant and unpleasant odors (Bensafi et al., 2001). The hypotheses were as follows:

1. Hedonic judgment of odors differs in kind from detection, intensity, and familiarity judgments.
2. The processing of unpleasant odors differs from the processing of neutral or pleasant odors.

Twelve odorants were chosen after perusal of a larger sample of 20 studied by Godinot (1999). There were 4 pleasant, 4 neutral, and 4 unpleasant odorants, according to the mean hedonic judgments of 40 subjects. Sixty-four subjects performed four different tasks: detection and intensity, pleasantness, and familiarity judgments. In contrast with our first response-time study, the judgments were not comparative, but were absolute judgments: Subjects were instructed to answer yes or no as quickly as possible by pressing one of two keys to indicate if an odor was detected (yes/no) and if it was intense (yes/no), pleasant (yes/no), or familiar (yes/no). Task order was randomized.

A three-way analysis of variance showed a significant effect of "task" ($p < .05$), and the interaction between "task" and the hedonic class of the odor was significant ($p < .05$). Comparison of means showed no response-time differences for the detection and intensity tasks, which is in accordance with the hypothesis that these judgments rely on a perceptive process. But detection and intensity responses were faster than hedonic and familiarity responses ($p < .05$), and familiarity judgments required more time than hedonic judgments ($p < .05$). But more interesting, pair comparisons showed that response times to the unpleasant odors were significantly shorter than response times to either the neutral or pleasant odors: The swiftness of responses to unpleasant odors accounted for the differences in pleasantness and familiarity judgments (Figure 9.4). Hedonic response times to unpleasant odorants did not differ from response times for intensity judgments.

That speed of processing seen only with the unpleasant odors could be interpreted as further evidence of differential processing of malodors. However, different response times are not sufficient to allow us to conclude that there are separate neural substrates. As shown by Royet et al. (1999), some judgments that require the same response times, such as familiarity and comestibility, depend on different neural networks. Our hypothesis is that the faster hedonic responses to unpleasant odorants may be explainable by parallel routes of processing: a

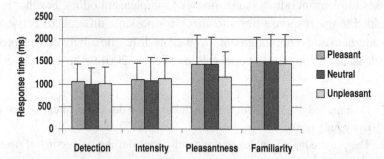

Figure 9.4. Mean response times for 64 subjects in four tasks (detection, intensity, pleasantness, and familiarity) and according to odor hedonic valence (pleasant, neutral, and unpleasant). The interaction between odor pleasantness and task was significant ($F = 2.260$; $p < .04$). Post-hoc comparisons showed that response times for unpleasant odors were significantly different from those for neutral and pleasant odors.

"quick and dirty" route for perception and reaction to potentially harmful odorant stimuli, and a slower, cognitively more complex route for neutral and pleasant stimuli. Brain imaging and autonomic recordings will be helpful in testing this hypothesis.

Concerning the hedonic dimension of odors, it is not entirely superposable onto the hedonic judgments we considered: Emotional responses and reactions imply that there are also autonomic responses, attentional shifts, and unconscious processes. It is important that the differences we found in hedonic and intensity judgments were not odor-specific, but task-specific: Hedonic judgment, which has been shown to activate the hypothalamic area (Zatorre et al., 2000), may be primed by unconscious processes.

In a more general vein, we have gathered evidence that the hedonic dimension is not to be found in an external and objective frame of reference, but rather in a complex of different categories of meaningful odors. These meanings are mainly ego-centered. This is in accordance with two lines of research: One states that a hedonic ecology cannot consider objects and goals of desire as independent of the perceiver (Rosch, 1996), nor decisions as independent of the hedonic context (Tversky and Kahneman, 1981). The other demonstrates that categorization itself is influenced by emotional states, and ultimately it considers that any categorization implies hedonics (Isen and Daubman, 1984; Niedenthal, Halberstadt, and Innes-Ker, 1999).

Finally, evidence of asymmetry in the processing of unpleasant and pleasant odors, both in language and in hedonic tasks, is in keeping with a research line called affective neuroscience, which works to elucidate different circuits underlying positive and negative emotions in the human brain (Davidson and Irwin, 1999). Thus, the complexity of olfactory hedonics certainly poses a challenge for psychophysics, but it also offers an unparalleled opportunity to improve our understanding of the relationship between cognition and emotion.

Acknowledgments

This research was funded by the French National Program in Cognition Sciences (*Catégorisation et invariance dans la perception des odeurs*) and by G.I.S. Sciences de la cognition (*Relations entre affectivité et cognition dans la perception des odeurs*). This chapter has benefited from discussions with G. Sicard, B. Schaal, S. David, D. Dubois, M. Mouélé, N. Godinot, R. Burnet, R. Gervais, R. Versace, V. Farget, M. Vigouroux, and, last but not least, M. Jones-Gotman and R. Zatorre.

References

Aggleton J P (1992). *The Amygdala. Neurobiological Aspects of Emotion, Memory and Mental Dysfunction*. New York: Wiley-Liss.

Alaoui-Ismaïli O, Robin O, Rada H, Dittmar A, & Vernet-Maury E (1997a). Basic Emotions Evoked by Odorants: Comparison between Autonomic Reponses and Self-evaluation. *Physiology and Behavior* 62:713–20.

Alaoui-Ismaïli O, Vernet-Maury E, Dittmar A, Delhomme G, & Chanel J (1997b). Odor Hedonics: Connection with Emotional Response Estimated by Autonomic Parameters. *Chemical Senses* 22:237–48.

Algom D & Cain W S (1991). Remembered Odors and Mental Mixtures: Tapping Reservoirs of Olfactory Knowledge. *Journal of Experimental. Psychology: Human Perception and Performance* 17:1104–19.

Armony J L, Servan-Schreiber D, Cohen J D, & LeDoux J (1997). Computational Modeling of Emotion: Explorations through the Anatomy and Physiology of Fear Conditioning. *Trends in Cognitive Sciences* 1:28–34.

Ayabe-Kanamura S, Schicker I, Laska M, Hudson R, Distel H, Kobayakawa T, & Saito S (1998). Differences in Perception of Everyday Odors: A Japanese-German Cross-Cultural Study. *Chemical Senses* 23:31–8.

Bensafi M, Rouby C, Farget V, Vigouroux M, & Holley A (2001). Are Pleasant and Unpleasant Odors Processed in the Same Way? *Chemical Senses* 26:786.

Berglund B, Berglund U, Engen T, & Ekman G (1973). Multidimensional Analysis of Twenty-one Odors. *Scandinavian Journal of Psychology* 14:131–7.

Berthoz A & Viaud-Delmon I (1999). Multisensory Integration in Spatial Orientation. *Current Opinion in Neurobiology* 9:708–12.

Boisson C (1997). La dénomination des odeurs: variations et régularités linguistiques. In: *Olfaction: Du Linguistique au Neurone*, ed. D Dubois & A Holley. *Intellectica* 1(24):29–49.

Brauchli P, Rüegg P B, Etzweiler F, & Zeier H (1995). Electrocortical and Autonomic Alteration by Administration of a Pleasant and an Unpleasant Odor. *Chemical Senses* 20:505–15.

Buck C D (1949). *A Dictionary of Selected Synonyms in the Principal Indo-European Languages, A Contribution to the History of Ideas*. University of Chicago Press.

Burnet R, Dubois D, & Rouby C (1997). Pleasantness of Odors: Perceptual or/and Semantic Processing? In: *Proceedings of the 19th Annual Conference of the Cognitive Science Society*, ed. M G Shafto & P Langley, p. 878. Hillsdale, NJ: Lawrence Erlbaum.

Carrasco M & Ridout J B (1993). Olfactory Perception and Olfactory Imagery: A Multidimensional Analysis. *Journal of Experimental Psychology: Human Perception and Performance* 19:287–301.

Craik F I M & Lockhart R S (1972). Levels of Processing: A Framework of Memory Research. *Journal of Verbal Learning and Verbal Behavior* 11:671–84.

Craik F I M & Tulving E (1975). Depth of Processing and the Retention of Words in Episodic Memory. *Journal of Experimental Psychology: General* 104:268–94.

Damasio A R (1994). *Descarte's Error: Emotion, Reason and Human Brain*. New York: Grosset/Putnam.

David S, Dubois D, & Rouby C (2000). Lexical Devices and the Construction of Objects: A Comparison between Sensory Modes. In: *LACUS Forum XXVI, The Lexicon*, ed. Linguistic Association of Canada and United States, pp. 225–35. Fullerton, CA: LACUS.

David S, Dubois D, Rouby C, & Schaal B (1997). L'expression des odeurs en français: analyse lexicale et représentation cognitive. In: *Olfaction: Du Linguistique au Neurone*, ed. D Dubois & A Holley. *Intellectica* 1(24):51–83.

Davidson R J & Irwin W (1999). The Functional Neuroanatomy of Emotion and Affective Style. *Trends in Cognitive Science* 3:11–21.

de Graaf C, van Staveren W, & Burema J (1996). Psychophysical and Psychohedonic Functions of Four Common Flavors in Elderly Subjects. *Chemical Senses* 21:293–302.

Distel H, Ayabe-Kanamura S, Martinez-Gomez M, Schicker I, Kobayakawa T, Saito S, & Hudson R (1999). Perception of Everyday Odors – Correlation between Intensity, Familiarity, and Strength of Hedonic Judgment. *Chemical Senses* 24:191–9.

Doty R L (1975). An Examination of the Relationships between the Pleasantness, Intensity, and Concentration of 10 Odorous Stimuli. *Perception and Psychophyics* 17:492–6.

Ehrlichman H, Brown S, Zhu J, & Warrenburg S (1995). Startle Reflex Modulation during Exposure to Pleasant and Unpleasant Odors. *Psychophysiology* 32:150–4.

Ehrlichman H, Brown-Kuhl S, Zhu J, & Warrenburg S (1997). Startle Reflex Modulation during Exposure to Pleasant and Unpleasant Odors in a Between-Subjects Design. *Psychophysiology* 34:726–9.

Eichenbaum H, Morton T H, Potter H, & Corkin S (1983). Selective Olfactory Deficits in Case H. M. *Brain* 106:459–72.

Ekman P, Levenson R W, & Friesen W V (1983). Autonomic Nervous System Activity Distinguishes among Emotions. *Science* 221:1208–10.

Engen T & Ross B M (1973). Long-Term Memory of Odors with and without Verbal Descriptions. *Journal of Experimental Psychology* 99:222–5.

Fulbright R K, Skudlarski P, Lacadie C M, Warrenburg S, Bowers A A, Gore J C, & Wexler B E (1998). Functional MR Imaging of Regional Brain Responses to Pleasant and Unpleasant Odors. *American Journal of Neuroradiology* 19:1721–6.

Gilbert A N, Fridlund A J & Sabini J (1987). Hedonic and Social Determinants of Facial Displays to Odors. *Chemical Senses* 12:355–63.

Godinot N (1999). Contribution à l'étude de la perception olfactive: qualité des odeurs et mélanges de composés odorants. Doctoral dissertation, Université Claude-Bernard Lyon 1.

Halgren E (1992). Emotional Neurophysiology of the Amygdala within the Context of Human Cognition. In: *The Amygdala. Neurobiological Aspects of Emotion, Memory and Mental Dysfunction*, ed. J P Aggleton, pp. 191–228. New York: Wiley-Liss.

Harper R, Land D G, Griffiths N M, & Bate-Smith E C (1966). Odour Qualities: A Glossary of Usage. *British Journal of Psychology* 59:231–52.

Henion K E (1971). Odor Pleasantness and Intensity: A Single Dimension? *Journal of Experimental Psychology* 90:275–9.

Hermann C, Ziegler S, Birnbaumer N, & Flor H (2000). Pavlovian Aversive and Appetitive Odor Conditioning in Humans: Subjective, Peripheral and Electrocortical Changes. *Experimental Brain Research* 132:203–15.

Herz R S, McCall C, & Cahill L (1999). Hemispheric Lateralization in the Processing of Odor Pleasantness versus Odor Names. *Chemical Senses* 24:691–5.

Hombert J M (1992). Terminologie des odeurs dans quelques langues du Gabon. *Pholia* 7:61–5.

Isen A M & Daubman K A (1984). The Influence of Affect on Categorization. *Journal of Personality and Social Psychology* 47:1206–17.

Kobal G, Hummel T, & Van Toller S (1992). Differences in Human Chemosensory Evoked Potentials and Somatosensory Chemical Stimuli Presented to Left and Right Nostrils. *Chemical Senses* 17:233–44.

Köster E P (1975). Human Psychophysics in Olfaction. In: *Methods in Olfactory Research*, ed. D G Moulton, A Turk, & J W Johnston, Jr, pp. 345–74. New York: Academic Press.

Linnaeus C (1765). *Odores medicamentorum. Amoenitates Academicae*, vol. 3, pp. 183–201. Stockholm: LarsSalvius.

Miltner W, Matjak M, Braun C, Diekmann H, & Brody S (1994). Emotional Qualities of Odors and Their Influence on the Startle Reflex in Humans. *Psychophysiology* 31:107–10.

Moncrieff R W (1966). *Odour Preferences*. New York: Wiley.

Morris J S, Frith C D, Perrett D I, Rowland D, Young A W, Calder A J, & Dolan R J (1996). A Differential Neural Response in the Human Amygdala to Fearful and Happy Facial Expressions. *Nature* 383:812–15.

Mouélé M (1997). L'apprentissage des odeurs chez les Waanzi: note de recherche. In: *L'odorat chez l'Enfant: Perspectives Croisées*, ed. B Schaal, *Enfance* 1:209–22.

Mouélé M, Hombert J M, Dubois D, Schaal B, Rouby C, & Sicard G (1997). Specific Odor Terminology: The Case of Li Wanzi (a Bantu Language of Central Africa). *Chemical Senses* 20:78.

Niedenthal P M, Halberstadt J B, & Innes-Ker A H (1999). Emotional Response Categorization. *Psychological Review* 106:337–61.

Otto T, Cousens G, & Herzog C (2000). Behavioral and Neuropsychological Foundations of Olfactory Fear Conditioning. *Behavioural Brain Reasearch* 110:119–28.

Parducci A (1995). *Happiness, Pleasure and Judgment: The Contextual Theory and Its Applications*. Hillsdale, NJ: Lawrence Erlbaum.

Rolls E (1999). *The Brain and Emotion*. Oxford University Press.

Rosch E (1996). The Environment of Minds: Towards a Noetic and Hedonic Ecology. In: *Cognitive Ecology*, ed. P M Friedman & E C Carterette, pp. 9–43. Orlando: Academic Press.

Rouby C, Jones-Gotman M, & Zatorre R (in press). Pleasantness and Intensity of Odors: Can They Be Dissociated?

Rouby C & Sicard G (1997). Des catégories d'odeurs? In: *Catégorisation et Cognition, de la perception au discours*, ed. D Dubois, pp. 59–81. Paris: Kimé.

Royet J P, Koenig O, Gregoire M C, Cinotti L, Lavenne F, Le Bars D, Costes N, Vigouroux M, Farget V, Sicard G, Holley A, Mauguière F, Comar D, & Froment J C (1999). Functional Anatomy of Perceptual and Semantic Processing. *Journal of Cognitive Neuroscience* 11:94–109.

Royet J P, Zald D, Versace R, Costes N, Lavenne F, Koenig O, & Gervais R (2000). Emotional Responses to Pleasant and Unpleasant Olfactory, Visual And Auditory Stimuli: A Positron Emission Tomography Study. *Journal of Neuroscience* 20:7752–9.

Rozin P & Vollmecke T A (1986). Food Likes and Dislikes. *Annual Review of Nutrition* 6:443–56.

Savic I & Gulyas B (2000). PET Shows that Odors Are Processed both Ipsilaterally and Contralaterally to the Stimulated Nostril. *Neuroreport* 11:2861–6.

Schaal B, Rouby C, Marlier L, Soussignan R, & Tremblay R E (1998). Variabilité et universaux de l'espace perçu des odeurs: approches inter-culturelles de l'hédonisme olfactif. In: *Géographie des odeurs*, ed. R Dulau & J R Pitte, pp. 25–47. Paris: L'Harmattan.

Schiffman S S (1974). Physicochemical Correlates of Olfactory Quality. *Science* 185:112–17.

Schiffman S S, Robinson D, & Erickson R P (1977). Multidimensional Scaling of Odorants: Examination of Psychological and Physiological Dimensions. *Chemical Senses* 2:375–90.

Soussignan R, Schaal B, Marlier L, & Jiang T (1997). Facial and Autonomic Responses to Biological and Artificial Olfactory Stimuli in Human Neonates: Re-examining Early Hedonic Discrimination of Odors. *Physiology and Behavior* 62:745–58.

Steiner J E (1979). Human Facial Expressions in Response to Taste and Smell Stimulations. In: *Advances in Child Development*, vol. 13, ed. L P Lipsitt & H W Reese, pp. 257–95. New York: Academic Press.

Tiberghien G (1984). *Initiation à la psychophysique*. Paris: Presses Universitaires de France.

Tversky A & Kahneman D (1981). The Framing of Decisions and the Psychology of Choice. *Science* 211:453–8.

von Haller A (1763). Liber XIV. Olfactus. *Elementa Physiologiae Corporis Humani*, Tome 5, pp. 125–85, Lausanne: Grasset.

Zald D H & Pardo J V (1997). Emotion, Olfaction, and the Human Amygdala: Amygdala Activation during Aversive Olfactory Stimulation. *Proceeding of the National Academy of Sciences USA* 94:4119–24.

Zatorre R, Jones-Gotman M, & Rouby C (2000). Neural Mechanisms Involved in Odor Pleasantness and Intensity Judgments. *Neuroreport* 11:2711–16.

Zwaardemaker H (1925). *L'odorat*. Paris: Doin.

10

Influences of Odors on Mood and Affective Cognition

Rachel S. Herz

In this chapter we shall review a wide range of research that shows how odors can influence mood, cognition, and behavior. This review, though not exhaustive, offers a comprehensive overview of the field. As background for this analysis, a discussion of odor-associative learning will first be given. The topics to be covered will then include the effects of odor exposure on (1) mood and specific emotions, (2) attitudes, work efficiency, and perceived health, (3) emotional memory, and (4) emotionally conditioned behavior. In addition to the behavioral evidence, neuroanatomic substantiation for the special relationship between odor and emotional associations will be presented.

1. Odor-associative Learning

The aim of the following paragraphs is to illustrate that almost all our responses to odors are learned, rather than innate. Evidence to support this idea comes primarily from research with infants and children. Although some work suggests that young children show adult-like preferences for certain odors (Schmidt and Beauchamp, 1988), most research with this age group indicates that children often do not differentiate between odors that adults find either very unpleasant or pleasant, such as butyric acid (rancid butter) versus amyl acetate (banana) (Engen, 1988; Schaal, Soussignan, and Marlier, Chapter 26, this volume), or they may have responses opposite to adult preferences, such as liking the smell of synthetic sweat and feces (Stein, Ottenberg, and Roulet, 1958; Engen, 1982). By age eight, however, most children's hedonic responses to odors mimic those of adults. Odor learning begins prior to birth. The olfactory system is functional in utero by 12 weeks of gestation (Schaal, Marlier, and Soussignan, 1998; Winberg and Porter, 1998), and there is considerable evidence that the specific odor characteristics of a mother's amniotic fluid (AF) are learned during gestation (Schaal and Rouby, 1990; Marlier, Schaal, and Soussignan, 1988; Winberg and Porter, 1998; Schaal et al., 1998, and Chapter 26, this volume). AF also contains flavor compounds from the mother's diet (Mennella, Johnson, and Beauchamp, 1995) that also can be incorporated into maternal milk (Mennella and Beauchamp, 1993). For example, it has been found that the infants of mothers who consume distinctive-smelling volatiles (e.g., garlic, alcohol, cigarette smoke) during pregnancy or lactation show preferences for those smells, as compared with infants who have not been exposed to those scents (Mennella and Beauchamp, 1991, 1993; Mennella et al., 1995). The consequences of odor learning go well beyond preferences for familiar smells and are not limited to biologically meaningful odorants. It has been shown that through simple exposure and reinforced conditioning (e.g., cuddling), preferences for odors such as cherry oil or the mother's perfume are readily developed by young infants (Balogh and Porter, 1986; Schleidt and Genzel, 1990; Sullivan et al., 1991).

Among adults, some of the most common instances of learned odor associations are the various hedonic responses to fine fragrances, which often have earned their associative connotations through their connections to significant others. Thus an individual's personal history with particular odorants tends to shape that individual's responses to those odors for life. The exception to this rule is that odors that cause strong trigeminal stimulation (e.g., ammonia) often are immediately repelling, because the irritation caused by trigeminal nerve activity at the moment of odor exposure produces a concomitant avoidance response. Thus there may be odors that an individual will automatically and instantly dislike. However, the proposition that there are odors that are automatically liked without prior experience seems unlikely. Even the smell of vanilla, which has been touted as universally appealing, is first experienced during suckling by both breast-fed and bottle-fed infants, for it is a primary odor compound in milk (Mennella and Beauchamp, 1996), and that context is certainly an instance of very salient emotional and nutritive reinforcement.

An area as yet unexplored is the extent to which individual differences in sensitivities to certain volatiles may differentially predispose people to like or dislike specific smells. It is known that anosmias to specific odors vary within populations and that this is largely determined genetically (Wysocki and Beauchamp, 1984). It is therefore possible that weakened or heightened sensitivities to certain chemicals might explain why some people enjoy the smell of skunk (and in the United States may even be members of the "Whiffy Club"), whereas most dislike it intensely. Empirical investigations into the relationships between individual physiological sensitivities and individual learning histories are clearly needed.

In addition to idiosyncratic experience, cultural modeling has been shown to play an important role in the development of odor preferences. We see this continually in ethnic differences in food preferences: "One man's meat is another man's poison." Evidence for culturally learned odor associations has been presented in two empirical studies examining olfactory hedonic responses: One study was conducted in the United Kingdom in the mid-1960s (Moncrieff, 1966), and the other in the United States in the late 1970s (Cain and Johnson, 1978). One might not think that those populations would have very different responses to flavors and smells, but the two studies demonstrated striking differences between the two cultures. Among the odors examined in both studies was methyl salicylate (wintergreen). In the U.K. study, that smell was given one of the lowest pleasantness ratings. By contrast, in the U.S. study, wintergreen was given the highest pleasantness rating among all the odors tested. Why would that be the case? The most likely explanation for the difference lies in cultural history. In the United Kingdom, the smell of wintergreen is associated with medicine, particularly so for the subjects in that 1966 study, who would have remembered the rub-on analgesics that were widely used during World War II, a time they may not have remembered fondly. Conversely, in the United States, the smell of wintergreen is exclusively a candy "mint" smell, and one that has very positive connotations. A similar pattern has been reported anecdotally for the smell of sarsaparilla, which in the United Kingdom is a disliked medicinal odor, but in the United States is the smell of root beer, a popular soft drink. It should always be kept in mind that because of the many idiosyncratic odor associations of individuals, it is often difficult to predict an individual's reaction to a specific odor, despite the existence of cultural norms.

Another factor that influences the type of association that an odor will evoke is the frequency with which the odor is encountered. For example, frequently experienced odors such as coffee lose their ability to elicit specific associations, and rather evoke general hedonic responses (e.g., pleasant, soothing, alerting), whereas an odor that is rarely encountered and has already been linked to a unique event may be able to elicit intense responses when smelled later: for example, the smell of eugenol (clove) and fear of the dentist's drill (Robin et al., 1998). The concept of odor-associative learning, vis-à-vis individual experience, culture, and odor–event specificity, provides the background for the following sections.

2. Effects of Odors on Mood and Specific Emotions

Many studies have shown that environmental manipulations such as exposure to pleasant or unpleasant music (Eich and Metcalf, 1989) or variations in indoor lighting (Baron, Rea, and Daniels, 1993) can have directional influences on emotional states. In a similar vein, exposures to pleasant and unpleasant odors have been found to impact moods. For example, in one study, pleasant fragrances used in a "real-life setting" were shown to improve mood for women and men during midlife and to alleviate some of the problems associated with menopause (Schiffman, et al., 1995b,c). In contrast, odors emanating from pig-holding facilities resulted in mood ratings that were significantly worse than those for control subjects (Schiffman et al., 1995a). Ludvigson and Rottman (1989) demonstrated that presentation of a pleasant odor (lavender) affected mood in a positive direction for subjects performing a stressful arithmetic task, though their performances on the task were inferior to those of subjects who did the task in a clove-odor or no-odor condition. Knasko (1995) found that exposure to an ambient smell of chocolate or baby powder caused people to look longer at photographic slides and to report being in a better mood, as compared with a no-odor condition. In contrast, exposure to the unpleasant odor of dimethyl disulfide caused subjects to report less pleasant mood ratings than did subjects exposed to lavender (Knasko, 1992). Thus it would seem that pleasant odors make people feel good, and unpleasant odors make them feel bad. In support of that contention, a study assessing a series of flower, herb, and fruit odors for their effects on emotion found that the scents that subjects liked best were those that they also found most relaxing (Nakano et al., 1992). Notably, it has been found that the mere suggestion of an ambient pleasant or unpleasant odor in a unscented room can produce the changes in mood that would be expected if such stimuli were present (Knasko, Gilbert, and Sabini, 1990). This implies that it is sufficient to have had past experience with the hedonic attributes of an odor for it to induce concordant mood changes. However, it also suggests a general caveat for any findings regarding odor and mood: It might be that any reported mood changes will have been influenced by demand characteristics, because subjects will ex-plicitly have been aware of the presence of an ambient odor. Experiments using ambient odors at concentrations near the detection threshold, with no explicit mention of odor having been made, would offer the most revealing approach. Within the odor–mood paradigm, only one study to date has tested with a barely detectable ambient odor (Kirk-Smith, Van Toller, and Dodd, 1983; Hermans and Baeyens, Chapter 8, this volume). In that experiment, subjects worked on a stressful task in the presence of a low concentration of trimethylundecylenic aldehyde (TUA) that they later reported being unable to detect. Nevertheless, at a subsequent nonchallenging task, those same subjects reported feeling anxious when that odor was present, as compared with subjects who had not been ex-posed to TUA during the stress task. That finding corroborates some results from studies in which the odor presence was not hidden and lends further support to the argument for the potency of odor-associative emotional learning.

In addition to self-reported mood changes, the physiological correlates of emotional states have been shown to be affected by odors. Miltner et al. (1994)

found that the presence of a negative odor (hydrogen sulfide) increased the amplitude of the eye-blink startle reflexes of subjects relative to a no-odor condition, whereas the presence of a positive odor (vanillin) decreased the startle reflex. Ehrlichman et al. (1997) reported similar results. Consistent with those observations, Vrana, Spence, and Lang (1988) have shown that the startle reflex is more pronounced during negative emotion, and muted during positive emotion. Changes in patterns of EEG theta rhythms have also been observed after exposure to different odorants, suggesting that various odors can differentially influence mental activity and mood (Lorig and Schwartz, 1988). Critical to the development of emotional associations to odor are first experiences, especially when they are emotionally salient. Reexposure to an odor that earlier was involved in such an episode can induce a spontaneous and intense emotional reaction commensurate with one's prior associational history with that odor. Initiation or reoccurrences of post-traumatic stress disorder often are triggered by an odor that was associated with the traumatic event (Kline and Rausch, 1985). For example, many Vietnam veterans have experienced physiological and psychological stress responses to odors associated with combat long after leaving the context of war (Kline and Rausch, 1985; McCaffrey et al., 1993). In less severe cases, patients who fear dental procedures show more stress-indicative autonomic responses to eugenol (clove odor, from dental cement) than do unafraid patients (Robin et al., 1998). Smells can also influence the cravings of drug addicts, and their contribution to the craving clearly is based on the associative-conditioning history that such odors have with particular drugs. For example, for a "crack" addict, the smell of a burning match can readily trigger a craving for cocaine (A. R. Childress, personal communication, 1995). Acquired odor–emotion associations have also been used therapeutically. Schiffman and Sieber (1991) trained subjects to associate an odor with muscle-relaxation exercises and found that the odor alone could later elicit relaxation responses. Odor–emotion associations have been used to increase patients' comfort and reduce anxiety during therapy (King, 1988). Thus, when odors have previously been associated with specific emotional states, they can have considerable capacity, when later encountered, to rekindle those emotions and influence both the cognitive and physiological parameters of experience.

On the basis of what is known about odor-associative learning, it seems evident that the directional effects on mood that are seen following exposure to certain odors are due to the hedonic connotations of those odors. An exception to that pattern, however, was recently reported by Chen and Haviland-Jones (1999), who found that an odor that was rated as smelling unpleasant could reduce feelings of depressive affect in subjects. In that study, college students rated their moods and then evaluated either (1) underarm odors from one of six donor groups (young boys, young girls, college women, college men, older women, older men) or (2) the smell of "home," using various hedonic dimensions. The subjects were later exposed to their target odors again, and their changes in mood were recorded. It was found that subjects who were exposed to the smell of older women rated it as unpleasant and intense (though not as unpleasant or intense as one of the other groups rated the smell of college males), but they also reported reduced feelings of depressed affect on the mood-assessment questionnaire. Reductions in depressed affect were also reported by subjects who smelled the odor of "home." Notably, the smell of older women was given a fairly high familiarity rating, so it may be that odors that are familiar, whether pleasant or unpleasant, are also comforting. Unfortunately, there was no determination of what associations were evoked by the smell of older women, or by any of the other odors tested. Therefore it is not possible to determine whether or not the subjects' prior personal associations (independent of the hedonic ratings) induced the observed effect. In the absence of such information, these data beg the question whether or not there was some biologically significant information carried by the smell of older women that was soothing.

"Aromatherapy" consists in application of the contention that various "natural" odors have intrinsic (essentially pharmacological) abilities to influence mood, behavior (cognition), and health. For example, some contend that the odor of jasmine oil has a stimulating effect, whereas lavender has a sedative effect, on mood and physical state (I. Imberger, personal communication, 1993). From an empirical standpoint there is nothing to suggest that the effects of these odors extend any further than the extent of one's associative learning, especially at a cultural level. That is, the claim that certain odors are experienced as relaxing, and others as stimulating, may be true, but that likely is due to the acquired reputations or meanings of the odors, not their intrinsic power; for example, in Western culture, lavender is commonly found in bath oils and soaps. Nevertheless, the findings of Chen and Haviland-Jones (1999) emphasize the need to determine whether or not any emotionally (or cognitively) significant information can be transmitted through exposure to such putatively biologically meaningful odorants (Jacob et al., Chapter 11, this volume).

3. Effects of Odors on Attitudes, Work Efficiency, and Perceived Health

The influence of odor hedonics on affective cognition is further illustrated by their effects on attitudes, performances, and perceptions of health. In a study to investigate how odor exposure would affect subjective evaluations, participants exposed to a malodor (1) judged paintings as less professional and inferior and

(2) assessed significantly lower ratings of well-being for people in photographs than did participants in a no-odor control group (Rotton, 1983). Baron (1983) reported that job applicants were evaluated more positively and credited with greater emotional competence when interviewed in a room that was scented with a pleasant odor prior to or during the interview, though certain complex interactions involving odor, gender, and style of dress were also observed. Pro-social behavior has been shown to be enhanced in the presence of positive ambient odors: Baron (1997) reported that assistive behavior toward strangers was much more likely to occur when subjects were exposed to pleasant ambient odors (baking cookies, roasting coffee) than in the absence of such odors, and respondents also reported significantly higher levels of positive affect in the presence of those smells. Results from several studies have demonstrated that pleasant odors have a positive influence on task performance: Baron (1990) showed that subjects who worked in the presence of a pleasant ambient odor reported higher self-efficacy, set higher goals, and were more likely to employ efficient work strategies than subjects who worked in a no-odor condition (Baron, 1990).

Similarly, Warm, Dember, and Parasuraman (1990) reported that pleasant odors enhanced vigilance during a tedious task, and Ehrlichman and Bastone (1992) found that the presence of a pleasant odor improved creative problem-solving relative to an unpleasant-odor condition. In studies examining the effects of pleasant ambient odors on anagram and word-completion tests (Baron and Bronfen, 1994; Baron and Thomley, 1994) it was found that in both low-stress and high-stress conditions subjects did better in the presence of pleasant odors. Notably, receipt of a small gift had similar positive effects on performance in those experiments. Many studies have shown that manipulations such as being given a small gift or being complimented can increase one's positive mood and improve behavior (Isen, 1984; Clark, 1991). Thus pleasant fragrances appear to be able to modify mood and behavior in a manner comparable to other mood manipulations.

Another way that odors have been shown to influence cognition is through the connotations they have for environments and for people's beliefs about the health risks or benefits of scented spaces. Historically, unpleasant odors long were viewed as carriers of disease, and good odors as curative (Levine and McBurney, 1986; Le Guérer, 1994). Today such negative perceptions are manifested as complaints about "multiple-chemical-sensitivity syndrome," in which severe allergic reactions and emotional distress are caused by the presence of malodorous environmental pollutants (Kjaergaard, Pederson, and Molhave, 1992), and "sick-building syndrome," in which individuals perceive that something in their indoor space is causing various health problems (Colligan and Murphy, 1982), and there is a sort of generic increase in symptom reporting in the presence of unfamiliar or unpleasant ambient odors (Neutra et al., 1991; Stahl and Lebedun, 1996). On the positive side, certain pleasant ambient odors are heralded in aromatherapy circles as natural healers (Lawless, 1991). Thus, contemporary beliefs are not so different from ancient beliefs. Whether excessively fearful, justifiably concerned, or indifferent to ambient odors, it is clear that the vast majority of people automatically make assumptions about the healthfulness or harmfulness of an environment on the basis of their hedonic evaluations of the ambient smells (Cain, 1987).

Notably, Dalton and colleagues (Dalton, 1996, 1999; Dalton et al., 1997) have shown that subjective beliefs about odors appear to be responsible for many instances of health complaints. In a study examining odor perception and cognitive bias, odors typically rated as healthful (wintergreen), harmful (alcohol), or ambiguous (balsam) were presented in three cognitive-bias conditions (healthful, harmful, and neutral). The subjects reported significantly more health symptoms and gave more intense odor ratings and higher irritation ratings to an odor (regardless of what it was) when it was presented in the "harmful" cognitive-bias context than when it was presented with a neutral or healthful bias (Dalton, 1999). There was also a synergistic effect between odorant and bias condition, such that the worst effects of odorant irritation were reported when alcohol was presented in the harmful context. In other studies (Dalton, 1996; Dalton et al., 1997), giving subjects false information about the health consequences of exposure (positive, negative, or neutral) altered their adaptations to isobornyl acetate (balsam) and acetone (nail-polish remover). Subjects who were in the positive- or neutral-bias context reported a decrease in perceived odor intensity (normal adaptation response) over time. In contrast, subjects who were in the negative-bias context showed initial adaptation, but then a sensitization effect that resulted in later odor ratings that in some cases were higher than the initial ratings.

Thus, just as with emotional perceptions, the physical aspects (health symptoms) of odor exposure can be mediated by subjects' beliefs, where the beliefs stem from the hedonic connotations of the odorants. This effect can be so extreme that the mere suggestion of an odor, without it actually being present, can cause people to report physical distress. For example, a tone played over the radio induced olfactory hallucinations in a number of listeners, some of whom reported discomfort from the "odor exposure" (O'Mahoney, 1978). Similarly, in a laboratory study, Knasko et al. (1990) found that feigned exposure to an unpleasant ambient odor produced increased ratings of discomfort and health symptoms. If such responses are not due to the actual presence of an odor stimulus, then they must be learned; and as can be seen, the response is consistent with the acquired hedonic connotation of the smell (real or imagined). The phenomenon of associative health effects produced by odors is further witnessed by the find-ing that subjects reported fewer health symptoms when exposed to baby powder than when exposed to chocolate or to no odor (Knasko, 1995). Such mood effects clearly are in line with the learned connotation of baby powder as comforting and clean.

4. Effects of Odor Exposure on Emotional Memory

The way in which odors elicit hedonically congruent mood and attitudinal states is by evoking memories of past experiences. In studies to examine the hedonic congruence between odor pleasantness and the valence of memories evoked, it was found that the personal memories that subjects recalled were significantly more positive when they were in the presence of a pleasant ambi-ent odor (almond) than when in the presence of an unpleasant odor (pyridine) (Ehrlichman and Halpern, 1988). Notably, the subjects' memories in that study, though personal, were not directly related to the ambient odors present, but rather were simply congruently toned with the hedonic valences of the odors. That is, the ambient odor influenced the subject's overall mood, and the subject's mood then influenced the tendency to remember more good or bad life events. This effect is an extension of the same associative mechanisms that determine hedonic influences of odors on moods, beliefs, and behaviors. Another, more direct way in which odors can influence memory is in their links to specific "autobiographical" events (e.g., the Proust phenomenon). Several investigators have used the autobiographical-memory method to explore the specific charac-teristics of odor-evoked memories. Those studies have shown that odor-evoked memories are reported as very evocative and that odors seem to be especially good reminders (see Chu and Downes, 2000, for a review). Most importantly, such studies have shown that odor-evoked memories are experienced as highly emotional, as measured by both self-reports and increased heart-rate responses (Laird, 1935; Herz and Cupchik, 1992; Herz, 1998a).

The most comprehensive method for assessing the special characteristics of odor-evoked memories is with a cross-modal approach, where memories elicited by stimuli presented in various sensory modalities are compared. In the first cross-modal olfactory-memory experiment of that type, Rubin, Groth, and Goldsmith (1984) showed participants 15 familiar stimuli (e.g., coffee, Band-Aids, cigarettes) to be assessed in olfactory, verbal, or picture form. For each item, the participant described the memory that was evoked and rated it on sev-eral scales: age of the memory, vividness, emotionality at the time of the event, emotionality at the time of recall, how many times it had been recalled, and when it was last recalled (prior to the experiment). From those measures, the only finding that was statistically reliable was that memories evoked by odors were thought of and talked about less often than memories evoked by words and pictures. There was a trend for odor-evoked memories to be more emotional, but that effect was not statistically significant.

Hinton and Henley (1993) compared subjects' free associations (via word, odor, and visual perceptions) to six familiar items (coffee, tobacco, carnations, oranges, Ivory soap, pine). Confirming the intuition that odors have a special capacity to elicit hedonic responses, they found that the free associations elicited by odors were more emotional, as evaluated by independent judges, than the free associations elicited by words or visual versions of the same items. Moreover, when the free associations resembled autobiographical recollections, they most often had been elicited by odor cues, thus suggesting that associations elicited by odors are more hedonically toned and personally involving than those elicited by other sensory stimuli.

Several cross-modal experiments from our laboratory have further demonstrated that memories evoked by odors are distinguished by their emotional potency, as compared with memories cued by other modalities (visual, tactile, verbal, music) (Herz and Cupchik, 1995; Herz, 1998b; Herz, 2001; Herz and Schooler, 2002). In one set of experiments, familiar objects (cues) were presented to participants in olfactory, verbal, visual, or tactile form (e.g., the smell of popcorn, the word "popcorn," seeing popcorn, or feeling popcorn) while the participants viewed emotionally evocative pictures. The participants were told that the experiment concerned the effects of different environmental cues on the perception of pictures. Memory was never mentioned. Two days later, however, when participants returned to the laboratory, they were given a surprise cued-recall test concerning their picture experiences, and the accuracy and emotionality of their memories were assessed. In each experiment it was found that the memories evoked by the various types of cues did not differ in accuracy (the same numbers of pictures were correctly recalled for each cue type); however, memories evoked by odor cues were significantly more emotional than memories prompted by any other cue. In a second series of experiments in which abstract odors, music, and visual images were compared as memory cues, it was found that odor-evoked memories caused higher heart rates during recall than did memories evoked by the other sensory cues, and again memory accuracy was unaffected by cue type. Most recently, we have combined autobiographical and cross-modal methods and have demonstrated that autobiographical memories elicited by odors are more emotional and evocative than recollections of the same personal events triggered by visual or verbal stimuli. Together, these findings show that odor-associated memory is distinctively more emotional than memory elicited through other sensory modalities.

5. Effects of Odors on Emotionally Conditioned Behavior

From the literature reviewed in this chapter, it is clear that odors have a special propensity to become associated with events and to cue recall. Perceptions of odors can also be emotionally conditioned. In a recent study it was shown that when a neutral odor was unobtrusively paired with toilet stimuli, subjects with negative attitudes toward "going to the toilet" rated the paired odor more negatively than they rated a non-toilet-paired control scent one week later (Baeyens et al., 1996; Hermans and Baeyens, Chapter 8, this volume). Thus, Pavlovian conditioning can modify perceptions of odors. Odors can also influence moods, either through an individual's specific associational history or because of the acquired cultural connotation of an odor. We also know that moods can influence behavior. For example, a positive affect is generally associated with an increase in work productivity and an inclination to help others (Isen, 1984; Clark, 1991), whereas negative feelings have been shown to reduce the propensity to assist (e.g., Underwood, Froming, and Moore, 1977) and to increase aggression (Baron and Bell, 1976). Following logically from that, one would predict that an odor experienced during an emotionally significant event could, when reintroduced, elicit an emotional response consistent with the past event and perhaps alter behavior accordingly.

To test that idea, we investigated whether or not an odor that had been associated with failure and frustration could, when later encountered, produce a detrimental effect on performance in an unrelated task. In that study (Epple and Herz, 1999), we gave five-year-olds the task of trying to solve an impossible maze (the "failure maze"). All subjects worked at the task while in a scented room. As the task was unsolvable, the result was failure and frustration, as subsequently confirmed by independent analyses of videotapes from the session. After the failure-maze task, the subjects were given a 20-minute break in an unscented area. They were then divided into three groups and taken to other experimental rooms that were scented with the same odor as in the failure-maze task, a novel odor, or no odor and were given a cognitively challenging test. It was found that subjects who performed the test in the presence of the same odor as in the failure-maze task did significantly worse than subjects in the other groups. Subjects in the novel-odor and no-odor groups did not differ in performance. Such findings show that subjects can become experientially conditioned to odors and that later exposure to such odors can have direct impacts on behavior. Although our study demonstrated a negative effect produced by emotional conditioning to an odor, many positive extrapolations are possible from such findings. For example, positive conditioning with odor manipulations could be implemented to improve scholastic performances for children with low self-confidence and poor academic orientation. Additionally, the underlying concepts and methods of odor-associated conditioning can be applied toward a variety of health-enhancing and positive-behavior-promoting techniques for both children and adults.

6. Neuroanatomic Connections between Olfaction and Associative Emotional Processing

The behavioral findings that indicate the highly associative and emotionally evocative properties of odors are being substantiated on neuroanatomical grounds. Olfactory afferent mechanisms have a uniquely direct connection with the neural substrates for emotional processing (amygdala-hippocampal complex) (Turner, Mishkin, and Knapp, 1980; Cahill et al., 1995). Only two synapses separate the olfactory nerve from the amygdala – critical for experiencing and expressing emotion (Aggleton and Young, 2000) and maintaining human emotional memory (Cahill et al., 1995). And only three synapses separate the olfactory nerve from the hippocampus, involved in the selection and transmission of information in working memory, in transfers between short-term and long-term memory, and in various declarative memory functions (Staubli, Ivy, and Lynch, 1984; Schwerdtfeger Buhl, and Gemroth, 1990; Eichenbaum, 1996). Additionally, the orbitofrontal cortex and amygdala have been shown to play major roles in stimulus-reinforced associative learning (Rolls, 1999), and olfactory processing has been shown to activate the orbitofrontal cortex and to some extent the amygdala (Zatorre et al., 1992; Phillips and Heining, Chapter 12, this volume). The fact that no other sensory system makes this kind of direct, dynamic contact with the neural substrates for emotion and memory provides strong support for the emotional distinctiveness of olfactory cognition.

The existence of a unique relationship between olfaction and emotional processing can also be defended on neuroevolutionary grounds. Olfaction was the first sense to evolve and is the most fundamental in terms of the information that it imparts. Even for the most simple creatures of one or a few cells the chemical sense enables the organisms to know what to approach (chemotaxis) for nutritional and reproductive reasons, that is, knowing what to approach for their survival. In more complex animals, such as mammals, olfactory information is the key form of communication for all of the most critical aspects of behavior – knowing which are one's kin, scenting out the reproductively available conspecifics, finding food, sniffing out the sources of danger – again, information that tells the animal what to approach and what to avoid in order to survive and thrive. We humans no longer rely on olfactory information as the key to our survival; rather, visual and verbal communications have become the major means by which we maneuver ourselves adaptively through the world, though olfaction still retains some of its basic functions.

The most immediate response we have to a smell is to like or dislike it, that is, approach or avoid. This basic hedonic choice is an affective response (Livesey, 1986). It is also the case that when one thinks abstractly about what emotions tell us, they are about what is good (to approach) and what is bad (to avoid). For example, positive emotions such as joy encourage our approach to the joyful stimulus and thus presumably foster our survival and reproductive potential, whereas negative emotions such as fear enable us to avoid or counter dangers, thus fostering our survival through avoidance of death. Thus, conceptually, emotions tell us the same things that smells tell our animal brethren, that is, what to approach and what to avoid for maximal reproductive advantage. In other words, emotions and olfaction are functionally and ultimately analogous; both are fundamentally geared to prompt either an approach or avoid response that will enable the organism to react appropriately to best ensure its survival.

It is also notable that during human phylogenetic development, the most ancient part of the brain, called the rhinencephalon (literally, "nose-brain"), which comprises the olfactory and limbic areas, developed from olfactory structures first, and the limbic structures (e.g., amygdala-hippocampal complex) then followed (McLean, 1969). In other words, the ability to experience and express emotion grew directly out of the ability to cognitively process smell. In sum, there appears to be neuroanatomical and neuroevolutionary convergence linking the perception of smell and the experience of emotion (Phillips and Heining, Chapter 12, this volume).

7. Conclusion

The literature reviewed in this chapter has shown that exposure to odors can alter moods, change attitudes, influence perceptions of health, affect task performance, and evoke memories. In general, pleasant-smelling odors induce positive moods, attitudes, and behavioral changes, and unpleasant odors induce negative moods, attitudes, and behavior. Almost all of these effects can be explained by associative learning and conditioning principles, where the acquired hedonic connotation of an odor is responsible for inducing congruent moods and behaviors; in more specific cases, a particular odor is tied to a past event and induces reactions consistent with its associational history. The neuroanatomical, phylogenetic, and functional relationships between olfaction and limbic-system structures provide strong support for the special and perhaps unique emotional potency and associative propensity of our perceptions of odors. Despite our current knowledge, our understanding of how odors influence mood and affective cognition is still in its infancy. Many topics await further exploration, and many new questions await answers. It is envisioned that neuroimaging techniques in conjunction with behavioral studies will greatly expand our capacity to deal with these questions and use wisely the answers we shall find.

References

Aggleton J P & Young A E (2000). The Enigma of the Amygdala: On Its Contribution to Human Emotion. In: *Cognitive Neuroscience of Emotion*, ed. R D Lane & L Nadel, pp. 106–28. Oxford University Press.

Baeyens F, Wrzesniewski A, de Houwer J, & Eelen P (1996). Toilet Rooms, Body Massages, and Smells: Two Field Studies on Human Evaluative Odor Conditioning. *Current Psychology: Developmental, Learning, Personality, Social* 15:77–96.

Balogh R D & Porter R H (1986). Olfactory Preference Resulting from Mere Exposure in Human Neonates. *Infant Behavior and Development* 9:395–401.

Baron R A (1983). "Sweet Smell of Success"? The Impact of Pleasant Artificial Scents on Evaluations of Job Applicants. *Journal of Applied Psychology* 68:709–13.

Baron R A (1990). Environmentally-induced Positive Affect: Its Impact on Self-efficacy, Task Performance, Negotiation, and Conflict. *Journal of Applied Social Psychology* 20:368–84.

Baron R A (1997). The Sweet Smell of Helping: Effects of Pleasant Ambient Fragrance on Prosocial Behavior in Shopping Malls. *Personality and Social Psychology Bulletin* 23:489–503.

Baron R A & Bell P A (1976). Aggression and Heat: The Influence of Ambient Temperature, Negative Affect, and a Cooling Drink on Physical Aggression. *Journal of Personality and Social Psychology* 33:245–55.

Baron R A & Bronfen M I (1994). A Whiff of Reality: Empirical Evidence Concerning the Effects of Pleasant Fragrances on Work-related Behavior. *Journal of Applied Social Psychology* 24:1179–203.

Baron R A, Rea M S, & Daniels S G (1993). Effects of Indoor Lighting on the Performance of Cognitive Tasks and Interpersonal Behaviors: The Potential Mediating Role of Positive Affect. *Motivation and Emotion* 16:1–33.

Baron R A & Thomley J (1994). A Whiff of Reality: Positive Affect as a Potential Mediator of the Effects of Pleasant Fragrances on Task Performance and Helping. *Environment and Behavior* 26:766–84.

Cahill L, Babinsky R, Markowilsch H J, & McGaugh J (1995). Amygdala and Emotional Memory. *Nature* 377:295–6.

Cain W S (1987). Indoor Air as a Source of Annoyance. In: *Environmental Annoyance: Characterization, Measurement and Control*, ed. H S Koelega, pp. 189–200. Amsterdam: Elsevier.

Cain W S & Johnson F Jr (1978). Lability of Odor Pleasantness: Influence of Mere Exposure. *Perception* 7:459–65.

Chen D & Haviland-Jones J (1999). Rapid Mood Change and Human Odors. *Physiology and Behavior* 68:241–50.

Chu S & Downes J J (2000). Odour-evoked Autobiographical Memories: Psychological Investigations of Proustian Phenomena. *Chemical Senses* 25:111–16.

Clark M S (1991). *Mood and Helping: Mood as a Motivator of Helping and Helping as a Regulator of Mood*. Newbury Park, CA: Sage.

Colligan M J & Murphy L (1982). A Review of Mass Psychogenic Illness in Work Settings. In: *Mass Psychogenic Illness: A Social Psychological Analysis*, ed. M Colligan, J Pennbaker, & L Murphy, pp. 35–56. Hillsdale, NJ: Lawrence Erlbaum.

Dalton P (1996). Odor Perception and Beliefs about Risk. *Chemical Senses* 21:447–58.

Dalton P (1999). Cognitive Influences on Health Symptoms from Acute Chemical Exposure. *Health Psychology* 18:579–90.

Dalton P, Wysocki C J, Brody M J, & Lawley H J (1997). The Influence of Cognitive Bias on the Perceived Odor, Irritation and Health Symptoms from Chemical Exposure. *International Archives of Occupational and Environmental Health* 69:407–17.

Ehrlichman H & Bastone L (1992). The Use of Odour in the Study of Emotion. In: *Fragrance: The Psychology and Biology of Perfume*, ed. S Van Toller & G H Dodd, pp. 143–59. London: Elsevier.

Ehrlichman H & Halpern J N (1988). Affect and Memory: Effects of Pleasant and Unpleasant Odors on Retrieval of Happy and Unhappy Memories. *Journal of Personality and Social Psychology* 55:769–79.

Ehrlichman H, Kuhl S B, Zhu J, & Warrenburg S (1997). Startle Reflex Modulation by Pleasant and Unpleasant Odors in a Between-Subjects Design. *Psychophysiology* 34:726–9.

Eich J E, & Metcalf J (1989). Mood Dependent Memory for Internal versus External Events. *Journal of Experimental Psychology: Learning, Memory and Cognition* 15:443–55.

Eichenbaum H (1996). Olfactory Perception and Memory. In: *The Mind–Brain Continuum*, ed. R Llinas & P Churchland, pp. 173–202. Cambridge, MA: MIT Press.

Engen T (1982). *The Perception of Odors*. New York: Academic Press.

Engen T (1988). The Acquisition of Odour Hedonics. In: *Perfumery: The Psychology and Biology of Fragrance*, ed. S Van Toller & G H Dodd, pp. 79–90. London: Chapman & Hall.

Epple G & Herz R S (1999). Ambient Odors Associated to Failure Influence Cognitive Performance in Children. *Developmental Psychobiology* 35:103–7.

Herz R S (1998a). An Examination of Objective and Subjective Measures of Experience Associated to Odors, Music and Paintings. *Empirical Studies of the Arts* 16:137–52.

Herz R S (1998b). Are Odors the Best Cues to Memory? A Cross-Modal Comparison of Associative Memory Stimuli. *Annals of the New York Academy of Sciences* 855:670–4.

Herz R S (2001). How Odor-evoked Memories Differ from Other Memory Experiences: Experimental Investigations into the Proustian Phenomenon. In: *Compendium of Olfactory Research*, ed. T Lorig, pp. 23–38. New York: Olfactory Research Fund.

Herz R S & Cupchik G C (1992). An Experimental Characterization of Odor-evoked Memories in Humans. *Chemical Senses* 17:519–28.

Herz R S & Cupchik G C (1995). The Emotional Distinctiveness of Odor-evoked Memories. *Chemical Senses* 20:517–28.

Herz R S & Schooler J W (2002). A Naturalistic Study of Autobiographical Memories Evoked to Olfactory versus Visual Cues. *American Journal of Psychology* 115.

Hinton P B & Henley T B (1993). Cognitive and Affective Components of Stimuli Presented in Three Modes. *Bulletin of the Psychonomic Society* 31:595–8.

Isen A M (1984). Toward Understanding the Role of Affect in Cognition. In: *Handbook of Social Cognition*, ed. A L Wyer & T K Scrull, pp. 179–236. Hillsdale, NJ: Lawrence Erlbaum.

King J R (1988). Anxiety Reduction Using Fragrances. In: *Perfumery: The Psychology and Biology of Fragrance*, ed. S Van Toller & G H Dodd, pp. 121–44. London: Chapman & Hall.

Kirk-Smith M D, Van Toller C, & Dodd G H (1983). Unconscious Odour Conditioning in Human Subjects. *Biological Psychology* 17:221–31.

Kjaergaard S, Pederson O F, & Molhave L (1992). Sensitivity of the Eyes to Airborne Irritant Stimuli. Influence of Individual Characteristics. *Archives of Environmental Health* 47:45–50.

Kline N A & Rausch J L (1985). Olfactory Precipitants of Flashbacks in Post-traumatic Stress Disorder: Case Reports. *Journal of Clinical Psychiatry* 46:383–4.

Knasko S C (1992). Ambient Odor's Effect on Creativity, Mood, and Perceived Health. *Chemical Senses* 17:27–35.

Knasko S (1995). Pleasant Odors and Congruency: Effects on Approach Behavior. *Chemical Senses* 20:479–87.

Knasko S C, Gilbert A N, & Sabini J (1990). Emotional State, Physical Well-being and Performance in the Presence of Feigned Ambient Odor. *Journal of Applied Social Psychology* 20:1345–57.

Laird D A (1935). What Can You Do With Your Nose? *Scientific Monthly* 41:126–30.

Lawless H (1991). Effects of Odors on Mood and Behavior: Aromatherapy and Related Effects. In: *The Human Sense of Smell*, ed. D G Laing, R L Doty, & W Breipohl, pp. 361–87. New York: Springer-Verlag.

Le Guérer A (1994). *Scent*. New York: Kodansha America.

Levine J M & McBurney D H (1986). The Role of Olfaction in Social Perception and Behaviour. In: *Physical Appearance, Stigma and Social Behaviour: The Ontario Symposium*, vol. 3, ed. C P Herman, M P Zanna, & E T Higgins, pp. 179–217. Hillsdale, NJ: Lawrence Erlbaum.

Livesey P J (1986). *Learning and Emotion: A Biological Synthesis. Vol. 1: Evolutionary Processes*. Hillsdale, NJ: Lawrence Erlbaum.

Lorig T S & Schwartz G E (1988). Brain and Odor: I. Alteration of Human EEG by Odor Administration. *Psychobiology* 16:281–4.

Ludvigson H W & Rottman T R (1989). Effects of Ambient Odors of Lavender and Cloves on Cognition, Memory, Affect and Mood. *Chemical Senses* 14:525–36.

McCaffrey R J, Lorig T S, Pendrey D L, McCutcheon N B, & Garrett J C (1993). Odor-induced EEG Changes in PTSD Vietnam Veterans. *Journal of Traumatic Stress* 6:213–24.

McLean P D (1969). *A Triune Concept of the Brain and Behavior*. New York: Bowker.

Marlier L, Schaal B, & Soussignan R (1998). Neonatal Responsiveness to the Odor of Amniotic and Lacteal Fluids: A Test of Chemosensory Continuity. *Child Development* 69:611–23.

Mennella J A & Beauchamp G K (1991). The Transfer of Alcohol to Human Milk: Effects on Flavor and Infants and the Infant's Behavior. *New England Journal of Medicine* 325:981–5.

Mennella J A & Beauchamp G K (1993). The Effects of Repeated Exposure to Garlic-flavored Milk on the Nursling's Behavior. *Pediatric Research* 34:805–8.

Mennella J A & Beauchamp G K (1996). The Human Infant's Responses to Vanilla Flavors in Mother's Milk and Formula. *Infant Behavior and Development* 19:13–19.

Mennella J A, Johnson A, & Beauchamp G K (1995). Garlic Ingestion by Pregnant Women Alters the Odor of Amniotic Fluid. *Chemical Senses* 20:207–9.

Miltner W, Matjak M, Braun C, Diekmann H, & Brody S (1994). Emotional Qualities of Odors and Their Influence on the Startle Reflex in Humans. *Psychophysiology* 31:107–10.

Moncrieff R W (1966). *Odour Preferences*. New York: Wiley.

Nakano Y, Kikuchi A, Matsui H, & Hatayama T (1992). A Study of Fragrance Impressions, Evaluation and Categorization. *Tohoku Psychologica Folia* 51:83–90.

Neutra R, Lipscomb J, Satin K, & Shusterman D (1991). Hypotheses to Explain the Higher Symptom Rates Observed Around Hazardous Waste Sites. *Environmental Health Perspectives* 94:31–5.

O'Mahoney M (1978). Smell Illusions and Suggestion: Reports of Smells Contingent on Tones Played on Radio and Television. *Chemical Senses and Flavor* 3:183–9.

Robin O, Alaoui-Ismaïli O, Dittmar A, & Vernet-Maury E (1998). Emotional Responses Evoked by Dental Odors: An Evaluation from Autonomic Parameters. *Journal of Dental Research* 77:1638–46.

Rolls E T (1999). *The Brain and Emotion*. Oxford University Press.

Rotton J (1983). Affective and Cognitive Consequences of Malodorous Pollution. *Applied Social Psychology* 4:171–91.

Rubin D C, Groth E, & Goldsmith D J (1984). Olfactory Cueing of Autobiographical Memory. *American Journal of Psychology* 97:493–507.

Schaal B, Marlier L, & Soussignan R (1998). Olfactory Function in the Human Fetus: Evidence from Selective Neonatal Responsiveness to the Odor of Amniotic Fluid. *Behavioral Neuroscience* 112:1438–49.

Schaal B & Rouby C (1990). Le développement des sens chimiques: Influences exogènes prénatales, conséquences postnatales. In: *Progrès en Néonatologie*, vol. 10, ed. J P Relier, pp. 182–201. Basel: Karger.

Schiffman S S, Miller E A, Suggs M S, & Graham B G (1995a). The Effects of Environmental Odors Emanating from Commercial Swine Operations on the Mood of Nearby Residents. *Brain Research Bulletin* 37:369–75.

Schiffman S S, Sattley-Miller E A, Suggs M S, & Graham B G (1995b). The Effect of Pleasant Odors and Hormone Status on Mood of Women at Midlife. *Brain Research Bulletin* 36:19–29.

Schiffman S S & Sieber J M (1991). New Frontiers in Fragrance Use. *Cosmetics and Toiletries* 106:39–45.

Schiffman S S, Suggs M S, & Sattley-Miller E A (1995c). Effect of Pleasant Odors on Mood of Males at Midlife: Comparison of African-American and European-American men. *Brain Research Bulletin* 36:31–37.

Schleidt M & Genzel C (1990). The Significance of Mother's Perfume for Infants in the First Weeks of Their Life. *Ethology and Sociobiology* 11:145–54.

Schmidt H J & Beauchamp G K (1988). Adult-like Odor Preferences and Aversions in Three Year Old Children. *Child Development* 59:1136–43.

Schwerdtfeger W L, Buhl E H, & Gemroth P (1990). Disynaptic Olfactory Input to the Hippocampus Mediated by Stellate Cells in the Entorhinal Cortex. *Journal of Comparative Neurology* 194:519–34.

Stahl S M & Lebedun M (1996). Mystery Gas: An Analysis of Mass Hysteria. *Journal of Health and Social Behavior* 15:44–50.

Staubli U, Ivy G, & Lynch G (1984). Hippocampal Denervation Causes Rapid Forgetting of Olfactory Information in Rats. *Proceedings of the National Academy of Sciences USA* 81:5885–7.

Stein M, Ottenberg M D, & Roulet N (1958). A Study of the Development of Olfactory Preferences. *Archives of Neurological Psychiatry* 80:264–6.

Sullivan R M, Taborsky-Barba S, Mendoza S, Itano A, Leon M, Cotman W, Payne T, & Lott I (1991). Olfactory Classical Conditioning in Neonates. *Pediatrics* 87:511–17.

Turner B H, Mishkin M, & Knapp M (1980). Organization of Amygdalopetal Projections from Modality-specific Cortical Association Areas in the Monkey. *Journal of Comparative Neurology* 191:515–43.

Underwood B, Froming W J, & Moore B S (1977). Mood, Attention, and Altruism. *Developmental Psychology* 13:541–2.

Vrana S R, Spence E L, & Lang P J (1988). The Startle Probe Response: A New Measure of Emotion? *Journal of Abnormal Psychology* 97:487–91.

Warm J S, Dember W N, & Parasuraman R (1990). Effects of Fragrances on Vigilance, Performance and Stress. *Perfumer and Flavorist* 15:16–17.

Winberg J & Porter R H (1998). Olfaction and Human Neonatal Behaviour: Clinical Implications. *Acta Paediatrica* 87:6–10.

Wysocki C J & Beauchamp G K (1984). Ability to Smell Androstenone Is Genetically Determined. *Proceedings of the National Academy of Sciences USA* 81:4899–902.

Zatorre R J, Jones-Gotman M, Evans A C, & Meyer E (1992). Functional Localization and Lateralization of Human Olfactory Cortex. *Nature* 360:339–40.

11

Assessing Putative Human Pheromones

Suma Jacob, Bethanne Zelano, Davinder J. S. Hayreh,
and Martha K. McClintock

Whether or not human pheromones exist and how they might influence human physiology and behavior have been debated for decades; for a review, see Preti and Wysocki (1999). In this chapter, we present the concepts and data that are shaping our understanding of pheromones and pheromone-like effects in humans and other species. This will include the semantic issues associated with use of the term "pheromone" and descriptions of experiments designed to determine the psychological effects of two putative human pheromones. The expectation that human pheromones can consistently elicit stereotyped behavior is unrealistic. We argue that it is more likely that airborne signals are context-dependent and have more general, modulatory effects that can best be captured conceptually in terms of "modulatory pheromones" or "social chemosignals" (McClintock, 2000). Research with humans also presents a unique opportunity to consider the functional role of an awareness or conscious experience of pheromones and social chemosignals, whereas animal research is limited to studies of overt behavior.

1. Definition of "Pheromone"

According to the classic definition (Karlson and Lüscher, 1959), pheromones are airborne chemical signals emitted by an individual that trigger specific neuroendocrine, behavioral, or developmental responses in other individuals of the same species. Early pheromone research began with insects (Karlson and Butenandt, 1959), and the term "pheromone" was coined to designate a group of externally released "active substances" that triggered specific behavioral responses and were similar to, but could not be called, hormones. An example is bombykol, a substance emitted by female silkworm moths. A male will follow the bombykol up its concentration gradient to find the female and mate (Schneider, 1974).

As pheromone research broadened, pheromones were divided into three general classes based on their functional effects. In one class of pheromones are the releaser pheromones (Wilson and Bossert, 1963), which trigger or "release" a stereotyped behavior, (such as the male moth following the trail of bombykol (Schneider, 1974), or mating behavior in cattle (Rivard and Klemm, 1990). In a second class are priming pheromones (Wilson and Bossert, 1963), which trigger a regulatory or developmental neuroendocrine change in the receiving organism (van der Lee and Boot, 1955, 1956; Bruce, 1960a,b; Vandenbergh, Witsett, and Lombardi, 1975). The third class comprises the signaling pheromones that indicate or signal to members of the same species information such as identity or reproductive status of an individual (Bronson, 1968). Recently, we have proposed the concept of a fourth class, modulating pheromones (McClintock, 2000), largely in response to the research with humans described later in this chapter. These modulate behavioral or psychological reactions to a particular context, without triggering specific behaviors or thoughts.

Although it may be simplest to categorize a pheromone by its clear and obvious behavioral or physiological effects, some have called for more specific criteria concerning a signal's cognitive attributes and developmental program, especially in regard to mammalian chemosensory communication. For example, it has been proposed that a pheromone should be chemically discrete (just one or a few compounds) and that pheromonal responses must be shaped minimally by olfactory qualities or learning and primarily by genetics (Beauchamp et al., 1976; Goldfoot, 1981).

The idea that olfaction and cognitive recognition do not play roles in pheromonal communication arose in part from the observation that in many mammals, pheromones do not always initiate their effects by binding at the main olfactory epithelium. Instead, they may bind at specialized chemoreceptors within the vomeronasal organ (Wysocki and Meredith, 1987; Keverne, 1999). The vomeronasal organ is the receptor organ of the vomeronasal system, and neural projections from the vomeronasal organ extend to the accessory olfactory bulb and to parts of the brain associated with emotional responses, social behavior, and reproduction, such as the limbic system and the anterior hypothalamus (Wysocki and Meredith, 1987). There are no known projections directly to the primary and secondary olfactory cortices, which are understood to be necessary for recognition and categorization of odors.

It has become clear, however, that some mammalian pheromones are indeed processed by the main olfactory system. Androstenone is the pig pheromone produced and released by the boar that causes the sow to assume her mating stance. Simply spraying this steroid from an aerosol can, without a boar present, is sufficient, along with a light touch, to trigger the behavior (Melrose, Reed, and Patterson, 1971). Nonetheless, the vomeronasal organ is not necessary for this response. Sows will respond after their vomeronasal organs have been plugged, implicating involvement of the main olfactory system in the response (Dorries, Adkins-Regan, and Halpern, 1997). Moreover, androstenone has a strong odor detectable by about 50% of human adults and presumably by sows (Wysocki, Dorries, and Beauchamp, 1989; Dorries et al., 1995). Thus, demonstrating involvement of the olfactory system is not evidence that a chemosignal is not a pheromone. Humans are the species in which questions concerning the roles of olfaction and cognition in pheromonal communication can best be explored.

2. Chemical Communication in Humans

The presence of the vomeronasal organ in human adults has been quite controversial and marked by conflicting reports. Reports of vomeronasal frequency in humans have been inconsistent, ranging from 39% (Johnson, Josephson, and Hawke, 1985) to 100% (Moran, Jafek, and Rowley, 1991), possibly because some investigators may mistakenly have identified other nasal structures as the vomeronasal organ (Jacob et al., 2000). At present, it can be concluded only that human adults have at least a remnant of a vomeronasal organ in the nose, but whether or not it is functional in adults remains debatable (Keverne, 1999; Preti and Wysocki, 1999; Smith et al., 1999). Recent claims for a functional vomeronasal organ have stemmed mainly from surface-potential changes recorded from the putative human vomeronasal epithelium during exposure to steroidal compounds (Berliner et al., 1996; Grosser et al., 2000). However, there has been valid criticism of the interpretation of those results as proof of functionality (Preti and Wysocki, 1999), and evidence for receptor potentials and neural transmission to the brain is needed. Furthermore, attempts to locate vomeronasal receptor genes in human vomeronasal tissue have thus far found only pseudogenes (Tirindelli, Mucignat-Caretta, and Ryba, 1998), although one, V1RL1, has been isolated from nasal epithelium (Rodriguez et al., 2000).

Regardless of whether or not humans have or need a functional vomeronasal system for processing pheromones, the important issue facing experimenters is what effects of pheromones to look for. At present, there is evidence in humans only for priming pheromones that regulate endocrine function (Stern and McClintock, 1998): Axillary secretions collected from women were found to modulate the duration of the follicular phase and the timing of the preovulatory surge in luteinizing hormone (LH) in recipient women who had the secretions applied just under their noses. We hypothesized that, as in other species, human pheromones might be produced by apocrine glands (active only during reproductive maturity), eccrine glands (which produce sweat that contains compounds found also in saliva and urine), exfoliated epithelial cells, hair, or bacterial action

(Nixon, Mallet, and Gower, 1988). We collected compounds from the axillae (armpits) because they contain all five of those potential sources.

After those compounds had been collected on pads, frozen, and then thawed, they did not have an odor that was detectable by the women in the study. It was also of central theoretical importance that all participants in our study be blind to the experiment's hypothesis and the source of the compounds. The study was presented to the subjects as being focused primarily on the development of noninvasive methods for detecting ovulation, and only secondarily on sensitivity to the odor of small amounts of "natural essences" (consent was obtained for a list of 30 compounds).

We do not yet know the structures of the compounds that regulate ovulation without being consciously detectable as odors. Axillary compounds, such as 5α-androst-16-en-3β-ol or (E)-3-methyl-2-hexenoic acid, at first would not seem likely to be the active compounds, because they have a strong odor. On the other hand, the latter compound is odorless when bound tightly with its carrier protein, which would be the case in natural secretions (Spielman et al., 1995). Thus it could be that the larger molecule is odorless, even though the smaller component that has a pheromonal action does have an odor.

Whereas we have shown that these chemosignals affect endocrine function without having detectable odors, there is some evidence for odors affecting human behavior and psychological states, presumably acting through the main olfactory pathway. For example, living within smelling distance of a pig farm has been shown to significantly depress mood (Schiffman et al., 1995), and there is some evidence for consistent emotional influences by low levels of odors (Kirk-Smith, Van Toller, and Dodd, 1983; Lorig et al., 1990; Schwartz et al., 1994). The power of odors to evoke emotional memories of past experiences in humans has often been noted (Proust, 1922; Degel and Köster, 1998), and the hedonic qualities of odors in everyday life are commonly recognized and inescapable (Tassinary, 1985).

Suggestive evidence for human signaling odors (if not for pheromones) comes from studies of the major histocompatibility complex (MHC). Ober and colleagues demonstrated that marriages between people with identical human leukocyte antigen (HLA; the human MHC) haplotypes are less frequent than expected (Ober et al., 1997, 1999) and that a high degree of HLA similarity between partners increases the chance of miscarriage (Ober et al., 1998).

Furthermore, studies have shown that people find unpleasant the odors from individuals whose HLA genes are similar to their own (Wedekind et al., 1995; Wedekind and Füri, 1997). Together, these intriguing findings suggest that people may use chemosignals in mate-choice decisions to reduce the fitness costs due to early fetal loss or miscarriage.

Whether this observed pattern of disassortative mating in humans is mediated by olfaction, as it is in rodents (Yamazaki et al., 1979; Boyse, Beauchamp, and Yamazaki, 1987), or mediated by some other mode of action has yet to be determined. If social chemosignals do play a role (regardless of which pathway mediates their effects), we do not suggest that they induce an instinctive attraction at a level below conscious control or awareness. However, it is conceivable to think that, in appropriate social contexts, pheromones or other chemosignals could modulate behavior and attraction and be one of the many influencing factors that lead to decisions and actions with long-lasting effects.

To date, research on specific behavioral effects (rather than physiological effects) of putative human pheromones has been conceptually limited to identifying releaser pheromones. The focus has been on sociosexual behaviors similar to those in some animal models, such as sex attraction in silkworm moths, hamsters, and rhesus monkeys (Michael, Keverne, and Bonsall, 1971; Schneider, 1974; Singer et al., 1986; Singer, 1991; Wood, 1998) and mating reflexes in pigs and rats (Gower, 1972; Sachs, 1997). Despite the fact that chemical communication among non-human animals is often complex and context-dependent (see Izard, 1983, for a review), the entertainment media and perfume industry have championed the idea that dramatic sexual behaviors or drives can be triggered in humans (see Table 2 of Preti and Wysocki, 1999).

In general, the search for human pheromones involved in sexual behavior has produced mixed results. For example, when Morris and Udry (1978) studied married couples in their homes, they failed to find that aliphatic acids increased the frequency of intercourse. Cutler and colleagues, however, reported that an unidentified compound increased sexual interactions of men (Cutler, Friedman, and McCoy, 1998). Unfortunately, that study did not address whether or not the behaviors were influenced by learned associations with consciously perceived odors, nor whether it was the women themselves who were affected or their male partners. However, even if the effects are not as dramatic as the perfume industry touts them to be, the idea that there may be detectable influences on behavior is certainly intriguing.

The expectation that discrete behaviors and states, specifically sexual intercourse or sexual desire, can be triggered in humans exceeds what can be predicted from our knowledge of human cognition and behavior. The inherent multimodal, multidimensional complexities of human behavior render such a strict model inappropriate, and the elaborate neocortical cognitive systems of the human nervous system preclude the initiation of complex behaviors by a simple signal. Just as in many other mammals (Signoret, 1976, 1991; Goldfoot et al., 1980; Sorensen, 1996), human chemical communication likely is inextricably linked with other sensory and emotional systems.

Rather than focusing on releaser effects and a limited range of observable behaviors, what seems more justifiable and productive is investigation into the more fundamental and well-studied aspects of psychological state and function. Using methodologies currently employed in psychological and physiological research to test on-line fluctuations would make investigations into the behavioral effects of putative human pheromones more flexible, sensitive, and comparable to extant knowledge in related fields. Specifically, pheromonal influences might be studied by measuring variations in common state measures such as mood and arousal, with corresponding concurrent psychophysiological measures (e.g., phasic and/or tonic autonomic nervous system variations or evoked-response measures).

The hypothesis that pheromonal effects can be detected in mood and autonomic measures is supported by the mammalian literature. The vomeronasal and olfactory inputs project to regions of the amygdala and hypothalamus in other mammals (Wysocki and Meredith, 1987; Risold and Swanson, 1995). Both of these areas of the brain play roles in causing changes in emotional tone (Aggleton, 1993) and in regulation of physical states such as body temperature, autonomic nervous system tone, appetite, and reproductive function. Thus, investigating changes in emotional, physical, and body states not only is a viable research avenue but also can provide a fundamental foundation for characterizing the full range of potential-circuit-based effects produced by putative human pheromones. We have employed such an approach in the research described next.

3. Psychological Effects of Androstadienone and Estratetraenol

In our laboratory we have recently been investigating a tantalizing development in the human pheromone field. Claims that two steroids, 4,16-androstadien-3-one (androstadienone) and 1,3,5,(10),16-estratetraen-3-ol (estratetraenol), are human pheromones have generated much debate. Androstadienone is found in men's sweat (Labows, 1988), semen (Kwan et al., 1992), and axillary hair (Nixon et al., 1988; Rennie et al., 1990), and estratetraenol has been isolated from the urine of pregnant women in the third trimester (Thysen, Elliott, and Katzman, 1968). Perfume marketers and their associated scientists claim behavioral releaser effects for these compounds, including stimulation of specific social cognitions or thoughts, such as increased friendliness, sociability, self-confidence, and sense of well-being (Blakeslee, 1993; Taylor, 1994). More recent findings, using a 70-item modification of the Derogatis Inventory, suggest a decrease in negative emotional self-ratings (not specifically related to social qualities) after exposure to androstadienone (Grosser et al., 2000). It also was reported that these compounds generated electrical responses from the human vomeronasal organ, not the olfactory epithelium, and that subjects did not detect any odors (Monti-Bloch and Grosser, 1991; Monti-Bloch et al., 1994). Furthermore, those electrophysiological responses were sex-specific: Androstadienone significantly stimulated only the female vomeronasal organ, whereas estratetraenol significantly stimulated only the male vomeronasal organ.

Although the electrophysiological findings have elicited considerable skepticism (Preti and Wysocki, 1999), those data presented not only an interesting development in the field but also an opportunity to investigate the effects of specific compounds. Our interest was to test the claims of specific effects on specific social cognitions or thoughts, such as increased self-confidence in social situations, as well as the sex specificity of the steroids, a claim based to date only on electrophysiological evidence.

Our hypothesis was that sex-specific releasing effects were unlikely in humans. Our alternative hypothesis was that human behavioral pheromones or social chemosignals would only modulate basic drive states, attention, or subcortical affect systems. Chemosignals might do this in several ways: (1) by affecting particular neurotransmitter systems in the brain (similar to the psychoactive effects of specific drug classes), (2) by activating or inhibiting brain circuits involved in emotion or motivation, or (3) by influencing arousal, attention, or memory-biasing systems. The predicted primary effect on behavior would be a change in the underlying tone or valence for perceiving external stimuli, in other words, a mood. We further hypothesized that cortical inputs could easily override a behavioral response. In that scenario, neither behavior nor specific cognitions would be expected to be reflexively associated with chemosignal exposure, as is the case in classical releasing pheromone systems.

We conducted two experiments (Jacob and McClintock, 2000). In both, we exposed our subjects to no more than 900 micromoles of either steroid in a propylene glycol carrier, applied under the nose with a cotton swab so that subjects would be exposed to it continuously. In order to minimize the chance of falsely accepting negative results, our concentrations were higher than those used by the perfumers, who directly delivered steroids to the vomeronasal organ or olfactory epithelium. In the initial study, we masked the steroids with a carrier of propylene glycol, which has a weak odor and was used by Monti-Bloch and colleagues (Monti-Bloch and Grosser, 1991; Monti-Bloch et al., 1994). In the second, replication study, we masked any odor from the steroids with a strong odor of clove oil added to the propylene glycol carrier.

To determine whether or not the steroids had psychological and sex-specific effects, we used a 15–20-minute psychometric battery consisting of well-established and validated scales for assessing the subjective effects of pharmacological agents on mood and psychological state (de Wit and Griffiths, 1991).

The Profile of Mood States (POMS) (McNair, Lorr, and Droppleman, 1971) was selected because it is an established measure that is sensitive to mood changes related to olfactory cues (Schiffman et al., 1995), to drug-induced changes in mood (de Wit and Griffiths, 1991; Fischman and Foltin, 1991), to hormonal-state influences (Kraemer et al., 1990), and to normal transient mood shifts in a wide range of circumstances (Lieberman et al., 1982; Der and Lewington, 1990; Horswill et al., 1990; Cockerill, Nevill, and Lyons, 1991; Williams, Krahenbuhl, and Morgan, 1991). The POMS scale we used in this study had 72 items in an adjective checklist format. From that extensive checklist, eight factors have been derived empirically, and these clusters of items are called Anxiety, Depression, Anger, Vigor, Fatigue, Confusion, Friendliness, and Elation.

The pharmacological-state Addiction Research Center Inventory (ARCI) (Haertzen, 1974a,b) questionnaire was originally developed to provide empirically derived scales to distinguish the sensations and perceptions uniquely associated with specific drugs or classes of drugs. The ARCI has five empirically derived scales: the morphine-benzedrine group (MBG) scale, which reflects drug euphoria; the lysergic acid diethylamide (LSD) scale, which reflects drug dysphoria and mental confusion; the pentobarbital-chlorpromazine-alcohol group (PCAG) scale, which measures level of sedation; the amphetamine (A) scale, which measures amphetamine-like stimulant effects; and the benzedrine group (BG) scale, which measures stimulant-like effects or intellectual efficacy.

A visual-analogue scale (VAS) is often used to assess momentary changes in affect (Folstein and Luria, 1973). We chose an established scale used in psychopharmacological research (Zacny, Bodker, and de Wit, 1992; Brauer and de Wit, 1997; de Wit, Clark, and Brauer, 1997; Kirk, Doty, and de Wit, 1998) that measures participants' responses to six adjectives ("stimulated," "high," "anxious," "sedated," "down," and "hungry").

In the initial experiment, we studied 10 women and 10 men in our laboratory. In each of three randomized, counterbalanced, and double-blind sessions, we applied a solution (estratetraenol, androstadienone, or propylene glycol carrier control) and then tested the participants' psychological responses. After an initial hour in our testing room, participants returned to their everyday lives and self-administered the test-battery questionnaires approximately 2, 4, and 9 hours after we had applied the test solutions.

We used similar methods in the replication experiment with the strong odor mask. However, we focused only on women and their responses to androstadienone within the first 2 hours (in contrast to the 9 hours covered in the first study). In order to increase statistical sensitivity, we measured psychological responses in terms of changes from their responses to a baseline battery administered prior to exposure to chemosignals, but after acclimation to the testing environment. All questionnaires were filled out in the testing room, except for one that was filled out 2 hours after application, roughly 1 hour after the subject had left the laboratory.

We found no support for a releaser effect nor for the marketing claims that androstadienone and estratetraenol have exclusively sex-specific effects and that they make individuals feel more social, self-confident, and friendly. At least in the context of everyday life in and around the university or within a laboratory setting, these steroids did not act as simple, sex-specific behavior releasers, cognitive releasers, or mental-state releasers. Nonetheless, in both settings, we did identify effects on general emotional states and demonstrated that these compounds were not exclusively mimicking psychoactive drugs in the major pharmacological classes (e.g., alcohol or opiates). Furthermore, at the low concentrations of androstadienone and estratetraenol that we presented, most people could not verbally describe a difference between the odor of the carrier with the steroids and its odor without the steroids (Figure 11.1). Subjects also did not distinguish between the strongly masked steroid and control presentations in the replication study. Thus, our data indicate that these steroids do not have to be detected consciously and identified as odors in order to exert their psychological effects.

Contrary to the sex-exclusive claims based on surface-potential activity of the vomeronasal organ (Monti-Bloch and Grosser, 1991; Berliner, 1994), both men and women responded to each steroid on a number of our measures. Compared with the control context, women responded positively to both steroids, but men

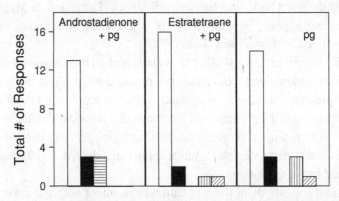

Figure 11.1. Smell descriptors and associations from 20 participants presented with androstadienone with propylene glycol (pg), estratetraene with pg, and pg alone. Unfilled bars indicate no smell was reported, solid bars indicate chemical descriptors, horizontal lines indicate musky descriptors, vertical lines indicate sweet/sour/bitter descriptors, and diagonal lines indicate floral descriptors. (Adapted from Jacob and McClintock, 2000.)

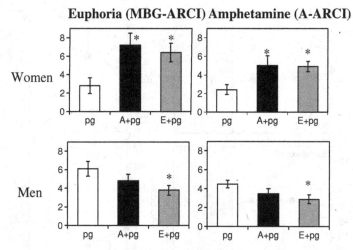

Figure 11.2. Absence of sex-exclusive psychological responses. Women reported initial euphoric and amphetamine-like responses to the estratetraenol (E) and propylene glycol mixture (E + pg; *p = .05), as well as to the androstadienone (A) and pg mixture (A + pg). Men reported decreased euphoria (MBG-ARCI) and amphetamine-like responses to estratetraenol (*p = .05).

responded negatively, especially in response to estratetraenol (Figure 11.2). There are several possible explanations for the lack of sex-specific results. Differences in methodologies and procedures between our study and previous ones could account for the difference in results. We used a repeated-measures design and controlled heavily for the context of the subjects' experience (e.g., odor-controlled room, trained testers, scripted procedures). In addition, our study looked at effects of long-term exposure to the steroids, as opposed to the immediate effects tested in previous studies (Monti-Bloch and Grosser, 1991; Berliner, 1994). We may have presented the steroids in a mode or concentration that affected other sensory processing systems – both steroids might produce their psychological effects via one or more systems that do not involve the vomeronasal organ. Given the possibility of multiple sensory systems responding to the stimuli (i.e., vomeronasal organ and olfactory epithelium), we may have obtained results that represent an interaction of the detection signals. Finally, psychological measures could simply be more sensitive to an overall response than are certain end-organ surface potentials.

The observed effects of androstadienone do not support a simple story of how these compunds affect attitudes related to social interactions. Androstadienone did not trigger feelings of self-assurance, sociability (VAS measures), or friendliness (POMS) at any time (Figure 11.3). It did, however, modulate women's mood states by preventing deterioration in general mood. Under the control condition,

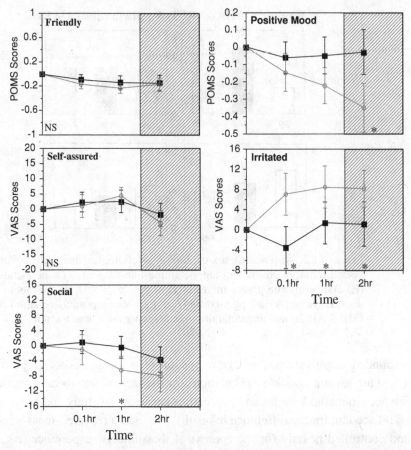

Figure 11.3. Time courses for subscales related to mood in women in the second replication study. Filled squares, androstadienone condition; open circles, propylene glycol odor-carrier condition; NS, no significant effects; *significant contrasts, $p = .05$. Included is a composite POMS Positive Mood score.

there was a gradual increase in negative mood that did not occur or was prevented under the androstadienone condition (see VAS, "Irritated," Figure 11.3). Similarly, there was a decrease in positive mood under the control condition that was prevented in the same subjects under the androstadienone condition (see POMS, "Positive Mood," Figure 11.3). As discussed in previous reports related to these data (Jacob and McClintock, 2000), the overall effect observed was determined most simply by calculating factors using factor analysis and varimax rotation. Androstadienone produced an overall positive shift in mood states in both studies, and estratetraenol had this effect in the first study. Inspection of individual measures from the initial study also suggested a drop in positive mood and an increase in negative mood at 9 hours after application. This suggests a rebound or "crash" from the influence of the steroids. This rebound pattern could be consistent with a withdrawal from a positive state. Alternatively, it could result from exposure to the steroids without an appropriate sociosexual context or in a manner not normally encountered in everyday life (e.g., concentration or time course at which the steroids evaporated and presented themselves to sensory epithelia). Characterizing the mechanisms of these longer-term psychological responses will require further study.

4. Putative Behavioral Pheromones?

Determining whether or not androstadienone and estratetraenol are pheromones will require more data. These steroids are not simply odors, because they exert their effects without being consciously detected or cognitively related to a particular odor or valenced memory. Although they do have characteristic odors at high concentrations, recognizable by a subset of people, their effects at low concentrations do not depend on the recipient being able to describe a particular odor, memory, or association. Thus, it is clear that the functional effects of these steroids do not necessitate that they be smelled consciously, an insight not possible with animal research.

How conscious versus unconscious odor detection contributes to the definition of a pheromone in higher mammals, and humans in particular, is something that can be determined only through additional research and discussion. To clarify existing terminology and concepts of chemical signals, we shall need to ascertain the functions of relevant compounds in a variety of contexts. We shall also need to ascertain their mechanisms of action. Where are their receptor sites? Are their effects neural or neuroendocrine? Are the primary and secondary olfactory cortices involved, and what other neural circuits ultimately process the chemical information? Finally, how do the various sensory systems integrate, if at all?

Because these steroids occur naturally in humans, they are strong candidates for being social signals, if not modulator pheromones, among humans. However, we do not yet know if these steroids, which have been tested only in a pure form, are produced, detected, and functional at natural levels and are social chemosignals that have evolved over time to mediate social interactions within the human physical environment. Moreover, there is always the possibility that these particular steroids are not themselves social signals, but merely agonists binding at a receptor for a different natural compound.

In sum, our data do demonstrate that these chemosignals are not classic behavior-releasing pheromones, because they are not prepotent stimuli driving a stereotyped behavior across a wide range of contexts and social environments. Instead, these steroids modulate emotional reactivity or mood associated with a social context, such as participating in our experiment. They may exert these effects by modulating the limbic system and prefrontal areas that regulate mood. They may also modulate attention, perception, and sensory integration involved in a variety of cognitive tasks or ongoing behaviors. Such effects would, of course, be highly context-dependent.

It must also be recognized that we cannot exclude the possibility that our observations reflect responses derived from a learned association with androstadienone. This raises the question whether or not all human "pheromone" effects can be completely dissociated from any learning or cognitive integration, as in more restrictive definitions of pheromones (Beauchamp et al., 1976). Determination of whether or not these particular steroids are "modulating pheromones," in parallel with releaser and primer pheromones, must await data on their bioavailability during human social interactions and a consensus on the use of the term "pheromone." In a natural setting, androstadienone and estratetraenol will be present alongside many other chemical and nonchemical signals adding to the overall sensory landscape of a particular social environment and context. What roles these consciously undetectable compounds play in the larger scheme of human chemical communication is an intriguing question that requires future research at the interface of chemosensory cognition, neurophysiology, and psychology.

Acknowledgments

This work was supported by the Mind-Body Network of the John D. and Catherine T. MacArthur Foundation, the NIH MERIT award R37 MH41788 to Martha K. McClintock, and the Olfactory Research Fund's Tova Fellowship to Suma Jacob.

References

Aggleton J P (1993). The Contribution of the Amygdala to Normal and Abnormal Emotional States. *Trends in Neuroscience* 16:328–33.

Beauchamp G K, Doty R L, Moulton D G, & Mugford R A (1976). The Pheromone Concept in Mammalian Chemical Communication: A Critique. In: *Mammalian Olfaction, Reproductive Processes, and Behavior*, ed. R L Doty, pp. 144–57. New York: Academic Press.

Berliner D L (1994). Fragrance Compositions Containing Human Pheromones. U.S. patent 5,278,141.

Berliner D L, Monti-Bloch L, Jennings-White C, & Diaz-Sanchez V (1996). The Functionality of the Human Vomeronasal Organ (VNO): Evidence for Steroid Receptors. *Journal of Steroid Biochemistry and Molecular Biology* 58:259–65.

Blakeslee S (1993). Human Nose May Hold an Additional Organ for a Real Sixth Sense. *New York Times*, pp. C3.

Boyse E A, Beauchamp G K, & Yamazaki K (1987). The Genetics of Body Scent. *Trends in Genetics* 3:97–102.

Brauer L H & de Wit H (1997). High Dose Pimozide Does Not Block Amphetamine-induced Euphoria in Normal Volunteers. *Pharmacology, Biochemistry and Behavior* 56:265–72.

Bronson F H (1968). Pheromonal Influences on Mammalian Reproduction. In: *Perspectives in Reproduction and Sexual Behavior*, ed. M Diamond, pp. 341–61. Bloomington: Indiana University Press.

Bruce H M (1960a). A Block to Pregnancy in the Mouse Caused by Proximity of Strange Males. *Journal of Reproduction and Fertility* 1:96–103.

Bruce H M (1960b). Further Observations of Pregnancy Block in Mice Caused by Proximity of Strange Males. *Journal of Reproduction and Fertility* 2:311–12.

Cockerill I M, Nevill A M, & Lyons N (1991). Modeling Mood States in Athletic Performance. *Journal of Sports Science* 9:205–12.

Cutler W B, Friedman E, & McCoy N L (1998). Pheromonal Influences on Sociosexual Behavior in Men. *Archives of Sexual Behavior* 27:1–13.

Degel J & Köster E P (1998). Implicit Memory for Odors: A Possible Method for Observation. *Perceptual and Motor Skills* 86:943–52.

Der D F & Lewington P (1990). Rational Self-directed Hypnotherapy: A Treatment for Panic Attacks. *American Journal of Clinical Hypnosis* 32:160–7.

de Wit H, Clark M, & Brauer L H (1997). Effects of *d*-Amphetamine in Grouped versus Isolated Humans. *Pharmacology, Biochemistry and Behavior* 57:333–40.

de Wit H & Griffiths R R (1991). Testing the Abuse Liability of Anxiolytic and Hypnotic Drugs in Humans. *Drug and Alcohol Dependence* 28:83–111.

Dorries K M, Adkins-Regan E, & Halpern B P (1995). Olfactory Sensitivity to the Pheromone, Androstenone, Is Sexually Dimorphic in the Pig. *Physiology and Behavior* 57:255–9.

Dorries K M, Adkins-Regan E, & Halpern B P (1997). Sensitivity and Behavioral Responses to the Pheromone Androstenone Are Not Mediated by the Vomeronasal Organ in Domestic Pigs. *Brain, Behavior and Evolution* 49:53–62.

Fischman M W & Foltin R W (1991). Utility of Subjective-Effects Measurements in Assessing Abuse Liability of Drugs in Humans. *British Journal of Addiction* 86:1563–70.

Folstein M F & Luria R (1973). Reliability, Validity, and Clinical Application of the Visual Analogue Mood Scale. *Psychological Medicine* 3:479–86.

Goldfoot D A (1981). Olfaction, Sexual Behavior, and the Pheromone Hypothesis in Rhesus Monkeys: A Critique. *American Zoologist* 21:153–64.

Goldfoot D A, Westerborg-Van Loon H, Groeneveld W, & Slob A K (1980). Behavioral and Physiological Evidence of Sexual Climax in the Female Stump-tailed Macaque (*Macaca arctoides*). *Science* 208:1477–9.

Gower D B (1972). 16-Unsaturated C19 Steroids: A Review of Their Chemistry, Biochemistry and Possible Physiological Role. *Journal of Steroid Biochemistry* 3:45–103.

Grosser B I, Monti-Bloch L, Jennings-White C, & Berliner D L (2000). Behavioral and Electrophysiological Effects of Androstadienone, a Human Pheromone. *Psychoneuroendocrinology* 25:289–99.

Haertzen C (1974a). *Addiction Research Center Inventory*. Washington, DC: National Institute of Mental Health.

Haertzen C (1974b). *An Overview of the Addiction Research Center Inventory (ARCI): An Appendix and Manual of Scales*. Rockville, MD: National Institute on Drug Abuse.

Horswill C A, Hickner R C, Scott J R, Costill D L, & Gould D (1990). Weight Loss, Dietary Carbohydrate Modifications, and High Intensity Physical Performance. *Medicine and Science in Sports and Exercise* 22:470–6.

Izard M K (1983). Pheromones and Reproduction in Domestic Animals. In: *Pheromones and Reproduction in Mammals*, ed. J G Vandenbergh, pp. 253–85. New York: Academic Press.

Jacob S & McClintock M K (2000). Psychological State and Mood Effects of Steroidal Chemosignals in Women and Men. *Hormones and Behavior* 37:57–78.

Jacob S, Zelano B, Gungor A, Abbott D, Naclerio R, & McClintock M K (2000). Location and Gross Morphology of the Nasopalatine Duct in Human Adults. *Archives of Otolaryngology – Head and Neck Surgery* 126:741–8.

Johnson A, Josephson R, & Hawke M (1985). Clinical and Histological Evidence for the Presence of the Vomeronasal (Jacobson's) Organ in Adult Humans. *Journal of Otolaryngology* 14:71–9.

Karlson P & Butenandt A (1959). Pheromones (Ectohormones) in Insects. *Annual Review of Entomology* 4:39–58.

Karlson P & Lüscher M (1959). "Pheromones": A New Term for a Class of Biologically Active Substances. *Nature* 183:55–6.

Keverne E B (1999). The Vomeronasal Organ. *Science* 286:716–20.

Kirk J M, Doty P, & de Wit H (1998). Effects of Expectancies on Subjective Responses to Oral Delta-9-tetrahydrocannabinol. *Pharmacology, Biochemistry and Behavior* 59:287–93.

Kirk-Smith M D, Van Toller C, & Dodd G H (1983). Unconscious Odour Conditioning in Human Subjects. *Biological Psychology* 17:221–31.

Kraemer R R, Dzewaltowski D A, Blair M S, Rinehardt K F, & Castracane V D (1990). Mood Alteration from Treadmill Running and Its Relationship to Beta-endorphin, Corticotropin, and Growth Hormone. *Journal of Sports Medicine and Physical Fitness* 30:241–6.

Kwan T K, Trafford D J, Makin H L J, Mallet A I, & Gower D B (1992). GC-MS Studies of 16-Androstenes and Other Studies of C19 Steroids in Human Semen. *Journal of Steroid Biochemistry and Molecular Biology* 43:549–56.

Labows J (1988). Odor Detection, Generation and Etiology in the Axilla. In: *Antiperspirants and Deodorants*, ed. C Felger & K Laden, pp. 321–43. New York: Marcel Dekker.

Lieberman H R, Corkin S, Spring B J, Growdon J H, & Wurtman R J (1982). Mood, Performance and Pain Sensitivity: Changes Induced by Food Constituents. *Journal of Psychiatric Research* 17:135–45.

Lorig T S, Herman K B, Schwartz G E, & Cain W S (1990). EEG Activity during Administration of Low-Concentration Odors. *Bulletin of the Psychonomic Society* 28:405–8.

McClintock M K (2000). Human Pheromones: Primers, Releasers, Signalers or Modulators? In: *Reproduction in Context*, ed. K Wallen & J E Schneider, pp. 355–420. Cambridge, MA: MIT Press.

McNair D M, Lorr M, & Droppleman L F (1971). *Profile of Mood States*. San Francisco: Educational and Industrial Testing Service.

Melrose D R, Reed H C B, & Patterson R L S (1971). Androgen Steroids Associated with Boar Odour as an Aid to the Detection of Oestrus in Pig Artificial Insemination. *British Veterinary Journal* 127:497–501.

Michael R P, Keverne E B, & Bonsall R W I (1971). Pheromones: Isolation of Male Sex Attractants from a Female Primate. *Science* 172:964–6.

Monti-Bloch L & Grosser B I (1991). Effect of Putative Pheromones on the Electrical Activity of the Human Vomeronasal Organ and the Olfactory Epithelium. *Journal of Steroid Biochemistry and Molecular Biology* 39:573–82.

Monti-Bloch L, Jennings-White C, Dolberg D S, & Berliner D L (1994). The Human Vomeronasal System. *Psychoneuroendocrinology* 19:673–86.

Moran D T, Jafek B W, & Rowley J C III (1991). The Vomeronasal (Jacobson's) Organ in Man: Ultrastructure and Frequency of Occurrence. *Journal of Steroid Biochemistry and Molecular Biology* 39:545–52.

Morris N M & Udry J R (1978). Pheromonal Influences on Human Sexual Behaviour: An Experimental Search. *Journal of Biosocial Science* 10:147–57.

Nixon A, Mallet A I, & Gower D B (1988). Simultaneous Quantification of Five Odorous Steroids (16-Androstenes) in the Axillary Hair of Men. *Journal of Steroid Biochemistry and Molecular Biology* 29:505–10.

Ober C, Hyslop T, Elias S, Weitkamp L R, & Hauck W W (1998). Human Leukocyte Antigen Matching and Fetal Loss: Results of a 10-Year Prospective Study. *Human Reproduction* 13:33–8.

Ober C, Weitkamp L R, & Cox N (1999). HLA and Mate Choice. In: *Chemical Signals in Vertebrates,* vol. 8, ed. R E Johnston, D Müller-Schwarze, & P W Sorensen, pp. 189–97. New York: Plenum Press.

Ober C, Weitkamp L R, Cox N, Dytch H, Kostyes D, & Elias S (1997). HLA and Mate Choice in Humans. *American Journal of Human Genetics* 61:497–504.

Preti G & Wysocki C J (1999). Human Pheromones: Releasers or Primers? In: *Chemical Signals in Vertebrates*, vol. 8, ed. R E Johnston, D Müller-Schwarze, & P W Sorensen, pp. 315–31. New York: Plenum Press.

Proust M (1922). *Swann's Way*. New York: Holt.

Rennie P J, Holland K T, Mallet A I, Watkins W J, & Gower D B (1990). 16-Androstene Content of Apocrine Sweat and Microbiology of the Human Axilla. In: *Chemical Signals in Vertebrates*, ed. D W MacDonald, D Müller-Schwarze, & S N Natynczuk, vol. 5, pp. 55–60. Oxford University Press.

Risold P Y & Swanson L W (1995). Evidence for a Hypothalamothalamocortical Circuit Mediating Pheromonal Influences on Eye and Head Movements. *Proceedings of the National Academy of Sciences USA* 92:3898–902.

Rivard G & Klemm W R (1990). Sample Contact Required for Complete Bull Response to Oestrous Pheromone in Cattle. In: *Chemical Signals in Vertebrates*, vol. 5, ed. D W McDonald, D Müller-Schwarze, & S E Natynczuk, pp. 627–33. Oxford University Press.

Rodriguez I, Greer C A, Mok M Y, & Mombaerts P (2000). A Putative Pheromone Receptor Gene Expressed in Human Olfactory Mucosa. *Nature Genetics* 26:18–19.

Sachs B (1997). Erection Evoked in Male Rats by Airborne Scent from Estrous Females. *Physiology and Behavior* 62:921–4.

Schiffman S S, Sattely-Miller E A, Suggs M S, & Graham B G (1995). The Effect of Pleasant Odors and Hormone Status on Mood of Women at Midlife. *Brain Research Bulletin* 36:19–29.

Schneider D (1974). The Sex-Attractant Receptor of Moths. *Scientific American* 231:28–35.

Schwartz G E, Beil I R, Dikman Z V, Fernandez M, Kline J P, & Peterson J M (1994). EEG Responses to Low-Level Chemicals in Normals and Cacosmics. *Toxicology and Industrial Health* 10:633–43.

Signoret J P (1976). Chemical Communication and Reproduction in Domestic
    Mammals. In: *Mammalian Olfaction, Reproductive Process, and Behavior,*
    ed. R L Doty, pp. 243–56. New York: Academic Press.
Signoret J P (1991). Sexual Pheromones in the Domestic Sheep: Importance and
    Limits in the Regulation of Reproductive Physiology. *Journal of Steroid
    Biochemistry and Molecular Biology* 39:639–45.
Singer A G (1991). A Chemistry of Mammalian Pheromones. *Journal of Steroid
    Biochemistry and Molecular Biology* 39:627–32.
Singer A G, Macrides F, Clancy A N, & Agosta W C (1986). Purification and Analysis
    of a Proteinaceous Aphrodisiac Pheromone from Hamster Vaginal Discharge.
    Journal of Biological Chemistry 261:13323–6.
Smith T D, Siegel M I, Burrows A M, Mooney M P, Burdi A R, Fabrizio P A,
    & Clemente F R (1999). Histological Changes in the Fetal Human Vomeronasal
    Epithelium during Volumetric Growth of the Vomeronasal Organ. In: *Advances in
    Chemical Signals in Vertebrates,* vol. 8, ed. R E Johnston, D Müller-Schwarze,
    & P W Sorensen, pp. 583–91. New York: Plenum Press.
Sorensen P W (1996). Biological Responsiveness to Pheromones Provides
    Fundamental and Unique Insight into Olfactory Function. *Chemical Senses*
    21:245–56.
Spielman A, Zeng X N, Leyden J, & Preti G (1995). Proteinaceous Precursors of
    Human Axillary Odor: Isolation of Two Novel Odor-binding Proteins. *Experientia*
    51:40–6.
Stern K & McClintock M K (1998). Regulation of Ovulation by Human Pheromones.
    Nature 392:177–9.
Tassinary L G (1985). Odor Hedonics: Psychophysical, Respiratory and Facial
    Measures. Unpublished dissertation, Dartmouth College.
Taylor R (1994). Brave New Nose: Sniffing out Human Sexual Chemistry. *Journal of
    NIH Research* 6:47–51.
Thysen B, Elliott W H, & Katzman P A (1968). Identification of
    Estra-1,3,5,(10),16-tetraen-3-ol (Estratetraenol) from the Urine of Pregnant
    Women. *Steroids* 11:73–87.
Tirindelli R, Mucignat-Caretta C, & Ryba N J P (1998). Molecular Aspects of
    Pheromonal Communication via the Vomeronasal Organ of Mammals. *Trends in
    Neuroscience* 21:482–6.
Vandenbergh J G, Witsett J M, & Lombardi J R (1975). Partial Isolation of a
    Pheromone Accelerating Puberty in Female Mice. *Journal of Reproductive
    Fertility* 43:515–23.
van der Lee S & Boot L M (1955). Spontaneous Pseudopregnancy in Mice. *Acta
    Physiologica et Pharmacologica Neerlandica* 4:442–4.
van der Lee S & Boot L M (1956). Spontaneous Pseudopregnancy in Mice II. *Acta
    Physiologica et Pharmacologica Neerlandica* 5:213–4.
Wedekind C & Füri S (1997). Body Odour Preferences in Men and Women: Do They
    Aim for Specific MHC Combinations or Simply Heterozygosity? *Proceedings of
    the Royal Society of London. Series B: Biological Sciences* 264:1471–9.
Wedekind C, Seebeck T, Bettens F, & Paepke A J (1995). MHC-dependent Mate
    Preferences in Humans. *Proceedings of the Royal Society of London. Series B:
    Biological Sciences* 260:245–9.
Williams T J, Krahenbuhl G S, & Morgan D W (1991). Mood State and Running
    Economy in Moderately Trained Male Runners. *Medicine and Science in Sports
    and Exercise* 23:727–31.

Wilson E O & Bossert W H (1963). Chemical Communication among Animals. *Recent Progress in Hormone Research* 19:673–710.

Wood R I (1998). Integration of Chemosensory and Hormonal Input in the Male Syrian Hamster Brain. *Annals of the New York Academy of Sciences* 855:362–72.

Wysocki C J, Dorries K M, & Beauchamp G K (1989). Ability to Perceive Androstenone Can Be Acquired by Ostensibly Anosmic People. *Proceedings of the National Academy of Sciences USA* 86:7976–8.

Wysocki C J & Meredith M (1987). *The Vomeronasal System*. New York: Wiley.

Yamazaki K, Yamaguchi M, Baranoski L, Bard J, Boyse E A, & Thomas L (1979). Recognition among Mice: Evidence from the Use of a Y-Maze Differentially Scented by Congenic Mice of Different Major Histocompatibility Types. *Journal of Experimental Medicine* 150:755–60.

Zacny J P, Bodker B K, & de Wit H (1992). Effects of Setting on the Subjective and Behavioral Effects of *d*-Amphetamines in humans. *Addictive Behaviors* 17:27–33.

12

Neural Correlates of Emotion Perception: From Faces to Taste

Mary L. Phillips and Maike Heining

Is there a relationship between the brain regions involved in perception of others' emotions and those important for olfaction and taste? The aim of this chapter is to discuss the nature of emotions, and in particular the studies that have examined brain regions important in the perception of distinct emotions, such as fear and disgust, and then to demonstrate that similar brain regions are involved in the perception of odors and flavors and emotive stimuli presented in other sensory modalities.

1. What Is the Relationship among Olfaction, Taste, and Emotion?

1.1. Emotions

What are emotions, and why do we have them? Dualist, or "feeling," theories proposed by Descartes and, in later years, by James (1890) describe emotions as epiphenomena, or nonfunctional feelings, separate from the physiological changes or behavior seen in response to provoking stimuli. Behaviorist theories, such as that of Skinner (1974), define emotions in terms of reinforced patterns of behavior. Cognitive theories dating from Aristotle emphasize the importance of cognitions as causal to emotions, with theorists such as Lyons (1992) describing the appraisal or interpretation of events, which then leads to physiological changes, as central to the formation of an emotion. Ekman (1992) has described emotions as "having evolved through their adaptive value in dealing with fundamental life-tasks." He argues that emotions are characterized by several unique features, including distinctive facial expressions, distinctive physiology, their presence in other primates, and distinctive antecedent events (Ekman, 1992). Intact perception and experience of emotion would thus appear to be vital, in evolutionary terms, for survival in the social environment.

How many different emotions are there? One theory (Cannon, 1927; Lyons, 1992) argues against separate, basic emotions, and instead suggests that a general level of arousal will be interpreted by an individual in terms of the events and evaluations with which it is associated. Davidson (1992) has proposed a single emotion continuum built on primitive adaptive responses: the range from approach (positive) to withdrawal (negative). The other type of theory (Darwin, 1965; Ekman, Friesen, and Ellsworth, 1982) argues for the existence of separate, basic emotions, proposing six: sadness, happiness, anger, surprise, fear, and disgust. This theory has become more popular in recent years as studies have begun to elucidate the neurobiological substrates for different emotions, as discussed later.

1.2. Olfaction, Taste, Fear, and Disgust

Odors and flavors can be classified as either pleasant or unpleasant, and enjoyed or avoided, respectively. Two specific emotions that therefore would appear to be of relevance to olfaction and taste are fear, important in prompting appropriate avoidance of unpleasant and threatening stimuli in the environment (LeDoux, 1996), and disgust (literally, "bad taste"). Disgust has, indeed, been defined in terms of a food-related emotion. Darwin (1965) wrote that disgust was "something offensive to the taste," and later authors described the emotion as "revulsion at the prospect of (oral) incorporation of an offensive object" (Rozin and Fallon, 1987). Often the objects of disgust have been identified as waste products of human and animal bodies (Angyal, 1941). In addition, the concept of disgust can be expanded to involve causes such as *violation of body borders* at points other than the mouth (Rozin and Fallon, 1987; Rozin, Lowery, and Ebert, 1994). This core concept of disgust has been further elaborated to include *animal-origin disgust*, encompassing the tendency of humans to emphasize the human–animal boundary, avoiding unnecessary contact with animals (Rozin and Fallon, 1987; Rozin et al., 1994). Other causes of disgust include *interpersonal contamination*, with disgust elicited by physical contact with unpleasant or unknown people (Rozin et al., 1994), and, finally, disgust can arise from the *moral or sociocultural domain* (Miller, 1997), with disgust at certain beliefs or behaviors, such as sexual abuse of children, acting as powerful means of transmitting social values (Rozin and Fallon, 1987). It has been argued that other complex emotions, such as shame, guilt, and embarrassment, are derived from the basic emotion of disgust, with the focus of disgust being on the self (Power and Dalgleish, 1997).

Are there similarities in the neural correlates of taste and olfaction and fear and disgust? The nature of the neural substrates underlying perceptions of different emotions will be reviewed, followed by a discussion of recent studies examining the neural correlates of olfaction and taste.

2. The Neurology of Emotion Perception: Current Issues

Despite increased interest in studies of the neurobiological basis of emotion perception in recent years (LeDoux, 1996), there are several areas in need of further study.

2.1. Do Different Emotions Have Specific Neural Substrates?

Lesion studies have implicated both the left hemisphere (Young et al., 1993) and the right hemisphere (Adolphs et al., 1996) in the perception of facial expressions. The valence hypothesis links positive (approach-related) emotions with the left, and negative (withdrawal-related) emotions with the right, frontotemporal brain regions (Davidson, 1995; Davidson and Irwin, 1999), and further evidence for that comes from a recent functional imaging study (Canli et al., 1998). Other authors have provided evidence indicating roles for the orbitofrontal cortex (e.g., Rolls, 1990; Hornak, Rolls, and Wade, 1996; Angrilli et al., 1999) and the retrosplenial cortex (Maddock, 1999) in emotion perception.

Lesion studies and functional neuroimaging studies have also been successful in demonstrating roles for different brain regions in neural responses to different specific emotions in humans. Many of those studies have employed as stimuli the facial expressions from the series of Ekman and Friesen (1976), in which subjects view depictions of faces displaying expressions of fear, disgust, anger, sadness, happiness, and surprise, in addition to a neutral expression. Those facial expressions can be transformed with computer software to depict different intensities of the emotions (Calder et al., 1997). There has been an assumption, which has not yet been refuted, that when viewing a facial expression, subjects empathize with the emotion displayed. The neural response to a specific facial expression therefore reflects not only recognition by the subject of that emotion in another but also the experience, in part, of the emotion.

With the use of such stimuli, it has been demonstrated that the amygdala is critical to perception of fearful facial expressions (Adolphs et al., 1994, 1995; Young et al., 1995; Breiter et al., 1996; Calder et al., 1996; Morris et al., 1996). Adolphs et al. (1994) demonstrated that patients with bilateral amygdala lesions were impaired in their ability to perceive facial expressions of fear, as well as their ability to match facial expressions. In a functional neuroimaging study employing positron-emission tomography (PET), Morris et al. (1996) demonstrated a positive correlation between the rate of amygdala blood flow and the intensity of fear depicted in a facial expression.

Although there have been fewer studies examining the nature of neural responses to other specific emotions, the anterior insula and components of a cortico-striatal-thalamic circuit implicated in perception of offensive stimuli in primates (Alexander, Crutcher, and DeLong, 1990), but not the amygdala, have been demonstrated to be important in the perception of facial expressions of disgust (Phillips et al., 1997). That study was the first to demonstrate differential neural responses to stimuli depicting different specific emotions.

Other recent studies have employed facial expressions to examine the neural correlates of perception of other basic emotions. Such studies have demonstrated the importance of the medial frontal cortex and the amygdala (Schneider et al., 1997; Blair et al., 1999) for recognition of sad facial expressions, and the orbitofrontal cortex and anterior cingulate gyrus for recognition of angry facial expressions (Blair et al., 1999). Studies examining the neural correlates of happiness have been inconclusive. Several different brain regions have been implicated in the neural response to this emotion, such as widespread decreases in blood flow over the whole brain (George et al., 1995), amygdala activation (Schneider et al., 1997), and activation in bilateral posterior cingulate gyri and medial frontal cortex and the left anterior cingulate gyrus (Phillips et al., 1998a). Such findings suggest that perception of a given positive emotion may not be dependent on activity in a discrete brain region, but instead may be associated with simultaneous activities in several different brain regions. Furthermore, positive emotions such as happiness may be processed in terms of the environmental context in which they exist, so that happiness occurring in the context of fear might be associated with activity in brain regions different from these activated during the processing of happiness occurring in an otherwise neutral context. It is possible, therefore, that happiness may be a more complicated and less "basic" emotion than previously thought. That problem will be the subject of future studies.

Other types of visual stimuli have been employed in studying the neural correlates of emotion perception. Emotive scenes from the International Affective Picture System (IAPS) (Lang, Bradley, and Cuthbert, 1997) have been employed in several studies. The IAPS comprises scenes rated as depicting different intensities of positive (pleasant), negative (aversive), and neutral emotions. When subjects have viewed those scenes, activation has been demonstrated predominantly in bilateral visual cortical regions (e.g., Lane et al., 1997; Lang et al., 1998; Lane, Chua, and Dolan, 1999). Other studies have employed emotive film excerpts, demonstrating in response to those stimuli activation in occipitotemporal cortex and limbic structures (Reiman et al., 1997) and medial prefrontal cortex (Paradiso et al., 1997). The results of those studies indicate that regions important for visual-object processing are activated to a greater extent by emotionally salient stimuli than by neutral, complex visual stimuli, despite efforts to control for the degree of visual complexity (luminance, color, detail) contained in the two types of stimuli. Some authors have suggested that that may reflect a modulation of activation in visual regions by regions important for emotion perception, such as the amygdala (Morris et al., 1998).

2.2. Are Specific Neural Substrates for Perception of Different Basic Emotions Supramodal?

It is important in evolutionary terms that an appropriate behavioral response be made to an emotionally salient stimulus, regardless of the sensory modality in which the stimulus is presented. There have been fewer studies examining neural responses to emotional stimuli presented in nonvisual modalities. One study to investigate perception and evaluation of the emotions projected by spoken words has demonstrated increased blood flow in the frontal cortex and cerebellum (Imaizumi et al., 1997). Impaired perception of fearful and angry vocalizations has been reported in a patient with bilateral amygdala lesions (Scott et al., 1997), and perception of fearful vocalizations has been reported to be associated with interactions between the amygdala and brainstem regions (Morris, Scott, and Dolan, 1999). In another study, amygdala activation to both fearful faces and sounds was reported in the same group of individuals, with the findings also suggesting a possible general role in emotional behavior for the superior temporal gyrus (Phillips et al., 1998b). A recent study (Blood et al., 1999) has shown a correlation between increasing dissonance or consonance of music and activity in areas of the brain involved in emotion processing, including the right parahippocampal gyrus and precuneus, bilateral orbitofrontal cortex, medial subcallosal cingulate gyrus, and right frontal polar regions.

Processing of painful tactile stimuli is thought to occur in many regions of the brain, but has been associated in particular with increased activity in primary and secondary sensory cortices, the anterior cingulate gyrus, the thalamus, and the insula (Casey et al., 1994; Oshiro et al., 1998). Other types of unpleasant sensory stimulation, including nonpainful and painful gastric stimulation, have been shown to activate bilateral central sulcal regions, the insula, and the frontoparietal operculum, with painful gastric stimulation associated with activation in bilateral insulae and the anterior cingulate gyrus (Aziz et al., 1997). Those findings are further support for the roles of limbic, frontal, and sensory cortical regions in the perception of emotionally salient stimuli, but further studies are needed to increase our understanding of the nature of the neural responses to different types of emotional stimuli presented in different sensory modalities.

2.3. Emotional Learning

Some authors have examined the neural correlates of the processes of learning and recalling emotionally salient events by asking subjects to view and subsequently remember events with either an emotional or nonemotional significance. Activation in the amygdala has been associated with the ability to remember emotionally salient stimuli and events (Cahill et al., 1996; Phelps and Anderson, 1997; Hamann et al., 1999), but further studies are needed to clarify the nature of the different neural systems underlying the learning and recall of emotional material.

Other studies have employed conditioning paradigms to determine the neural correlates of emotional learning. It has been well established from studies with non-human primates that fear conditioning involves the amygdala (e.g., LeDoux, 1996; Quirk, Armony, and LeDoux, 1997). In some functional imaging studies, human subjects have been allowed to develop a conditioned fear response to a stimulus such as an angry face paired with the nonconditioning stimulus (often a burst of white noise), with subsequent examination of the neural correlates of their perceptions of the conditioning stimulus (Buchel et al., 1998; LaBar et al., 1998). The results indicate that the amygdala is important for fear conditioning. Those and other studies (e.g., Breiter et al., 1996) have also highlighted the phenomenon of habituation of the amygdala response to fearful stimuli over time. Further studies employing more sophisticated methods of analysis, in which the temporal nature of the responses of specific brain regions implicated in the perception of specific emotions (time-series analysis and event-related fMRI paradigms), will help to clarify our understanding of the neural networks involved in emotion conditioning and learning.

3. Neural Correlates of Olfaction and Taste

The neural correlates of perceptions of negative emotions, particularly fear and disgust, have been shown to include the amygdala and insula, in addition to sensory and medial frontal cortices. The neural correlates of perceptions of pleasant and unpleasant odors and flavors might be predicted to include those structures. There have been fewer studies examining the neural correlates of olfaction and taste, and especially few examining neural responses to emotionally salient odors and flavors. In this review, we attempt in particular to compare the neural correlates of perceptions of emotionally salient odors and flavors with the neural responses to the basic emotions of fear and disgust.

3.1. Neural Correlates of Olfaction

Receptor cells from the olfactory epithelium project across the cribriform plate and innervate the olfactory bulb; the primary projection of the olfactory bulb neurons are to the piriform cortex, olfactory tubercle, anterior olfactory nucleus, amygdala, and entorhinal cortex (Price, 1987). The olfactory system is in fact the only sensory system that bypasses the thalamus and has direct projections to cortical areas. Lesion studies have highlighted the importance of the orbitofrontal cortex and temporal lobes in olfactory identification (Jones-Gotman and Zatorre, 1988) and discrimination tasks (Zatorre and Jones-Gotman, 1991), and the greater importance of the right, rather than left, anterior temporal cortex in olfactory memory (Rausch, Serafetinides, and Crandall, 1977). Furthermore, the role of the amygdala in olfaction is indicated by the demonstration that amygdalotomy is a successful treatment for olfactory hallucinations in patients with seizure disorders (Chitanondh, 1966).

Neuroimaging studies have highlighted the roles of the orbitofrontal cortex (e.g., Zatorre et al., 1992; Sobel et al., 1998) and the right frontal lobe in olfaction, with the latter activated to a greater extent in females than in males (Yousem et al., 1999). The roles of the piriform cortex, orbitofrontal cortex, amygdala, and entorhinal/hippocampal region in olfaction have been emphasized in a recent review of PET and fMRI studies of the human olfactory system (Zald and Pardo, 2000).

With regard to examination of the neural correlates of perceptions of emotionally salient odors, activation of the left medial frontal lobe, inferior frontal cortex, and bilateral insulae has been demonstrated during olfaction per se, with pleasant odors producing increased left insula activation, as compared with unpleasant odors (Fulbright et al., 1998). Another study has reported increased blood flow in the left orbitofrontal cortex and bilateral amygdalae in response to perception of unpleasant (although not specifically disgusting) odorants (Zald and Pardo, 1997). Perceptions of familiar odorants have also been associated with right orbitofrontal cortical activation (Royet et al., 1999). Although such studies have been few, the findings to date indicate that many of the brain regions associated with emotion perception are also involved in perception of olfactory stimuli: the orbitofrontal cortex, amygdala, and insula. Because odors are rarely devoid of emotional salience, it is probable that the involvement of the orbitofrontal cortex, amygdala, and insula demonstrated in the neural response to olfactory stimulation reflects, at least in part, processing of the emotional component of such a stimulus.

3.2. *Neural Correlates of Taste Perception*

There have been few lesion studies examining taste perception in subjects with lesions in higher cortical areas, rather than the brainstem or midbrain. Some epilepsy patients experience gustatory hallucinations, which can be induced by stimulation of the parietal or rolandic operculum (Hausser-Haw and Bancaud, 1987). According to a comprehensive review of studies investigating neural substrates for taste perception (Small et al., 1999), the insula, parietal and frontal opercula, and orbitofrontal cortex have been demonstrated to play important roles, predominantly within the right hemisphere. Those authors have suggested that orbitofrontal activity may be dependent on the motivational features of specific taste tasks, whereas the right hemisphere predominance for taste perception may be a result of the specialization of the left hemisphere for language. One study has also reported activation of the anterior cingulate gyrus and thalamus, in addition to the areas mentioned earlier (Faurion et al., 1998). Others have emphasized the importance of activation in the insula and perisylvian region and have reported functional lateralization of taste perception related to handedness in the inferior part of the insula (Cerf et al., 1998; Faurion et al., 1999). In one study in which neural responses to aversive gustatory stimulation were examined, amygdala and orbitofrontal cortex activation was demonstrated specifically in response to such stimulation (Zald et al., 1998).

The findings in such studies indicate that there is significant overlap between regions important for perception of distinct flavors and those involved in emotion perception. Flavors, like odors, frequently have emotional significance, and it is probable that they too are processed to a large extent in terms of their emotional component.

3.3. *Examination of the Overlap among Neural Responses to Emotional, Olfactory, and Gustatory Stimuli*

The results from the studies just discussed indicate that similar brain regions are activated by emotional stimuli presented in the visual modality and by odors and flavors – both pleasant and unpleasant. The potential overlap between brain regions activated by emotional stimuli presented in different sensory modalities, including the olfactory and gustatory modalities, was investigated directly in a recent study (Francis et al., 1999). Neural responses to a pleasant touch (velvet), a pleasant olfactory stimulus (vanilla), and a pleasant gustatory stimulus (glucose) were examined in the same group of four subjects. Similar (medial) regions of the orbitofrontal cortex were activated by all three types of stimuli, particularly right-sided. The pleasant olfactory and gustatory stimuli also activated similar regions of the bilateral anterior insulae. Those authors suggested that their findings provided evidence of a role for the orbitofrontal cortex in particular in the learning and representation of rewards, and they argued that emotions per se can be defined in terms of states elicited by reward (or punishing) stimuli. The findings of that and other studies therefore highlight the role of the orbitofrontal cortex and insula in emotion processing. Furthermore, odors and flavors appear to activate these regions in particular, suggesting that these stimuli are processed primarily in terms of their emotional salience.

4. Conclusion: From Faces to Olfaction and Taste

With the advent of functional neuroimaging techniques, it has become possible to examine the neural correlates of sensory and emotion perception. The earlier studies concentrated on investigation of neural substrates for perceptions of visual stimuli depicting facial expressions and unpleasant and pleasant scenes. Later studies have employed techniques to present emotionally salient stimuli in other sensory modalities: auditory, tactile, olfactory, and gustatory. It is clear from these studies that similar regions, in particular the insula, amygdala, and primary sensory and orbitofrontal cortical regions, are implicated in the perception of aversive stimuli presented in all five sensory modalities. The orbitofrontal cortex and insula are also activated by several different types of odors and flavors. This suggests that emotion processing and perception of odors and flavors have similar neural bases and that olfactory and gustatory stimuli seem to be processed to a significant extent in terms of their emotional content, even if not presented in an emotional context. The aim of future studies in this area will be to determine the nature of the specific neural systems involved in olfaction, taste, and emotion perceptions by employing analysis techniques to allow examination of the temporal and functional relationships between the different brain regions identified in these processes.

References

Adolphs R, Damasio H, Tranel D, & Damasio A R (1996). Cortical Systems for the Recognition of Emotion in Facial Expressions. *Journal of Neuroscience* 16:7678–87.

Adolphs R, Tranel D, Damasio H, & Damasio A R (1994). Impaired Recognition of Emotion in Facial Expressions following Bilateral Damage to the Human Amygdala. *Nature* 372:669–72.

Adolphs R, Tranel D, Damasio H, & Damasio A R (1995). Fear and the Human Amygdala. *Journal of Neuroscience* 15:5879–92.

Alexander G E, Crutcher M D, & DeLong M R (1990). Basal Ganglia–Thalamocortical Circuits: Parallel Substrates for Motor, Oculomotor, "Prefrontal" and "Limbic" Functions. *Progress in Brain Research* 85:119–46.

Angrilli A, Palomba D, Cantagallo A, Maietti A, & Stegagno L (1999). Emotional Impairment after Right Orbitofrontal Lesion in a Patient without Cognitive Deficits. *NeuroReport* 10:1741–6.

Angyal A (1941). Disgust and Related Aversions. *Journal of Abnormal and Social Psychology* 36:393–412.

Aziz Q, Andersson J L, Vakind S, Sundin A, Hamdy S, Jones A K, Foster E R, Langstrom B, & Thompson D G (1997). Identification of Human Brain Loci Processing Esophageal Sensation Using Positron Emission Tomography. *Gastroenterology* 113:50–9.

Blair R J R, Morris J S, Frith C D, Perrett D I, & Dolan R J (1999). Dissociable Neural Responses to Facial Expressions of Sadness and Anger. *Brain* 122:883–93.

Blood A J, Zatorre R J, Bermudez P, & Evans A C (1999). Emotional Responses to Pleasant and Unpleasant Music Correlate with Activity in Paralimbic Brain Regions. *Nature Neuroscience* 2:382–7.

Breiter H C, Etcoff N L, Whalen P J, Kennedy W A, Rauch S L, Buckner R L, Strauss M M, Hyman S E, & Rosen B R (1996). Response and Habituation of the Human Amygdala during Visual Processing of Facial Expression. *Neuron* 17:875–87.

Buchel C, Morris J, Dolan R J, & Friston K J (1998). Brain Systems Mediating Aversive Conditioning: An Event-related fMRI Study. *Neuron* 20:947–57.

Cahill L, Haier R J, Fallon J, Alkire M T, Tang C, Keator D, Wu J, & McGaugh J L (1996). Amygdala Activity at Encoding Correlated with Long-Term, Free Recall of Emotional Information. *Proceedings of the National Academy of Sciences USA* 93:8016–21.

Calder A J, Young A, Rowland D, & Perrett D (1997). Computer-enhanced Emotion in Facial Expressions. *Proceedings of the Royal Society of London* 264B:919–25.

Calder A J, Young A W, Rowland D, Perrett D I, Hodges J R, & Etcoff N L (1996). Facial Emotion Recognition after Bilateral Amygdala Damage: Differentially Severe Impairment of Fear. *Cognitive Neuropsychology* 13:699–745.

Canli T, Desmond J E, Zhao Z, Glover G, & Gabrieli J D E (1998). Hemispheric Asymmetry for Emotional Stimuli Detected with fMRI. *NeuroReport* 9:3233–9.

Cannon W B (1927). The James-Lange Theory of Emotions. A Critical Examination and an Alternative Theory. *American Journal of Psychology* 39:106–24.

Casey K L, Minoshima S, Berger K L, Koeppe R A, Morrow T J, & Frey K A (1994). Positron Emission Tomographic Analysis of Cerebral Structures Activated Specifically by Repetitive Noxious Heat Stimuli. *Journal of Neurophysiology* 71:802–7.

Cerf B, Le Bihan D, Van den Moortele P F, MacLeod P, & Faurion A (1998). Functional Lateralization of Human Gustatory Cortex Related to Handedness Disclosed by fMRI Study. *Annals of the New York Academy of Sciences* 855:575–8.

Chitanondh H (1966). Stereotaxic Amygdalotomy in the Treatment of Olfactory Seizures and Psychiatric Disorders with Olfactory Hallucination. *Confinia Neurologica* 27:181–96.

Darwin C (1965). *The Expression of the Emotions in Man and Animals.* University of Chicago Press. (Originally published 1872.)

Davidson R J (1992). Prolegomenon to the Structure of Emotion: Gleanings from Neuropsychology. *Cognition and Emotion* 6:245–68.

Davidson R J (1995). Cerebral Asymmetry, Emotion and Affective Style. In: *Brain Asymmetry*, ed. R J Davidson & K Hugdahl, pp. 361–87. Cambridge, MA: MIT Press.

Davidson R J & Irwin W (1999). The Functional Neuroanatomy of Affective Style. *Trends in Cognitive Sciences* 3:11–21.

Ekman P (1992). An Argument for Basic Emotions. *Cognition and Emotion* 6:169–200.

Ekman P & Friesen W P (1976). *Pictures of Facial Affect.* Palo Alto: Consulting Psychologists.

Ekman P, Friesen W V, & Ellsworth P (1982). What Emotion Categories or Dimensions Can Observers Judge from Facial Behaviour? In: *Emotion in the Human Face,* 22nd ed., ed. P Ekman, pp. 39–55. Cambridge University Press.

Faurion A, Cerf B, Le Bihan D, & Pillias A M (1998). fMRI of Taste Cortical Areas in Humans. *Annals of the New York Academy of Sciences* 855:535–45.

Faurion A, Cerf B, Van den Moortele P F, Lobel E, MacLeod P, & Le Bihan D (1999). Human Taste Cortical Areas Studied with Functional Magnetic Resonance Imaging: Evidence of Functional Lateralization Related to Handedness. *Neuroscience Letters* 277:189–92.

Francis S, Rolls E T, Bowtell R, McGlone F, O'Doherty J, Browning A, Clare S, & Smith E (1999). The Representation of Pleasant Touch in the Brain and Its Relationship with Taste and Olfactory Areas. *NeuroReport* 10:453–9.

Fulbright R K, Skudlarski P, Lacadie C M, Warrenburg S, Bowers A A, Gore J C, & Wexler B E (1998). Functional MR Imaging of Regional Brain Responses to Pleasant and Unpleasant Odors. *American Journal of Neuroradiology* 19:1721–6.

Hamann S B, Ely T D, Grafton S T, & Kilts C D (1999). Amygdala Activity Related to Enhanced Memory for Pleasant and Aversive Stimuli. *Nature Neuroscience* 2:289–93.

Hausser-Haw C & Bancaud J (1987). Gustatory Hallucinations in Epileptic Seizures. *Brain* 110:339–59.

Hornak J, Rolls E T, & Wade D (1996). Face and Voice Expression Identification in Patients with Emotional and Behavioural Changes following Ventral Frontal Lobe Damage. *Neuropsychologia* 34:247–61.

Imaizumi S, Mori K, Kritani S, Kawashima R, Sugiura M, Fukuda H, Itoh K, Kato T, Nakamura A, Hatano K, Kojima S, & Nakamura K (1997). Vocal Identification of Speaker and Emotion Activates Different Brain Regions. *NeuroReport* 8:2809–12.

James W (1890). *The Principles of Psychology,* 2 vols. New York: Holt.

Jones-Gotman M & Zatorre R J (1988). Olfactory Identification Deficits in Patients with Focal Cerebral Excisions. *Neuropsychologia* 26:387–400.

LaBar K S, Gatenby C, Gore J C, LeDoux J E, & Phelps E A (1998). Human Amygdala Activation during Conditioned Fear Acquisition and Extinction: A Mixed-Trial fMRI Study. *Neuron* 20:937–45.

Lane R D, Chua P M L, & Dolan R J (1999). Common Effects of Emotional Valence, Arousal and Attention on Neural Activation during Visual Processing of Pictures. *Neuropsychologia* 17:989–97.

Lane R D, Reiman E M, Ahern G L, Scwartz G E, & Davidson R J (1997). Neuroanatomical Correlates of Happiness, Sadness and Disgust. *American Journal of Psychiatry* 154:926–33.

Lang P J, Bradley M M, & Cuthbert B N (1992). A Motivational Analysis of Emotion: Reflex–Cortex Connections. *Psychological Science* 3:44–9.

Lang P J, Bradley M M, & Cuthbert B N (1997). *International Affective Picture System (IAPS).* NIMH Center for the Study of Emotion and Attention, University of Florida, Gainesville.

Lang P J, Bradley M M, Fitzsimmons J R, Cuthbert B N, Scott J D, Moulder B, & Nangia V (1998). Emotional Arousal and Activation of the Visual Cortex: An fMRI Analysis. *Psychophysiology* 35:199–210.

LeDoux J (1996). *The Emotional Brain.* London: Weidenfeld & Nicolson.

Lyons W (1992). An Introduction to the Philosophy of Emotions. In: *International Review of Studies on Emotion*, vol. 2, ed. K T Strongman, pp. 295–313. Chichester: Wiley.

Maddock R J (1999). The Retrosplenial Cortex and Emotion: New Insights from Functional Neuroimaging of the Human Brain. *Trends in Neurosciences* 22:310–16.

Miller S B (1997). *The Anatomy of Disgust*. Cambridge, MA: Harvard University Press.

Morris J S, Friston K J, Buchel C, Frith C D, Young A W, Calder A J, & Dolan R J (1998). A Neuromodulatory Role for the Human Amygdala in Processing Emotional Facial Expressions. *Brain* 12:47–57.

Morris J S, Frith C D, Perrett D I, Rowland D, Young A W, Calder A J, & Dolan R J (1996). A Differential Neural Response in the Human Amygdala to Fearful and Happy Facial Expressions. *Nature* 383:812–15.

Morris J S, Scott S K, & Dolan R J (1999). Saying It with Feeling: Neural Responses to Emotional Vocalizations. *Neuropsychologia* 37:1155–63.

Oshiro Y, Fuijita N, Tanaka H, Hirabuki N, Nakamura H, & Yoshiya I (1998). Functional Mapping of Pain-Related Activation with Echo-Planar MRI: Significance of the SII-Insular Region. *NeuroReport* 9:2285–9.

Paradiso S, Robinson R G, Andreasen N C, Downhill J E, Davidson R J, Kirchner P T, Watkins G L, Boles Ponto L L, & Hichwa R D (1997). Emotional Activation of Limbic Circuitry in Elderly Normal Subjects in a PET Study. *American Journal of Psychiatry* 154:384–9.

Phelps E A & Anderson A K (1997). What Does the Amygdala Do? *Current Biology* 7:311–13.

Phillips M L, Bullmore E, Howard R, Woodruff P W R, Wright I, Williams S C R, Simmons A, Andrew C, Brammer M, & David A S (1998a). Investigation of Facial Recognition Memory and Happy and Sad Facial Expression Perception: An fMRI Study. *Psychiatry Research: Neuroimaging* 83:127–38.

Phillips M L, Young A W, Scott S K, Calder A J, Andrew C, Giampietro V, Williams S C R, Bullmore E T, Brammer M, & Gray J A (1998b). Neural Responses to Facial Expressions of Fear and Disgust. *Proceedings of the Royal Society, London, B* 265:1809–17.

Phillips M L, Young A W, Senior C, Calder A J, Perrett D, Brammer M, Bullmore E T, Andrew C, Williams S C R, Gray J, & David A S (1997). A Specific Neural Substrate for Perception of Facial Expressions of Disgust. *Nature* 389:495–8.

Power M & Dalgleish T (1997). *Cognition and Emotion. From Order to Disorder*. Hove: Psychology Press, Erlbaum, Taylor and Francis.

Price J L (1987). The Central and Accessory Olfactory Systems. In: *Neurobiology of Taste and Smell*, ed. T E Finger & W L Silver, pp. 179–204. New York: Wiley.

Quirk G J, Armony J L, & LeDoux J E (1997). Fear Conditioning Enhances Different Temporal Components of Tone-evoked Spike Trains in Auditory Cortex and Lateral Amygdala. *Neuron* 19:613–24.

Rausch R, Serafetinides E A, & Crandall P H (1977). Olfactory Memory in Patients with Anterior Temporal Lobectomy. *Cortex* 13:445–52.

Reiman E M, Lane R D, Ahern G L, Schwartz G E, Davidson R J, & Friston K J (1997). Neuroanatomical Correlates of Externally and Internally Generated Human Emotion. *American Journal of Psychiatry* 154:918–25.

Rolls E T (1990). A Theory of Emotion and Its Application to Understanding the Neural Basis of Emotion. *Cognition and Emotion* 4:161–90.

Royet J P, Koenig O, Grégoire M C, Cinotti L, Lavenne F, Le Bars D, Costes N, Vigouroux M, Farget V, Sicard G, Holley A, Mauguière F, Comar D, & Froment J C (1999). Functional Anatomy of Perceptual and Semantic Processing for Odors. *Journal of Cognitive Neuroscience* 11:4–109.

Rozin P & Fallon A E (1987). A Perspective on Disgust. *Psychological Review* 94:23–41.

Rozin P, Lowery L, & Ebert R J (1994). Varieties of Disgust Faces and the Structure of Disgust. *Journal of Personality and Social Psychology* 66:870–81.

Schneider F, Grodd W, Weiss U, Klose U, Mayer K R, Nagele T, & Gur R C (1997). Functional MRI Reveals Left Amygdala Activation during Emotion. *Psychiatry Research: Neuroimaging* 76:75–82.

Scott S K, Young A W, Calder A J, Hellawell D J, Aggleton J P, & Johnson M (1997). Impaired Recognition of Fear and Anger following Bilateral Amygdala Lesions. *Nature* 385:254–7.

Skinner B F (1974). *About Behaviourism*. New York: Knopf.

Small D M, Zald D H, Jones-Gotman M, Zatorre R J, Pardo J V, Frey S, & Petrides M (1999). Human Cortical Gustatory Areas: A Review of Functional Neuroimaging Data. *NeuroReport* 10:7–14.

Sobel N, Prabhakaran V, Desmond J E, Glover G H, Goode R L, Sullivan E V, & Gabrieli J D E (1998). Sniffing and Smelling: Separate Subsystems in the Human Olfactory Cortex. *Nature* 392:282–5.

Young A W, Aggleton J P, Hellawell D J, Johnson M, Brooks P, & Hanley J R (1995). Face Processing after Amygdallotomy. *Brain* 118:15–24.

Young A W, Newcombe F, de Haan E H F, Small M, & Hay D C (1993). Face Perception after Brain Injury. Selective Impairments affecting Identity and Expression. *Brain* 116:941–59.

Yousem D M, Maldjian J A, Siddiqi F, Hummel T, Alsop D C, Geckle R J, Bilker W B, & Doty R L (1999). Gender Effects on Odor-stimulated Functional Magnetic Resonance Imaging. *Brain Research* 818:480–7.

Zald D H, Lee J T, Fluegel K W, & Pardo J V (1998). Aversive Gustatory Stimulation Activates Limbic Circuits in Humans. *Brain* 121:1143–54.

Zald D H & Pardo J V (1997). Emotion, Olfaction, and the Human Amygdala: Amygdala Activation during Aversive Olfactory Stimulation. *Proceeding of the National Academy of Sciences USA* 94:4119–24.

Zald D H & Pardo J V (2000). Functional Neuroimaging of the Human Olfactory System in Humans. *International Journal of Psychophysiology* 36:165–81.

Zatorre R J & Jones-Gotman M (1991). Human Olfactory Discrimination after Unilateral Frontal or Temporal Lobectomy. *Brain* 114:71–84.

Zatorre R J, Jones-Gotman M, Evans A C, & Meyers E (1992). Functional Localization and Lateralization of Human Olfactory Cortex. *Nature* 360:339–40.

Section Four
Memory

Concerning memory processes in olfaction, most of the authors herein note that the evidence is sparse, mainly because the standard procedures used in the fields of vision and hearing are difficult to transfer to olfaction. Early research in olfaction, mimicking studies of the processing of visual information, was oriented toward episodic and semantic functions. However, recently there has been increasing interest in olfaction specificity, therefore shifting the focus to implicit forms of odor memory and to priming effects.

Chapter 13, by Sylvie Issanchou and colleagues, explains how the study of odor memory specificity in everyday life led them to contrast (1) incidental (nonintentional) learning and implicit recollection and (2) consciously learned and consciously recollected memories of odors in laboratory studies. They examine the ecological validity of traditional laboratory experiments and of incidental-learning paradigms in the study of olfaction by trying to get at the experimental conditions that can best predict memory performances in everyday life. They prefer priming and same/different judgments to identification tasks. However, Issanchou and associates, like Maria Larsson (Chapter 14) and Mats Olsson and his colleagues (Chapter 15), note that in spite of their efforts, it has been unexpectedly difficult to show priming effects in olfaction, as compared with priming in other modalities, the challenge being to figure out whether priming effects depend on perceptual processes or on conceptual (naming) processes.

Maria Larsson, in questioning whether priming is conceptually driven (mediated by subvocal naming) or simply perceptually driven, finds that the evidence for priming effects in olfaction is still sparse and presents a mixed pattern of results. In sorting out the different memory processes traditionally identified in visual (and verbal) memory research, she finds that olfactory research has yielded puzzling evidence and that the links either to neurophysiological bases or to language (naming) remain much less straightforward than in the case of visual information.

In the introductory overview of their Chapter 15, Mats Olsson and his colleagues report on three experiments published in German by Wippich et al. (1993) that therefore may not have come to the general awareness of the international scientific community. They further present new data on olfactory repetition priming. That experimental paradigm was designed to avoid the problem that priming has tended to be too dependent on the processing of odor names, whereas the attention of participants should be turned toward the quality of odors during both study and test phases. However, in spite of their efforts, and in line with previous research, they find no statistically reliable evidence regarding perceptual repetition priming. They are therefore led to conclude that, when observed, priming effects rely mainly on verbal priming.

Steven Nordin and Claire Murphy, in Chapter 16, address the question whether or not studies of odor memory in populations with specific neurological or geriatric disorders, such as Alzheimer's disease, can provide a better understanding of the neural processes underlying odor memory. More precisely, they examine whether or not specific olfactory memory impairments rely on specific neurological abnormalities. A secondary issue in their chapter is the potential for use of olfactory memory tests for early diagnosis of Alzheimer's disease.

In Chapter 17, Johannes Lehrner and Peter Walla point out that although a considerable literature has been accumulated on odor perception, memory, and naming in studies of adults or newborns (see Schaal and colleagues, Chapter 26), less attention has been paid to the developmental processes of olfactory functions in childhood and young adulthood. The observation that consistent use of odor labels is strongly dependent on the relevance of the label given to a particular odor, with "veridical naming" producing the highest level of consistent naming, led them to conclude that more research is needed on semantic processing of odor memory in childhood.

In short, the reviews and the original studies of memory for odors presented here reveal many contrasts and a somewhat puzzling picture. The difficulties encountered in getting at reliable data similar to those obtained in the visual domain suggest the importance of olfactory memory specificity: incidentally learned and unnamed in everyday conditions. However, that contrasts with the fact that perceptual priming is not observed, but that evaluative and even hedonic (semantic?) priming seems possible. This last remark suggests that memory for odors relies on principles that are at variance with the principles of memory for visual and auditory stimuli and rather would seem to be connected to emotional and affective values, as discussed in the preceding section.

13

Testing Odor Memory: Incidental versus Intentional Learning, Implicit versus Explicit Memory

Sylvie Issanchou, Dominique Valentin, Claire Sulmont, Joachim Degel, and Egon Peter Köster

1. Introduction

How good is human memory for odors? Tests of memory involve two phases: an exposure or learning phase and a testing phase, separated by a retention period. During the exposure phase, stimuli are presented, and, depending on the instructions and experimental conditions, the odors can be memorized *incidentally* or *intentionally*. During the testing phase, the same odor stimuli are presented again, generally accompanied by new stimuli, and memory can be tested *explicitly* or *implicitly* (i.e., intentional retrieval by the subject may or may not be involved). Compared with other kinds of memory, such as verbal memory, pictorial memory, and face memory, there have been very few studies of odor memory. Moreover, most of those few studies investigated *consciously* learned and *consciously* recollected memories of odors. But in everyday life, odors are generally learned *incidentally*. Rarely does anyone decide "I should memorize this odor" (Baeyens et al., 1996; Haller et al., 1999, Sulmont, 2000). In other words, whereas in everyday life odor learning is nonintentional and its recollection is usually implicit, in laboratory studies odor memory has been evaluated using intentional learning and explicit recollection. That raises the question of the ecological validity of traditional laboratory experiments to test odor memory. Indeed, following Neisser (1976), we must wonder if that type of approach has not been ignoring some of the main features of odor memory as they occur in ordinary life.

To examine the effects of experimental paradigms on memory performance for odors is the main goal of this chapter. More precisely, we are interested in finding the experimental conditions that can best predict memory performance in everyday life. We shall begin by defining the testing procedures that are generally referred to as explicit and implicit memory tasks. Then we shall review the literature on odor memory to examine the effects of experimental conditions on memory performance. Previous work has suggested that the processes governing odor memory are different from those governing other types of nonverbal memory, or at least that odor memory has a variety of important distinguishing characteristics (Herz and Engen, 1996). Thus, we shall, when possible, compare the results obtained for odors with the results obtained with other nonverbal stimuli. In agreement with Murphy et al. (1991), we have chosen human faces as comparative nonverbal stimuli. Though there are obvious differences between these two types of stimuli, it is our opinion that odors and faces have some important common features. Like faces, odors provide social signals, elicit emotions, are difficult to describe, seem to be perceived holistically, elicit context-dependent memory effects, and are quite resistant to forgetting. To some extent these common features may have their origins in the way we acquire knowledge about these two types of stimuli. Like other nonverbal stimuli, such as music or paintings, odors and faces are learned through repeated exposures without explicit learning intention. During those exposures, sets of features useful for distinguishing among the stimuli are assembled little by little. For example, whereas at first two burgundy wines such as a Pommard and a Chambolle may seem rather similar, the difference between them will become increasingly clear over several exposures. In general, this type of learning, called perceptual learning (Gibson, 1969), is characterized by difficulty in precisely describing the features used to distinguish between stimuli, difficulty in generalizing to new stimuli, and the phenomenon of recognition/naming dissociation. Who has not experienced the feeling of recognizing a face, a melody, or an odor, but then being unable to recall anything else about it? When it applies to an odor, this phenomenon has been called the tip-of-the-nose effect by Lawless and Engen (1977).

What makes identifying faces different from identifying odors, however, is our great expertise in identifying faces. This great expertise results from the social importance of faces. Indeed, the face not only provides the most distinctive and useful information for identifying a known person but also enables us to infer the gender, the age, the emotional state, and perhaps even the health condition of an unfamiliar person. Earlier work on odors (Engen and Ross, 1973) showed that we are far from being able to identify large numbers of odors. Yet it seems that even though we are not very good at identifying odors, our capacity to discriminate between odors is as impressive as our capacity to discriminate faces (Holley, 1999). One plausible explanation for the discrepancy between our good discrimination performance and our poor recognition/identification performance is that although we are physiologically well equipped for perceiving odors, we do not use all our capacities, because the sense of smell is not crucial in day-to-day life. Congenitally anosmic individuals are not always aware of their loss (Engen, Gilmore, and Mair, 1991), whereas prosopagnosic individuals are severely impaired (Wacholtz, 1996).

2. Explicit versus Implicit Memory

It is now a well-known fact that even deeply amnesic patients are capable of learning. Thus, the often-described amnesic patient H.M. (Milner, Corkin, and Teuber, 1968) was able to learn new motor skills such as tracking a moving target. His performance in that task improved gradually with practice, even though he was not aware of having performed the task before. That type of memory has been called *implicit memory*, because subjects do not need to remember having encountered the situation before to perform better. The tasks used to evaluate implicit memory differ greatly from the explicit memory tasks usually used in a laboratory. Explicit memory tasks require conscious recollection of previously experienced events "through some sort of directed controlled search into stored information" (Perruchet and Baveux, 1989). In contrast, conscious or voluntary recollection of previous experiences is not necessary in implicit memory tasks. In an implicit memory task, memory is demonstrated by a change in perfor-mance, shorter latency or greater accuracy, on tasks involving either previously experienced items (i.e., repetition priming effect) or items closely associated to previously experienced items (i.e., semantic priming effect).

2.1. Explicit Tasks

The explicit tasks most commonly used are recall and recognition tasks. In a recall task, subjects are exposed to a series of stimuli (acquisition phase). Some time later, they are asked to recall as many of the stimuli as possible with (cued recall) or without (free recall) the help of a list of cues. Recall tasks typically are used to probe verbal memory. In that case, the stimuli would be a list of words, and the cues could be, for example, the first two letters of each word. Strict recall tasks would be quite difficult to use with faces and odors, as they would require that subjects re-create the stimuli (i.e., draw previously seen faces or recompose odors). Such tasks would be possible only for painters or perfumers. However, it is worth mentioning that such procedures have been used by Tuorila, Theunissen, and Ahlström (1996) and by Vanne, Tuorila, and Laurinen (1998) for testing quantitative memory for taste. Those authors asked their subjects to reproduce the subjective taste intensity of a tastant solution previously tasted, by mixing portions of a pure medium (with no tastant) and a high concentration of the tastant. Both studies showed that subjects tended to recall tastes as more intense than they had actually been. It is also worth mentioning that some authors have used a "recall" task with odors and faces, but in fact have asked their subjects either to provide a verbal description of the stimulus or to recall not the stimulus itself but the name of the stimulus. That last task, however, is possible only with familiar stimuli for which subjects already have names. That paradigm has been used, for example, by Hanley, Pearson, and Howard (1990) with photographs and by Annett and Lorimer (1995) for familiar odors of common household substances. For faces, subjects were first presented with a series of famous faces and then asked to write down the names of as many faces as they could remember. For odors, authors asked their subjects to write down the names, or brief descriptions, of as many of the previously smelled odors as they could remember. Then participants were presented with each odor again and asked to provide a verbal label or a brief description for it. Because of the idiosyncratic nature of odor labeling, that procedure was chosen to facilitate the scoring of recall responses. It is clear that such recall tasks with odors and faces are not equivalent to the recall task with words, because subjects are not asked to recall the stimuli per se, but to give the names of the stimuli.

The recognition task is the experimental paradigm most frequently used to study explicit memory for odors and faces. The acquisition phase is identical with that of a recall task, but the testing phase differs in that subjects have to recognize the target stimuli among distractor stimuli. This is usually done in one of two different ways: a two-alternative forced choice (2AFC) or a yes/no task. In a 2AFC task, stimuli are presented by pairs composed of an "old" stimulus (presented during the learning stage) and a "new" one. Subjects are asked to determine which stimulus in a pair is the old one. In a yes/no task, old or new stimuli are presented one at a time, and subjects are asked to answer "yes" or "no" to the question "Were you presented with this stimulus in the first part of the experiment?"

2.2. Implicit Tasks

Implicit memory tests were first designed to study implicit memory for words. The general principle of these tests is as follows: First subjects are presented with a list of words. Then they are asked to perform a task such as word-stem completion, word-fragment completion, tachistoscopic identification, or anagram resolution (Table 13.1). Half of the words presented during this task will have been in the list presented in the first part of the experiment (targets); the other half will be distractors. Implicit memory is demonstrated if the performance on the second task is greater for targets than for distractors. That type of test has been extended to the study of visual memory using picture-completion or picture-clarification tasks (Table 13.2). The general idea of "completion,"

Table 13.1. *Classic paradigms for implicit word-memory tasks*

Test	Learning stage	Retention	Retrieval stage	Retrieval task	Dependent variable
Word-fragment completion	Visual or auditory presentation of the target stimuli *autumn* *orange* *tuning*	Distractor task	Visual or auditory presentation of the target stimuli and of the new stimuli *a - - m -* *f - o - - -* *o - - n - -* *- u n - -* *v i - - - -* *ca - - - -*	Give the first word that comes to mind	Number of completed words (new and old)
Anagram	Visual presentation of the target stimuli *autumn* *orange* *tuning*	Distractor task	Visual presentation of the target stimuli and of the new stimuli *antumu* *fwoler* *oeangr* *gunint* *vinlio* *ctrroa*	Give the first solution that comes to mind	Number of solved words (new and old)
Tachistoscopic identification	Brief visual exposure to the target stimuli	Distractor task	Words embedded within a mask that gradually vanished	Give the first solution that comes to mind	Number of identified words (new and old)
Perceptual clarification	Visual presentation of the target stimuli	Distractor task	Words embedded within a mask that gradually vanished	Give the word as soon as it has been identified	Latency (new old words)

Table 13.2. *Classic paradigms for implicit picture-memory tasks*

Test	Learning stage	Retention	Retrieval stage	Retrieval task	Dependent variable
Picture completion	Identification of target stimuli presented from the most incomplete version to the complete version or	Distractor task	Visual presentation of the target stimuli and of the new stimuli	Same as learning stage	Gain between the first and the second presentations for the perceptual identification threshold (i.e., the level of fragmentation at which the picture was identified)
	Identification of target stimuli presented at one level of fragmentation	Distractor task	Visual presentation of the target stimuli and of the new stimuli	Identification of target stimuli presented from the most incomplete version to the complete version	Difference in perceptual identification threshold for old and new stimuli

however, is rather difficult to apply to odors and faces. Thus slightly different paradigms are generally used for those stimuli. They rely on the assumption that the presentation of a first stimulus sharing some properties (visual, olfactory, or semantic) with a second stimulus should affect the processing of the second stimulus more than would the presentation of an irrelevant stimulus.

Bruce and Valentine (1985) were the first to use that type of paradigm to study implicit memory for faces. They presented subjects with a series of famous-face pictures or names. The subjects' task was to identify the faces from their pictures or to read their names. After a 20-minute break they were asked to identify (exp. 1) or recognize (exp. 2) a second series of face images. That second series was composed of faces whose pictures had been presented in the first series, faces whose names had been presented in the first series, and new faces (control). Implicit memory was demonstrated when performance was better for faces from the first list than for new faces. Since that first work, numerous variations of that procedure have been used, but the general principle remains the same (see Bruyer, 1990, for a review). All such variations have converged to the finding that face identification is facilitated by previous presentation of either the faces themselves or their names, whereas face recognition (i.e., familiarity judgment) is facilitated only by previous presentation of the faces, not by previous presentation of their names.

In comparison, as discussed by Mair, Harrison, and Flint (1995), implicit memory for odors has received scant attention in the literature. Schab and Crowder (1995) were the first to report experiments testing implicit memory for odors. Their paradigm was similar to that used with faces. During the learning phase, subjects were exposed to a series of jars filled either with an odorant or with water. As each jar was presented, they were given the name of the odor they were actually (odor-and-name condition) or supposedly (name-only condition) smelling. During the testing phase, subjects were presented with a second series of jars containing an odorant. That series included the odorants presented during the learning phase in the odor-and-name condition and in the name-only condition, as well as new odorants. Their task was to identify (exp. 1) or detect (exp. 2 and 3) the odors, or judge their pleasantness (exp. 4). As before, implicit memory was demonstrated if performance was better for odors presented in the first series than for new odors. The data showed that, as was the case for faces, a name priming effect was observed in the identification task (i.e., a significant difference between the name-only condition and the control condition was observed) but not in the detection task. Because that effect was even larger in the name-plus-odor condition, the authors concluded that there was odor priming.

However, it is important to note that the paradigm used by Schab and Crowder differed on one point from that used by Bruce and Valentine (1985): In Schab and Crowder's experiments, odors were always presented with their names in the learning stage, but Bruce and Valentine's experiment included a learning condition in which faces were presented alone. As pointed out by Degel and Köster (1998), it is not clear that, given such conditions, a "pure" implicit memory for odors was tested. In fact, Schab and Crowder's experiments would allow to study the impact of verbal mediation in odor identification. As indicated by the authors themselves, the superior performances in identification scores and reaction times for the odor-and-name learning condition compared with the name-only learning condition may have been caused by a strengthening of the association between a perceptual representation and its related verbal label or by an increased availability of a verbal label because of prior activation of the perceptual representation in combination with its verbal label. That could explain why a priming effect was observed only in their first experiment, when the test task was an identification task. Nevertheless, because our identification capacity for odors is not crucial in day-to-day life, the identification task should be replaced by another task based on discrimination. In particular, it would seem possible to adapt the paradigm used for faces to odors by presenting an odors-only condition during the learning stage, and not an odors-and-labels condition.

In an attempt to minimize the effect of verbal processing, Olsson and Cain (1995) used a subvocal identification paradigm to demonstrate odor priming. Subjects were asked to indicate when they had last smelled six mono-rhinally presented odors. After a 10-minute break, they were presented with the same odors as well as six new odors. Their task was to press a button when they realized what they were smelling, without giving a verbal label. A positive priming effect (i.e., primed odors were identified faster than control ones) was observed when odors were presented to the left nostril. However, as pointed out later by Olsson (1999), even though subvocal identification was used, the testing task was a verbal task, and consequently that was not "a process-pure test of priming." In a later paper, Olsson (1999) reported another repetition priming experiment based on latency of identity rejection. A positive priming effect was observed for odors that could not be identified. On the contrary, a negative priming effect was observed for odors that could be identified. Thus, it seems that identification may interfere with the encoding or retrieval of odor memory. However, the task used during the learning stage (i.e., to judge when they had last smelled the odor), because of its emphasis on remembering, almost certainly would have precluded true measures of implicit memory in the final phase.

Recently, Degel and Köster (1999) reported a new method to analyze implicit memory for odors. During the learning stage, subjects completed a creativity test, a letter-counting concentration test, and a mathematical test in several weakly odorized rooms without being aware of the odors. In that experiment the presentation of the stimuli during the learning stage was somewhat similar to a tachistoscopic word presentation in that subjects were not aware of the stimuli. After a 30-minute retention time, subjects were shown photographic slides of different surroundings, including the room they had been in, and were asked to rate how well each of 12 odors, including the one they had been exposed to, fitted with each context shown in the pictures. The hypothesis was that if there was an effect of implicit memory for odors, the ratings of fit between odors and contexts would be positively influenced by odors previously experienced unconsciously and subliminally in the test rooms. The data showed that subjects who had worked in a room with a given odor subsequently assigned a higher fit for that odor to the picture of that room than did subjects who had worked in rooms with other odors, but that was true only for people who could not identify the odor by its name.

In summary, explicit and implicit tasks differ on several points. In an implicit memory test, learning is unintentional, retrieval is involuntary, and memory is tested indirectly, as subjects are not required to remember past events. In an explicit task, retrieval is voluntary, and memory is tested directly, whereas learning can be intentional or incidental. The notion of the intentionality versus the nonintentionality of learning and retrieval plays a crucial role in subjects' performances. The task at the learning stage determines the encoding process, which also has a key role in memory performance. We shall illustrate these points in the following sections.

3. Effects of Experimental Conditions on Memory Performance

3.1. Intentional versus Incidental Learning

Since the frequently cited study by Craik and Lockhart (1972), we know that all methods of encoding information are not equally useful. In particular, laboratory studies have shown that the simple intention to remember something is not enough if that intention is not followed by an effective elaborative encoding. Real-life experiences suggest that the intention to learn is not even a necessary condition. Indeed, most of the experiences that we recall from our day-to-day lives were not encoded with the deliberate intention to remember them. Accordingly, Engen and Ross (1973), in a study reporting three memory experiments with odors, the first one using intentional learning, and the other two incidental learning, noted that "when due allowance was made for experimental differences, the retention percentages found for incidental learning (experiments II and III) appear to be in line with the percentages found for intentional memory (in experiment I)." Likewise, Ayabe-Kanamura, Kikuchi, and Saito (1997), in

Table 13.3. *Impact of intentionality on recognition performances for faces and odors*

Type of stimulus	Group	Number of subjects	Type of learning	d' mean
Faces	1	25	incidental	1.46[a]*
	2	27	intentional	1.32[a]
Odors	1	25	incidental	1.69[a]
	2	27	intentional	1.62[a]

*d' values followed by the same letter within a cell are not significantly different.

studying the effects of verbal cues on recognition memory for unfamiliar odors, observed no difference in performance between intentional and incidental learning. Although both studies should be interpreted with caution, as different sets of odors were used in the intentional and incidental conditions, they suggest that odor can be learned incidentally. That conclusion has been supported more recently in a study by Senouci (1999) to evaluate the effect of learning intentionality on memory for unfamiliar odors and unfamiliar faces: During the learning phase, subjects were presented with a series of pairs of odors and a series of pairs of faces. Their task was to decide which odor, or which face, in each pair they preferred. In addition, half of the subjects were asked to memorize the stimuli. After a delay of a week, all subjects participated in a recognition task. The data show that, for both odors and faces, no significant difference was observed between the incidental and intentional learning conditions (Table 13.3).

3.2. Encoding Processes

Warrington and Ackroyd (1975) were the first to demonstrate that subjects' knowledge of the fact that a subsequent recognition test was to be taken had no effect on memory performance for faces. Those authors showed that the intention to learn per se was not a determining factor, but that how the subjects processed the information was most important. In particular, they showed that memory for faces was better following a judgment-of-pleasantness condition than following either a judgment-of-height condition or an intention-to-learn condition. Since that first work, many studies have been carried out to examine the effect of the type of processing on facial recognition (see Coin and Tiberghien, 1997, for a review). Those studies showed that the more elaborate and rich one's memory representations are, the better the memory performance. In line with that conclusion, Hanley et al. (1990) showed that subjects who were asked to name familiar faces performed better at a recognition test than did subjects who had been asked to make occupation or familiarity decisions. They hypothesized that the better performance in the naming task was due to the fact that naming a face gives rise to the elaboration of pictorial, semantic, and lexical codes.

A positive effect of naming has also been shown by most recent studies on odor memory. Lyman and McDaniel (1986, 1990) showed that recognition of odors was higher when subjects were asked to name odors at the inspection phase, as compared with control groups who were asked to remember or simply smell the odors. Several authors did not compare performances obtained when some subjects were asked to identify the odors and some were not, but rather tested only one group of subjects (who were asked to describe odors at the learning stage) and compared the results obtained for odors that were correctly identified or consistently named to the results obtained for odors that were not correctly identified or were inconsistently named. Whatever the type of learning – incidental for Lehrner (1993), and intentional for Rabin and Cain (1984) and Lehrner, Glück, and Laska (1999a) – they showed that correctly identified odors and consistently named odors were better recognized than incorrectly identified and inconsistently named odors, and respectively, they also found that consistency in labeling had a positive effect on odor recognition.

Sulmont (2000) also observed a positive effect of labeling consistency when subjects had to perform paired preference tests at learning, indicating that subjects could have spontaneously tried to identify odors. That positive effect of labeling is consistent with the results obtained by Jehl, Royet, and Holley (1992), who compared the recognition performances of subjects previously familiarized with the odors in different labeling conditions (no label, veridical label, chemical name, label personally generated by the subject) (on odor labeling, see Dubois and Rouby, Chapter 4, this volume). Indeed, those authors observed the highest recognition performances for the veridical label and the personally generated label, and the worst score for the no-label condition.

Finally, Lesschaeve and Issanchou (1996), who analyzed odor descriptions given by subjects in terms of precision (i.e., categorization level: orchard, fruit, apple) rather than in terms of accuracy (compared with a veridical name), also observed a positive effect of precision in labeling on odor recognition. From all those results, it appears that the key point for improving recognition is not the presence of a label but the "quality" of the label. To help subjects memorize an odor, the label must be appropriate (i.e., it must evoke personal experiences and/or evoke a relevant source of that odor for the subject). As an explanation, Lehrner et al. (1999b) suggested that the way odors are processed depends not only on the task but also on the way the task is performed. They assumed that odors that could not be correctly named were processed on a more perceptual and lower level than were correctly named odors. They found no age effect in recognition performances for incorrectly named odors, but young adults performed better than did children and the elderly for correctly named odors. Those authors concluded that there are two different forms of human odor memory: a perceptual memory and a semantic memory.

Most authors agree that naming is not necessary for memory, but enhances it, and most would agree that that can be accounted for by a dual-coding theory (Paivio, 1986). Lyman and McDaniel (1990) observed that recognition performance was higher when odors were presented with their names or with photographs of their sources than when only odors were presented, and the highest recognition was obtained when odors were presented together with their names and photographs of their sources. Consequently, those authors suggested an alternative to the dual-coding theory: a general elaborative-network model in which encoding representations from different modalities would provide more retrieval paths than would encoding in a single modality. They suggested that the key point was not the verbal coding, but the multimodality coding. In order to test the relevance of the dual-coding theory in olfactory memory, Perkins and McLaughlin Cook (1990) used a suppression paradigm: During the acquisition stage, subjects were presented odors in one of four different conditions. Subjects in the visual-suppression group had to play a computer game. Subjects in the verbal-suppression group had to listen through headphones to a random sequence of numbers and had to repeat each number as soon as they heard it. Subjects in the visual-plus-verbal-suppression group had to perform both tasks. There was no suppression for the control group. Recognition was tested 2.5 minutes and one week after the learning stage. As expected by the authors, recognition performance after one week decreased in the following order: control, visual, verbal, and visual-plus-verbal. After a delay of 2.5 minutes, only the verbal condition was significantly different from the control condition. However, as pointed out by the authors, the effects seen for those suppression tasks may simply have reflected the relative complexity of the total task.

In contrast to the foregoing, it should be noted that verbalization can also have a negative effect on memory performance. Fallshore and Schooler (1996), in an experiment on recognition performance for own-race faces versus other-race faces, observed that verbalization at the learning stage impaired recognition in the case of other-race faces. From those findings and from other data on music and visual forms, Melcher and Schooler (1996) hypothesized that verbal overshadowing occurs when the level of perceptual expertise and the level of verbal expertise differ. Their hypothesis was confirmed in an experiment conducted on wine, using subjects who did not drink wine, untrained wine drinkers, and trained

ANTARCTIC
EARTH SCIENCE
Fourth International Symposium

SYMPOSIUM SPONSORS

    Scientific Committee on Antarctic Research
    Australian Academy of Science
    University of Adelaide
    International Union of Geological Sciences
    Australian Academy of Technological Sciences
    Geological Society of Australia

SCAR STEERING COMMITTEE

    R. J. Adie (U.K.)
    C. Craddock (U.S.A.)
    F. Davey (N.Z.)
    J. C. Dooley (Australia)
    G. Grikurov (U.S.S.R.)
    R. L. Oliver (Australia), Chairman
    R. J. Tingey (Australia)

ADELAIDE ORGANIZING COMMITTEE

    D. H. Boyd
    C. M. Fanning
    C. G. Gatehouse
    F. Jacka
    J. B. Jago, Secretary
    P. R. James
    P. G. Law
    D. Moller, Treasurer
    R. L. Oliver, Chairman
    A. J. Parker
    E. A. Rudd
    C. C. Von Der Borch
    H. Worner

SUPPORTING ORGANISATIONS

    Ansett Airlines
    Australian Academy of Science
    Australian Academy of Technological Sciences
    Australian Consolidated Industries
    Australian Mineral Development Laboratories
    Broken Hill Proprietary Limited
    C.R.A. Exploration
    Department of Administrative Services
    Esso Australia Limited
    Ian Potter Foundation
    International Union of Geological Sciences
    Newmont Holdings
    Santos Limited
    Scientific Committee on Antarctic Research
    South Australian Brewing Holdings Limited
    South Australian Department of Mines and Energy
    South Australian Institute of Technology
    The Opal Mine
    University of Adelaide
    Utah Foundation
    Western Mining Corporation Limited

wine experts. Degel and Köster (1999, 2001) also reported a negative effect of verbal knowledge in observing that implicit memory for odors was found only in subjects who were unable to give the right names for the odors, but in their case, explicit verbalization did not come into play. As mentioned in Section 1.2, Olsson (1999) also observed a negative impact of odor identification on implicit memory.

3.3. Implicit versus Explicit Recollection

As pointed out by Mandler (1980), most examples of remembering involve both conscious and subconscious processes. First, it must be noted that even if the task instructions for tests of indirect measures of memory do not make any reference to the old/new character of the stimuli (i.e., presented or not presented during the learning phase), it is possible that uninformed subjects will realize early in the test session that some items have previously been presented, as noted by several investigators (e.g., Light and Singh, 1987). Consequently, as pointed out by Schacter (1990), even though subjects are not required to think back to the stimuli presented during the exposure phase, it is possible that some of them will adopt intentional, explicit retrieval strategies, thus turning the indirect test into a direct test.

3.4. Match or Mismatch between Encoding Operations and Retrieval Operations

As mentioned by Schab and Crowder (1995), dissociations between indirect tests could also be due to the degree to which the types of processing at encoding and at retrieval match. In the domain of face memory, Wells and Hryciw (1984), for example, observed that trait judgments (e.g., honesty–dishonesty) at the encoding stage yielded higher performances in full-face recognition than did feature judgments (e.g., narrow nose–wide nose). Even more interesting, they observed that feature judgments produced better reconstructions than did trait judgments. Those authors explained their findings by suggesting that making trait judgments involves holistic encoding processes and that face recognition involves a holistic process rather than a feature-based process. Likewise, in a second experiment, Lyman and McDaniel (1990) observed that when subjects were presented with a set of names of common objects, asking them to imagine the odors of those objects facilitated later odor recognition, and asking them to visually imagine those objects facilitated later visual recognition. The authors explained that effect in terms of the concept of "transfer-appropriate processing." According to that concept, the usefulness of different types of processing will depend on the degree to which the processing performed at encoding matches the demands of the retrieval task.

4. Similarity between Experimental Conditions and Real-Life Conditions

Like Neisser (1976), we think that memory research should have ecological validity. This means that we should examine naturally occurring memory phenomena and develop experimental designs that will allow us to place subjects in conditions as representative as possible of everyday situations. So, before concluding, we would like to point out the important differences between odor memory as it is studied in the laboratory and odor memory as it occurs in real life. How do we learn odors in everyday life? As mentioned by Köster (Chapter 3, this volume), we generally do not learn to name odors. The only people who actively learn to recognize odors and to match odors and labels are perfumers, flavorists, and members of sensory panels. For those of us who are not professionally involved in the world of odors, there are nevertheless some particular odors that we probably learn to identify in everyday life. The learning of odor names generally is limited to cases where odors are associated with toxic or dangerous substances, such as leaking gas and decaying products, and some taints, such as cork taint in wines, or rancid butter. Only a few of the names for the odors of commonly used foods are learned, more as a convenience for referring to the foods than as a reference to their odors, and as a consequence we find it difficult to identify those odors when they are presented without their foods. It is thus surprising that most research on odor memory has been focused on the impact of naming. Moreover, odors are learned incidentally, and subjects are not always aware of the presence of odorants.

It is also important to keep in mind that odors are remembered very differently, as compared with faces, for example. In everyday life, we have occasions to recognize faces, but also to identify a person's face, and occasionally we even need to recall faces when we "try to describe a face verbally to someone else, when we try to draw a face from memory, or when we try to picture a face by forming a mental image of it" (Cohen, 1989). As pointed out by Cohen, memory for people is "a crucial element in everyday life, both in social interaction and in work and family life." As mentioned earlier, odor identification usually is performed only in particular occupations and activities, such as perfumers' and flavorists' work and sensory analysis, but odor identification is also carried out spontaneously by wine amateurs who regularly try to identify origins, and vintages. But except in rare cases, odor identification is not performed in everyday life.

Odor recall is even more unusual. Odor recognition tends to occur in a spontaneous way, as when one perceives a familiar perfume at the cinema. The memory phenomenon that has most frequently been reported is that an odor cue can evoke memories of events that long ago were associated with the presence of that odor. Perhaps the most famous example was provided by Proust, who described how the flavor of a *madeleine* soaked in tea reminded him of his aunt's house. It is clear that the association between a particular odor and a particular context is learned incidentally and that such memories are retrieved automatically, without any effort. Moreover, several features of this phenomenon are worth noting. The first is the remarkable duration of the effect, with memories being evoked from months, years, even decades earlier (Degel and Köster, 1998; Aggleton and Waskett, 1999; Haller et al., 1999). The second feature is the precision and detail of the autobiographical events associated with the odor. It should be noted that such features are not exclusive to memories prompted by odors, but also can be observed with word and picture evocations (Rubin, Groth, and Goldsmith, 1984). However, it appears that memories evoked by odors elicit the strongest affective reactions. Herz and Cupchik (1995) reported that memory accuracy in recalling paintings was no better with an olfactory cue than with a verbal label, but odor-evoked memories were more emotional than verbally cued memories. So the majority of the experiments would seem to confirm the autobiographical reports, and there is also evidence for the strong emotional content of odor-evoked memories, which is one of the particular features of the Proust phenomenon. However, in all those laboratory experiments the retention interval was short (24 or 48 hours) compared with the retention interval that is typical for autobiographical accounts.

Aggleton and Waskett (1999) reported a study conducted in a real-world setting. In a test to determine if reexposure to the unique combination of odors present in a museum would aid subjects' recall of their visit to the museum, they asked subjects who had visited the museum in different odor conditions to complete questionnaires about the contents of the museum. Their findings indicated that the odors that had been present in the museum at the time of encoding could improve recall memory of various displays that had been seen, on average six years earlier, in the same odor context. Likewise, a study conducted by Haller et al. (1999) with 133 German adults revealed an influence of early experience with vanillin on food preferences later in life. In Germany, bottled milk for infants and babies had been flavored with vanilla for many years. When preferences for ketchup without or with vanilla were measured, it was found that those who had been bottle-fed as infants preferred ketchup with vanilla, whereas the breast-fed group preferred ketchup without it. That result demonstrates the effectiveness and the duration of unintentional learning in olfaction.

What do we learn about odors in everyday life? Though we do not learn to name odors, we learn to associate hedonic values to odors through various mechanisms that all involve memory. The first such mechanism is the mere exposure effect. Zajonc (1968) has suggested that repeated exposures constitute a sufficient condition for increased liking of stimuli: For foods, that effect has been demonstrated in adults by Pliner (1982) and in children by Birch and Marlin (1982), and their findings have recently been confirmed by Sulmont (2000). However, a detailed analysis of such results has revealed that the effect of repeated exposures on hedonic values is very dependent on both the stimulus and the subject. In many instances, repeated exposures to novel stimuli can even lead to a decrease in appreciation (Lévy and Köster, 1999). The second type of mechanism that can induce changes in our hedonic valuations of odors is associative conditioning. For example, Zellner et al. (1983) demonstrated that after pairing a neutral flavor with sugar (a hedonically positive test), subjects assessed increased hedonic value to the combination at the end of the presentation and one week later. Baeyens et al. (1990) reported a negative taste-flavor conditioning effect, but no significant taste-flavor conditioning and no color-flavor conditioning when the conditioned stimulus was hedonically positive. Those results reveal an asymmetry between negative and positive conditioning and show that not all associations are effective in inducing a hedonic shift. Social conditioning can also occur and can have a positive or negative effect on the hedonic value of a food, as demonstrated by the following experiments: Birch, Zimmerman, and Hind (1980) reported that if a neutral food was given to children either as a reward contingent on a pro-social behavior or paired with an adult's positive attention, a positive hedonic shift was observed, as compared with children who received the food in a non-social-affective context. To the contrary, Birch, Marlin, and Rotter (1984) reported a negative shift in preferences for foods eaten to obtain a reward. In the three preceding experiments, a hedonic reaction evoked by a stimulus (the unconditioned stimulus, US) was transferred to a previously neutral stimulus (conditioned stimulus, CS) presented with the US. To be effective, such simultaneous presentation of the US and the CS must be repeated several times. As reported by Baeyens et al. (1990), subjects had no explicit knowledge about the CS–US contingency relationship. Conditioning can also occur when a flavor is associated with positive or negative post-ingestion consequences. For example, Booth, Mather, and Fuller (1982) found a positive hedonic shift for a flavor paired with a high caloric content and offered to hungry subjects, as compared with a flavor presented an equal number of times, but with low caloric content. Using a questionnaire filled out by 198 subjects, Pelchat and Rozin (1982) found that when nausea follows ingestion of a food, people tend to develop a dislike for the taste of that food, but a negative hedonic shift is not reported for ingestion that is followed by other negative consequences, such as headache, diarrhea, or rash. Compared with the previously described cases of conditioning, this nausea conditioning is unique, because a single negative experience is sufficient.

5. Conclusion

Several features of odor memory as it occurs in the real world have been described, and they should be taken into account by anyone attempting to design better experimental procedures to study odor memory in the laboratory. First, odors are learned unintentionally, and in most cases without one's awareness. Consequently, future experiments with odor memory should take place in a context such that the subject's attention is not specifically drawn to the odor at learning time. That is, no task related to the odors should be given at the learning stage. When studying environmental odors, each odor should be diffused in a room in which subjects perform tasks or activities unrelated to the odor, as in the study of Degel and Köster (1999), or dissolved into the instruction sheets, as in the study by Kirk-Smith, Van Toller, and Dodd (1983). As the human olfactory system is not well oriented toward description, tasks presented to evaluate retrieval should avoid verbalization and preferably should be based on priming for hedonic judgments or for different–same judgments, rather than on identification. The previously discussed reports of conditioning effects seem to reveal an asymmetry between positive association and negative association in odor learning, the latter being more efficient, and possibly more resistant to change. Such possible asymmetry should be considered in future experiments on odor learning with conditioning. The major difficulty in trying to create a truly natural experimental condition is in reproducing associations involving emotions that will be as strong as they originally were when they occurred naturally. Then one must test those long-term memories so common in real life. However, we think that combining "field" studies and laboratory studies should lead to a better understanding of how odor memory works and what importance it has in food-preference dynamics.

References

Aggleton J P & Waskett L (1999). The Ability of Odours to Serve as State-dependent Cues for Real-World Memories: Can Viking Smells Aid Recall of Viking Experiences? *British Journal of Psychology* 90:1–7.

Annett J M & Lorimer A W (1995). Primacy and Recency in Recognition of Odours and Recall of Odour Names. *Perceptual and Motor Skills* 81:787–94.

Ayabe-Kanamura S, Kikuchi T, & Saito S (1997). Effect of Verbal Cues on Recognition Memory and Pleasantness Evaluation of Unfamiliar Odors. *Perceptual and Motor Skills* 85:275–85.

Baeyens F, Eelen P, van den Bergh O, & Crombez G (1990). Flavor–Flavor and Color–Flavor Conditioning in Humans. *Learning and Motivation* 21:434–55.

Baeyens F, Wrzesniewski A, de Houwer J, & Eelen P (1996). Toilet Rooms, Body Massages, and Smells: Two Field Studies on Human Evaluative Odor Conditioning. *Current Psychology: Developmental, Learning, Personality, Social* 15:77–96.

Birch L L & Marlin D W (1982). I Don't Like It; I Never Tried It: Effects of Exposure on Two-Year-Old Children's Food Preferences. *Appetite* 3:353–60.

Birch L L, Marlin D W, & Rotter J (1984). Eating as the "Means" Activity in a Contingency: Effects on Young Children's Food Preference. *Child Development* 55:431–9.

Birch L L, Zimmerman S I, & Hind H (1980). The Influence of Social-Affective Context on the Formation of Children's Food Preferences. *Child Development* 51:856–61.

Booth D A, Mather P, & Fuller J (1982). Starch Content of Ordinary Foods Associatively Conditions Human Appetite and Satiation, Indexed by Intake and Eating Pleasantness of Starch-paired Flavours. *Appetite* 3:163–84.

Bruce V & Valentine T (1985). Identity Priming in the Recognition of Familiar Faces. *British Journal of Psychology* 76:373–83.

Bruyer R (1990). *La reconnaissance des visages*. Neuchâtel (Switzerland): Delachaux et Niestle SA.

Cohen G (1989). *Memory in the Real World*. Hove: Lawrence Erlbaum.

Coin C & Tiberghien G (1997). Encoding Activity and Face Recognition. *Memory* 5:545–68.

Craik F & Lockhart R (1972). Levels of Processing: A Framework for Memory Research. *Journal of Verbal Learning and Verbal Behaviour* 11:671–84.

Degel J & Köster E P (1998). Implicit Memory for Odors: A Possible Method for Observation. *Perceptual and Motor Skills* 86:943–52.

Degel J & Köster E P (1999). Odors: Implicit Memory and Performance Effects. *Chemical Senses* 24:317–25.

Degel J & Köster E P (2001). Implicit Learning and Implicit Memory for Odors: The Influence of Odor Identification and Retention Time. *Chemical Senses* 26:267–80.

Engen T, Gilmore M M, & Mair R G (1991). Odor Memory. In: *Smell and Taste in Health and Disease*, ed. T V Getchell, R L Doty, L M Bartoshuk, & J B Snow, pp. 315–28. New York: Raven Press.

Engen T & Ross B M (1973). Long-Term Memory of Odors with and without Verbal Descriptions. *Journal of Experimental Psychology* 100:221–7.

Fallshore M & Schooler J W (1996). The Verbal Vulnerability of Perceptual Expertise. *Journal of Experimental Psychology: Learning, Memory and Cognition* 21:1608–23.

Gibson E J (1969). *Principles of Perceptual Learning and Development*. New York: Appleton-Century-Crofts.

Haller R, Rummel C, Henneberg S, Pollmer U, & Köster E P (1999).The Influence of Early Experience with Vanillin on Food Preference Later in Life. *Chemical Senses* 24:465–7.

Hanley J R, Pearson N A, & Howard L A (1990). The Effects of Different Types of Encoding Task on Memory for Famous Faces and Names. *Quarterly Journal of Experimental Psychology* 42A:741–62.

Herz R S & Cupchik G C (1995). The Emotional Distinctiveness of Odor-evoked Memories. *Chemical Senses* 20:517–28.

Herz R S & Engen T (1996). Odor Memory: Review and Analysis. *Psychonomic Bulletin and Review* 3:300–13.

Holley A (1999). *Eloge de l'odorat*. Paris: Odile Jacob.

Jehl C, Royet J P, & Holley A (1992). Role of Verbal Encoding Processes in Olfactory Memory. *Chemical Senses* 17:845.

Kirk-Smith M D, Van Toller C, & Dodd G H (1983). Unconscious Odour Conditioning in Human Subjects. *Biological Psychology* 17:221–31.

Lawless H & Engen T (1977). Associations to Odors: Interference, Mnemonics, and Verbal Labeling. *Journal of Experimental Psychology: Human Learning and Memory* 3:52–9.

Lehrner J P (1993). Gender Differences in Long-Term Odor Recognition Memory: Verbal versus Sensory Influences and the Consistency of Label Use. *Chemical Senses* 18:17–26.

Lehrner J P, Glück J, & Laska M (1999a). Odor Identification, Consistency of Label Use, Olfactory Threshold and Their Relationships to Odor Memory over the Human Lifespan. *Chemical Senses* 24:337–46.

Lehrner J P, Walla P, Laska M, & Deecke L (1999b). Different Forms of Human Odor Memory: A Developmental Study. *Neuroscience Letters* 275:17–20.

Lesschaeve I & Issanchou S (1996). Effects of Panel Experience on Olfactory Memory Performance: Influence of Stimuli Familiarity and Labeling Ability of Subjects. *Chemical Senses* 21:699–709.

Lévy C & Köster E P (1999). The Relevance of Initial Hedonic Judgements in the Prediction of Subtle Food Choices. *Food Quality and Preference* 10:185–200.

Light L L & Singh A (1987). Implicit and Explicit Memory in Young and Older Adults. *Journal of Experimental Psychology: Learning, Memory and Cognition* 13:531–41.

Lyman B J & McDaniel M A (1986). Effects of Encoding Strategy on Long-Term Memory for Odours. *Quarterly Journal of Experimental Psychology* 38:753–65.

Lyman B J & McDaniel M A (1990). Memory for Odors and Odor Names: Modalities of Elaboration and Imagery. *Journal of Experimental Psychology: Learning, Memory and Cognition* 16:656–64.

Mair R G, Harrison L M, & Flint D L (1995). The Neuropsychology of Odor Memory. In: *Memory for Odors*, ed. F R Schab & R G Crowder, pp. 39–69. Mahwah, NJ: Lawrence Erlbaum.

Mandler G S (1980). Recognizing: The Judgment of Previous Occurrence. *Psychological Review* 87:252–71.

Melcher J M & Schooler J W (1996). The Misremembrance of Wines Past: Verbal and Perceptual Expertise Differentially Mediate Verbal Overshadowing of Taste Memory. *Journal of Memory and Language* 35:231–45.

Milner B, Corkin P, & Teuber H (1968). Further Analysis of the Hippocampal Amnesic Syndrome: Fourteen Year Follow-up Study of H.M. *Neuropsychologia* 6:215–34.

Murphy C, Cain W S, Gilmore M M, & Skinner R B (1991). Sensory and Semantic Factors for Recognition Memory for Odors and Graphic Stimuli: Elderly versus Young Persons. *American Journal of Psychology* 104:161–92.

Neisser U (1976). *Cognition and Reality: Principles and Implications of Cognitive Psychology*. New York: Freeman.

Olsson M J (1999). Implicit Testing of Odor Memory: Instances of Positive and Negative Repetition Priming. *Chemical Senses* 24:347–50.

Olsson M J & Cain W S (1995). Early Temporal Events in Odor Identification. *Chemical Senses* 20:753.

Paivio A (1986). *Mental Representations: A Dual Coding Approach*. Oxford University Press.

Pelchat M L & Rozin P (1982). The Special Role of Nausea in the Acquisition of Food Dislikes by Humans. *Appetite* 3:341–51.

Perkins J & McLaughlin Cook N (1990). Recognition and Recall of Odours: The Effect of Suppressing Visual and Verbal Encoding Processes. *British Journal of Psychology* 81:221–6.

Perruchet P & Baveux P (1989). Correlational Analyses of Explicit and Implicit Memory Performance. *Memory and Cognition* 17:77–86.

Pliner P (1982). The Effects of Mere Exposure on Liking for Edible Substance. *Appetite* 3:283–90.

Rabin M D & Cain W S (1984). Odor Recognition: Familiarity, Identifiability and Encoding Consistency. *Journal of Experimental Psychology: Learning, Memory and Cognition* 10:316–25.

Rubin G R, Groth R E, & Goldsmith D J (1984). Olfactory Cueing of Autobiographical Memory. *American Journal of Psychology* 97:493–507.

Schab F R & Crowder R G (1995). Implicit Measures of Odor Memory. In: *Memory for Odors*, ed. F R Schab & R G Crowder, pp. 71–91. Mahwah, NJ: Lawrence Erlbaum.

Schacter D L (1990). Introduction to "Implicit Memory: Multiple Perspectives." *Bulletin of the Psychonomic Society* 28:338–40.

Senouci K (1999). *Influence de la nature de l'apprentissage sur la mémorisation des odeurs et des visages*. Diplôme d'Etudes Approfondies de Sciences de l'Alimentation, Option Science des Aliments. Université de Bourgogne.

Sulmont C (2000). Impact de la mémoire des odeurs sur la réponse hédonique au cours d'une exposition répétée. Thèse de doctorat, Université de Bourgogne.

Tuorila H, Theunissen M J M, & Ahlström R (1996). Recalling Taste Intensities in Sweetened and Salted Liquids. *Chemical Senses* 21:29–34.

Vanne M, Tuorila H, & Laurinen P (1998). Recalling Sweet Taste Intensities in the Presence and Absence of Other Tastes. *Chemical Senses* 23:295–301.

Wacholtz E (1996). Can We Learn from the Clinically Significant Face Processing Deficits, Prosopagnosia and Capgrass Delusion? *Neuropsychology Review* 6:203–57.

Warrington E & Ackroyd C (1975). The Effect of Orienting Tasks on Recognition Memory. *Memory and Cognition* 3:140–2.

Wells G L & Hryciw B (1984). Memory for Faces: Encoding and Retrieval Operations. *Memory and Cognition* 12:338–44.

Zajonc R B (1968). Attitudinal Effects of Mere Exposure. *Journal of Personality and Social Psychology. Monograph Supplement* 9:1–27.

Zellner D A, Rozin P, Aron M, & Kulish C (1983). Conditioned Enhancement of Humans' Liking for Flavors by Pairing with Sweetness. *Learning and Motivation* 14:338–50.

14

Odor Memory: A Memory Systems Approach

Maria Larsson

Research focusing on human memory suggests that memory is not a single or unitary faculty of the mind. Instead, it can be conceived of as a variety of distinct and dissociable processes and systems that are subserved by particular constellations of neural networks that mediate different forms of learning (e.g., Gabrieli et al., 1995; Tulving, 1995; McDonald, Ergis, and Winocur, 1999).

One categorical distinction within human memory is that between declarative memory and non-declarative memory (Tulving, 1995). According to this division, non-declarative memory is characterized by unintentional learning, or learning without awareness, and by inability to access conscious recall. This form of memory is manifested in multiple dissociable processes and is measured in terms of changes in performance (produced by conditioning and priming) in the learning of motor skills (e.g., Schacter, 1992). In contrast, declarative memory can be conceptualized as learning with awareness and refers to the acquisition and retention of information about events and facts, and is typically assessed by accuracy in tests of recall and recognition. The available evidence suggests that medial temporal/diencephalic structures are critical for the integrity of declarative memory, whereas non-declarative forms of memory rely on other brain areas, such as occipital structures (visual priming) (Gabrieli et al., 1995) and basal ganglia (procedural memory) (Heindel et al., 1989).

Although based on hypothetical constructs and still under considerable debate, the view that human memory is composed of at least five different systems has been highly influential (e.g., Nyberg and Tulving, 1996; Roediger, Buckner, and McDermott, 1999). According to the fivefold classification system described by Schacter and Tulving (1994), the non-declarative and declarative classes of human memory can be decomposed into five interrelated memory systems that have evolved both phylogenetically and ontogenetically in the following order: procedural memory, perceptual representation system (PRS), semantic memory, working memory, and episodic memory. The latter three systems typically are described as declarative forms of memory, whereas the first two are examples of non-declarative expressions of memory. It is noteworthy that the system that evolved last, episodic memory, is the form of memory that has been proved to be the most sensitive to disturbances (e.g., aging, depression, dementia, health status), whereas the systems that developed earliest are considerably more robust. The highly developed neural complexity supporting the functions of episodic memory, as contrasted with the more localized mnemonic properties of the systems developed earlier (procedural, PRS), likely underlies the sensitivity of this form of memory. It is also of interest to note that the brain structures especially critical to the functioning of episodic memory have been shown to undergo gradual changes that start in early adulthood (Braak, Braak, and Mandelkov, 1993; Simic et al., 1997) and mimic life-span data, covering memory functioning from early to late adulthood (e.g., Nilsson et al., 1997).

During the past decade, several reviews of studies in olfactory memory have been presented, but none has explicitly tried to conceptualize the various expressions of olfactory memory in the realm of a memory-systems framework (Richardson and Zucco, 1989; Schab, 1991; Herz and Engen, 1996). Most research on odor memory has been oriented toward episodic and semantic memory functions, with little attention paid to non-declarative aspects of olfactory memory. However, increasing interest in implicit forms of odor memory has been noted, and this is particularly true for the principles governing odor conditioning (Otto, Cousens, and Rajewski, 1997; Robin et al., 1999). Also, researchers have been trying to determine whether or not priming effects can be shown for olfactory information (e.g., Olsson, 1999; Olsson et al., Chapter 15, this volume).

The main aim of this chapter is to organize some of the available knowledge on the behavioral and psychological manifestations of odor memory using a memory-systems approach. The review is highly selective, and the reports cited serve only as examples to be used in constructing a theoretical framework that potentially may further our understanding of the various expressions of olfactory memory. Also, we shall consider some neuropsychological evidence and some brain-imaging evidence indicating that various olfactory functions draw on different neural correlates. A classification scheme for the different olfactory functions as they relate to each memory system is provided in Table 14.1.

1. Non-declarative Memory

1.1. Procedural Memory

Procedural memory is a form of memory underlying the acquisition of skills and other aspects of knowledge that are not directly accessible to consciousness and

Table 14.1. *Classification scheme*

Memory system	Olfactory function
Procedural memory	Odor conditioning; aversions
Perceptual representation system	Odor priming
Semantic memory	Hedonics; familiarity; identification; metamemory
Working memory	Odor discrimination
Episodic memory	Odor-recognition memory

whose presence can be demonstrated only indirectly by action (e.g., walking, skiing, conditioned responses). Procedural memory has been subdivided into a number of other systems, including (1) skills and habits, (2) simple associative learning or conditioning, and (3) nonassociative forms of learning such as habituation and sensitization (Schacter and Tulving, 1994).

Procedural memory may have appeared early during evolution and is shared in various forms by most living organisms (Tulving, 1993). In contrast to other forms of memory, non-declarative memory is little influenced by the passage of time, which is reflected in the fact that we can adequately perform tasks that we have not carried out in years, and we can experience conditioned responses to stimuli that we may not have encountered for decades (Robin et al., 1999).

Procedural memory comes into play in one of the most prominent features of the olfactory sensory system: the production of odor and taste aversions. The development of aversions occurs as a response to the ingestion of a sickness-provoking substance, which from an evolutionary perspective makes sense in that the aversion works as self-protection against food poisoning (Borison, 1989). It is of interest to note that the ability to learn to avoid ingesting toxic substances is present in simple invertebrates, which implies that relatively primitive neural networks can mediate this form of learning (Sahley, Gelperin, and Rudy, 1981).

However, there are situations in which elicitation of conditioned aversions can be problematic. For instance, patients undergoing anti-cancer treatment often experience nausea and vomiting, which tends to make them averse to further treatment. That can eventually result in the anticipatory syndrome in which the patient experiences nausea and vomiting when thinking of the treatment, or when visiting the hospital environment and being exposed to the sights, sounds, and smells associated with the treatment (Andrews and Sanger, 1993). Cancer patients can also develop an aversion to food that is consumed in close association with the treatment. Conditioned odor/taste aversions can be difficult to extinguish because of the lengthy association between the particular food and previous illness (Bernstein, 1978; Garcia et al., 1985; Hursti et al., 1992).

In a similar vein, recent evidence suggests that the principles governing acquisition of conditioned emotional (fear) responses to auditory and visual information are also valid for olfactory information, although it is not clear that odor-conditioned fear is acquired more rapidly or is more robust than other forms of conditioned fear (Otto et al., 1997). Robin et al. (1999) examined negative conditioning for the odorant eugenol. Eugenol is a substance commonly used in tooth fillings and in restorative dentistry, and it is responsible for the typical odor of dental offices. When fearful and nonfearful dental patients were exposed to eugenol, it was seen that the evoked responses of the autonomic nervous system were associated with positive basic emotions (happiness, surprise) in nonfearful subjects, whereas fearful subjects displayed negative basic emotions (fear, anger, disgust). That finding provides evidence of the potential for olfactory stimuli to serve as conditioning elicitors of negative or positive responses, depending on the previous associations with specific odors.

1.2. Perceptual Representation System

"Repetition priming" refers to the unconscious facilitation of performance following prior exposure to a target item or a related stimulus (Schacter, 1992). Like tasks in procedural memory, priming tasks have been referred to as "implicit," because the test instructions do not inform the participants to actively think back to a previous study episode. Considerable research efforts have been devoted to exploration of lexical and visual priming, but the available evidence on the possibility of priming for olfactory information is sparse and presents a mixed pattern of results (Schab and Crowder, 1995; Olsson, 1999; Olsson et al., Chapter 15, this volume).

In a series of experiments attempting to explore the potential for odor priming, Schab and Crowder (1995) examined the effects of previous exposure to an odor and/or exposure to its name on odor-detection and odor-identification thresholds. Their data yielded only weak support for the idea of odor priming, although presentation of a suprathreshold odor enhanced subsequent ability to identify that particular odor. Likewise, Olsson and Cain (1995) reported that primed odors were identified faster than control odors and that the priming effect was more pronounced when the olfactory information was presented in the left nostril rather than in the right nostril. Although the possibility of subvocal naming was not controlled for, that pattern of results might indicate that the priming effect was related to subvocal naming, suggesting a conceptually driven priming effect rather than a perceptually driven priming effect.

2. Declarative Memory

2.1. Semantic Memory

"Semantic memory" or "generic memory" refers to our general knowledge of the world. It encompasses the meanings of words, concepts, and symbols and their associations, as well as rules for manipulating these concepts and symbols (Tulving, 1993). As opposed to episodic memory, the information in semantic memory is stored without reference to the temporal and spatial context that was present at the time of its storage. Semantic memory also involves knowledge about one's own memory proficiency and one's own memory processes. This aspect of semantic memory has been termed "metamemory" (Flavell and Wellman, 1977).

Semantic memory in an olfactory context concerns a person's knowledge or experience with a specific odorant and typically is exemplified in odor identification and perceptions of familiarity. However, semantic memory also comes into play in perceptual experiences of hedonic odor properties and in tip-of-the-nose states. The following sections describe a number of olfactory functions and states tapping reservoirs of prior knowledge about odors.

Hedonic Valence. A fundamental question in olfactory research is whether hedonic responses to odors are present at birth or whether affective reponses are acquired postnatally, in interaction with experience (Menella and Beauchamp, 1997). The available data show a mixed pattern of findings that most likely is related to difficulties in assessing affective reactions in pre-verbal children, as well as to problems in selecting test odorants (i.e., whether affective responses are the results of olfactory or trigeminal stimulation).

Engen (1982) postulated that all odor preferences are learned relatively late in the pre-school period. That position is based on reports indicating that children less than five years old do not show systematic, adult-like odor-preference patterns. For example, Stein, Ottenberg, and Roulet (1958) reported that four-year-old children were as likely to say that they liked the smell of amyl acetate as they were to say that they liked the smell of synthetic sweat and feces. That response pattern was reversed among six-year-olds, who reported aversive sensations for feces and sweat smells, and pleasant sensations for amyl acetate. In opposition to those findings, several studies have shown that affective odor reactions may begin in early infancy, suggesting that some odors may be inherently pleasant or unpleasant (Steiner, 1979; Strickland, Jessee, and Filsinger, 1988). Further research into the developmental aspects of semantic memory is needed. Much more research into the relationships among genetic mechanisms, physiological maturation, and various experiential factors is of critical importance if we are to achieve a better understanding of the development of odor preferences.

Odor Familiarity. In contrast to identification tasks, which require explicit retrieval of verbal descriptors of odorants, familiarity ratings require only that a subject indicate a perceived familiarity on a scale varying from low to high. Thus familiarity can be regarded as a continuum covering the subject's implicit level of odor knowledge. A low familiarity rating may reflect an extremely vague perception, with no distinct semantic cues elicited by the olfactory experience. A medium rating may involve moderately meaningful associations, whereas a high familiarity rating presumably reflects the experience of having access to more specific knowledge about the odor. Occasionally, a high familiarity rating may simply reflect knowledge of the odor name, which would make this measure equivalent to identification. That presumably is true in some cases, but considering the high incidence of tip-of-the-nose experiences for olfactory stimuli, an item rated as highly familiar may truly reflect access to a more general or idiosyncratic knowledge of the odor.

Odor Identification. Performance at odor identification can be assessed by free identification or in a multiple-choice format. It is widely accepted that naming odors spontaneously is extremely difficult. Reviews indicate that unaided free identification of odors by young laypersons varies between 22% and 57%, with set sizes ranging from 7 to 80 items (e.g., Cain, 1979; Richardson and Zucco, 1989; Chobor, 1992). In contrast, when subjects have access to a number of response alternatives, they perform better (Doty, Shaman, and Dann, 1984; Larsson et al., 1999). Most likely that superiority is related to a decrease in cognitive demands: The provision of label alternatives reduces the effort an individual has to invest in searching for an appropriate label, which ultimately yields a more supportive task situation.

Related studies oriented toward life-span changes in olfactory functioning have indicated that aging effects are minor in perceptions of familiarity, whereas aging has deleterious effects on odor-identification ability (Larsson and Bäckman, 1993; Lehrner et al., 1999). This dissociative pattern of results suggests that increasing age is associated with retrieval impairment regarding the specificity of information (i.e., the verbal label), although semantic knowledge, as reflected in perceptions of familiarity, remains unaffected by aging. Similar findings have been reported for visual information (Kempler and Zelinski, 1994).

A somewhat neglected area of interest is the relationship between general semantic functioning (e.g., vocabulary, information) and proficiency in odor identification. Consequently, we examined this issue in a recent study (Larsson, Finkel, and Pedersen, 2000) in which we found evidence of that odor-identification aptitude is strongly and positively related to performance in other semantic memory tasks (i.e., analogies, vocabulary, information). We noted that odor identification proved to be unrelated to performance in other cognitive functions, such as visuospatial functioning, short-term-memory functioning, and episodic-memory functioning. This observation strengthens the notion that se-mantic memory (i.e., crystallized intelligence) and odor identification tap the same cognitive domain. Further support for this was provided in a follow-up study addressing the heritability aspects of various odor functions. There the re-sults indicated that the relationship between odor identification and measures of verbal ability was primarily genetically mediated (Finkel, Pedersen, and Larsson, 2001).

Metamemory. The general definition given for "metamemory" is that it involves knowledge of and monitoring of one's own memory proficiency and one's own memory processes (Flavell and Wellman, 1977). Studies examining the relationship between predicted and actual olfactory performances are sparse, although there is some evidence that subjects tend to overestimate their odor-memory performances and odor-identification abilities (Schab, 1991).

One element of metamemory knowledge is the commonly experienced feeling of knowing that underlies tip-of-the-tongue experiences (i.e., to know a word, but not be able to recall it) (Brown and McNeill, 1966; Brown, 1991). The olfactory analogue of this phenomenon was described by Lawless and Engen (1977) as the tip-of-the-nose experience. However, in contrast to the former, persons in the tip-of-the-nose state typically cannot answer any questions about the name of the odor, such as the initial letter, the number of syllables, or the general configuration of the word. Subjects can, however, answer questions about the odor's quality, such as its taxonomic category, or say something about objects associated with it (Lawless and Engen, 1977).

2.2. Short-Term Memory/Working Memory

Short-term memory can be divided into two components: primary memory and working memory. Both types of memory are indices of short-term memory, because the items to be remembered reside in consciousness. Where they differ has to do with whether or not additional processing operations are performed on the information to be remembered. Specifically, items in primary memory are maintained passively, whereas items in working memory are manipulated in some fashion (Baddeley, 1986).

It is well established that working memory can be subdivided into a central executive system and at least two slave systems – one specialized for verbal information, and the other for nonverbal (visual) information (Baddeley, 1992) – that depend on slightly different cortical regions (Vallar and Papagno, 1995; Owen, 1997).

Although earlier research suggested that olfactory memory consisted only of long-term-memory components (e.g., Engen, 1982, 1991), today the general picture suggests the presence of a viable short-term/working-memory olfactory system (see White, 1998, for a review), although its storage capacity and processing efficiency remain unclear. Two olfactory abilities that immediately involve working memory are discrimination and odor matching. In these tasks, the subject must be able to form a transient representation of the target odor and use that representation for comparison with the subsequent odorant. It has been noted that odor discrimination seems to be particularly affected by age, which is in keeping with age effects in working memory for verbal and visual information (Salthouse, 1995; Larsson et al., 1999). It is important to remember that some of the earlier research that focused on short-term retention for olfactory information may have confounded output from short-term memory with actual retrieval from episodic memory. For example, in traditional memory research, a simple and effective way of discriminating items residing in short-term memory from items in long-term memory (i.e., episodic memory) is to evaluate recency effects in recall tasks (Tulving and Colotla, 1970). Odor memory is assessed by means of recognition with a fixed test order, making the interpretation of any recency effects difficult. Recency effects in recognition memory should be treated with caution, because it is highly likely that they reflect a temporal gradient in episodic memory, thus suggesting that the most recently consolidated episodic information will be best remembered.

As noted earlier, working memory is composed of at least two slave systems; verbal and nonverbal. As compared with the verbal system, the nonverbal system in working memory is yet relatively unexplored. Speculatively, it is possible that working memory may be composed of an additional number of modality-specific short-term stores (e.g., olfactory working memory). Whether that is true or whether all incoming olfactory information is manipulated and processed in either or both of the slave systems remains to be addressed in future research.

2.3. Episodic Memory

Episodic memory deals with the acquisition and retrieval of information that was acquired in a particular place at a particular time. It involves traveling back in time to remember personally experienced events through conscious recollective processes (Tulving, 1993). That orientation toward the past makes episodic memory phenomenologically different from other varieties of memory. In the laboratory, episodic memory typically is assessed by asking persons to recall or recognize some information encountered earlier in an experimental setting.

It is well known that episodic-memory performance is a context-dependent phenomenon related to encoding/retrieval factors (e.g., encoding activity, recall or recognition), the nature of the information to be remembered (e.g., visual, verbal), and subject-related factors (e.g., health status), as well as to the multiple interactions among those factors (Jenkins, 1979). Earlier work suggested that episodic memory for odors differed fundamentally from episodic retention of verbal and visual information. That dissimilarity was reflected in a flat forgetting function, no influence of semantic factors, and a negligible impact of retroactive interference (e.g., Lawless and Cain, 1975; Engen, 1987). However, that view has changed, as later research has revealed that episodic memory for odors is to a large extent governed by the same principles that govern information acquired via other modalities (see Larsson, 1997, for a review).

Addressing the interdependence between the memory systems, it is of interest to highlight some of the interplay observed among working memory, semantic memory, and proficiency in episodic-memory odor recognition. Episodic memory for odors has been proved to vary according to the degree and type of elaboration (imagery, verbal elaboration) the subjects engage in during encoding or during exposure to an odorant that relates to working-memory processing (Lyman and McDaniel, 1990; Larsson and Bäckman, 1993). Also, recognition memory for odors is sensitive to the degree of familiarity and identifiability of the olfactory information, which are instances of semantic memory (Rabin and Cain, 1984; Larsson, 1997).

2.4. Neuropsychological and Physiological Evidence

As noted at the outset, one implication of the memory-systems theory is that different memory functions are subserved by certain constellations of neural networks responsible for different forms of learning. Although the available evidence speaking to this issue is sparse, there have been some clinical observations and neuroimaging studies suggesting that different olfactory functions are indeed mediated by and dependent on different underlying substrates (Eichenbaum et al., 1983; Savic et al., 2000).

The most striking illustration that olfactory functions can be dissociated by cerebral damage was manifested in H.M., a patient with bilateral medial temporal lobe resection (Eichenbaum et al., 1983). In contrast to a normal capacity to detect odors, to make intensity discriminations, and to adapt to odors, H.M. was unable to discriminate the quality of odors, whether tested by same–different judgments or matching-to-sample tasks. Moreover, although H.M. was successful in naming common objects using visual and tactile cues, he could not identify them by smell. Those findings suggest that structures in the medial temporal lobe play critical roles in quality discrimination and odor identification.

Relatedly, Savic et al. (1997) examined episodic odor recognition and odor-quality discrimination in two groups of epileptic patients and a healthy control group. The epilepsia groups included one group with mesial temporal lobe seizures (MTLS) and a group of patients suffering from neocortical seizures (NS). The results showed that MTLS patients were selectively impaired in odor discrimination relative the NS group and controls, whereas odor-memory performances were equal for the two epileptic groups, but lower than for controls. That pattern of results indicates that different pathological states can induce dissociable impairments in higher olfactory functions such as quality discrimination and episodic memory for odors, which ultimately implies that the two functions are mediated by different neurological structures.

In order to investigate whether or not odor perception, discrimination, and episodic-memory odor recognition are mediated by different sets of brain structures, and to explore the neurobiological correlates of those specific tasks, we used positron-emission tomography (PET) to measure regional cerebral blood flow in a group of healthy adults (Savic et al., 2000). The different tasks varied in degree of complexity and differed in terms of the cognitive demands imposed by the tasks (from low to high demands): passive smelling of odorless air, passive smelling of single odors, intensity discrimination, quality discrimination, and episodic-memory odor recognition. In agreement with earlier findings, passive smelling of olfactory information activated the amygdala-piriform cortex, right orbitofrontal cortex, right thalamus, and insula (Zatorre et al., 1992; Sobel et al., 1998). Depending on the task, different subsets of those regions were recruited, along with other areas interconnected with them. Both discrimination tasks engaged the left insula and right cerebellum. However, discrimination of quality proved also to involve structures hierarchically above those engaged in intensity discrimination: the thalamus, cingulum, orbitofrontal and prefrontal cortex, the frontal operculum, caudatus, insula, and subiculum. Several studies have shown that the prefrontal cortex is a major area of activation in tasks tapping working memory, which implies that odor-quality discrimination draws on the same neural structures that have been shown to support working-memory functions (Cabeza and Nyberg, 2000). The absence of a prefrontal effect for intensity discrimination is less clear, but may be related to the fact that the task is more perceptual/sensory-mediated (stronger/weaker), whereas discrimination by quality is more cognitively loaded (similar/dissimilar).

Furthermore, episodic memory for odors, which was considered to be the task posing the highest cognitive demands, also proved to show the most extensive activation pattern. It activated the same areas as the quality-discrimination task, with the exception of caudatus, subiculum, and insula. However, episodic memory for odors activated the piriform cortex and two remote neocortical areas: the temporal and parietal cortices. That outcome is in accordance with a number of studies addressing neuroanatomical correlates of episodic-memory functioning for other types of sensory information (e.g., Buckner et al., 1995; Cabeza and Nyberg, 2000). Thus, the activation pattern of episodic memory for odors maps well to what is known for verbal and visual episodic memory.

3. Concluding Comments

According to systems theory, different kinds of information about an event are "stored" in different memory systems and subsystems and used as needed (Tulving, 1995). Whether that is also valid for olfaction or whether odor memory is best described as a form of memory governed by its own premises remains to be explored in future research. Here, the cognitive neuroscience approach will provide valuable insights into the organization and representation of olfactory memory in the human brain.

Acknowledgments

This work was supported by a post-doctoral fellowship from The Swedish Council for Research in the Humanities and the Social Sciences (F1484/1997) to Dr. Maria Larsson. I am grateful to Drs. Timo Hursti and Mats J. Olsson for providing me with materials and stimulating discussions.

References

Andrews P L R & Sanger G J (1993). The Problem of Emesis in Anti-cancer Therapy: An Introduction. In: *Emesis in Anti-cancer Therapy: Mechanisms and Treatment*, ed. P L R Andrews & G J Sanger, pp. 1–8. London: Chapman & Hall.
Baddeley A (1986). *Working Memory*. Oxford University Press.
Baddeley A (1992). Working Memory. *Science* 255:556–9.
Bernstein I L (1978). Learned Taste Aversions in Children Receiving Chemotherapy. *Science* 200:1302–3.

Borison H L (1989). Area Postrema: Chemoreceptor Circumventricular Organ of the Medulla Oblongata. *Progress in Neurobiology* 32:351–90.

Braak E, Braak H, & Mandelkov E M (1993). Cytoskeletal Alterations Demonstrated by the Antibody AT8: Early Signs for the Formation of Neurofibrillary Tangles in Man. *Society for Neuroscience Abstracts* 19:228.

Brown A S (1991). A Review of the Tip-of-the-Tongue Experience. *Psychological Bulletin* 109:204–23.

Brown R, & McNeill D (1966). The "Tip of the Tongue" Phenomenon. *Journal of Verbal Learning and Verbal Behavior* 5:325–37.

Buckner, R L, Petersen S E, Ojemann J G, Miezin F M, Squire L R, & Raichle M E (1995). Functional Anatomical Studies of Explicit and Implicit Memory Retrieval Tasks. *Journal of Neuroscience* 15:12–29.

Cabeza R & Nyberg L (2000). Imaging Cognition II: An Empirical Review of 275 PET and fMRI Studies. *Journal of Cognitive Neuroscience* 12:1–47.

Cain W S (1979). To Know with the Nose: Keys to Odor Identification. *Science* 203:467–70.

Chobor K L (1992). Human Olfaction in Infancy and Early Childhood. In: *Science of Olfaction*, ed. M J Serby & K L Chobor, pp. 381–95. New York: Springer-Verlag.

Doty R L, Shaman P, & Dann M (1984). Development of the University of Pennsylvania Smell Identification Test: A Standardized Microencapsulated Test of Olfactory Function. *Physiology and Behavior* 32:489–502.

Eichenbaum H, Morton T H, Potter H, & Corkin S (1983). Selective Olfactory Deficits in Case H.M. *Brain* 106:459–72.

Engen T (1982). *The Perception of Odors*. Toronto: Academic Press.

Engen T (1987). Remembering Odors and Their Names. *American Scientist* 75:497–503.

Engen T (1991). *Odor Sensation and Memory*. New York: Praeger.

Finkel D, Pedersen N L, & Larsson M (2001). Olfactory Functioning and Cognitive Abilities: A Twin Study. *Journal of Gerontology: Psychological Sciences* 56B:226–33.

Flavell J H & Wellman H M (1977). Metamemory. In: *Perspectives on the Development of Memory and Cognition*, ed. R V Kail Jr & J W Hagen, pp. 3–34. Hillsdale, NJ: Lawrence Erlbaum.

Gabrieli J D E, Fleischman D A, Keane M M, Reminger S L, & Morrell F (1995). Double Dissociation between Memory Systems Underlying Explicit and Implicit Memory in the Human Brain. *Psychological Science* 6:76–82.

Garcia J, Laister P S, Bermudez-Rattoni F, & Deems D A (1985). A General Theory of Aversions Learning. *Annals of the New York Academy of Sciences* 443:8–21.

Heindel W C, Salmon D P, Shults C W, Walicke P A, & Butters N (1989). Neuropsychological Evidence for Multiple Implicit Memory Systems: A Comparison of Alzheimer's, Parkinson's, and Huntington's Disease Patients. *Journal of Neuroscience* 9:582–7.

Herz R S & Engen T (1996). Odor Memory: Review and Analysis. *Psychonomic Bulletin and Review* 3:300–13.

Hursti T, Fredrikson M, Börjeson S, Fürst C J, Peterson C, & Steineck G (1992). Association between Personality Characteristics and Extinction of Conditioned Nausea after Chemotherapy. *Journal of Psychosocial Oncology* 10:59–77.

Jenkins J J (1979). Four Points to Remember: A Tetrahedral Model of Memory Experiments. In: *Levels of Processing in Human Memory*, ed. L S Cermak & F I M Craik, pp. 429–46. Hillsdale, NJ: Lawrence Erlbaum.

Kempler D & Zelinski E M (1994). Language in Dementia and Normal Aging. In: *Dementia and Normal Aging*, ed. F A Huppert, C Brayne, & D W O'Connor, pp. 331–65. Cambridge University Press.

Larsson M (1997). The Influence of Semantic Factors in Episodic Recognition of Common Odors: A Review. *Chemical Senses* 22:623–33.

Larsson M & Bäckman L (1993). Semantic Activation and Episodic Odor Recognition in Young and Older Adults. *Psychology and Aging* 8:582–8.

Larsson M & Bäckman L (1997). Age-related Differences in Episodic Odour Recognition: The Role of Access to Specific Odour Names. *Memory* 5:361–78.

Larsson M, Finkel D, & Pedersen N L (2000). Odor Identification: Influences of Age, Cognition, and Personality. *Journal of Gerontology: Psychological Sciences* 55B:304–10.

Larsson M, Semb H, Winblad B, Amberla K, Wahlund L O, & Bäckman L (1999). Odor Identification in Normal Aging and Early Alzheimer's Disease: Effects of Retrieval Support. *Neuropsychology* 13:47–53.

Lawless H T & Cain W S (1975). Recognition Memory for Odors. *Chemical Senses and Flavor* 1:331–7.

Lawless H T & Engen T (1977). Associations to Odors: Interference, Memories, and Verbal Labeling. *Journal of Experimental Psychology* 3:52–9.

Lehrner J P, Walla P, Laska M, & Deecke L (1999). Different Forms of Human Odor Memory: A Developmental Study. *Neuroscience Letters* 272:17–20.

Lyman B J & McDaniel M A (1990). Memory for Odors and Odor Names: Modalities of Elaboration and Imagery. *Journal of Experimental Psychology: Learning, Memory, and Cognition* 16:656–64.

McDonald R M, Ergis A M, & Winocur G (1999). Functional Dissociation of Brain Regions in Learning and Memory: Evidence for Multiple Systems. In: *Memory: Systems, Process, or Function?*, ed. J K Foster & M Jelicic, pp. 66–103. Oxford University Press.

Menella J A & Beauchamp G K (1997). Human Flavor Perception. In: *Tasting and Smelling*, ed. G K Beauchamp & L Bartoshuk, pp. 199–221. Orlando: Academic Press.

Nilsson L G, Bäckman L, Erngrund K, Nyberg L, Adolfsson R, Bucht G, Karlsson S, Widing M, & Winblad B (1997). The Betula Prospective Cohort Study: Memory, Health, and Aging. *Aging, Neuropsychology, and Cognition* 4:1–32.

Nyberg L & Tulving E (1996). Classifying Human Long-Term Memory: Evidence from Converging Dissociations. *European Journal of Cognitive Psychology* 8:163–83.

Olsson M J (1999). Implicit Testing of Odor Memory: Instances of Positive and Negative Repetition Priming. *Chemical Senses* 24:347–50.

Olsson M J & Cain W S (1995). Early Temporal Events in Odor Identification (Abstract). *Chemical Senses* 20:753.

Otto T, Cousens G, & Rajewski K (1997). Odor-guided Fear Conditioning in Rats: Acquisition, Retention, and Latent Inhibition. *Behavioral Neuroscience* 111:1257–64.

Owen A M (1997). Tuning in to the Temporal Dynamics of Brain Activation Using Functional Magnetic Resonance Imaging (fMRI). *Trends in Cognitive Sciences* 1:123–5.

Rabin M D & Cain W S (1984). Odor Recognition: Familiarity, Identifiability, and Encoding Consistency. *Journal of Experimental Psychology: Learning, Memory, and Cognition* 10:316–25.

Richardson J T E & Zucco G M (1989). Cognition and Olfaction: A Review. *Psychological Bulletin* 105:352–60.

Robin O, Alaoui-Ismaïli O, Dittmar A, & Vernet-Maury E (1999). Basic Emotions Evoked by Eugenol Odor Differ according to the Dental Experience. A Neurovegetative Analysis. *Chemical Senses* 24:327–35.

Roediger H L, Buckner R L, & McDermott K B (1999). Components of Processing. In: *Memory: Systems, Process, or Function?*, ed. J K Foster & M Jelicic, pp. 31–65. Oxford University Press.

Sahley C L, Gelperin A, & Rudy J (1981). One-Trial Associative Learning Modifies Food Odor Preferences of a Terrestrial Mollusc. *Proceedings of the National Academy of Sciences USA* 78:640–2.

Salthouse T A (1995). Aging, Inhibition, Working Memory, and Speed. *Journal of Gerontology: Psychological Sciences* 50B:297–306.

Savic I, Bookheimer S Y, Fried I, & Engel J Jr (1997). Olfactory Bedside Test: A Simple Approach to Identify Temporo-orbitofrontal Dysfunction. *Archives of Neurology* 54:162–8.

Savic I, Gulyàs B, Larsson M, & Roland P (2000). Olfactory Functions Are Mediated by Parallel and Hierarchical Processing. *Neuron* 26:735–45.

Schab F R (1991). Odor Memory: Taking Stock. *Psychological Bulletin* 109:242–51.

Schab F R & Crowder R G (1995). Implicit Measures of Odor Memory. In: *Memory for Odors*, ed. R G Crowder & F R Schab, pp. 71–92. Hillsdale, NJ: Lawrence Erlbaum.

Schacter D L (1992). Understanding Implicit Memory: A Cognitive Neuroscience Approach. *American Psychologist* 47:559–69.

Schacter D L & Tulving E (1994). What Are the Memory Systems of 1994? In: *Memory Systems 1994*, ed. D L Schacter & E Tulving, pp. 1–38. Cambridge, MA: MIT Press.

Simic G, Kostovic I, Winblad B, & Bogdanovich N (1997). Volume and the Number of Neurons of the Human Hippocampal Formation in Normal Aging and Alzheimer's Disease. *Journal of Comparative Neurology* 379:482–94.

Sobel N, Prabhakaran V, Desmond J E, Glover G H, Goode R L, Sullivan E V, & Gabrieli J D (1998). Sniffing and Smelling: Separate Subsystems in the Human Olfactory Cortex. *Nature* 392:282–6.

Stein M, Ottenberg P, & Roulet N (1958). A Study of the Development of Olfactory Preferences. *Archives of Neurology and Psychiatry* 80:264–6.

Steiner J E (1979). Human Facial Responses in Response to Taste and Smell Stimulation. *Advances in Child Development and Behavior* 13:257–95.

Strickland M, Jessee P O, & Filsinger E E (1988). A Procedure for Obtaining Young Children's Reports of Olfactory Stimuli. *Perception and Psychophysics* 44:379–82.

Tulving E (1993). Human Memory. In: *Memory Concepts – 1993. Basic and Clinical Aspects,* ed. P Andersen, O Hvalby, O Paulsen, & B Hökfelt, pp. 27–45. New York: Elsevier.

Tulving E (1995). How Many Memory Systems Are There? *American Psychologist* 40:385–98.

Tulving E & Colotla V (1970). Free Recall of Trilingual Lists. *Cognitive Psychology* 1:86–98.

Vallar G & Papagno C (1995). Neuropsychological Impairments of Short-Term Memory. In: *Handbook of Memory Disorders*, ed. A D Baddeley & B A Wilson, pp. 135–65. New York: Wiley.

White T L (1998). Olfactory Memory: The Long and Short of It. *Chemical Senses* 23:433–41.

Zatorre R J, Jones-Gotman M, Evans A C, & Meyer E (1992). Functional Localization of the Human Olfactory Cortex. *Nature* 360:339–41.

15

Repetition Priming in Odor Memory

Mats J. Olsson, Maria Faxbrink, and Fredrik U. Jönsson

1. Introduction

1.1. Implicit Memory

Tests for implicit memory have been pursued extensively for about two decades, and the findings from such memory experiments are commonly considered to have confirmed the phenomenon of implicit memory. Schacter (1987) defined "implicit memory" as facilitation of task performance because of prior experience, in the absence of conscious or intentional recollection (explicit memory). That rather broad definition might seem to encompass the definitions of implicit learning, conditioned learning, motor-skills learning, and perceptual adaptation (Roediger and McDermott, 1993), but will not be considered to do so here. Instead, our focus will be on the branch of implicit memory known as repetition priming. In our terminology, repetition priming is tested when the stimuli presented are the same (identical) or of the same type (essentially the same, but varying in some way) at priming and at testing. In this type of experiment, the response to a repeated stimulus is facilitated without the influence of explicit memory from the first encounter.

Implicit memory can also be measured using cross-modal priming, where the sensory modalities used differ between priming and testing. This review will cover cases of cross-modal priming in which odor names are presented at priming, and odors at testing. Several recent studies have examined cross-modal priming using the olfactory and visual modalities, assessing the influence of processing an odor on the later processing of a visual stimulus (e.g., Hermans, Baeyens, and Eelen, 1998; Grigor et al., 1999; Pauli et al., 1999; Sarfarazi et al., 1999). Those (visual) priming studies will not be covered in this review. Other studies of related interest have been conducted by Degel and Köster (1998, 1999) to address implicit learning (Buchner and Wippich, 1998) associated with surreptitious presentation of odors.

A common way to probe repetition priming in the visual domain has been with tests of word-stem or word-fragment completion (e.g., Warrington and Weiskrantz, 1974; Tulving, Schacter, and Stark, 1982). In such tests, participants are exposed to a list of words in the priming phase. In a later testing phase they are presented with word stems or fragments that they are asked to complete with the first word that comes to mind. Although participants' attention is directed away from the earlier priming list, they nevertheless tend to select words from that list. In fact, even in the absence of episodic recognition (i.e., explicit memory for the words on the priming list) (Tulving et al., 1982), participants use the priming words significantly more often than would be expected. In a similar vein, amnesic patients have performed well on this type of test, whereas they have done very poorly on tests of explicit memory (Graf, Squire, and Mandler, 1984).

Repetition priming has most often been interpreted as involving perceptual processes (e.g., Jacoby and Dallas, 1981). It has also been proposed that priming largely reflects the operation of a separate perceptual representation system (Tulving and Schacter, 1990). But even for versions of repetition priming tests that have been considered to be of the perceptual type, there are now findings that imply that these tests are driven by conceptual processing as well as by perceptual processing (e.g., Keane and Gabrieli, 1991). The issue of what type of priming a certain test is tapping is important and will emerge as one of the key issues in our empirical review of repetition priming in olfaction.

1.2. Objectives

Although there is considerable interest in implicit memory, only a few studies have been performed using odors as the stimuli to be remembered. One possible reason for that is that repetition priming for odors is difficult to investigate with many of the standard procedures used in the fields of vision and hearing (Issanchou et al., Chapter 13, this volume). The objectives of this review are therefore (1) to review the empirical evidence concerning implicit memory for odors as assessed in priming experiments, (2) to present new data on olfactory repetition priming, (3) to provide a state-of-the-art assessment, and (4) to suggest what further research is needed in these areas.

2. Review of Priming Experiments

Basically, three groups of researchers have performed experiments on priming in olfaction. We shall review the experiments of each group in order to follow the lines of reasoning behind their work.

2.1. The Schab and Crowder Experiments

When Schab and Crowder (1995) published their study on implicit memory for odors, there was nothing on this topic in English. They designed their experiments with reference to how implicit-memory tests are typically structured in the visual realm, and they hypothesized that odors would lend themselves to priming much as did visual, auditory, and tactile stimuli. In four experiments they investigated the effects of previous exposure on identification scores and reaction times; olfactory detection thresholds, identification thresholds, and latency of pleasantness ratings.

In the priming phase of their first experiment, Schab and Crowder exposed participants to 10 odor names alone and 10 other odor names together with their respective odors. The stimuli were at all times presented in jars (for the name-only condition, the jars did not contain any odorants). To try to offset that potential problem, participants were told that some odors could be very faint. Five minutes later, the participants attempted to identify the 20 experimental odors and 10 control odors. Both identification scores and reaction times (RTs) were measured. Exposure to name and odor at priming led to significantly better identification and shorter RTs in the test phase than did the name-only condition (which by itself yielded significant priming). Schab and Crowder argued that this "name priming" was a common finding in lexical research, but did not necessarily demonstrate odor-specific priming. More interesting, the name-and-odor condition showed higher facilitation, both for identification scores and for RTs, than did the name-only condition. Those findings were interpreted as showing that "(a) the activation of an odor's perceptual representation, in conjunction with the activation of its verbal label, enhances subsequent odor identification by temporarily strengthening the association between the odor's perceptual representation and its verbal code or (b) the presentation of an odor with its corresponding verbal label strengthens the accessibility of that verbal label until the subsequent odor identification task more than does presentation of the label only" (p. 75).

In the detection experiment (exp. 2), thresholds for nine different odors were measured after previous exposure to the name or to the name and the odor. The data showed that neither of those conditions generated enhanced sensitivity as compared with control odors. Because in the experiment on detection they did not obtain the same results as in the first experiment on identification, Schab and Crowder suggested that if participants had to make decisions on odor quality, one would observe priming effects.

To test that hypothesis, Schab and Crowder performed an identification threshold experiment (exp. 3). The experiment entailed three separate parts and sought

to determine if the concentration at which odors were identifiable would decrease because of previous exposure to names or to names plus odors. In the
priming phase (part 1), six odors distributed over two study conditions were
rated for pleasantness and familiarity. The two conditions were name and name
plus odor. In the name-only condition, the participants "smelled" a labeled "odor"
consisting of de-ionized water, which they were told was "very faint." Shortly
after that, the participants were asked to identify the six experimental odors and
three control odors presented at 10 concentrations each. They were to select
among four odor names, one of which was correct. The data showed that the
odor-identification scores for the experimental conditions were not significantly
different from those for the control condition, in which only new odors were
presented.

At that point, Schab and Crowder argued that the difference between the priming condition and test task could have been detrimental to the aim of observing
priming. That idea rests on the theory of transfer-appropriate processing (Morris,
Bransford, and Franks, 1977), stressing that memory performance is dependent
on whether or not the encoded information is appropriate for the memory test
used to measure retention. The importance of transfer-appropriate processing for
priming has been pointed out in a number of studies (Roediger and McDermott,
1993). In part 2, Schab and Crowder tested participants for identification thresholds twice. The first time, two stimuli were presented by name only, and two
were presented as names plus odors. Although the means lined up according
to the hypothesis that the name-plus-odor condition would generate lower identification thresholds than the name-only condition (which in turn would show
lower thresholds than the control condition), the differences fell short of being
statistically significant.

Encouraged by the pattern of results in part 2, Schab and Crowder replicated
that part with more participants (20), this time achieving a significant main effect
for condition. However, a pairwise comparison of the name-only and name-
plus-odor conditions yielded a small and nonsignificant difference, as in part 2.
Because the pairing of the odor and the name did not significantly increase the
priming effect, Schab and Crowder concluded that they had not shown "odor
priming," but simply "name priming."

In their final experiment, Schab and Crowder returned to their testing of
repetition priming with odors above threshold, this time measuring only RTs.
Participants rated the perceived intensities of 24 common odors, half of which
were presented as odors along with their names, and the other half as names
alone, but with the instruction to try to imagine the corresponding odor. At
testing five minutes later, participants rated those odors and 12 control odors
for pleasantness. RTs were also measured for these ratings. The data revealed

that the means for the three conditions were almost identical and hence did not demonstrate any priming at all.

Schab and Crowder concluded that their efforts to demonstrate priming of odor memory "paint a surprisingly bleak picture" and that the phenomenon was "not yet on a solid factual basis." Regarding those statements, a comment might be made concerning their decision not to use a condition in which only the odor would be presented in the priming phase. They discussed that extensively and concluded that an odor-only condition would have yielded ambiguous results with regard to the type of priming found (odor priming or name priming), because it would have been impossible to control the participants' thought processes. On the other hand, using the difference between name priming and name-plus-odor priming as a measure of "odor priming" is questionable (Olsson and Fridén, 2001). The fact that identification rates for familiar odors seldom exceed 50% (de Wijk, Schab, and Cain, 1995) means that inclusion of odor names along with the odors in the priming phase would have introduced new information to the participants that could have been used later in the test phase. Therefore it is doubtful that the name-plus-odor condition taps only implicit memory.

2.2. The Wippich Experiments

We shall review three studies by Wippich and colleagues (Wippich, Mecklenbräuker, and Trouet, 1989; Wippich, 1990; Wippich, Mecklenbräuker, and Banning, 1993). They were published in German and therefore may not have come to the general awareness of the international scientific community. We realize, of course, that more studies on this topic can be found in other languages.

Wippich et al. (1989) wanted to study the commonly observed dissociation between implicit memory and explicit memory for different levels of processing. Typically, implicit-memory tests are not sensitive to levels of processing, whereas explicit-memory tests are (Jacoby and Dallas, 1981; Graf et al., 1984). The priming phase of the first experiment by Wippich et al. (1989) entailed three conditions: (1) estimation of the subjective duration of exposure (*Darbietungsdauer*) (which was 3 sec at all times), (2) condition 1 plus an attempt to identify the odor, and (3) condition 2 plus a rating of pleasantness. In the test phase shortly thereafter, odors were presented in pairs (one old and one new odor), and participants judged each odor with regard to subjective duration. In addition, for half of the pairs they judged which of the two pair members was the more pleasant. The data indicated no difference between new and old odors for any of the study conditions, either for ratings of duration or ratings of pleasantness. In other words, no priming was observed.

In a second experiment, Wippich et al. (1989) used two priming conditions: Group 1 judged the lingering time for presented (3-sec) odors, and group 2 made an additional attempt to identify the odors. The identification was cued with one correct name and one distractor name. In the test phase, directly following the priming phase, all participants judged each odor for lingering time, pleasantness, and identity. Identification was once again cued with a correct name and a distractor name. The results showed that neither the ratings of lingering time (*Nachwirkungsdauer*) nor the ratings of pleasantness were significantly affected by previous exposure in any of the groups. With regard to odor-identification scores, the following results were obtained: Odors that had been smelled in the priming phase (group 1) were better identified than control odors. Odors that had been smelled before and whose identification had been attempted with the aid of the two cues (group 2) were not identified significantly better at testing than were the control odors that had not been smelled before but whose names had appeared as distractors in the priming phase. The results for identification RTs did not prove significant, but there was a tendency in group 2 toward shorter RTs for old odors than for new odors. The authors concluded that it appeared that verbal mediation was necessary in order for priming to be observed.

In order to differentiate between odor priming and name priming, Wippich (1990) exposed participants either to odors or to their names. For each exposure, participants were asked either to relate an autobiographical memory or to judge the odor/name for edibility. Three dependent measures were registered in the test phase in response to new and old odors: RTs for relating an autobiographical memory, scores and RTs for identification. The data showed that identification of previously presented odors (without names) was better and faster than identification of new odors. In the case in which only names had been presented in the priming phase, there was no significant priming effect. At that point, the authors concluded that repetition priming for naming odors was not purely a verbal-priming effect, but also presupposed olfactory processing in the priming phase. With regard to RTs for autobiographical memory, a priming effect was observable only in the condition in which autobiographical memories had been reported in the priming phase, not for the condition in which edibility had been judged. Another main result was that the level of processing did not seem to matter for measures of implicit memory, but it did for some measures of explicit memory.

A third study (Wippich et al., 1993) sought further proof for olfactory-specific priming. In the priming phase, there were three encoding conditions: exposure to single odors, to their names, or to pairs of odors. For the group that was presented names, the instructions were that they should try to imagine the odor corresponding to a name and rate the degree of clarity at which they could evoke that phenomenon. The group that was presented with single odors was similarly asked to rate the degrees of clarity with which they perceived the odors. The group that was presented with pairs judged whether they were the same or different. The test phase, directly following the priming phase, was identical for all participants. Odors were presented in pairs, with the two members of each pair being identical or different, and the test task was to judge whether they were the same or different. Participants who responded "different" were asked to judge which odor was the more pleasant. For the group that had received odor pairs in the priming phase, the pairs in the test phase appeared as either old pairs (identical or different) or new pairs (identical or different). If old, the order of the odors was reversed relative to the order in the priming phase. For the group that studied odor pairs, the data showed that old pairs were more quickly judged for whether they were composed of identical or different odors than were new pairs. That effect was observed to the same extent whether different or identical odors were repeated. Participants who were exposed to single odors or to their names in the priming phase were only nonsignificantly faster to judge old pairs as compared with new pairs.

With regard to preference judgments, priming effects were observed. The group that had been exposed to single odors and had rated them for perceptual clarity in the priming phase preferred old odors over new ones in significantly more than 50% of cases. In a related experiment, Cain and Johnson (1978) showed that unpleasant odors were rated as less unpleasant after previous exposure; on the other hand, pleasant odors were judged less pleasant. Cain and Johnson discussed their results using the notion of "affective habituation." It should be noted that the odors used by Wippich et al. (1993), according to the few examples that were given, were on the positive side in terms of preference ratings. Therefore, those two studies may be somewhat at odds.

As noted, the group that studied odor pairs judged old pairs more quickly than new pairs with respect to whether they were composed of identical or different odors, whereas the group that had been presented single odors in the priming phase did not. Two observations regarding that pattern of results were discussed by Wippich et al. (1993). First, in accordance with a transfer-appropriate-processing account of priming effects, the findings suggest that in order to observe priming, task-specific processes, not only the activation of odor representations, need to be repeated between priming and testing. Second, an alternative interpretation of that pattern of results would be that the presentation of one member of a pair triggered the episodic memory of the other member. The short RTs when judging old pairs, therefore, might possibly be attributable to explicit-memory processes.

On the basis of their efforts to show repetition priming in olfaction (Wippich et al., 1989, 1993; Wippich, 1990), those authors, in their third paper, concluded that it had been unexpectedly difficult to show priming in olfaction as compared with priming in other modalities. It will be remembered that Schab and Crowder (1995) had a similar comment, suggesting that the observed priming in their experiments may have been verbally mediated. However, Wippich et al. (1993) concluded that priming occurs not only as a function of name activation but also as a function of odor activation in the priming phase.

2.3. The Olsson Experiments

Following the Schab and Crowder (1995) experiments, Olsson and Cain (1995, in press) attempted to design a repetition priming experiment that would not draw on the processing of odor names as heavily as had the study by Schab and Crowder. In addition, the design entailed only mono-rhinal smelling. Data on the nostril used (left or right) at priming and testing were factorially combined. Mono-rhinal testing is supposed to reflect functional characteristics of the ipsilateral cerebral hemisphere (Brodal, 1981; Doty et al., 1997).

In the priming phase used by Olsson and Cain (exp. 1), participants estimated how long it had been since they had last smelled each of six very familiar odors. In doing that, participants had no access to names or visual cues to the identity of each odor. Five minutes later, they smelled 12 odors (6 old and 6 new) and indicated by pressing a response key when they thought they knew the identity of an odor. They were not required to know the names of the odors at the time of pressing the key; a visualization of the odorous object itself would qualify. Measured that way, the findings revealed faster RTs for identification of old odors than for new ones.

When the data were broken down by nostril, the following pattern emerged. Which nostril had been used at priming did not matter at all. Which was used as the test nostril, however, did matter: Priming was evident only when testing was via the left nostril, irrespective of the side used at priming, and therefore it was believed to reflect left-hemisphere (LH) functioning. Because the LH is associated with verbal processing, the left-nostril priming we observed may be also. That is, the observed priming may again represent name priming rather than odor priming.

In a follow-up experiment, Olsson (1999) tried to draw even less on verbal processes. The idea was to restrict the requirements of verbal processing by intervening in the naming process to see if odor priming in the LH was still evident. It should be noted that most theories of object naming would agree that at least three stages are present in the naming process: object identification, name activation, and response generation (Johnson, Paivio, and Clark, 1996). In Olsson's experiment, the priming task required participants to rate how long it had been since they had last smelled the odors (because Olsson and Cain had found priming only when tested from the LH, participants used the left nostril throughout the whole experiment). In the test phase, participants were introduced to a comparison odor (orange or coffee). They were then presented with old and new odors. The task was to decide, as quickly as possible, whether or not the old and new odors were identical with the comparison odor. The general idea was that only early (perceptual) processing should be necessary in order to reject the proposition that a test odor was identical with the comparison odor. Priming was then supposed to reveal itself as faster rejections for old odors than for new ones. The results showed that old and new odors were rejected equally fast, suggesting that the original experiment by Olsson and Cain had tapped some type of conceptual priming rather than perceptual priming. However, the results proved more complicated than first realized. After the priming test, participants had been asked to identify each odor, and as is commonly observed in odor-identification experiments, their correct identifications had been low (48%). A post-hoc analysis of the RT data, dividing the trials into two categories, one for trials that were followed by successful identification and one for trials followed by incorrect identification, showed that unidentifiable odors had been associated with shorter latencies than had control odors and hence had exhibited repetition priming. Identifiable odors, on the other hand, had been associated with slower RTs. That latter result was discussed in terms of negative repetition priming (Tipper, 1985), that is, an expression of memory in which the processing of a stimulus is in some sense inhibited because of a previous encounter with it.

Another attempt to reduce the impact of verbal processing in the quest to show repetition priming in olfaction is seen in a study by Olsson and Fridén (2001). They argued that perceptions of edibility would be less laden with verbal processing and that judgments thereof could be made without prior identification. In their first experiment, participants were asked to judge the edibility of 24 bi-rhinally presented odors, half of which were edible. Ten minutes later, they repeated such judgments for 48 odors, 24 old and 24 new, that were presented mono-rhinally. The results indicated that edibility judgments for old odors were significantly faster, but not more accurate. There were no differences in performance for the two nostrils.

A second experiment followed the lines of the first, with one major change. The participants attempted to identify the odors in the priming phase and judged old and new odors for edibility in the test phase. The results showed that priming, in terms of faster RTs, was observable only when testing was via the right nostril. It can be noted that testing via the left nostril was associated with more correct judgments of edibility, as compared with the right nostril.

The main finding of Olsson and Fridèn (2001) that pertains to our review is that edibility judgments were faster when repeated for the same odors. The authors reported two observations that support the notion that edibility judgments are not conditional on name activation: First, the reaction times were quite fast for edibility judgments. The RTs for judging whether or not an odor was edible were at least 30–35% shorter than those for making a decision on the odor source (Olsson and Cain, in press). Second, there was no correlation across odors between the primability and identifiability of odors. On the other hand, Olsson and Fridén discussed the possibility that the observed priming could be based on categorization of odors into some number of edible and inedible subcategories, such as spices, cleaning products, and so forth (see Dubois and Rouby, Chapter 4, this volume, on this point). In other words, this type of priming may tap into conceptual processes other than the strictly perceptual ones.

3. An Experiment Probing Perceptual Repetition Priming

One of the critical points in the search for evidence of olfactory repetition priming concerns the development of methods that can measure odor priming, or perceptual priming, rather than name priming. We shall present some original data from an experiment aimed at investigating such perceptual priming: The design was based on an experiment by Wippich et al. (1993) that in our opinion seems adequate for tapping perceptual priming. Although their data showed nominal, but not significant, priming effects, it seemed possible that a more focused experimental effort might. The idea was to see if exposure to odors in conjunction with an orienting task focusing on odor quality during the priming phase would influence perceptions of odor quality at a later test, thereby facilitating judgments on perceptual clarity and same/different judgments of odor pairs.

3.1. Method

Participants. Twenty male and 20 female participants, primarily students from Uppsala University, took part in the experiment, either for course credit or for a movie ticket. They ranged in age from 19 to 52 years (mean = 26.8). Their olfactory functioning was normal, according to self-assessments. Smokers ($n = 6$) were asked to refrain from smoking for one hour before the experiment. Anyone suffering from a cold or a temporary nasal obstruction was excluded.

Stimuli. The stimulus set consisted of 48 different odorous items that were placed in amber-tinted glass jars and occluded from visual inspection by cotton pads. The odors represented common everyday objects, both edible and inedible, such as coffee, cinnamon, shoe polish, and so forth.

Design and Procedure. In the priming phase, participants judged the similarities of 24 odors to a comparison odor (pine) on a scale from zero (very dissimilar) to 100 (identical). To suppress verbal processing of the odors during the priming phase, participants were required to perform a verbal-fluency test (Benton and Hamsher, 1976) between the odor presentations. To prevent verbal processing during the retention interval between the priming and test phases, participants engaged in a working-memory test (Skandinaviska Testförlaget, 1967).

In the test phase, participants were presented with 32 odor pairs, half of which were made up of old odors, and the other half of new odors. Identical pairs and different pairs were equally represented among the new and old pairs. At the presentation of the first member of a pair, participants sampled the odor on hearing a tone signal and then indicated when they had perceived the odor quality "clearly" by pressing a timer button. For the second member of the pair, participants indicated whether the odor was the same as or different from the first member of the pair. For that judgment, both correctness and RTs were noted.

3.2. Results and Discussion

Three analyses of variance (ANOVA) were performed to investigate repetition priming. First, the RT for perceptual clarity in response to the first member of a pair in the priming phase was analyzed in a two-way ANOVA (Old/New × Sex), with repeated measures on the first factor. The results showed that old and new odors were perceived with clarity equally fast (1,450 msec and 1,440 msec, respectively) [$F(1, 38) = 0.17$, $p =$ ns]. The dependent variable for the second analysis was the RT for giving a same/different judgment (for pairs whose two members were different) when presented the second odor of a pair. The reason for omitting the responses to pairs of identical odors from the analysis was that in those trials the same odor was presented twice, and therefore the second presentation could not be considered "new." An ANOVA (Old/New × Sex) yielded no significant difference between old odors (1,230 msec) and new odors (1,257 msec) [$F(1, 38) = 1.29$, $p =$ ns], and hence no indication of repetition priming. The third test of priming concerned correctness scores for the same/different judgments. For the same reasons as before, that analysis was performed only for pairs whose two members differed. An ANOVA (Old/New × Sex) yielded no significant difference between old (95.5%) and new (95.7%) pairs

$[F(1, 38) = 0.03, p = \text{ns}]$. It should be noted, however, that performance levels were quite high, indicating that ceiling effects may have prevented detection of a difference between old and new pairs. On the other hand, performance levels for new and old pairs were close to identical. Women performed nominally faster and better than men in all three analyses, but those differences were not statistically significant.

To conclude, a repetition priming experiment designed to tap perceptual priming was performed. Task requirements oriented the participants toward the quality rather than the identity of odors at both priming and testing. In addition, distraction tasks prevented additional processing of odor identities or other semantic processing. The effects of repetition were assessed in three ways. No statistically reliable evidence could be found for perceptual repetition priming. Our findings are in line with previous results from Wippich et al. (1993).

4. Discussion

This review of the literature indicates that several priming effects for a number of tasks in olfaction have been observed. Moreover, such priming can be seen to be lateralized between the cerebral hemispheres, as assessed by mono-rhinal testing. However, several authors have concluded that repetition priming in olfaction has received weaker empirical support than would have been expected on the basis of findings for other modalities. Moreover, the effects that were observed could have been due to verbal processing rather than perceptual processing. Given that background, we conducted an experiment aimed at probing further into perceptual priming. Our results supported the notion that strictly perceptual processing (in this case, attending to sensory quality) may not suffice to generate repetition priming effects in olfaction.

A second major issue regarding the existence and nature of olfactory repetition priming concerns the issue of external validation. Typically, investigators in olfaction have adopted procedures similar to those that have been developed and validated for probing repetition priming in the field of visual memory. However, as can be noted in this review, there is no proof that the methods used to probe olfactory repetition priming really tap the operations of implicit memory. Therefore, we need tests to show independence between explicit memory and implicit memory for odors at the level of individual items (cf., Tulving et al., 1982). Another way to validate the existence of olfactory priming would be to show that amnesics exhibit olfactory priming in the absence of normal performance on other explicit tests of odor memory (Cave and Squire, 1992). A third possibility concerns assessment of the effects of surreptitiously presented odors on subsequent performance (Degel and Köster, 1999).

5. Conclusions

This review of different experimental conditions attempting to assess repetition priming in olfaction has led to the following conclusions: Response facilitation due to repetition (priming) in olfaction can be observed for several tasks. However, several authors have argued that such priming experiments have had limited success as compared with tests of priming in other modalities, especially vision. Experimental procedures that target odor identity/naming typically are more successful in exhibiting priming than are other procedures. This indicates that the observed effects concern "name priming" rather than "odor priming" or "perceptual priming."

Other experimental procedures, drawing less on the ability to identify odors by their names, have also been developed. Repeated edibility judgments, for instance, have shown reliable priming effects in terms of shorter RTs. Other procedures designed to focus on strictly perceptual processing (e.g., comparing odor qualities) have found little to no priming.

Transfer-appropriate processing shows more priming than transfer-inappropriate processing. In other words, priming is more likely to be observed if the priming and test tasks are repeated or otherwise overlap with respect to information or processing.

Results from several experimental conditions indicate that priming effects differ for the two hemispheres (nostrils). This is an interesting finding that could further our understanding of functional cerebral lateralization, as well as the processes/neural substrates that are associated with priming in olfaction.

It should be noted that there has been no validation of the methods used to probe repetition priming in olfaction, that is, whether or not they tap the operations of implicit memory, thereby avoiding "leakage" between priming and testing via explicit memory. Such methods have been developed within the field of visual memory, and we need to find ways to extend them to olfaction.

Acknowledgments

This work was supported by The Bank of Sweden Tercentenary Foundation and The Alrutz Fund.

References

Benton A L & Hamsher K (1976). *Multilingual Aphasia Examination.* Iowa City, IA: University of Iowa Press.
Brodal A (1981). *Neurological Anatomy in Relation to Clinical Medicine*, 3rd ed. Oxford University Press.

Buchner A & Wippich W (1998). Differences and Commonalities between Implicit Learning and Implicit Memory. In: *French Handbook of Implicit Learning*, ed. M A Staedler & P A French, pp. 3–46. Thousand Oaks, CA: Sage.

Cain W S & Johnson F Jr (1978). Lability of Odor Pleasantness: Influence of Mere Exposure. *Perception* 7:459–65.

Cave C B & Squire L R (1992). Intact and Long-Lasting Repetition Priming in Amnesia. *Journal of Experimental Psychology: Learning, Memory, and Cognition* 18:509–20.

Degel J & Köster E P (1998). Implicit Memory for Odors: A Possible Method for Observation. *Perceptual and Motor Skills* 86:943–53.

Degel J & Köster E P (1999). Odors: Implicit Memory and Performance Effects. *Chemical Senses* 24:317–25.

de Wijk R A, Schab F R, & Cain W S (1995). Odor Identification. In: *Memory for Odors*, ed. F R Schab & R G Crowder, pp. 1–12. Mahwah, NJ: Lawrence Erlbaum.

Doty R L, Bromley S M, Moberg P J, & Hummel T (1997). Laterality in Human Nasal Chemoreception. In: *Cerebral Asymmetries in Sensory and Perceptual Processing*, ed. S Christman, pp. 497–552. Amsterdam: North Holland.

Graf P, Squire L R, & Mandler G (1984). The Information that Amnesic Patients Do Not Forget. *Journal of Experimental Psychology: Learning, Memory, and Cognition* 10:164–78.

Grigor J, Van Toller S, Behan J, & Richardson A (1999). The Effect of Odour Priming on Long Latency Visual Evoked Potentials of Matching and Mismatching Objects. *Chemical Senses* 24:137–44.

Hermans D, Baeyens F, & Eelen P (1998). Odours as Affective-Processing Context for Word Evaluation. A Case of Cross-Modal Affective Priming. *Cognition and Emotion* 12:601–13.

Jacoby L L & Dallas M (1981). On the Relation between Autobiographical Memory and Perceptual Learning. *Journal of Experimental Pschology: General* 3:306–40.

Johnson C J, Paivio A, & Clark J M (1996). Cognitive Components of Picture Naming. *Psychological Bulletin* 120:113–39.

Keane M M & Gabrieli D E (1991). Evidence for a Dissociation between Perceptual and Conceptual Priming in Alzheimer's Disease. *Behavioral Neuroscience* 105:326–42.

Morris C D, Bransford J D, & Franks J J (1977). Levels of Processing versus Transfer Appropriate Processing. *Journal of Verbal Learning and Verbal Behavior* 16:519–33.

Olsson M J (1999). Implicit Testing of Odor Memory: Instances of Positive and Negative Priming. *Chemical Senses* 24:347–50.

Olsson M J & Cain W S (1995). Early Temporal Events in Odor Identification (abstract). *Chemical Senses* 20:753.

Olsson M J & Cain W S (in press). Implicit vs Explicit Memory for Odors: Hemispheric Differences.

Olsson M J & Fridén M (2001). Evidence of Odor Priming: Edibility Judgments Are Primed Differently between the Hemispheres. *Chemical Senses* 26:117–23.

Pauli P, Bourne L E Jr, Diekmann H, & Birbaumer N (1999). Cross-Modality Priming between Odors and Odor-congruent Words. *American Journal of Psychology* 112:175–86.

Roediger H L III & McDermott K B (1993). Implicit Memory in Normal Human Subjects. In: *Handbook of Neuropsychology*, vol. 8, ed. F Boller & J Grafman, pp. 63–131. New York: Elsevier.

Sarfarazi M, Cave B, Richardson A, Behan J, & Sedgwick E M (1999). Visual Event Related Potentials Modulated by Contextually Relevant and Irrelevant Olfactory Primes. *Chemical Senses* 24:145–54.

Skandinaviska Testförlaget (1967). *Kontorstesten*. Stockholm: Skandinaviska Testförlaget.

Schab F R & Crowder R (1995). Implicit Measures of Odor Memory. In: *Memory for Odors*, ed. F R Schab & R G Crowder, pp. 71–92. Mahwah, NJ: Lawrence Erlbaum.

Schacter D L (1987). Implicit Memory: History and Current Status. *Journal of Experimental Psychology: Learning, Memory, and Cognition* 13:501–18.

Tipper S P (1985). The Negative Priming Effect: Inhibitory Priming by Ignored Objects. *Quarterly Journal of Experimental Psychology* 37A:571–90.

Tulving E & Schacter D L (1990). Priming and Human Memory Systems. *Science* 247:301–6.

Tulving E, Schacter D L, & Stark H A (1982). Priming Effects in Word-Fragment Completion Are Independent of Recognition Memory. *Journal of Experimental Psychology: Learning, Memory, and Cognition* 8:336–42.

Warrington E K & Weiskrantz L (1974). The Effect of Prior Learning and Subsequent Retention in Amnesic Patients. *Neuropsychologica* 12:419–28.

Wippich W (1990). Erinnerungen an Gerüche: Benennungsmasse und autobiographische Erinnerungen zeigen Geruchsnachwirkungen an. *Zeitschrift für experimentell und angewandte Psychologie* 37:679–95.

Wippich W, Mecklenbräuker S, & Banning R (1993). Sensorische Geruchsnachwirkungen bei indirekten und direkten Behaltensprüfungen. *Schweizerische Zeitschrift für Psychologie* 52:193–204.

Wippich W, Mecklenbräuker S, & Trouet J (1989). Implizite und explizite Erinnerungen an Gerüche. *Archiv für Psychologie* 141:195–211.

16

Odor Memory in Alzheimer's Disease

Steven Nordin and Claire Murphy

1. Introduction

The neuropsychology of olfactory cognition is a relatively young scientific domain that in recent years has in many respects advanced from explorative research to more focused and hypothesis-driven investigations. That advancement can be attributed in part to important studies of certain neurological and geriatric disorders. A question of interest here is whether or not research in odor memory in populations with those disorders can provide a better understanding of the neural processes underlying odor memory. For example, there is evidence that Korsakoff patients experience rapid forgetting in the visual, auditory, and tactile modes, but not in the olfactory realm (Jones, Moscowitz, and Butters, 1975; Mair, Harrison, and Flint, 1995). That has been used as part of the evidence that odor memory may be a specialized subsystem of memory (Herz and Eich, 1995). If so, it might be expected that a pattern quite opposite to that of Korsakoff patients would be possible. Thus, memory impairment for olfaction may, under certain neuropathologic conditions, be more severe than memory deficits in other sensory modalities.

The notion of olfactory-specific memory impairment can be linked to the possibility of odor memory being particularly vulnerable to certain cortical neuropathologic conditions. In a rather extensive review of the neuropsychological literature involving human olfaction, Mair et al. (1995) concluded that, in contrast to the situation for other sensory modalities, there is no evidence for morphologically distinct centers mediating memory for and perception of odors. Thus, impairment of odor memory is strongly associated with impairment in the ability to perceive the quality of an odor. The general association between impairments in perception and memory for odors suggests that relatively limited cortical areas serve a relatively wide range of olfactory functions. That, in turn, might imply that odor memory could be more sensitive than other sensory systems to focal neuropathologic changes, because both perceptual functions (prerequisites for the encoding of odors) and odor memory per se would be affected.

Patients with Alzheimer's disease (AD) constitute an interesting population with regard to questions concerning olfactory-specific memory impairment and an understanding of the neural processes underlying odor memory. Of particular interest are cases of AD in its very early stages, because it is assumed that cortical abnormalities at an early stage of the disease progression are relatively focal and limited to the entorhinal and transentorhinal areas (Braak and Braak, 1997).

The main issue to be addressed in this chapter is the question of what can be learned about the neuropsychology of odor memory from studies of patients with AD. At present, several criteria are used to assess AD (McKhann et al., 1984), though unfortunately a determination of disease status can be made definitively only by neurological examination at autopsy. More sensitive tools clearly are needed for diagnosis in the early stages of the disease progression. Because of early, pronounced neuropathologic changes in olfactory brain areas in AD, this sensory system is of particular interest for the development of diagnostic tools. Consequently, a secondary issue of this chapter is the potential for using memory-based olfactory tests for early diagnosis of AD. To approach these two issues, we shall review the state of research into the neural and behavioral abnormalities of the olfactory system at various stages of the disease. Finally, our review will be integrated with the most recent scientific findings to address our two issues and to shed light on further research needs.

2. Neuropathology of AD

AD is a neurological disorder characterized by progressive memory loss, and the most common cause of dementia. About 5% of people over the age of 65 years suffer from dementia (Jorm, Korten, and Henderson, 1987), and AD accounts for 50% to 60% of the dementing diseases (Katzman, 1986). Although cognitive symptoms dominate, other symptoms such as anxiety, depression, delusions, and hallucinations plague many of these patients (Folstein and Bylsma, 1994). The most prominent cognitive deficit typically associated with AD is amnesia, with additional deficits in language, abstract reasoning, certain executive functions, and visuospatial abilities (Bondi, Salmon, and Butters, 1994b).

The neuropathology of AD is characterized by the presence of neuritic plaques (NP), neurofibrillary tangles (NFT), and cell loss. As mentioned earlier, it has been suggested that regions of importance for olfactory processing are among those most affected, and they may even be sites of initial involvement in the disease (e.g., Pearson et al., 1985; Braak and Braak, 1997). NPs, NFTs, and neuropil threads have been found in the anterior olfactory nucleus (Ohm and

Braak, 1987; Price et al., 1991), and NFTs, neuropil threads (Esiri and Wilcock, 1984), and axonal loss (Davies, Brooks, and Lewis, 1993) have been reported in the olfactory bulb. Signs of abnormalities in the olfactory epithelium have been less consistent; some researchers have reported changes (Talamo et al., 1989), whereas others have reported no significant abnormalities (Davies et al., 1993). Such findings tend to focus the search for olfactory deficits on more central regions of this sensory system.

Van Hoesen and Solodkin (1994) pointed out that olfactory brain areas appear uniquely affected, given the relative sparing of other sensory areas. For example, the entorhinal cortex and periamygdaloid nucleus, which are parts of the olfactory cortex, are especially damaged. The work of Price et al. (1991) has confirmed the presence of tangles in areas that mediate olfactory function, particularly the anterior olfactory nucleus, entorhinal cortex, and amygdala, in AD patients with very mild dementia. Reyes, Deems, and Suarez (1993) reported NPs and especially NFTs in the entorhinal and prepiriform cortices and in the periamygdaloid nucleus and concluded that there was greater evidence of pathologic changes in the olfactory cortex than in the hippocampus, which is the area characteristically associated with damage in AD. Hyman (1997) reported the most severe lesions to be in the entorhinal and perirhinal cortices, the CA1/subicular area of the hippocampus, the amygdala, and the association cortices. Braak and Braak (1992) have argued that the lesions in the entorhinal and transentorhinal areas effectively disconnect the hippocampus from the isocortex, preventing the transfer of information essential to memory function.

3. Odor Detection and Discrimination

Although the focus of this chapter is on odor memory, it is important to understand that more sensory-based olfactory functions required for memory performance are also affected in AD. Losses in absolute detection sensitivity have been demonstrated for a large number of substances (Richard and Bizzini, 1981; Knupfer and Spiegel, 1986; Doty, Reyes, and Gregor, 1987; Rezek, 1987; Murphy et al., 1990; Serby, Larson, and Kalkstein, 1991; Lehrner et al., 1997; Nordin et al., 1997). That poor performance in odor detection most likely is due to neurological rather than rhinological status (Feldman et al., 1991) and to sensory loss rather than lack of comprehension of the detection task per se (e.g., Nordin, Monsch, and Murphy, 1995). The loss in detection sensitivity is further illustrated by its association with the degree of dementia (Murphy et al., 1990; Nordin et al., 1997) and by an association between a rapid progression in dementia and a swift decline in sensitivity (Nordin et al., 1993). A patient's odor threshold has, interestingly, also been found to correlate with a family history of AD (Schiffman,

Clark, and Warwick, 1990). Obviously AD patients are at considerable risk with respect to safety issues (detection of smoke, gas leaks, spoiled food, etc.), because many anosmic AD patients are unaware of their olfactory loss (Doty et al., 1987; Nordin et al., 1995).

Another sensory function underlying performance on tests of odor memory is quality discrimination. Impairment in that olfactory function has been shown in AD when using same-or-different (Koss, 1986) and match-to-sample paradigms (Kesslak et al., 1988; Buchsbaum et al., 1991; Kesslak, Nalcioglu, and Cotman, 1991). The roles of detection and discrimination sensitivity for accurate odor memory will be discussed next.

4. Odor Memory

Considering the severe olfactory neuropathologic abnormalities found in the medial temporal lobe in AD, it is of particular interest to study behavioral impairment of memory for odors in this population. Thus far, much research has been directed toward explicit memory (see Issanchou et al., Chapter 13, this volume). Hence the memory domains to be reviewed here are of the episodic and/or semantic nature. Episodic memory is commonly defined as memory for personal episodes, and it is the type of memory for which the most pronounced auditory and visual deficits have been reported in AD (Welsh et al., 1991). Poor performances in these two sensory modalities on tests of recognition memory, inability to learn new information over repeated learning trials, and rapid forgetting suggest impairment in storing (i.e., encoding) and retaining new information in AD (e.g., Delis et al., 1991; Welsh et al., 1991). Such findings suggest that various dimensions of episodic odor memory are of potential interest with respect to AD.

Impairment in semantic processing in AD has also been well documented for visual and auditory material. It includes identification by naming and verbal fluency with respect to both categories and letters (Butters et al., 1987; Chertkow and Bub, 1990). This impairment may well reflect breakdown of semantic networks in AD, as has been demonstrated by applying various graphic scaling techniques (Chan et al., 1993, 1995), including multidimensional scaling (MDS). Findings from our laboratory obtained with the technique of MDS, with triads of odor stimuli being compared for similarity, suggest a breakdown in the semantic network for odors, but not colors, in term of associations between odors (Chan et al., 1998). Not surprisingly, our patients also showed losses in the ability to identify odors. Another memory function that requires intact semantic memory, but also episodic memory, is verbal recall of previously presented odors. Preliminary findings from our laboratory suggest that AD patients perform particularly poorly on this task and show basically no ability to improve across repeated trials. This includes immediate recall as well as short- and long-delay recall, both with and without semantic cues (Razani et al., 1996). Analyses of the data suggest that impaired odor identification underlies much of the failure of recall.

4.1. Recognition memory

The ability to recognize previously experienced odors is predominantly a matter of episodic memory, but also is influenced by semantic knowledge of the odor item (e.g., Larsson and Bäckman, 1997). Deficits in recognition memory for odors have been documented in mild to moderate cases of AD for both short-term memory (retention intervals of 5–30 sec) (Knupfer and Spiegel, 1986) and long-term memory (10–20 min) (Moberg et al., 1987; Murphy, Lasker, and Salmon, 1987; Lehrner et al., 1997). Preliminary data from our laboratory also suggest that AD patients have limited trust in their recognition memory for odors even when they in fact remember correctly (Wilhite et al., 1993).

We have examined the question of when in the disease progression a decline in recognition memory for odors begins (Murphy, Nordin, and Jinich, 1999). That was addressed by studying 78 patients with probable AD, categorized as very mild, mild, or moderate in degree of dementia, and 78 controls. Common household odors were used as stimuli, as were visual stimuli of common faces and abstract, uncommon symbols for comparison. Ten stimuli for each modality were presented for inspection. After intervals averaging 20 min, half of the stimuli were presented again, together with the same number of new stimuli and the instruction to tell which stimuli had been presented before. In addition, odor-detection sensitivity for *n*-butyl alcohol was assessed for comparison. The results showed that whereas the earliest declines in odor sensitivity and visual memory were found in patients with mild dementia, a decline in odor memory was found in patients with very mild dementia. That suggested that odor-recognition memory may be affected earlier in the disease progression than are visual-recognition memory and odor sensitivity.

To further our understanding of the early decline in odor-recognition memory in AD, we shifted our focus to persons with the diagnosis of questionable AD. These persons are rare, but ideal for studying the early changes in AD. They show impairment in cognitive functions, similar to patients with probable AD, but, in contrast, they do not yet show changes in everyday functioning as reported by significant others (Bondi et al., 1994a). Using the same test procedures as in the study described earlier (Murphy et al., 1999), we investigated 16 persons diagnosed with questionable AD and 16 controls (Nordin

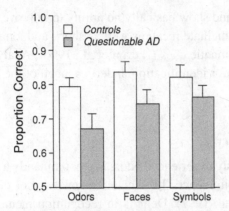

Figure 16.1. Mean (±SE) proportions of correct responses for odors, faces, and symbols on tests of recognition memory in questionable AD and normal controls. (From Nordin and Murphy, 1996. Reprinted by permission of the American Psychological Association.)

and Murphy, 1996). Six of the questionable-AD subjects were diagnosed again one to three years after the odor-memory testing, and at that time all six were given the diagnosis of probable AD. Thus, those subjects can be considered to represent preclinical AD. The results are presented in Figure 16.1 and suggest a relative olfactory-specific impairment in recognition memory. Thus, persons with questionable AD performed significantly poorer than did controls for odors, but showed only a tendency toward poorer performance for visual material.

4.2. Familiarity

Familiarity with a certain object or event can be considered as a form of remote memory – memories for stimuli encoded at some unspecified previous time. Besides episodic memory (e.g., "This is familiar; it is something I had for breakfast some time ago"), familiarity may also involve semantic memory (e.g., "This is familiar; it is some kind of spice"). Research in the visual and auditory domains in AD patients has demonstrated deficits in retrieval from remote memory (e.g., Beatty et al., 1988; Mitrushina et al., 1994). In a study to compare familiarity performances for visual and olfactory materials, test stimuli of odors, faces, and symbols were used when studying 32 mild to moderately demented patients with probable AD and 32 controls (Niccoli-Waller et al., 1999). The subjects were instructed to rate the stimuli for familiarity on a visual-analogue scale. It was found that AD patients rated odors, but not faces or symbols, as significantly less familiar than did controls, indicative of deficits in remote odor memory.

Follow-up testing one year later in 12 of the AD patients and 12 of the controls again showed the same result.

4.3. Identification

The ability to identify odors by name depends on semantic memory, but also requires a great deal of perceptual processing such as quality discrimination (e.g., Cain and Potts, 1996). This is especially true when response alternatives are available, which eases the cognitive load. Difficulties in odor identification by naming have been demonstrated in AD when assessed by a test to match odors to names (Corwin et al., 1985; Knupfer and Spiegel, 1986), without presentation of response alternatives (Rezek, 1987; Bacon Moore, Paulsen, and Murphy, 1999). However, the most commonly used means to study odor identification in AD has been the University of Pennsylvania Smell Identification Test (UPSIT), which is a lexical-based test using four written response alternatives for each test odor (Doty, Shaman, and Dann, 1984). Results from such studies consistently show losses in the ability to name odors (Warner et al., 1986; Doty et al., 1987, 1991; Kesslak et al., 1991; Serby et al., 1991; Moberg et al., 1997).

The studies of odor identification cited earlier relied on lexical functioning as the response mode, such as the interpretation or formulation of words. Therefore, we wondered if odor identification that did not rely on lexical function would also show impairments (Morgan, Nordin, and Murphy, 1995). Eighteen patients with probable AD at mild to moderate stages of dementia were compared with 18 controls, and 8 persons diagnosed with questionable (preclinical) AD were compared with 8 controls. The participants were given the UPSIT and the San Diego Odor Identification Test (SDOIT) (Anderson, Maxwell, and Murphy, 1992). The latter is an eight-item test (of which six were used for our purpose) developed for children, with pictures as the response alternatives from which to choose. Thus, the participant would point at the picture representing the odor item, which eliminated lexical demands. To verify earlier findings of relatively spared visual identification in AD (Doty et al., 1987, 1991), the Picture Identification Test (PIT) (Vollmecke and Doty, 1985) was administered. It is identical in content and format with the UPSIT except that pictures, instead of odors, serve as stimulus items. The results are presented in Figure 16.2 and show impairments in both lexical-based (UPSIT) and picture-based (SDOIT) odor identifications for both probable and questionable AD. With the same items represented by pictures (PIT), the AD groups, in contrast, performed at the same level as their control groups.

These findings suggest that the impairments in odor identification commonly found in AD cannot be attributed predominantly to lexical difficulties. That

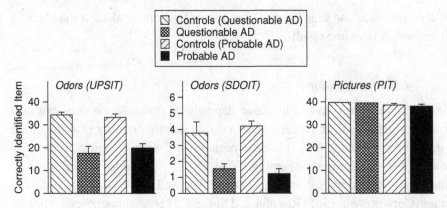

Figure 16.2. Mean (±SE) numbers of correctly identified odors and pictures in questionable AD, probable AD, and normal controls. (From Morgan et al., 1995. Reprinted by permission of Swets & Zeitlinger.)

conclusion is supported by the findings of Larsson et al. (1999), who compared performances in odor identification when support was provided (either written response alternatives or real-life objects as alternatives, or no support). The AD patients performed very similarly on both tasks with support, but considerably better with support than without support, and significantly poorer than controls on all three tasks. The degrees of improvement due to support were equally large in AD patients and controls. Such similarity in support-induced improvements, also reported by Peabody and Tinklenberg (1985) and Rezek (1987), was interpreted by the authors as showing that degeneration of olfactory knowledge plays a more important role than lexical-access problems in AD.

In a recent study (Devanand et al., 2000) it was demonstrated that a large proportion of patients with mild cognitive impairment have difficulty in identifying odors (UPSIT) and that those patients with such difficulties are more likely to develop AD than are other patients with mild cognitive impairment. Interestingly, it was reported that the combination of poor odor identification and lack of awareness of one's olfactory deficit predicted the time of development of AD.

Still another approach to investigation of early, possible AD is by means of typing apolipoprotein E (ApoE). This protein is found in increased amounts in patients with AD, specifically in plaques and tangles (Namba et al., 1991), suggesting that ApoE is a genetic risk factor for AD (Corder et al., 1993). Pairs of three primary alleles, 2, 3, and 4, determine an individual's ApoE status. About 25% of the general population will have at least one 4-allele, yet 80% of patients with familial AD and 64% of patients with the sporadic form of the disease have one or both of the 4-alleles (Sing and Davignon, 1985). In a study of healthy elderly people, the participants were categorized as allele-positive (having at least one 4-allele: 2/4, 3/4, 4/4) or allele-negative (2/2, 2/3, 3/3) and tested on odor identification (SDOIT) (Murphy et al., 1998). The results showed that the allele-positive elderly performed significantly poorer than the allele-negative participants on that olfactory task. Further support for a genetic link between odor identification and AD has been provided by data suggesting that performance on odor identification (UPSIT) is impaired in first-degree relatives of patients with probable AD (Serby et al., 1996).

4.4. Verbal Fluency

Some well-documented difficulties in verbal fluency, as seen when subjects are instructed to generate names of items with respect to a given category or an initial letter, motivated a study of name generation for items specifically related to the olfactory domain (Bacon Moore et al., 1999). Forty patients with probable AD and 40 controls were asked to generate (within 60 sec) as many names as possible for (1) odors belonging to a given category, (2) odors beginning with the letters P, C, and S, and (3) odors that came to mind when presented with two odorants. The results demonstrated impairments in AD for all three measures, and in accordance with findings from conventional testing of verbal fluency (e.g., Monsch et al., 1994), category fluency was more impaired than letter fluency. Those findings also support the claim that the poor odor identification in AD is better explained by cognitive than by sensory impairment. Furthermore, the difficulties in generating names for odorants may reflect a breakdown in the semantic network for odors.

5. Discussion and Perspective

From our review of the neuropathology of AD, the very early changes appear to be found predominantly in the entorhinal cortex, but also in the amygdala, piriform and prepiriform cortices, and to some extent in the anterior olfactory nucleus. As the disease progresses, the pathologic processes also involve the hippocampus (CA1 and subiculum) and the olfactory bulb. Whereas the olfactory epithelium may be affected to some extent, there are no indications of neurodegeneration in the orbitofrontal cortex.

In the behavioral arena, as described in our review, the very early olfactory impairments in AD that have been demonstrated thus far concern the abilities to recognize and identify odors, with the latter being the more severe. Early declines in odor identification have been demonstrated in questionable AD, in patients with mild cognitive impairment, and in first-degree relatives of AD patients, often by means of ApoE typing. In mild to moderate cases of AD, various other sensory- and cognition-based olfactory functions are affected, including absolute detection, quality discrimination, familiarity, verbal fluency for odor items, and, probably, any type of semantic memory requiring an intact semantic network for odors. To what extent these functional impairments are present at a very early stage of AD remains to be studied.

The question of peripheral versus central contributions to poor task performance may in this context determine to what extent losses in odor-detection sensitivity can explain poor performances on tests of odor memory in AD. Although a decline in detection sensitivity may contribute, there is strong support for the view that semantic odor memory, rather than odor sensitivity, is predominantly affected in early AD. Thus, deficits in odor identification have been found in AD in the absence of significant odor-detection deficits (St. Clair et al., 1985; Rezek, 1987; Koss et al., 1988; Larsson et al., 1999). Impairment in identification ability has also been found at an earlier disease stage than impairment in detection ability (Serby et al., 1991), and the former has more commonly been found to correlate with global cognitive status (Knupfer and Spiegel, 1986; Kesslak et al., 1988, 1991; Serby et al., 1991; Larsson et al., 1999). Finally, identification deficits appear to be present before detection deficits in persons at risk for AD due to ApoE 4-allele status (Bacon et al., 1998; Murphy et al., 1998). In accordance with this, impairment in recognition memory for odors has been found in AD in the absence of significant odor-detection deficits (e.g., Nordin and Murphy, 1996; Murphy et al., 1999).

5.1. Neuropsychology of Odor Memory

A difficulty in developing an understanding of the neuropsychology of odor memory based on findings in AD is that a relatively wide range of areas already show pathologic involvement by the time a diagnosis of probable AD can be made. Therefore, very early cases of AD play an important role in our understanding, for the neuropathologic changes at that stage are still relatively focal. The combination of very early abnormality in the entorhinal cortex and a very early decline in odor identification, as shown in many studies to be prominent at more advanced stages of AD, suggests that the entorhinal cortex plays an important role in the ability to identify odors. Although this cortical region may play a key role for odor identification, other areas may be important as well. In MRI studies of patients who had undergone temporal-lobe resection, Jones-Gotman et al. (1997) discussed the possibility of the piriform cortex being a critical site for odor identification, with periamygdaloid damage causing disconnection from the piriform cortex and thus affecting odor identification.

The theory of the entorhinal and transentorhinal areas disconnecting the hippocampus from the isocortex early in AD, preventing the transfer of information essential to memory function (Braak and Braak, 1992), directs further attention to the entorhinal cortex, but also to the hippocampus. There is, unfortunately, a lack of documentation of any relationship between olfactory performance in early AD and neuroimaging data. However, some information about the more advanced progression of AD is available. For example, lower metabolic activity, as compared with controls, has been found in the parahippocampal gyrus (entorhinal cortex) in AD while performing an odor-discrimination task (match-to-sample task, with a 10-sec load on working memory). Furthermore, Kesslak et al. (1991) used MRI to show significant atrophy in the entorhinal cortex and hippocampus in AD patients post mortem. The volumes for those areas were also found to correlate with performances on quality discrimination and identification of odors, but poorly with visual discrimination.

High correlations with atrophy in the entorhinal cortex and hippocampus for both quality discrimination and identification (Kesslak et al., 1991) support the notion that relatively limited cortical areas serve a relatively wide range of olfactory functions (Mair et al., 1995). That implies that the entorhinal cortex and possibly the hippocampus may serve both quality discrimination and identification of odors. Regarding the hippocampus, Otto and Eichenbaum (1991) suggested, with reference to animal models, that this structure is necessary to encode relationships between odors, thus affecting discrimination. The claim that poor odor identification in AD cannot be explained simply by loss in discrimination sensitivity is supported by findings of breakdown in semantic networks for odors (Chan et al., 1998; Bacon Moore et al., 1999).

The demonstrated impairment in recognition memory for odors in very early AD (Figure 16.1) raises the question whether or not the entorhinal cortex also plays a major role in episodic odor memory. Although further investigations clearly are needed to answer this question, such a role seems reasonable in view of the well-established opinion that the entorhinal-hippocampal-subicular complex is a key structure in the encoding of memory, independent of type of sensory modality (e.g., Kandel, Schwartz, and Jessell, 2000).

Documentation of past research involving neuropsychological approaches to an understanding of odor familiarity is sparse. However, Royet et al. (1999) have used positron-emission tomography to show selective activation in the right medial orbitofrontal area in normal subjects during familiarity judgments. Considering the spared orbitofrontal areas in AD, and yet a decline in familiarity judgment in AD (Nordin and Murphy, 1996; Niccoli-Waller et al., 1999), those additional regions probably are involved in this process. Hypothetically, the task of asking oneself whether or not a certain odor is familiar may activate the orbitofrontal cortex, whereas activation in other areas, presumably in the medial temporal lobe, may be critical for successful retrieval of the memory trace (generating an experience of familiarity). Future studies will have to untangle this question.

5.2. Memory-based Olfactory Tests for Early Diagnosis of AD

A first prerequisite for the introduction of an olfactory test for diagnosis of AD in clinical settings is that there be both a neuroanatomical rationale and a functional rationale for such an introduction. In this respect it can be concluded that the very early and predominant abnormalities are found in cerebral areas that are crucial for basic olfactory processing: the prepiriform and piriform cortices (primary olfactory cortex), entorhinal cortex and amygdala (secondary olfactory cortex), and the anterior olfactory nucleus. A second prerequisite is that the olfactory test either detect the disease with more accuracy than conventional tests or that it be faster to administer with equal accuracy. In this context it is of interest to note that identification testing with the six-item SDOIT ($\sim$5 min) has shown an 83% correct classification rate (the ability of the test to correctly classify subjects as healthy or diseased) for persons with questionable AD and for controls. With the 40-item UPSIT (15–20 min), the correct rate increased to 100% (Morgan et al., 1995).

Despite the need for further investigation into whether or not odor memory is more affected than memory for other sensory modalities in AD, a few studies have already compared odor memory and visual memory. Recognition memory and familiarity (Nordin and Murphy, 1996; Niccoli-Waller et al., 1999) as well as identification (Morgan et al., 1995) and semantic networks (Chan et al., 1998) have been found to be more affected for odorous material than for visual material.

From our review of studies of odor memory, it appears evident that odor-identification testing is a strong candidate for use in a battery of conventional tests for early diagnosis of AD. Although it is tempting to suggest the use of this test to clinical practitioners, more research is needed. In particular, this includes direct comparisons on different odor-memory modalities (recognition, identification, recall, etc.) in AD.

It is important to point out that olfactory changes have also been demonstrated in other forms of dementia, the best documented being Parkinson's disease, Huntington's disease, and Down's syndrome. Attempts have been made to provide overviews of the research conducted on olfactory similarities and differences in the various forms of dementia (e.g., Harrison and Pearson, 1989: Mesholam et al., 1998; Murphy, 1999). However, far more research is needed before a clear picture will emerge regarding the impairment patterns for different olfactory functions and dementia causes. Such information will prove most valuable for differential diagnosis.

Acknowledgments

Supported by NIH grants AG08203 and AG04085 (to Claire Murphy), by the Bank of Sweden Tercentenary Foundation (1998-0270:01), and by the Swedish Council for Research in the Humanities and Social Sciences (to Steven Nordin). The excellent assistance of Anna Bacon Moore, Jodi Harvey, Samuel Jinich, Stacy Markison, Charlie D. Morgan, Caprice Niccoli-Waller, and Jill Razani is gratefully acknowledged.

References

Anderson J, Maxwell L, & Murphy C (1992). Odorant Identification Testing in the Young Child. *Chemical Senses* 17:590.

Bacon A W, Bondi M W, Salmon D P, & Murphy C (1998). Very Early Changes in Olfactory Functioning due to Alzheimer's Disease and the Role of Apolipoprotein E in Olfaction. *Annals of the New York Academy of Sciences* 855:723–31.

Bacon Moore A, Paulsen J S, & Murphy C (1999). A Test of Odor Fluency in Patients with Alzheimer's and Huntington's Disease. *Journal of Clinical and Experimental Neuropsychology* 21:341–51.

Beatty W W, Salmon D P, Butters N, Heindel W C, & Granholm E L (1988). Retrograde Amnesia in Patients with Alzheimer's Disease or Huntington's Disease. *Neurobiology of Aging* 9:181–6.

Bondi M W, Monsch A U, Galasko D, Butters N, Salmon D P, & Delis D C (1994a). Preclinical Cognitive Markers of Dementia of the Alzheimer Type. *Neuropsychology* 8:374–84.

Bondi M W, Salmon D P, & Butters N (1994b). Neuropsychological Features of Memory Disorders in Alzheimer's Disease. In: *Alzheimer's Disease*, ed. R D Terry, R Katzman, & K L Bick, pp. 41–63. New York: Raven Press.

Braak H & Braak E (1992). The Human Entorhinal Cortex: Normal Morphology and Lamina-specific Pathology in Various Diseases. *Neuroscience Research* 15:6–31.

Braak H & Braak E (1997). Frequency of Stages of Alzheimer-related Lesions in Different Age Categories. *Neurobiology of Aging* 18:351–7.

Buchsbaum M S, Kesslak J P, Lynch G, Chui H, Wu J, Sicotte N, Hazlett E, Teng E, & Cotman C W (1991). Temporal and Hippocampal Metabolic Rate during an Olfactory Memory Task Assessed by Positron Emission Tomography in Patients with Dementia of the Alzheimer Type and Controls. Preliminary studies. *Archives of General Psychiatry* 48:840–7.

Butters N, Granholm E, Salmon D P, Grant I, & Wolfe J (1987). Episodic and Semantic Memory: A Comparison of Amnesic and Demented Patients. *Journal of Clinical and Experimental Neuropsychology* 9:479–97.

Cain W S & Potts B C (1996). Switch and Bait: Probing the Discriminative Basis of Odor Identification via Recognition Memory. *Chemical Senses* 21:35–44.

Chan A, Salmon D, Nordin S, Murphy C, & Razani J (1998). Abnormality of Semantic Network in Patients with Alzheimer's Disease: Evidence from Verbal, Perceptual,

274   *Steven Nordin and Claire Murphy*

and Olfactory Domains. *Annals of the New York Academy of Sciences* 855:681–5.

Chan A S, Butters N, Salmon D P, Johnson S A, Paulsen J S, Swenson M R, & Swerdlow N (1995). Comparison of the Semantic Networks in Patients with Dementia and Amnesia. *Neuropsychology* 9:177–86.

Chan A S, Butters N, Salmon D P, & McGuire K A (1993). Dimensionality and Clustering in the Semantic Network of Patients with Alzheimer's Disease. *Psychology and Aging* 8:411–19.

Chertkow H & Bub D (1990). Semantic Memory Loss in Dementia of Alzheimer's Type: What Do Various Measures Measure? *Brain* 133:397–417.

Corder E H, Saunders A M, Strittmatter W J, Schmechel D E, Gaskell P C, Small G W, Rose A D, Haines J L, & Pericak-Vance M A (1993). Gene Dose of Apolipoprotein E Type 4 Allele and the Risk of Alzheimer's Disease in Late Onset Families. *Science* 261:921–3.

Corwin J, Serby M, Conrad P, & Rotrosen J (1985). Olfactory Recognition Deficit in Alzheimer's and Parkinsonian Dementias. *IRCS Medical Science* 13:260.

Davies D C, Brooks J W, & Lewis D A (1993). Axonal Loss from the Olfactory Tracts in Alzheimer's Disease. *Neurobiology of Aging* 14:353–7.

Delis D C, Massman P J, Butters N, Salmon D P, Cermak L S, & Kramer J H (1991). Profiles of Demented and Amnesic Patients on the California Verbal Learning Test: Implications for the Assessment of Memory Disorders. *Psychological Assessment: A Journal of Consulting and Clinical Psychology* 3:19–26.

Devanand D P, Michaels-Marston K S, Liu X, Pelton G H, Padilla M, Marder K, Bell K, Stern Y, & Mayeux R (2000). Olfactory Deficits in Patients with Mild Cognitive Impairment Predict Alzheimer's Disease at Follow-up. *American Journal of Psychiatry* 157:1399–405.

Doty R L, Perl D P, Steele J C, Chen K M, Pierce J D, Reyes P, & Kurland L T (1991). Olfactory Dysfunction in Three Neurodegenerative Diseases. *Geriatrics (Suppl. 1)* 46:47–51.

Doty R L, Reyes P F, and Gregor T (1987). Presence of Both Identification and Detection Deficits in Alzheimer's Disease. *Brain Research Bulletin* 18:597–600.

Doty R L, Shaman P, & Dann M (1984). Development of the University of Pennsylvania Smell Identification Test: A Standardized Microencapsulated Test of Olfactory Function. *Physiology and Behavior* 32:489–502.

Esiri M M & Wilcock G K (1984). The Olfactory Bulbs in Alzheimer's Disease. *Journal of Neurology, Neurosurgery and Psychiatry* 47:56–60.

Feldman J I, Murphy C, Davidson T M, Jalowayski A A, & Galindo de Jaime G (1991). The Rhinologic Evaluation of Alzheimer's Disease. *Laryngoscope* 101:1198–202.

Folstein M F & Bylsma F W (1994). Noncognitive Symptoms of Alzheimer's Disease. In: *Alzheimer's Disease*, ed. R D Terry, R Katzman, & K L Bick, pp. 27–40. New York: Raven Press.

Harrison P J & Pearson R C A (1989). Olfaction and Psychiatry. *British Journal of Psychiatry* 155:822–8.

Herz R S & Eich E (1995). Commentary and Envoi. In: *Memory for Odors,* ed. F R Schab & R G Crowder, pp. 159–75. Mahwah, NJ: Lawrence Erlbaum.

Hyman B T (1997). The Neuropathological Diagnosis of Alzheimer's Disease: Clinical-Pathological Studies. *Neurobiology of Aging* 18:S27–S32.

Jones B P, Moskowitz H R, & Butters N (1975). Olfactory Discrimination in Alcoholic Korsakoff Patients. *Neuropsychologia* 13:173–9.

Jones-Gotman M, Zatorre R J, Cendes F, Olivier A, Andermann F, McMackin D, Staunton H, Siegel A M, & Wieser H G (1997). Contribution of Medial versus

Lateral Temporal-Lobe Structures to Human Odour Identification. *Brain* 120:1845–56.

Jorm A F, Korten A E, & Henderson A S (1987). The Prevalence of Dementia: A Quantitative Integration of the Literature. *Acta Psychiatrica Scandinavica* 76:465–79.

Kandel E R, Schwartz J H, & Jessell T M (2000). *Principles of Neural Science.* New York: McGraw-Hill.

Katzman R (1986). Alzheimer's Disease. *New England Journal of Medicine* 314:964–73.

Kesslak J P, Cotman C W, Chui H C, van den Noort S, Fang H, Pfeffer R, & Lynch G (1988). Olfactory Tests as Possible Probes for Detecting and Monitoring Alzheimer's Disease. *Neurobiology of Aging* 9:399–403.

Kesslak J P, Nalcioglu O, & Cotman C W (1991). Quantification of Magnetic Resonance Scans for Hippocampal and Parahippocampal Atrophy in Alzheimer's Disease. *Neurology* 41:51–4.

Knupfer L & Spiegel R (1986). Differences in Olfactory Test Performance between Normal Aged, Alzheimer and Vascular Type Dementia Individuals. *International Journal of Geriatric Psychiatry* 1:3–14.

Koss E (1986). Olfactory Dysfunction in Alzheimer's Disease. *Developmental Neuropsychology* 2:89–99.

Koss E, Weiffenbach J, Haxby J, & Friedland R P (1988). Olfactory Detection and Identification Performance Are Dissociated in Early Alzheimer's Disease. *Neurology* 38:1228–32.

Larsson M & Bäckman L (1997). Age-related Differences in Episodic Odour Recognition: The Role of Access to Specific Odour Names. *Memory* 5:361–78.

Larsson M, Semb H, Winblad B, Amberl K, Wahlund L O, & Bäckman L (1999). Odor Identification in Normal Aging and Early Alzheimer's Disease: Effects of Retrieval Support. *Neuropsychology* 13:47–53.

Lehrner J P, Brücke T, Dal-Bianco P, Gatterer G, & Kryspin-Exner I (1997). Olfactory Functions in Parkinson's Disease and Alzheimer's Disease. *Chemical Senses* 22:105–10.

McKhann G, Drachman D, Folstein M, Katzman R, Price D, & Stadlan E M (1984). Clinical Diagnosis of Alzheimer's Disease: Report of the NINCDS-ADRDA Work Group under the Auspices of Department of Health and Human Services Task Force on Alzheimer's Disease. *Neurology* 34:939–44.

Mair R G, Harrison L M, & Flint D L (1995). The Neuropsychology of Odor Memory. In: *Memory for Odors,* ed. F R Schab & R G Crowder, pp. 39–69. Mahwah, NJ: Lawrence Erlbaum.

Mesholam R I, Moberg P J, Mahr R N, & Doty R L (1998). Olfaction in Neurodegenerative Disease: A Meta-analysis of Olfactory Functioning in Alzheimer's and Parkinson's Disease. *Archives of Neurology* 55:84–90.

Mitrushina M, Drebing C, Uchiyama C, Satz P, Van-Gorp W, & Chervinsky A (1994). The Pattern of Deficits in Different Memory Components in Normal Aging and Dementia of Alzheimer's Type. *Journal of Clinical Psychology* 50:591–6.

Moberg P J, Doty R L, Mahr R N, Mesholam R I, Arnold S E, Turetsky B I, & Gur R E (1997). Olfactory Identification in Elderly Schizophrenia and Alzheimer's Disease. *Neurobiology of Aging* 18:163–7.

Moberg P J, Pearlson G D, Speedie L J, Lipsey J R, Strauss M E, & Folstein S E (1987). Olfactory Recognition: Differential Impairments in Early and Late Huntington's and Alzheimer's Disease. *Journal of Clinical and Experimental Neuropsychology* 9:650–64.

Monsch A U, Bondi M W, Butters N, Paulsen J S, Salmon D P, Brugger P, & Swenson M R (1994). A Comparison of Category and Letter Fluency in Alzheimer's Disease and Huntington's Disease. *Neuropsychology* 8:25–30.

Morgan C D, Nordin S, & Murphy C (1995). Odor Identification as an Early Marker for Alzheimer's Disease: Impact of Lexical Functioning and Detection Sensitivity. *Journal of Clinical and Experimental Neuropsychology* 17:793–803.

Murphy C (1999). Loss of Olfactory Function in Dementing Disease. *Physiology and Behavior* 66:177–82.

Murphy C, Baco A W, Bondi M W, & Salmon D P (1998). Apolipoprotein Status Is Associated with Odor Identification Deficits in Nondemented Older Persons. *Annals of the New York Academy of Sciences* 855:744–50.

Murphy C, Gilmore M M, Seery C S, Salmon D P, & Lasker B R (1990). Olfactory Thresholds Are Associated with Degree of Dementia in Alzheimer's Disease. *Neurobiology of Aging* 11:465–9.

Murphy C, Lasker B R, & Salmon D P (1987). Olfactory Dysfunction and Odor Memory in Alzheimer's Disease, Huntington's Disease, and Normal Aging. *Society for Neuroscience Abstracts* 13:387.

Murphy C, Nordin S, & Jinich S (1999). Very Early Decline in Recognition Memory for Odors in Alzheimer's Disease. *Aging, Neuropsychology, and Cognition* 6:229–40.

Namba Y, Tomonaga M, Kawasaki H, Otomo E, & Ikeda K (1991). Apolipoprotein E Immunoreactivity in Cerebral Amyloid Deposits and Neurofibrillary Tangles in Alzheimer's Disease. *Brain Research* 541:163–6.

Niccoli-Waller C A, Harvey J, Nordin S, & Murphy C (1999). Remote Odor Memory in Alzheimer's Disease: Deficits as Measured by Familiarity. *Journal of Adult Development* 6:131–6.

Nordin S, Almkvist O, Berglund B, & Wahlund L O (1997). Olfactory Dysfunction for Pyridine and Dementia Progression in Alzheimer's Disease. *Archives of Neurology* 54:993–8.

Nordin S, Monsch A U, & Murphy C (1995). Unawareness of Smell Loss in Normal Aging and Alzheimer's Disease: Discrepancy between Self-reported and Diagnosed Smell Sensitivity. *Journal of Gerontology: Psychological Sciences* 50B:P187–92.

Nordin S & Murphy C (1996). Impaired Sensory and Cognitive Olfactory Function in Questionable Alzheimer's Disease. *Neuropsychology* 10:113–19.

Nordin S, Murphy C, Nijjar R, & Quinoñez C (1993). Odor Detection and Recognition Memory in Alzheimer's Disease. *Chemical Senses* 18:607–8.

Ohm T G & Braak H (1987). Olfactory Bulb Changes in Alzheimer's Disease. *Acta Neuropathologica (Berlin)* 73:365–9.

Otto T & Eichenbaum H (1991). Olfactory Learning and Memory in the Rat: A "Model System" for Studies of the Neurobiology of Memory. In: *Science of Olfaction*, ed. M Serby & K L Chobor, pp. 213–44. New York: Springer-Verlag.

Peabody C A & Tinklenberg J R (1985). Olfactory Deficits and Primary Degenerative Dementia. *American Journal of Psychiatry* 142:524–5.

Pearson R C A, Esiri M M, Hiorns R W, Wilcock G K, & Powell T P S (1985). Anatomical Correlates of the Distribution of the Pathological Changes in the Neocortex in Alzheimer's Disease. *Proceedings of the National Academy of Sciences USA* 82:4531–4.

Price J L, Davis P B, Morris J C, & White D L (1991). The Distribution of Tangles, Plaques and Related Immunohistological Markers in Healthy Aging and Alzheimer's Disease. *Neurobiology of Aging* 12:295–312.

Razani J, Wilson D M, Acosta L, Nordin S, & Murphy C (1996). Odor Learning and Memory in Patients with Alzheimer's Disease. *Chemical Senses* 21:660–1

Reyes P F, Deems D A, & Suarez M G (1993). Olfactory-related Changes in Alzheimer's Disease: A Quantitative Neuropathologic Study. *Brain Research Bulletin* 32:1–5.

Rezek D L (1987). Olfactory Deficits as a Neurologic Sign in Dementia of the Alzheimer Type. *Archives of Neurology* 44:1030–2.

Richard J & Bizzini L (1981). Olfaction et démences: Premiers résultats d'une étude clinique et expérimentale avec le *N*-propanol. *Acta Neurologica (Belgium)* 81:333–51.

Royet J P, Koenig O, Gregoire M C, Cinotti L, Lavenne F, Le Bars D, Costes N, Vigouroux M, Farget V, Sicard G, Holley A, Mauguiere F, Comar D, & Froment J C (1999). Functional Anatomy of Perceptual and Semantic Processing for Odors. *Journal of Cognitive Neuroscience* 11:94–109.

Schiffman S S, Clark C M, & Warwick Z S (1990). Gustatory and Olfactory Dysfunction in Dementia: Not Specific to Alzheimer's Disease. *Neurobiology of Aging* 11:597–600.

Serby M, Larson P, & Kalkstein D (1991). The Nature and Course of Olfactory Deficits in Alzheimer's Disease. *American Journal of Psychiatry* 148:357–60.

Serby M, Mohan C, Aryan M, Williams L, Mohs R C, & Davis K L (1996). Olfactory Identification Deficits in Relatives of Alzheimer's Disease Patients. *Biological Psychiatry* 39:375–7.

Sing C F & Davignon J (1985). Role of the Apolipoprotein E Polymorphism in Determining Normal Plasma Lipid and Lipoprotein Variation. *American Journal of Human Genetics* 37:268–85.

St. Clair D M, Simpson J, Yates C M, & Gordon A (1985). Olfaction and Dementia: A Response. *Journal of Neurology, Neurosurgery, and Psychiatry* 48:849.

Talamo B R, Rudel R A, Kosik K S, Lee V M Y, Neff S, Adelman L, & Kauer J S (1989). Pathological Changes in Olfactory Neurons in Patients with Alzheimer's Disease. *Nature* 337: 736–9.

Van Hoesen G W & Solodkin A (1994). Cellular and System Neuroanatomical Changes in Alzheimer's Disease. *Annals of the New York Academy of Sciences* 747:12–35.

Vollmecke T & Doty R L (1985). Development of the Picture Identification Test (PIT): A Research Comparison to the University of Pennsylvania Smell Identification Test (UPSIT). *Chemical Senses* 10:413–14.

Warner M D, Peabody C A, Flattery J J, & Tinklenberg J R (1986). Olfactory Deficits and Alzheimer's Disease. *Biological Psychiatry* 21:116–18.

Welsh K A, Butters N, Hughes J P, Mohs R, & Heyman A (1991). Detection of Abnormal Memory Decline in Mild Cases of Alzheimer's Disease Using CERAD Neuropsychological Measures. *Archives of Neurology* 48:278–81.

Wilhite D, Acosta L, Quiñonez C, Nordin S, & Murphy C (1993). Confidence Ratings for Odor and Visual Stimuli in Alzheimer's Patients and Normal Controls. *Chemical Senses* 18:651

17

Development of Odor Naming and Odor Memory from Childhood to Young Adulthood

Johannes Lehrner and Peter Walla

The early experiences of children have been of considerable interest in memory research, because it is assumed that those experiences lay the foundation for their later behavioral and cognitive development. Implicit in that assumption is a capacity for long-term memory in children. Earlier, memories from the first year of life had been characterized by some researchers as being short-lived, highly generalized and diffuse, and devoid of place information (Nadel and Zola-Morgan, 1984; Nelson, 1984; Olson and Strauss, 1984; Schacter and Moscovitch, 1984; Mandler, 1990). More recently, however, there have been reports of intact memory functions in infants and very young children, using tests such as novelty-preference procedures, conditioning paradigms, auditory localization, and deferred-imitation procedures (reviewed by Rovee-Collier and Gerhardstein, 1997). Moreover, recent research indicates that young children can accurately recall specific events after delays of weeks and even months. Such long-term recall of events by young children would seem to depend on the temporal relationships among the components of the event, the familiarity or repeated experiencing of the event (Mandler, 1986), and the availability of cues or remainders of the event (Bauer, 1997).

Several explanations have been proposed for developmental differences in memory performance. The most important are different stores of resources in short-term memory (Cowan, 1997), different uses of mnemonic strategies (Guttentag, 1997), and different conditions of metamemory (Joyner and Kurtz-Costes, 1997). Further, it has been argued that developmental differences in memory performance may be attributable to age-related changes in the knowledge base (Kail, 1990). Still, memory performance in recalling events is inferior in children as compared with adults.

Memory functions in humans are not controlled by a single system. Rather, memory can be divided into several different and independent subsystems that serve distinct functions and are characterized by fundamentally different rules of operation (Tulving, 1985a,b; Sherry and Schacter, 1987; Squire, 1992). Evidence for the independence of memory systems has come mainly from clinical studies in which selective memory impairments have been reported (Tulving and Schacter, 1990). The distinction between memory based on conscious recollection (explicit memory) and memory resulting from unperceived influences (implicit memory) is fundamental. Explicit memory involves the capacity for conscious recollection of names, places, dates, and events. In contrast, implicit memory encompasses a variety of nonconscious abilities, including the capacity to learn habits and skills and receptivity to priming and some forms of classic conditioning. A defining feature of implicit memory is that the impact of experience is evidenced in a change in behavior or performance, though the experience that led to the change was not consciously accessible (Zola-Morgan and Squire, 1993). Recent studies of childhood changes in memory performance have revealed evidence for intact implicit memory but poorly developed explicit memory in children (Graf, 1990; Mitchell, 1993; Naito and Komatsu, 1993; Parkin, 1997). That is, studies of picture completion and word completion have reliably shown priming effects (Parkin and Streete, 1988; Naito, 1990; Russo et al., 1995). Studies to assess priming effects for recognition of faces revealed no apparent age effect on the implicit measures, despite age-related improvements in explicit measures (Lorsbach and Morris, 1991; Ellis, Ellis, and Hosie, 1993; Newcombe and Lie, 1995).

A considerable literature on odor perception in adults has been accumulated, particularly for odor naming and odor memory (Engen, 1982; Schab and Crowder, 1995). However, less attention has been paid to the developmental processes of olfactory functions in children. In this chapter, we shall delineate the current status of research on children's olfactory functions. We shall start with the development of odor naming and compare it to findings on the development of visual naming. We shall consider odor-memory functions in children, and recent findings in olfactory explicit/implicit memory will be discussed. Lastly, we shall provide a framework to describe the development of olfaction from childhood to young adulthood and suggest some new experimental approaches. Before we go on, it is important to note that developmental differences between children and adults in odor naming and memory cannot be attributed to sensory differences, because recent studies of olfactory sensitivity in children have documented comparable olfactory thresholds in children and young adults (Koelega and Köster, 1974; Perry et al., 1980; Cain et al., 1988, 1995; Dorries et al., 1989; Solbu, Jellestad, and Straetkvern, 1989; Lehrner, Glück, and Laska, 1999a).

1. Odor Naming

The literature on the development of childhood cognition in general, and naming in particular, has been reviewed by Schneider and Pressley (1989). One method of testing naming capacity is by means of confrontation naming. Recent neuropsychological data suggest that children have poorer naming capabilities for common objects than do young adults (La Barge, Edwards, and Knesevich, 1986; Cain et al., 1995). With advancing age, that effect disappears because of a steadily increasing lexicon.

For odor identification, which is a form of confrontation naming, similar findings have been reported. Prior studies using the multiple-choice University of Pennsylvania Smell Identification Test (UPSIT) found odor-naming performance to be poorer in children than in young adults (Doty et al., 1985); however, improvement was observed with increasing age from 4 to 17 years using both a shortened version of the UPSIT (Richman, Wallace, and Sheehe, 1995) and the University of California, San Diego, odorant-identification test with 20 common odorants and corresponding pictures and names adapted for use with children (Rothschild, Myer, and Duncan, 1995). In an odor-naming task with common household odors, similar results were obtained (De Wijk and Cain, 1994; Cain et al., 1995; Lehrner et al., 1999a).

Taken together, the available studies document gradual improvements in odor naming with increasing age. Because olfactory sensitivities are comparable for different ages, the inferior naming ability probably is due to children's poor semantic knowledge of odors. Whether this is due to limitations in cognitive-processing ability or to the fact that children will not have encountered odors as often as young adults, and thus will not have acquired an extensive odor-name lexicon, remains to be determined.

2. Stability of Odor Naming

The stability of odor-naming ability over time is a measure of retrieval/encoding consistency. Similar naming stabilities in children and young adults would suggest that cognitive operations for olfactory retrieval/encoding are also similar. Recent research with young adults suggests that there is something less than perfect consistency in label use after only a 10–15-minute time interval (Rabin and Cain, 1984; Lehrner, 1993). The only developmental study comparing children and young adults found inferior naming stability in children as compared with young adults, indicating developmental differences in olfactory retrieving/encoding (Lehrner et al., 1999a).

Consistency of label use and label quality

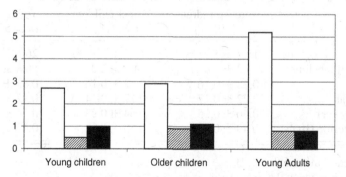

Figure 17.1. Numbers of consistently named odors across three age groups and three classes of label quality. White bars illustrate veridical naming: ANOVA $[F(2, 94) = 21.3; p < .001]$; single Scheffe post-hoc comparisons were significant between young adults and both child groups (all pairs $p < .001$). The child groups did not differ ($p > .9$). Cross-hatched bars indicate near misses: ANOVA $[F(2, 94) = 1.8; p > .16]$. Black bars indicate far misses: ANOVA $[F(2, 94) = 0.9; p > .41]$. (Data adapted from Lehrner et al., 1999a.)

For the purposes of this chapter, we analyzed the relationship between label quality (i.e., how much prior information a participant had about an odor) and consistency of labeling (Figure 17.1). Rabin and Cain (1984) reported a strong association between the consistency of labeling and the quality of labels in adults. As in the Rabin and Cain study, we found that veridical labeling produced the highest number of consistently labeled odors after a retention interval of 10 minutes. Statistical analysis revealed that our three age groups differed only for veridically named odors, with young adults performing best, whereas the two child groups did not differ.

These results indicate that consistent use of odor labels is strongly dependent on the quality of the label given to each odor, with veridical naming producing the highest level of consistent naming. That effect was particularly evident for the young adults, showing that if young adults had specific information about odors available to them, they could use that information to retrieve the specific odor name. In children, such consistency was still developing.

3. Explicit Memory for Odors

Memory research has documented the fact that explicit-memory performance improves with age in terms of quantity and quality. This has been shown for different explicit-memory procedures, such as free recall, cued recall, and recognition

Table 17.1. *Mean olfactory performances for three age groups*

Variable	Young children[a] (n = 20)	Older children[b] (n = 21)	Young adults[c] (n = 56)
Odor naming[d]	15.4 ± 5.5	17.7 ± 5.1	26.1 ± 6.3
Consistency of labeling[e]	4.2 ± 2.0	4.9 ± 2.2	6.8 ± 2.2
Hit rate[f]	0.78 ± 0.16	0.82 ± 0.16	0.86 ± 0.11
False-alarm rate[g]	0.22 ± 0.17	0.17 ± 0.14	0.12 ± 0.12
Odor memory (Pr)[h]	0.56 ± 0.25	0.64 ± 0.23	0.74 ± 0.19
Response criterion (Br)[i]	0.49 ± 0.22	0.53 ± 0.22	0.47 ± 0.20

[a] Ages 4–8 years (mean 6.8 ± 1.2).
[b] Ages 9–11 years (mean 9.6 ± 0.7).
[c] Ages 18–30 years (mean 24.9 ± 2.7).
[d] $[F(2, 94) = 31.2; p < .001]$. Single Scheffe post-hoc comparisons were significant between young adults and both child groups (all pairs $p < .001$). The child groups did not differ significantly ($p > .7$).
[e] $[F(2, 94) = 16.9; p < .001]$. Single Scheffe post-hoc comparisons were significant between young adults and both child groups (all pairs $p < .001$). The child groups did not differ significantly ($p > .5$).
[f] $[F(2, 94) = 3.2; p < .05]$. Post-hoc Scheffe tests showed that the hit rate was significantly higher for young adults than for young children ($p < .03$). No other differences were statistically significant (all $p > .5$).
[g] $[F(2, 94) = 3.6; p < .03]$. Post-hoc Scheffe tests showed that the false-alarm rate was significantly lower for young adults than for young children ($p < .04$). No other differences were statistically significant (all $p > .4$).
[h] Pr was derived by subtracting the false-alarm rate from the hit rate. $[F(2, 94) = 5.8; p < .005]$. Scheffe post-hoc tests showed significant differences between the young adults and the young-child group ($p < .008$). Older children did not differ from younger children ($p > .2$) or young adults ($p > .2$).
[i] $[F(2, 94) = 0.6; p > .5]$. The response criterion (Br) was calculated after Snodgrass and Corwin (1988). Accordingly, hit and false-alarm rates were corrected by adding 0.5 to each frequency and dividing by $n + 1$, where n is the number of old or new trials. For the computation of Br, the transformed values were used. A value of 0.5 indicates neutral bias, a value > 0.5 indicates liberal bias, and a value < 0.5 indicates conservative bias.
Source: Adapted from Lehrner et al. (1999a).

memory, with several factors influencing memory performance, both in children and in adults (Nelson, 1986). Younger children, as well as older children and adults, show a consistent pattern of superior memory for ordered events as compared with events that lack such relationships (Slackman and Nelson, 1984; Hudson and Nelson, 1986; Slackman, Hudson, and Fivush, 1986; Ratner, Smith, and Dion, 1986; Price and Godman, 1990). Familiarity, or repeated experiences of an event, influences recall for young as well as older children. Repeated experiences aid memory in terms of both the amount of information that can be remembered (Fivush, 1984; Hudson, 1986, 1990) and the duration of memories (Bauer, 1995). Reminding children of previously experienced events effectively serves to preserve event memories over longer periods of time (Sheffield and Hudson, 1994). Events of which children are verbally reminded at the time of retrieval also are well recalled after long delays (Fivush, 1984; Fivush, Hudson, and Nelson, 1984; Fivush and Hamond, 1990; Hudson, 1990; Hudson and Fivush, 1991). As discussed earlier, factors known to influence recall of events by children also influence recall by adults, thus demonstrating considerable continuity in mnemonic processes from young childhood to adulthood.

Studies of odor-recognition memory in children have rarely been performed. Hvastja and Zanuttini (1989), using children from three different age groups (mean ages 6 years, 8 years, and 10 years), found no association between odor memory and pleasant or unpleasant pictures. Jehl and Murphy (1998), using the child version of the California Olfactory Learning Test, documented better odor-recognition memory and lower false-alarm rates for children aged 10–14 years, as compared with children aged 8–10 years. Improved odor memory and lower false-alarm rates for older children (mean age 9.6 years), as compared with younger children (mean age 6.8 years), were also documented in another study (Table 17.1). Interestingly, performances reached adult levels by the age of 10 years. For the hit rate, false-alarm rate, and response criterion, the same conclusion can be drawn, indicating that 10-year-olds and young adults have similar capacities for recognition of odors. In contrast, 7-year-olds had not yet reached the adult processing level, although they did not differ for the response criterion.

4. Implicit Memory for Odors

Recent memory research suggests that recognition of repeated verbal stimuli may depend on two different memory processes: "familiarity" and "recollection." Familiarity judgments are made on the basis of a feeling, without specific information about the encoding episode, and thus relate to implicit memory. Recollection is seen as an elaborate, conceptually driven process that includes retrieval of contextual information and thus relates to explicit memory (Mandler, 1980). A method to investigate implicit/explicit memory is the level-of-processing approach, with a large body of evidence suggesting that different "levels" of processing lead to different memory performances (Craik and Lockhart, 1972). Semantic processing of words leads to good memory performance, whereas non-semantic processing leads to rather poor memory performance (Craik, 1983). Similar results have been obtained from odor-memory studies in young adults: It was found that (1) if the correct labels were provided by the experimenter

(Lyman and McDaniel, 1986, 1990) and (2) the quality of self-generated labels was high (Rabin and Cain, 1984) and (3) there was consistent labeling of odors (Rabin and Cain, 1984; Lehrner, 1993), those factors could enhance odor memory substantially, indicating that semantic processing of odors is important for retention of odor memory (Schab, 1991; Herz and Engen, 1996).

In an attempt to distinguish between explicit memory and implicit memory for odors in children, the combination of an odor-naming task and odor-recognition task was used to estimate the relationship between the depth or level of processing and retention of olfactory information (Lehrner et al., 1999b). Assuming that the name of an odor represents its semantic information, correct naming of an odor was interpreted as indicating a high (cognitive) level of processing. Odors that could not be correctly named were believed to be processed at a lower and more perceptual level. The applied analysis revealed different patterns depending on age. Odors that were not correctly named revealed no age effect. Recognition of odors that were correctly named depended strongly on age. The results indicated two forms of odor memory differently represented from childhood to young adulthood. Thus, the mechanism operating for odor-recognition memory may be similar to that for verbal-recognition memory, where the level of processing has a strong influence on memory performance.

5. Discussion and Perspective

As this review has shown, odor-naming capabilities progressively improve with increasing age. Whether poor odor-naming performance in children is determined by limited cognitive-processing ability or by the fact that children have not encountered odors as often as young adults, and thus have not acquired an extensive odor-name lexicon, is not clear at present. However, several lines of evidence indicate that poor odor-specific knowledge is responsible for poor odor naming.

In laboratory testing, children have steeper explicit paired-associate learning slopes for names of odors than do young adults, even though they start from a lower level of semantic knowledge of odors at the beginning. This indicates that children possess the capacity to make appropriate odor–name associations (Cain et al., 1995). In natural language acquisition, children add more words per month to their vocabularies than do adults, and they probably possess the capacity to learn more odor–name associations per month as well. However, according to Engen and Engen (1997), who reviewed naturalistic language studies, children are more concerned with the colors, sizes, and locations of objects than with their smells, and there are virtually no examples of children spontaneously using the perception verb "smell" to respond to odor experiences. In the future, the exact acquisition speed for odor–name associations with naturally occurring odors should be determined with naturalistic field studies and compared to performances involving the other senses.

What determines odor-name acquisition in children? Is it the frequency of encounters that offers more opportunities for learning, or is it the salience of specific odors? De Wijk and Cain (1994) and Cain et al. (1995) investigated that issue, assuming that the frequency of occurrence in English usage reflects the frequency of occurrence of items in everyday life. They found a significant positive correlation for odor naming and frequency in English usage, indicating that the more their opportunities to experience the odors, the better the naming of odors by children. Salience of odors seems to be less important, because children and young adults showed reasonable similarities in their profiles for identifying everyday olfactory items, indicating similar acquisitions.

Consistency of label use is inferior in children as compared with young adults. Laboratory analyses have indicated that prior knowledge of the veridical odor name determines the stability of naming to a large degree in adults, whereas selection of labels of poor quality produces similar naming stabilities in children and adults. Those results suggest some advantage in cognitive-processing ability for young adults, probably because of greater familiarity with the odors.

Taken together, the data indicate that acquisition of verbal labels is incidental, proceeding slowly (probably on a smell-by-smell basis) during child development, and is strongly dependent on the opportunities children are offered and on how well each child is encouraged to use those opportunities.

Odor memory gradually develops from childhood to adulthood, reaching an adult level at the age of 10 years. Although their performance is inferior to that of adults, children have already acquired a substantial odor vocabulary. They are capable of using that semantic knowledge to encode olfactory information, as evidenced by a significant positive correlation between odor memory and stability of odor naming (Lehrner et al., 1999a), indicating the importance of semantic encoding for odor memory in childhood. In future studies, it will be important to compare recognition memory for odors with recognition memory for other stimuli (e.g., pictures) in order to be able to compare different modalities directly and to assess the specific influence of verbal encoding on memory for olfactory stimuli compared with other stimuli in children.

A developmental odor-memory study has provided further support for the implicit/explicit-memory dissociation in children, documenting a fully functional implicit memory for odors and a still-developing explicit memory in children, with the former probably based more on perceptual processing, and the latter probably based more on semantic processing. Thus the hypothesis of distinct implicit memory and conscious or explicit memory in children can be extended to the olfactory modality.

Further developmental studies to investigate implicit memory for odors (e.g., priming studies) are needed, because recent reports have demonstrated (1) the importance of perinatal and postnatal olfactory information (Schaal et al., Chapter 26, this volume) and (2) the influence of neonatal experience with vanilla on food preferences later in life (Haller et al., 1999), indicating that implicit olfactory learning takes place in children.

Acknowledgments

Peter Walla was supported by the Austrian Science Fund and partly by the Austrian National Bank. We would like to thank Dr. Ross Cunnington for helpful comments on the manuscript.

References

Bauer P (1995). Recalling Past Events: From Infancy to Early Childhood. *Annals of Child Development* 11:25–71.
Bauer P (1997). Development of Memory in Early Childhood. In: *The Development of Memory in Childhood*, ed. N Cowan & C Hulme, pp. 83–113. Hove, East Sussex: Psychology Press.
Cain W S, Gent J F, Goodspeed R B, & Leonard G (1988). Evaluation of Olfactory Dysfunction in the Connecticut Chemosensory Clinical Research Center. *Laryngoscope* 98:83–8.
Cain W S, Stevens J C, Nickou C M, Giles A, Johnston I, & Garcia-Medina M R (1995). Life-Span Development of Odor Identification, Learning, and Olfactory Sensitivity. *Perception* 24:1457–72.
Cowan N (1997). The Development of Working Memory. In: *The Development of Memory in Childhood*, ed. N Cowan & C Hulme, pp. 163–200. Hove, East Sussex: Psychology Press.
Craik F I M (1983). On the Transfer of Information from Temporary to Permanent Memory. *Philosophical Transactions of the Royal Society (London)* 302:341–59.
Craik F I M & Lockhart R S (1972). Levels of Processing: A Framework for Memory Research. *Journal of Verbal Behavior* 11:671–84.
De Wijk R & Cain W S (1994). Odor Identification by Name and by Edibility: Life-Span Development and Safety. *Human Factors* 36:182–7.
Dorries K M, Schmidt H J, Beauchamp G K, & Wysocki C J (1989). Changes in Sensitivity to the Odor of Androstenone during Adolescence. *Developmental Psychobiology* 22:423–35.
Doty R L, Applebaum S, Zusho H, & Settle R G (1985). Sex Differences in Odor Identification Ability: A Cross Cultural Analysis. *Neuropsychologia* 23:667–72.
Ellis H D, Ellis D M, & Hosie J A (1993). Priming Effects in Children's Face Recognition. *British Journal of Psychology* 84:101–10.
Engen T (1982). *The Perception of Odors*. Toronto: Academic Press.
Engen T & Engen E (1997). The Relationship between Odor Perception and Language Development. *Enfance* 1:125–40.
Fivush R (1984). Learning about School: The Development of Kindergartners' School Scripts. *Child Development* 55:1697–709.

Fivush R & Hamond N R (1990). Autobiographical Memory across the Preschool Years: Toward Reconceptualizing Childhood Amnesia. In: *Knowing and Remembering in Young Children*, ed. R Fivush & J A Hudson, pp. 223–48. Cambridge University Press.

Fivush R, Hudson J A, & Nelson K (1984). Children's Long-Term Memory for a Novel Event: An Exploratory Study. *Merill-Palmer Quarterly*, 30:303–16.

Graf P (1990). Life-Span Changes in Implicit and Explicit Memory. *Bulletin of the Psychonomic Society* 28:353–8.

Guttentag R (1997). Memory Development and Processing Resources. In: *The Development of Memory in Childhood*, ed. N Cowan & C Hulme, pp. 247–74. Hove, East Sussex: Psychology Press.

Haller R, Rummel C, Henneberg S, Pollmer U, & Köster E P (1999). The Influence of Early Experience with Vanillin on Food Preference Later in Life. *Chemical Senses* 24:465–7.

Herz R S & Engen T (1996). Odor Memory: Review and Analysis. *Psychonomic Bulletin Review* 3:300–13.

Hudson J A (1986). Memories Are Made of This: General Event Knowledge and Development of Autobiographical Memory. In: *Event Knowledge: Structure and Function in Development*, ed. K Nelson, pp. 97–118. Hillsdale, NJ: Lawrence Earlbaum.

Hudson J A (1990). Constructive Processing in Children's Event Memory. *Developmental Psychology* 26:180–7.

Hudson J A & Fivush R (1991). As Time Goes by: Sixth Graders Remember a Kindergarten Experience. *Applied Cognitive Psychology* 5:246–360.

Hudson J A & Nelson K (1986). Repeated Encounters of a Similar Kind: Effects of Familiarity on Children's Autobiograpical Memory. *Cognitive Development* 1:253–71.

Hvastja L & Zanuttini L (1989). Odor Memory and Odor Hedonics in Children. *Perception* 18:391–6.

Jehl C & Murphy C (1998). Developmental Effects on Odor Learning and Odor Memory in Children. *Annals of the New York Academy of Sciences* 855:632–5.

Joyner M & Kurtz-Costes B (1997). Metamemory Development. In: *The Development of Memory in Childhood*, ed. N Cowan & C Hulme, pp. 275–300. Hove, East Sussex: Psychology Press.

Kail R V (1990). *The Development of Memory*. New York: Freeman.

Koelega H S & Köster E P (1974). Some Experiments on Sex Differences in Odor Perception. *Annals of the New York Academy of Sciences* 237:234–46.

La Barge E, Edwards D, & Knesevich J W (1986). Performance of Normal Elderly on the Boston Naming Test. *Brain and Language* 27:380–4.

Lehrner J (1993). Gender Differences in Long-Term Odor Recognition Memory: Verbal versus Sensory Influences and Consistency of Label Use. *Chemical Senses* 18:17–26.

Lehrner J, Glück J, & Laska M (1999a). Odor Identification, Consistency of Label Use, Olfactory Threshold and Their Relationships to Odor Memory over the Human Lifespan. *Chemical Senses* 24:337–46.

Lehrner J, Walla P, Laska M, & Deecke L (1999b). Different Forms of Human Odor Memory: A Developmental Study. *Neuroscience Letters* 272:17–20.

Lorsbach T C & Morris A K (1991). Direct and Indirect Testing of Picture Memory in Second and Sixth Grade Children. *Contempory Educational Psychology* 16:18–27.

Lyman B J & McDaniel M A (1986). Effects of Encoding Strategy on
    Long-Term Memory for Odors. *Quaterly Journal of Experimental Psychology*
    38A:753–65.
Lyman B J & McDaniel M A (1990). Memory for Odors and Odor Names: Modalities
    of Elaboration and Imagery. *Journal of Experimental Psychology: Learning,
    Memory and Cognition* 16:656–64.
Mandler G (1980). Recognizing: The Judgement of Previous Occurrence.
    Psychological Review 87:252–71. ·
Mandler J M (1986). The Development of Event Memory. In: *Human Memory and
    Cognitive Capabilities: Mechanisms and Performance*, ed. F Klix &
    H Hagendorf, pp. 459–67. New York: Elsevier.
Mandler J M (1990). Recall of Events by Preverbal Children. In: *The Development and
    Neural Basis of Higher Cognitive Function*, ed. A Diamond, pp. 485–516.
    New York Academy of Sciences.
Mitchell D B (1993). Implicit and Explicit Memory for Pictures: Multiple Views across
    the Lifespan. In: *Implicit Memory: New Directions in Cognition, Development,
    and Neuropsychology*, ed. P Graf & M Masson, pp. 167–90. Hillsdale, NJ:
    Lawrence Erlbaum.
Nadel L & Zola-Morgan S (1984). Infantile Amnesia: A Neurobiological Perspective.
    In: *Infant Memory*, vol. 9, ed. M Moscovitch, pp. 145–72. New York: Plenum
    Press.
Naito M (1990). Repetition Priming in Children and Adults: Age-related Dissociation
    between Implicit and Explicit Memory. *Journal of Experimental Child
    Psychology* 50:462–84.
Naito M & Komatsu S (1993). Processes Involved in Childhood Development of
    Implicit Memory. In: *Implicit Memory: New Directions in Cognition,
    Development, and Neuropsychology*, ed. P Graf & M Masson, pp. 231–60.
    Hillsdale, NJ: Lawrence Erlbaum.
Nelson K (1984). The Transition from Infant to Child Memory. In: *Infant Memory*,
    vol. 9, ed. M Moscovitch, pp. 103–30. New York: Plenum Press.
Nelson K (1986). *Event Knowledge*. Hillsdale, NJ: Lawrence Erlbaum.
Newcombe N & Lie E (1995). Overt and Covert Recognition of Faces in Children and
    Adults. *Psychological Science* 6:241–5.
Olson G M & Strauss M S (1984). The Development of Infant Memory. In: *Infant
    Memory*, vol. 9, ed. M Moscovitch, pp. 29–48. New York: Plenum Press.
Parkin A (1997). The Development of Procedural and Declarative Memory. In: *The
    Development of Memory in Childhood*, ed. N Cowan & C Hulme, pp. 113–37.
    Hove, East Sussex: Psychology Press.
Parkin A J & Streete S (1988). Implicit and Explicit Memory in Young Children and
    Adults. *British Journal of Psychology* 79:361–9.
Perry J D, Frisch S, Jafek B, & Jafek M (1980). Olfactory Detection Thresholds Using
    Pyridine, Thiophene, and Phenylethyl Alcohol. *Otolaryngology: Head and Neck
    Surgery* 88:778–82.
Price D W & Godman G S (1990). Visiting the Wizard: Children's Memory for a
    Recurring Event. *Child Development* 61:664–80.
Rabin M D & Cain W S (1984). Odor Recognition: Familiarity, Identifiability, and
    Encoding Consistency. *Journal of Experimental Psychology: Learning, Memory
    and Cognition* 10:316–25.
Ratner H H, Smith B S, & Dion S (1986). Development of Memory for Events. *Journal
    of Experimental Child Psychology* 41:411–28.

Richman R A, Wallace K, & Sheehe P R (1995). Assessment of an Abbreviated Odorant Identification Task for Screening: A Rapid Screening Device for Schools and Clinics. *Acta Paediatrica* 84:434–7.

Rothschild M A, Myer C M, & Duncan H J (1995). Olfactory Disturbance in Pediatric Tracheotomy. *Otolaryngology: Head and Neck Surgery* 111:71–6.

Rovee-Collier C & Gerhardstein P (1997). The Development of Infant Memory. In: *The Development of Memory in Childhood*, ed. N Cowan & C Hulme, pp. 5–39. Hove, East Sussex: Psychology Press.

Russo R P, Nichelli M, Gibertoni M, & Cornia C (1995). Developmental Trends in Implicit and Explicit Memory: A Picture Completion Study. *Journal of Experimental Child Psychology* 59:566–78.

Schab F R (1991). Odor Memory: Taking Stock. *Psychological Bulletin* 2:242–51.

Schab F R & Crowder R G (eds.) (1995). *Memory for Odors*. Hillsdale, NJ: Lawrence Erlbaum.

Schacter D L & Moscovitch M (1984). Infants, Amnesics, and Dissociable Memory Systems. In: *Infant Memory*, vol. 9, ed. M Moscovitch, pp. 173–216. New York: Plenum Press.

Schneider W & Pressley M (1989). *Memory Development between 2 and 20*. New York: Springer-Verlag.

Sheffield E G & Hudson J A (1994). Reactivation of Toddlers' Event Memory. *Memory* 2:447–65.

Sherry D F & Schacter D L (1987). The Evolution of Multiple Memory Systems. *Psychological Review* 94:439–54.

Slackman E A, Hudson J A, & Fivush R (1986). Actions, Actors, Links, and Goals: The Structure of Children's Event Representations. In: *Event Knowledge: Structure and Function in Development*, ed. K Nelson, pp. 47–69. Hillsdale, NJ: Lawrence Erlbaum.

Slackman E A & Nelson K (1984). Acquisition of an Unfamiliar Script in Story Form by Young Children. *Child Development* 55:329–40.

Snodgrass J G & Corwin J (1988). Pragmatics of Measuring Recognition Memory: Applications to Dementia and Amnesia. *Journal of Experimental Psychology: General* 117:34–50.

Solbu E H, Jellestad F K, & Straetkvern K O (1989). Children's Sensitivity to Odor of Trimethylamine. *Journal of Chemical Ecology* 16:1829–40.

Squire L R (1992). Memory and the Hippocampus: A Synthesis from Findings with Rats, Monkeys, and Humans. *Psychological Review* 99:195–231.

Tulving E (1985a). How Many Memory Systems Are There? *American Psychologist* 40:385–98.

Tulving E (1985b) Memory and Consciousness. *Canadian Psychologist* 26:1–12.

Tulving E & Schacter D L (1990). Priming and Human Memory Systems. *Science* 247:301–30.

Zola-Morgan S & Squire L R (1993). Neuroanatomy of Memory. *Annual Review of Neuroscience* 16:547–63.

Section Five

Neural Bases

As discussed in Sections 3 and 4, olfactory and gustatory cognitions have close links to emotion and memory, from both the psychological and physiological points of view. The chapters in this section will complement the preceding studies and focus on the neural bases supporting the neural code and chemical-sense cognition, while most often taking into account these emotional and memory dimensions. In Chapter 18, Gilles Sicard reviews recent findings on *neuroreceptor* identification and the patterns that have emerged in the neural representation of odors at the level of *neuroreceptors* and the *olfactory bulb*, emphasizing a potentially important role for *neural assemblies* in odor detection and recognition.

The chapters in this section deal with phenomena observed over different time scales and under different experimental conditions, varying from one study to the next, and the concepts of neural dynamics are not always easy to follow. Several sets of data presented in the following chapters will illustrate this point. For instance, in Chapter 19, Bettina Pause examines the effects of cognitive dimensions such as *attention, habituation*, and *short-term memory* on the characteristics of olfactory-related *evoked potentials* recorded from the scalp in humans. She reports on pronounced modulations of some evoked-potential components occurring within the first second after stimulus onset. In Chapter 20, Robert Zatorre reviews the findings on the contributions of functional imaging studies to our understanding of the processing of olfactory affective information.

Using deep *local-field potential* recordings in freely behaving rats, Nadine Ravel and colleagues, in Chapter 21, study the impact of olfactory learning on information processing at different levels of the olfactory pathways. For instance, it was found that, on one hand, habituation over a 1-min period to repeated olfactory presentations and, on the other hand, associative learning, which developed over weeks, could both strongly modify neural responses to stimuli. They report on the efficiency of synaptic transmission and the strength of functional coupling as revealed by oscillatory activities in neuron assemblies recorded at the level of the primary olfactory cortex and in plurimodal areas such as the *entorhinal cortex.*

Within the same conceptual perspective, Annick Faurion and colleagues examine how *familiarization* to repeated presentations of novel tastes can modulate neural reactivity, as can aversive conditioning. Such experiences are accompanied by changes in neural responses both at the periphery of the gustatory system in animals and in the *primary gustatory* centers in humans.

In Chapter 23, Edmund Rolls provides strong evidence that internal states such as food repletion and depletion induce modulation of single-cell responsiveness in primates at the level of the orbitofrontal cortex. That plurimodal associative structure is also found to be a convergence site for gustatory and olfactory information, taking into account the consequences of long-lasting associative learning. These chapters will illustrate in both animals and humans how recent and remote experiences determine neural responses to odorants and tastants at several central stages of information processing.

18

Odor Coding at the Periphery of the Olfactory System

Gilles Sicard

During the past decade, scientists have identified a large number of genes coding for olfactory receptor proteins in vertebrates, including humans, and in insects and nematodes (Buck and Axel, 1991; Selbie et al., 1992; Sengupta, Colbert, and Bargmann, 1994; Gao and Chess, 1999; Clyne et al., 1999). Much earlier, such entities had been hypothesized to exist, probably on the intuition that the nature of an odor was not "ethereal" but rather a material part of the odor source (Lucretius, *De rerum natura*, IV) and thus could interact directly with the detecting organism. During the twentieth century, that concept was commonly used by physiologists to discuss the coding of odors (Zwaardemaker, 1925; Guillot, 1948) and by pioneer chemists who postulated receptive sites for odor molecules (Amoore, 1967; Beets, 1982). Its recent implementation in identifying receptor proteins has emitted a "strong scent of success" (Lancet, 1991). However, when the data from molecular biology and the physiological properties of the olfactory system are compared, it becomes clear that the final word has not yet been spoken on olfactory coding.

When olfactory signals are detected and differentiated, they gain behavioral significance when recognized as representing particular odor sources. In terms of neurophysiology, such processes require highly organized neuronal circuitry, and an important finding in recent studies of receptor proteins is that the receptors themselves are involved in determining the neural space devoted to representation of the chemical environment. This chapter will describe anatomical and functional aspects of the olfactory message in the peripheral olfactory system of vertebrates, that is, features of the sensory neurons located in the olfactory mucosa and their projections to the glomeruli and the second-order neurons of the olfactory bulb. The output message of the olfactory bulb will also be considered.

Receptors recognize, and olfactory cells respond to, defined sets of molecular stimuli, to which we refer, using terminology borrowed from visual physiology, as "molecular receptive fields" (Mori and Shepherd, 1994). Molecular receptive fields are not definable spatially, and in defining them we must content ourselves with selectivity profiles obtained from testing a limited sample of odorants. In addition, we know that knowledge of the specificity of the receptors alone, without an understanding of their relationships to one another, leaves our description of the proximal part of the olfactory system incomplete. Although a comparison of data from recent advances in the molecular biology of olfactory receptors and data from electrophysiological characterization of the olfactory receptor neurons reveals some contradictory elements, the rough sketch seems coherent and provides a basis from which to further understand olfactory perception and to suggest important ideas and new experiments.

1. Receptors for Odorants

1.1. Olfactory Receptor Families and Specificity

Olfactory receptors selectively respond to environmental signals and represent a form of structural memory. As a result of evolution, several hundred receptors have been selected for olfactory discrimination in vertebrates. They were discovered by scientists working on the assumption that they would belong to the G-protein-coupled receptor family, would have seven trans-membrane domains, and would show a certain degree of homology in their amino acid sequences. The genes coding olfactory receptors constitute the largest known gene family, and in fact the largest family contributing to a single physiological function. However, on the basis of differences in their amino acid sequences, they can be grouped into subfamilies.

Whereas orthologous genes (i.e., genes with nearly identical sequences and functions) are found across different animal species (Krautwurst, Yau, and Reed, 1998), the total repertoire of genes available for olfactory receptor coding varies among species. For instance, in the rat, a repertoire of 1,000 genes has been estimated (Buck and Axel, 1991). The number of genes is more restricted for fish and, for example, has been estimated at about 100 for catfish (Ngai et al., 1993). In humans, the repertoire is also smaller, believed to be between 200 and 750 (Ressler, Sullivan, and Buck, 1994; Mombaerts, 1999). Interestingly, half of these are pseudogenes, which cannot encode a complete functional receptor, thus suggesting that the human olfactory receptor repertoire represents a phylogenetic decline in the sense of smell (Rouquier et al., 1998; Sharon et al., 1999). In the dolphin Stenella coeruleoalba, which has no sense of smell, only nonfunctional pseudogenes have been found (Freitag et al., 1998). Although we have no direct evidence of this, the level of olfactory gene expression may vary between individuals of a given species, including humans. Suggestive evidence comes from studies demonstrating differences between individuals in olfactory detection thresholds, and particularly from studies on specific hyposmia that reveal substance-specific olfactory deficits in some subjects.

Further, it has been shown in mice, rabbits, and humans that repeated presentations of an odorant are able to increase behavioral sensitivity to it, and in mice and rabbits that the electrophysiological response of the olfactory mucosa to the odorant is concomitantly increased (Wysocki, Dorries, and Beauchamp, 1989; Wang, Wysocki, and Gold, 1993; Semke, Distel, and Hudson, 1995). This observation is consistent with an exciting hypothesis: that the level of olfactory gene expression may depend on the chemical environment.

A first dimension of receptor specificity related to animal ecology is the partitioning of olfactory stimuli in hydrophilic and lipophilic chemicals. Indeed, it has been shown that aquatic vertebrates express an olfactory receptor subfamily, class I, that differs structurally from class II, found in terrestrial animals. Interestingly, amphibians express both classes (Freitag et al., 1998).

Regarding their chemical selectivity, most olfactory receptors currently must be considered to be orphan receptors, as we do not know their ligands. However, different approaches have been developed to describe their ranges of chemical selectivity, and further investigations are in progress (Raming et al., 1993; Zhao et al., 1998; Krautwurst et al., 1998; Touhara et al., 1999). In a few cases, transfections of olfactory receptor genes have been shown to confer chemosensitivity on different host cells. Thus, in the rat, overexpression of gene I7 in the olfactory mucosa enhanced the electrical responses to a small number of the 74 substances tested (i.e., to linear C7–C10 saturated aldehydes), with the best response to octanal (Zhao et al., 1998). Interestingly, in the mouse, an orthologous receptor responded best to heptanal, differing in the putative receptor pocket only by an isoleucine (6-carbon) residue in place of a valine (5-carbon) residue (Krautwurst et al., 1998). Such tight specificity is also illustrated by the OR17-40 human receptor, which was reported to recognize only 1 of the 100 odorants tested (Hatt, Gisselmann, and Wetzel, 1999).

Those observations seem to suggest that each odor receptor is tightly tuned to a limited set of structurally related molecules. In contrast, another study of mammalian olfactory receptors concluded that a given receptor recognized different molecular species (Malnic et al., 1999), although the structural difference between the molecules tested was a change in only a single functional group. The same study indicated that a given molecule could interact with several receptors. Thus, combinatorial coding of chemical stimuli may be occurring at the receptor level, a form of coding that previously had been attributed to the neural representation of odor stimuli only at the sensory neuron level!

1.2. Receptor Distribution

With regard to olfactory coding, two kinds of distributional questions are of interest: (1) How many different receptors are expressed by a given receptor neuron? (2) Is there a spatial patterning of the receptors in the olfactory mucosa?

Because only 0.1% of olfactory neurons were found to express a particular receptor gene in a repertoire evaluated to approximately 1,000, each neuron presumably expresses only one receptor type (Nef et al., 1992; Strotmann et al., 1992; Ressler et al., 1993). That assumption is consistent with allelic inactivation of olfactory gene expression (Chess et al., 1994) and with results from studies in which mRNAs from single olfactory neurons have been amplified (Malnic et al., 1999; Touhara et al., 1999). However, on the basis of single olfactory cell mRNA amplification and in situ hybridization, a recent study indicates that two different mRNA receptor sequences can be expressed by small samples of receptor neurons (Rawson et al., 2000). How many receptors can be expressed by a single olfactory neuron? That, together with the question of the specificity of receptor–ligand interactions, is a critical issue for understanding the nature of signal encoding by the olfactory system. Assuming tightly tuned receptors and exclusive expression of a single receptor type by a given cell, the system could simply operate by labeled lines. On the other hand, broad receptor tuning, or alternatively, mixed expression of receptors in one neuron, would necessitate that the stimulus be represented by an across-fiber pattern of activation. Because the stimulus itself is generally a combination of odorant molecules, a combinatorial representation by several receptors seems a valid model for thinking about olfactory coding more generally.

The second question is whether or not there is a spatial pattern of receptors in the sensory sheet and whether or not that would be of advantage for olfactory coding. A heterogeneous distribution of odorant receptors has been shown in the olfactory mucosa in mammals. Thus, in rats, neurons expressing mRNA of a given receptor appear to be restricted to four large zones extending in roughly parallel bands along the anteroposterior axis of the olfactory sensory sheet. Within those zones the distribution of receptor types appears to be random (Ressler et al., 1993; Vassar, Ngai, and Axel, 1993), although OR37 is apparently restricted to a smaller area, suggesting a different type of control for its spatial expression (Strotmann et al., 1999). Contrasting with the former data, such regional segregation is not found in chickens or fish (Leibovici et al., 1996; Vogt et al., 1997), where the repertoire of olfactory receptors is possibly 10 times smaller. Nevertheless, specific changes in sensitivity to odorants across the olfactory mucosa have been recorded, both in rodents and in amphibians (Daval, Leveteau, and MacLeod, 1970; Mackay-Sim and Kesteven, 1994), which could be a consequence of receptor segregation. It remains to evaluate the implications of the particular spatial arrangement for olfactory coding. The data suggest at least two comments: (1) Apparent differences among species would seem to imply that peripheral odor coding principles vary among vertebrates. (2) In other respects, as partial obstructions of the turbinates by mucus can frequently occur, the spatial representation of a given chemical can change, and the role of a peripheral chemotopy in odor representation remains questionable.

1.3. Receptor Neuron Projections

The discovery that olfactory receptor mRNAs are also expressed in the axons and terminals of the sensory neurons has shed new light on another topographic aspect of the olfactory coding system. The olfactory mucosa of the rat or mouse contains millions of sensory neurons, which send their axons to the olfactory bulb to reach about 2,000 glomeruli. The glomeruli are well-delineated anatomical structures. They are the sites where the receptor neurons synapse with dendrites of the secondary neurons (i.e., the mitral and tufted cells) and with dendrites of local interneurons such as the periglomerular cells. The primary dendrites of only about 20 secondary neurons contribute to an individual glomerulus. Consequently, there is a strong convergence of the receptor neurons on (1) the glomeruli and (2) the secondary neuronal population (Figure 18.1).

As shown by in situ hybridization studies in rats and mice, neurons expressing a given receptor project to just two glomeruli, one on the lateral aspect and one on the medial aspect of the olfactory bulb (Ressler et al., 1994; Vassar et al., 1994; Strotmann et al., 1994). Thus, in addition to the numerical convergence, there is a functional convergence based on receptor type, meaning that the total activation of a given receptor type is collected in the glomerular layer, resulting in an amplification of the stimulus. The patterns of projections are similar across individuals. In other words, the pattern of glomerular activation can be considered as a topographically organized map of the receptors activated. The mechanisms leading to that highly organized pattern of projections are not known, but it has been suggested that receptor proteins could serve as surface signals in the process of finding the target glomeruli, although other kinds of signals may also operate, such as morphogen gradients, which are generally considered to provide an explanation for axon targeting (O'Leary, Yates, and McLaughlin, 1999; Matsuzaki et al. 1999).

In mice, receptor neurons expressing P2 gene products converge on a pair of glomeruli. When the P2 receptor is exchanged with the M12 receptor, the sites of convergence shift, suggesting that the receptors themselves are involved in the axonal guidance (Mombaerts et al., 1996). Whereas glomerular convergence is precise in intact animals, following lesions of the olfactory nerve and

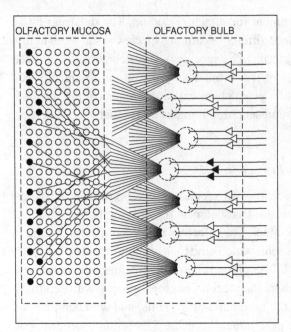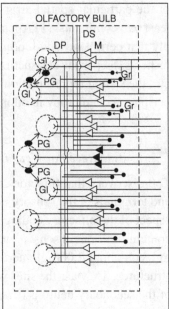

Figure 18.1. Olfactory receptor convergence rule and glomerular units in the olfactory bulb. Receptor neurons expressing a single receptor type (plain circles) are widely distributed in the olfactory mucosa. Their axons converge onto one or two glomeruli (Gl) in the olfactory bulb, where they form excitatory synapses with the primary dendrites (DP) of the relay neurons of the olfactory bulb mitral cells (M) and tufted cells. In the rat, a given relay neuron contacts only one glomerulus. Associated with a glomerulus are the dendrites of the periglomerular cells (PG), a first group of local interneurons that are also activated by the receptor neurons. They exert an inhibitory influence on neighboring glomeruli. Deeper in the bulb the secondary dendrites (DS) of the relay neurons spread radially (diameter 800 μm) and synapse with a large number of granular cells (Gr), the second group of local interneurons. In addition, the granular cells receive collaterals of mitral/tufted cell axons, with the mitral–granular cell circuits forming feedback loops.

regeneration the target of the regenerating sensory axons is no longer so well defined (Hudson, Distel, and Zippel, 1990; Costanzo, 2000). As a consequence, the spatial representation of odors in the olfactory bulb should be modified. That could explain why deficits in olfactory sensitivity (hyposmia) following nasal viral infections in humans often are associated with alterations in the perceived quality of stimuli (parosmia).

2. Functional Data

By means of direct recordings of the responses of receptor neurons, the structure of the transmitted message emerging from receptor–odorant interactions can be revealed.

2.1. Receptor Neuron Responsiveness

The selectivity profiles for olfactory receptor neurons in amphibians were characterized early in a number of studies (Gesteland et al., 1963; Duchamp et al., 1974), which showed that by considering only the output of a single receptor neuron, the stimulus could not be unambiguously defined. More recently, electrophysiological recordings have shown that the receptor neurons are able to respond to structurally different molecules, including aliphatic, aromatic, and spherical structures. For instance, in the frog, olfactory receptor neurons responded, on average, to 7 of 20 stimuli, although a significant percentage (19%) failed to respond to any of those (Sicard and Holley, 1984). Some of the nonresponding neurons, however, could have been tightly tuned and thus responsive to odorants not included in the test sample. Such neurons could express receptors as tightly tuned as that described by Zhao et al. (1998).

Although receptor neurons have been found to be more selective in the mouse than in the frog (Sicard, 1986), they can nevertheless respond to several structurally different odorants (Figure 18.2). Broad selectivity profiles for olfactory receptor neurons have been confirmed by recent investigations in the mouse and rat (Sato et al., 1994; Bozza and Kauer, 1998; Duchamp-Viret, Chaput, and Duchamp, 1999). If it is assumed that only one receptor gene is expressed by a given receptor neuron, then it should follow that some of the receptors have broad selectivity. The problem remains that whereas most direct investigations of receptor function suggest a high selectivity, narrow tuning is not consistent with the bulk of the physiological data. Is it possible that the extreme sensitivity of the olfactory transduction mechanisms is difficult to reproduce in host cells? If that should be the case, then we need to consider that besides the best ligand, other odorants may also activate the receptor, but with less efficiency. The assumption that other types of receptors co-exist with the identified heptahelical receptor family constitutes an alternative hypothesis to explain the actual molecular receptive field of the olfactory receptor cells (Gibson and Garbers, 2000).

2.2. Activity in the Bulbar Glomeruli

The olfactory glomerulus, containing no cell bodies but only fibers, is anatomically an isolated structure: Locally released factors might diffuse and exert influences over the whole glomerular neuropil, and thus the function of a given glomerulus could be controlled independently of its neighbors. Glomeruli were very early considered to be functional units (Leveteau and MacLeod, 1966). In the rat, indirect evidence for that was provided by the finding that the responses of the bulbar output neurons of a given glomerulus tended to be more similar

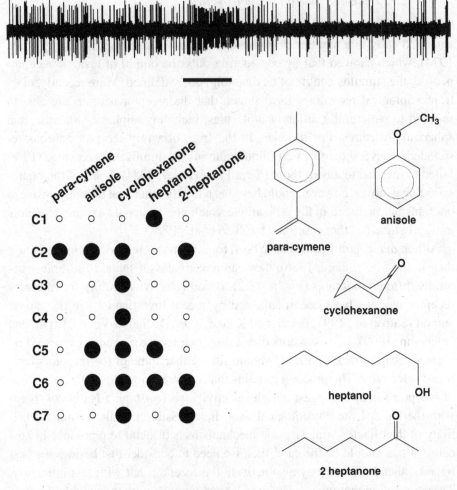

Figure 18.2. Responses of olfactory receptor neurons in the mouse. Upper part: Electrophysiological response of a receptor neuron to 2-heptanone (horizontal bar: 2-sec stimulation) recorded in an anesthetized mouse. Lower part: Examples of the structural diversity of the molecules recognized by seven receptor neurons recorded in the mouse. The large dots indicate an increase in the firing rate following chemical stimulation, and the small circles the absence of response to the stimulus: para-cymene, citrusy odor; anisole, powerful and harsh, very sweet odor, rather chemical; cyclohexanone, minty-camphoraceous, cool and solvent-like odor; heptanol, green-fatty, winey, and sap-like odor; 2-heptanone, fruity-spicy, light and volatile odor (Arctander, 1969).

to each other than to those of output neurons of different glomeruli (Buonviso and Chaput, 1990). Thus, assuming that converging receptor neurons express identical receptors, the glomeruli should have a homogeneous and unique input as well as a homogeneous and unique output.

In fact, local neuronal circuits and centrifugal fibers from distant areas of the brain modulate the activity of the output neurons. The first group of interneurons is composed of the periglomerular cells, which are distributed around the glomerular neuropil. They form inhibitory feedback loops with mitral/tufted cells and also inhibit neighboring glomeruli via their short axons. These connections tend to decrease activity in the proximity of activated glomeruli and are thought to enhance the contrast between highly activated glomeruli and their neighbors (Duchamp and Sicard, 1984a; Young and Wilson, 1999). These inhibitory circuits may also serve to protect the olfactory system against saturation during intense stimulation and to preserve, or further enhance, the signal-discrimination ability at that level (Duchamp-Viret, Duchamp, and Sicard, 1990). As the secondary dendrites of mitral cells can extend over large areas (Mori, Kishi, and Ojima, 1983) and support mitral–granular cell loops, we can also suppose that the connections take part in shaping the representation of the stimulus linking more distant glomeruli.

As observed in 2-deoxyglucose studies, the spatial patterns of glomerular activation elicited in the olfactory bulb by single substances are extensive, which appears to confirm – considering the rule by which the receptor neurons project to the bulb – that a given chemical can interact with a large number of receptor types. The 2-deoxyglucose method has also revealed a further type of convergence involving neighboring glomeruli. For instance, stimulation with propionic or isovaleric acid results in 2-deoxyglucose labeling of several glomeruli in a circumscribed zone of the dorsomedial olfactory bulb (Slotnick et al., 1987; Sicard, Royet, and Jourdan, 1989; Johnson et al., 1999), and an activation of neighboring glomeruli by structurally related molecules has also been reported in a recent optical recording study (Rubin and Katz, 1999). On the other hand, because the glomerular targets of neurons expressing a given receptor are found twice – one on the medial and one on the lateral side of the bulb – rules other than receptor-based convergence must operate to organize the distribution of input fibers to the glomerular layer.

The glomerular activation map is a highly structured representation of the stimulus. Nevertheless, interactions between glomeruli (i.e., receptor inputs) have been shown. Thus, even if epithelioglomerular convergence depends on receptor specificity, the spatial representation of an odor in the glomerular layer cannot be simply reduced to a list of activated receptor types.

2.3. Selectivity Profiles of Mitral Cells

Assuming that tangential intra-bulbar connections exert only modulatory influences, the selectivity of the secondary neurons should derive directly from that of the afferent receptor neurons. Indeed, the selectivity profiles of mitral/tufted cells do not differ markedly from the selectivity profiles of the receptor neurons, despite additional bulbar processing (Duchamp and Sicard, 1984b). In the frog, we found the selectivity profiles of mitral cells to be similar to those of the receptor neurons, that is, neurons in the olfactory bulb were activated, on average, by 6.4, and in the olfactory mucosa by 8.2, of the 19 odorants tested (Duchamp, 1982; Revial et al., 1982).

Similar observations have been reported in mammals, in the rabbit (Chaput and Holley, 1985), and in the rat, in which, for instance, single mitral cells were found to respond to limonene, carvone, amyl acetate, ethyl butyrate, and ethyl acetate (Wellis, Scott, and Harrison, 1989). More recently, Motokizawa (1996) observed that 69% of the secondary neurons recorded in the olfactory bulb of the rat responded to all 5 of the 5 odorants delivered and concluded that mitral/tufted cells respond to a broad spectrum of odorants. Once again, the chemical structures of the 5 effective stimuli (acetophenone, limonene, isoamyl acetate, 6-methyl-5-hepten-2-one, and butyl alcohol) were very different. If the convergence rule for the peripheral fibers is valid, the broad chemical range in the selectivity profiles for the secondary neurons would seem to confirm broad response profiles in the afferent receptor neurons.

3. Use of Neural Space in Odor Coding

Because of the broad response profiles for individual receptor neurons, the response of one neuron alone would not be able to distinguish between stimuli belonging to its profile, and thus to discriminate stimuli would require that groups of fibers from different receptor types be activated. Consequently, the specificity of the transmitted message should be represented by an assembly of neurons. Taking the selectivity profiles for single neurons into account, at the single-neuron level the peripheral message is only slightly altered by bulbar processing. From the olfactory bulb, a multifiber representation (i.e., an across-fiber pattern) of the stimulus is transmitted to cortical levels via the output neurons. The discrimination ability of the transmitted message is conserved across large concentration ranges (Van Drongelen, Holley, and Doving, 1978; Duchamp-Viret et al., 1990).

Theoretically, to represent olfactory stimuli adequately, no additional properties of the message need to be invoked other than the convergence of receptor types in the glomerular layer and the spatial dimension that implies. As mentioned earlier, a peripheral across-fiber pattern would lead to the activation of a glomerular assembly, resulting in amplification of receptor input and in an increase in the signal-to-noise ratio. Conceptually, that would mean that the treatment of the olfactory information would be massively parallel. Besides amplification, the neuronal arrangement would favor the integration of receptor inputs by means of lateral and modulatory feedback mechanisms, which would result in the treatment of incoming information according to its spatial and temporal properties. Some aspects of the spatial treatment have been mentioned earlier, and increasing efforts are being made to address the temporal aspects (Adrian, 1942; Freeman, 1978; Laurent et al., 1996). In many animal species the oscillatory behavior of the bulbar network is a major indicator of the temporal dimension of olfactory coding (Delanay et al., 1994; Stopfer et al., 1997; Kay and Freeman, 1998; Kashiwadani et al., 1999; Kay and Laurent, 1999; Buonviso et al., 2001; Lam et al., 2000). It can be affected by selective attention, learning, and memory (Freeman and Schneider, 1982; Stopfer and Laurent, 1999; Chabaud et al., 2000; Ravel et al., Chapter 21, this volume) and thus indicates already at the periphery high-level processing of the sensory information.

Considering the lack of effect of large bulbar lesions on olfactory-guided behavior (Hudson and Distel, 1987; Slotnick et al., 1987; Lu and Slotnick, 1998), and assuming that such lesions dramatically modify the spatial representation of chemical stimuli, some degree of flexibility must be added to the concept of spatial coding. Thus, one might consider redundancy to account for the limited effects of bulbar lesions. Such redundancy clearly would be supported by the broad selectivity of the olfactory receptors and parallel inputs. That, in turn, leads to the question whether or not amplification is the only role for glomerular convergence. Probably not; for instance, spatial convergence could contribute to the stabilization of the neural representation of odors during the turnover of the receptor neuron population. Nevertheless, confirmation of a functional role for the spatial distribution of information in the olfactory glomerular layer and, subsequently, in the bulbar output neurons needs more experimental investigation.

The notion of a neuronal assembly as a prerequisite for detection and discrimination of odor stimuli by the olfactory system – especially by synchronization of activations – is opening up an important area of investigation. The olfactory system is able to distinguish between elements simultaneously present in the olfactory scene, distinguishing foreground from background, describing traits or components of odorous sources, and exploring temporal clues in the olfactory environment. In that regard, a basic task might be the formation of associations between receptor types signaling the smell of a particular object. Although the extraction of such information could be performed more centrally in the olfactory system, some of the features of odor coding at the input level and in the olfactory bulb network that would be prerequisites for that have already been mentioned.

Thus it seems, given the broad selectivity profiles for receptor neurons, that the strategy of the olfactory system does not consist in developing receptors and corresponding labeled lines for each odor stimulus. Rather, with every stimulus the system is required to perform pattern recognition. The multiplicity of the receptors can be considered as an adaptation to the discontinuity of olfactory stimuli, involving a vast number of shapes that cannot be represented by a small number of physical dimensions. The analysis of the peripheral olfactory system to date indicates that a combinatorial strategy operates at all levels, including receptor selectivity and across-fiber patterning. This is most probably necessary to ensure high adaptability in olfactory coding; that is, for any odorant, whatever its origin or however new it may be, a specific – spatial and temporal – neural representation can be "formed."

References

Adrian E D (1942). Olfactory Reactions in the Brain of the Hedgehog. *Journal of Physiology (London)* 100:459–73.

Amoore J E (1967). Specific Anosmia: A Clue to the Olfactory Code. *Nature* 214:1095–8.

Arctander S (1969). *Perfume and Flavor Chemicals.* Montclair, NJ: Arctander.

Beets M G J (1982). Odorants and Stimulant Structure. In: *Fragrance Chemistry*, ed. E T Theimer, pp. 77–122. New York: Plenum Press.

Bozza T C & Kauer J S (1998). Odorant Response Properties of Convergent Olfactory Receptor Neurons. *Journal of Neuroscience* 18:4560–9.

Buck L B & Axel R (1991). A Novel Multigene Family May Encode Odorant Receptors: A Molecular Basis for Odor Recognition. *Cell* 65:175–87.

Buonviso N, Amat C, Fumanal G, Bertrand B, Farget V, & Vigouroux M (2001). Oscillations and Coherent Activities in the Rat OB. *Chemical Senses* 26:816.

Buonviso N & Chaput M A (1990). Response Similarity to Odors in Olfactory Bulb Output Cells Presumed to Be Connected to the Same Glomerulus: Electrophysiological Study Using Simultaneous Single-Unit Recordings. *Journal of Neurophysiology* 63:447–54

Chabaud P, Ravel N, Wilson D A, Mouly A M, Vigouroux M, Farget V, & Gervais R (2000). Exposure to Behaviourally Relevant Odour Reveals Differential Characteristics in Central Rat Olfactory Pathways as Studied through Oscillatory Activities. *Chemical Senses* 25:561–73.

Chaput M A & Holley A (1985). Responses of Olfactory Bulb Neurons to Repeated Odor Stimulations in Awake Freely-breathing Rabbits. *Physiology and Behavior* 34:249–58.

Chess A, Simon I, Cedar H, & Axel R (1994). Allelic Inactivation Regulates Olfactory Receptor Gene Expression. *Cell* 78:823–84.

Clyne P J, Warr C G, Freeman M R, Lessing D, Kim J, & Carlson J R (1999). A Novel Family of Divergent Seven-Transmembrane Proteins: Candidate Odorant Receptors in *Drosophila. Neuron* 22:327–38.

Costanzo R M (2000). Rewiring of the Olfactory Bulb: Changes in Odor Maps following Recovery from Nerve Section. *Chemical Senses* 25:199–205.

Daval G, Leveteau J, & MacLeod P (1970). Electro-olfactogramme local et discrimination olfactive chez la grenouille. *Journal de Physiologie (Paris)* 62:477–88.

Delaney K R, Gelperin A, Fee M S, Flores J A, Gervais R, Tank D W, & Kleinfeld D (1994). Waves and Stimulus-modulated Dynamics in an Oscillating Olfactory Network. *Proceedings of the National Academy of Sciences USA* 91:669–73.

Duchamp A (1982). Electrophysiological Responses of Olfactory Bulb Neurons to Odor Stimuli in the Frog. *Chemical Senses* 7:191–209.

Duchamp A, Revial M F, Holley A, & MacLeod P (1974). Odor Discrimination by Frog Olfactory Receptors. *Chemical Senses* 1:213–33.

Duchamp A & Sicard G (1984a). Odor Discrimination by Olfactory Bulb Neurons: Statistical Analysis of Electrophysiological Responses and Comparison with Odor Discrimination by Receptor Cells. *Chemical Senses* 9:1–14.

Duchamp A & Sicard G (1984b). Influence of Stimulus Intensity on Odor Discrimination by Olfactory Bulb Neurons as Compared with Receptor Cells. *Chemical Senses* 8:355–66.

Duchamp-Viret P, Duchamp A, & Sicard G (1990). Olfactory Discrimination over a Wide Concentration Range. Comparison of Receptor Cell and Bulb Neuron Abilities. *Brain Research* 517:256–62.

Duchamp-Viret P, Chaput M A, & Duchamp A (1999). Odor Response Properties of Rat Olfactory Receptor Neurons. *Science* 284:2171–4.

Freeman W J (1978). Spatial Properties of an EEG Event in the Olfactory Bulb and Cortex. *Electroencephalography and Clinical Neurophysiology* 44:586–605.

Freeman W J & Schneider W (1982) Changes in Spatial Patterns of Rabbit Olfactory EEG with Conditioning to Odors. *Psychophysiology* 19:44–56.

Freitag J, Ludwig G, Andreini I, Rossler P, & Breer H (1998). Olfactory Receptors in Aquatic and Terrestrial Vertebrates. *Journal of Comparative Physiology* [A] 183:635–50.

Gao Q & Chess A (1999). Identification of Candidate *Drosophila* Olfactory Receptors from Genomic DNA Sequence. *Genomics* 60:31–9.

Gesteland R C, Lettvin J Y, Pitts W H, & Rojas A (1963). Chemical Transmission in the Nose of the Frog. *Journal of Physiology (London)* 181:525–59.

Gibson A D & Garbers D L (2000). Guanylyl Cyclases as a Family of Putative Odorant Receptors. *Annual Review of Neurosciences* 23:417–39.

Guillot M (1948). Anosmie partielle et odeurs fondamentales. *Comptes Rendus de l'Académie des Sciences, Paris* 226:1307–9.

Hatt H, Gisselmann G, & Wetzel C H (1999). Cloning, Functional Expression and Characterisation of a Human Olfactory Receptor. *Cell and Molecular Biology* 45:285–91.

Hudson R & Distel H (1987). Regional Autonomy in the Peripheral Processing of Odor Signals in Newborn Rabbits. *Brain Research* 421:85–94.

Hudson R, Distel H, & Zippel H P (1990). Perceptual Performance in Peripherally Reduced Olfactory Systems. In: *Chemosensory Information Processing*, ed. D Schild, pp. 259–69. Berlin: Springer-Verlag.

Johnson B A, Woo C C, Hingco E E, Pham K L, & Leon M (1999). Multidimensional Chemotopic Responses to *n*-Aliphatic Acid Odorants in the Rat Olfactory Bulb. *Journal of Comparative Neurology* 409:529–48.

Kashiwadani H, Sasaki Y F, Uchida N, & Mori K (1999). Synchronized Oscillatory Discharges of Mitral/Tufted Cells with Different Molecular Receptive Ranges in the Rabbit Olfactory Bulb. *Journal of Neurophysiology* 82:1786–92.

Kay L M & Freeman W J (1998). Bidirectional Processing in the Olfactory-Limbic Axis during Olfactory Behavior. *Behavioral Neuroscience* 112:541–53.

Kay L M & Laurent G (1999). Odor and Context-dependent Modulation of Mitral Cell Activity in Behaving Rats. *Nature Neuroscience* 2:1003–9.

Krautwurtz D, Yau K W, & Reed R R (1998). Identification of Ligands for Olfactory Receptors by Functional Expression of a Receptor Library. *Cell* 95:917–26.

Lam Y W, Cohen L B, Wachowiak M, & Zochowski M R (2000). Odors Elicit Three Different Oscillations in the Turtle Olfactory Bulb. *Journal of Neuroscience* 15:749–62.

Lancet D (1991). The Strong Scent of Success. *Nature* 351:275–6.

Laurent G, Wehr M, Macleod K, Stopfer M, Leitch B, & Davidowitz H (1996). Dynamic Encoding of Odors with Oscillating Neuronal Assemblies in the Locust Brain. *Biological Bulletin* 191:70–5.

Leibovici M, Lapointe F, Aletta P, & Ayer-Le Lièvre C (1996). Avian Olfactory Receptors: Differentiation of Olfactory Neurons under Normal and Experimental Conditions. *Developmental Biology* 175:118–31.

Leveteau J & MacLeod P (1966). La discrimination des odeurs par les glomérules olfactifs du lapin (Etude électrophysiologique). *Journal de Physiologie (Paris)* 58:717–29.

Lu X C M & Slotnick B M (1998). Olfaction in Rats with Extensive Lesions of the Olfactory Bulbs: Implications for Odor Coding. *Neuroscience* 84:849–66.

Mackay-Sim A & Kesteven S (1994). Topographic Patterns of Responsiveness to Odorants in the Rat Olfactory Epithelium. *Journal of Neurophysiology* 71:150–60.

Malnic B, Hirono J, Sato T, & Buck L B (1999). Combinatorial Receptor Codes for Odors. *Cell* 96:713–23.

Matsuzaki O, Bakin R E, Cai X, Menco B P, & Ronnett G V (1999). Localization of the Olfactory Cyclic Nucleotide-gated Channel Subunit 1 in Normal, Embryonic and Regenerating Olfactory Epithelium. *Neuroscience* 94:131–40.

Mombaerts P (1999). Odorant Receptor Genes in Humans. *Current Opinion in Genetics and Development* 9:315–20.

Mombaerts P, Wang F, Dulac C, Chao S K, Nemes A, Mendelsohn M, Edmondson J, & Axel R (1996). Visualizing an Olfactory Sensory Map. *Cell* 87:675–86.

Mori K, Kishi K, & Ojima H J (1983). Distribution of Dendrites of Mitral, Displaced Mitral, Tufted and Granule Cells in the Rabbit Olfactory Bulb. *Journal of Comparative Neurology* 219:339–55.

Mori K & Shepherd G M (1994). Emerging Principles of Molecular Signal Processing by Mitral/Tufted Cells in the Olfactory Bulb. *Cell Biology* 5:65–74.

Motokizawa F (1996). Odor Representation and Discrimination in Mitral/Tufted Cells of the Rat Olfactory Bulb. *Experimental Brain Research* 112:24–34.

Nef P, Hermans-Borgmeyer I, Artieres-Pin H, Beasley L, Dionne V E, & Heinemann S F (1992). Spatial Pattern of Receptor Expression in the Olfactory Epithelium. *Proceedings of the National Academy of Sciences USA* 89:8948–52.

Ngai J, Dowling M M, Buck L B, Axel R, & Chess A (1993). The Family of Genes Encoding Odorant Receptors in the Channel Catfish. *Cell* 72:657–66.

O'Leary D M D, Yates P A, & McLaughlin T (1999). Molecular Development of Sensory Maps: Representing Sights and Smells in the Brain. *Cell* 96:255–69.

Raming K, Krieger J, Strotmann J, Boekhoff I, Kubick S, Baumstark C, & Breer H (1993). Cloning and Expression of Odorant Receptors. *Nature* 361:353–6.

Rawson N E, Eberwine J, Dotson R, Jackson J, Ulrich P, & Restrepo D (2000). Expression of mRNAs Encoding for Two Different Olfactory Receptors in a Subset of Olfactory Neurons. *Journal of Neurochemistry* 75:185–95.

Ressler K, Sullivan S, & Buck L B (1993). A Zonal Organization of Odorant Receptor Gene Expression in the Olfactory Epithelium. *Cell* 73:597–609.

Ressler K J, Sullivan S L, & Buck L B (1994). Information Coding in the Olfactory System. Evidence for a Stereotyped and Highly Organized Epitope Map in the Olfactory Bulb. *Cell* 79:1245–55.

Revial M F, Sicard G, Duchamp A, & Holley A (1982). New Studies on Odor Discrimination in the Frog's Olfactory Receptor Cells. I. An Experimental Study. *Chemical Senses* 7:175–90.

Rouquier S, Friedman C, Delettre C, van den Engh G, Blancher A, Crouau-Roy B, Trask B J, & Giorgi D (1998). A Gene Recently Inactivated in Human Defines a New Olfactory Receptor Family in Mammals. *Human Molecular Genetics* 7:1337–45.

Royet J P, Sicard G, Souchier C, & Jourdan F (1987). Specificity of Spatial Patterns of Glomerular Activation in the Mouse Olfactory Bulb: Computer-assisted Image Analysis of 2-DG Autoradiograms. *Brain Research* 417:1–11.

Rubin B D & Katz L C (1999). Optical Imaging of Odorant Representations in the Mammalian Olfactory Bulb. *Neuron* 23:499–511.

Sato T, Hirono J, Tonoike M, & Takebayashi M (1994). Tuning Specificities to Aliphatic Odorants in Mouse Olfactory Receptor Neurons and Their Local Distribution. *Journal of Neurophysiology* 72:2980–9.

Selbie L A, Townsend-Nicholson A, Iismaa T I, & Shine J (1992). Novel G Protein-coupled Receptors: A Gene Family of Putative Human Olfactory Receptor Sequences. *Molecular Brain Research* 13:159–63.

Semke E, Distel H, & Hudson R (1995). Specific Enhancement of Olfactory Receptor Sensitivity Associated with Foetal Learning of Food Odors in the Rabbit. *Naturwissenschaften* 82:148–9.

Sengupta P, Colbert H A, & Bargmann C I (1994). The *C. elegans* Gene odr-7 Encodes an Olfactory-specific Member of the Nuclear Receptor Superfamily. *Cell* 79:971–80.

Sharon D, Glusman G, Pilpel Y, Khen M, Gruetzner F, Haaf T, & Lancet D (1999). Primate Evolution of an Olfactory Receptor Cluster: Diversification by Gene Conversion and Recent Emergence of Pseudogenes. *Genomics* 61:24–36.

Sicard G (1986). Electrophysiological Recordings from Olfactory Receptor Cells in Adult Mice. *Brain Research* 397:405–8.

Sicard G & Holley A (1984). Receptor Cell Responses to Odorants: Similarities and Differences among Odorants. *Brain Research* 292:283–96.

Sicard G, Royet J P, & Jourdan F (1989). A Comparative Study of 2-Deoxyglucose Patterns of Glomerular Activation in the Olfactory Bulbs of C57BL/6J and AKR/J Mice. *Brain Research* 481:325–34.

Slotnick B M, Graham S, Laing D G, & Bell G A (1987). Detection of Propionic Acid Vapor by Rats with Lesions of Olfactory Bulb Areas Associated with High 2-DG Uptake. *Brain Research* 417:343–6.

Stopfer M, Bhagavan S, Smith B H, & Laurent G (1997) Impaired Odor Discrimination on Desynchronization of Odor-encoding Neural Assemblies. *Nature* 390:70–4.

Stopfer M & Laurent G (1999). Short-Term Memory in Olfactory Network Dynamics. *Nature* 402:664–8.

Strotmann J, Hope R, Conzelmann S, Feinstein P, Mombaerts P, & Breer H (1999). Small Subfamily of Olfactory Receptor Genes: Structural Features, Expression Pattern and Genomic Organization. *Gene* 236:281–91.

Strotmann J, Wanner I, Helfrich T, Beck A, Meinken C, Kubick S, & Breer H (1994). Olfactory Neurons Expressing Distinct Odorant Receptor Subtypes Are Spatially Segregated in the Nasal Neuroepithelium. *Cell and Tissue Research* 276:429–38.

Strotmann J, Wanner I, Krieger J, Raming K, & Breer H (1992). Expression of Odorant Receptors in Spatially Restricted Subsets of Chemosensory Neurons. *NeuroReport* 3:1053–6.

Touhara K, Sengoku S, Inaki K, Tsuboi A, Hirono J, Sato T, Sakano H, & Haga T (1999). Functional Identification and Reconstitution of an Odorant Receptor in Single Olfactory Neurons. *Proceedings of the National Academy of Sciences USA* 96:4040–5.

Van Drongelen W, Holley A, & Doving K B (1978). Convergence in the Olfactory System: Quantitative Aspects of Odor Sensitivity. *Journal of Theoretical Biology* 71:39–49.

Vassar R, Chao S K, Sitcheran R, Nunez J M, Vosshall L B, & Axel R (1994). Topographic Organization of Sensory Projections to the Olfactory Bulb. *Cell* 79:981–91.

Vassar R, Ngai J, & Axel R (1993). Spatial Segregation of Odorant Receptor Expression in the Mammalian Olfactory Epithelium. *Cell* 74:309–18.

Vogt R G, Lindsay S M, Byrd C A, & Sun M (1997). Spatial Patterns of Olfactory Neurons Expressing Specific Odor Receptor Genes in 48-Hour-old Embryos of Zebrafish *Danio rerio*. *Journal of Experimental Biology* 200:433–43.

Wang H W, Wysocki C J, & Gold G H (1993). Induction of Olfactory Receptor Sensitivity in Mice. *Science* 260:998–1000.

Wellis D P, Scott J W, & Harrison T A (1989). Discrimination among Odorants by Single Neurons of the Rat Olfactory Bulb. *Journal of Neurophysiology* 61:1161–77.

Wysocki C J, Dorries K M, & Beauchamp G K (1989). Ability to Perceive Androstenone Can Be Acquired by Ostensibly Anosmic People. *Proceedings of the National Academy of Sciences USA* 86:7976–8.

Young T A & Wilson D A (1999). Frequency-dependent Modulation of Inhibition in the Rat Olfactory Bulb. *Neuroscience Letters* 276:65–7.

Zhao H, Ivic L, Otaki J M, Hashimoto M, Mikoshiba K, & Firestein S (1998). Functional Expression of a Mammalian Odorant Receptor. *Science* 279:237–42.

Zwaardemaker H (1925). *L'odorat*. Paris: Doin.

19

Human Brain Activity during the First Second after Odor Presentation

Bettina M. Pause

Voltage fluctuations above the intact human scalp are called chemosensory event-related potentials (CSERPs) (Evans et al., 1993) when these variations are caused by experimental manipulations of odor presentations. CSERPs function as indicators of the speed, strength, and local distribution of neuronal brain activity related to odor perception. CSERPs are very time-sensitive, which allows odor perception to be separated into different processing stages within the first second after odor presentation (Pause and Krauel, 2000). The aim of this chapter is to describe the biological and psychological meanings of the different processing stages. Between 300 and 500 msec after odor presentation, the specific features of the olfactory environment are encoded. This process is accompanied by a first distinct wave that appears within the CSERP; it is negatively charged and therefore is called the N1 component. The N1 component seems to reflect a pre-attentive level of stimulus encoding, but also depends on the level of the subject's alertness. The next prominent wave within the CSERP is the positively charged P3 component, appearing between 700 and 1,200 msec after odor presentation. Like the N1 component, the P3 is sensitive to the attentional investment of the subject, but in addition it is sensitive to the probability of the odor occurrence and to the subjective significance of the odor. The extraction of the significance of the olfactory event is an evaluative process that depends on cognitive and emotional resources. Within this process, the olfactory information is compared to earlier olfactory experiences stored in short- and long-term memory, and its specific importance for the subject is identified.

This chapter aims to show the usefulness of CSERP analysis in understanding odor processing. In the first part, CSERP indicators of the mechanisms involved in general odor perception are presented. In the second part, a description of how CSERPs might help to explain individual differences in odor perception is given. As odor perception varies considerably between and within subjects, the perception of odors will be understood as a function of the individual rather than as a function of the odor itself. Within this model, differences in olfactory perception between subjects can be explained by biological and psychological traits, personal characteristics that remain stable over time. On the other hand, differences in odor perception within one subject can be explained by his or her biological and psychological state (i.e., the internal circumstances in the specific situation).

1. The CSERP as an Indicator of General Odor Perception

About 400 msec after an odor presentation, a first wave appears within the CSERP, called the N1 component (Figure 19.1). In response to an auditory or visual stimulus, the N1 usually appears much sooner (e.g., 100–150 msec after presentation) (Näätänen and Picton, 1987). Therefore, its late appearance at about 400 msec seems to be a characteristic unique to the olfactory modality. However, a possible explanation for the late brain response is the finding that

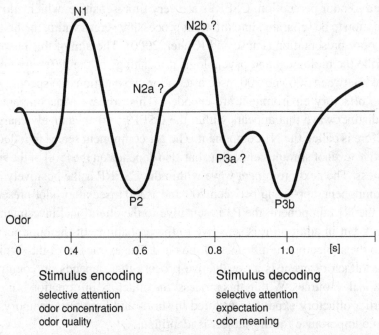

Figure 19.1. The components of the CSERP and their relation to stimulus encoding and decoding. This fictitious example does not refer to a real-case CSERP, but resembles a true CSERP in its time structure. The components are measured in microvolts and usually are detected in a voltage range from 1 to 20. Whereas the olfactory N1, P2, and P3b have been at the focus of CSERP research, further investigation will be required to understand the functional significance of the olfactory N2 and P3a.

the olfactory receptor neuron responds to olfactory stimulation with a latency of about 300 msec (Firestein and Werblin, 1989). The N1 component in response to acoustic stimuli comprises several subcomponents (Näätänen, 1992) and varies with stimulus characteristics (e.g., stimulus intensity) as well as with the internal conditions of the subject (e.g., selective attention).

In CSERP research it is still a matter of debate whether or not the amplitude of the N1 component varies with odor concentration. One of the difficulties is choosing a relatively pure olfactory stimulus that does not activate the trigeminal system (Hummel, 2000). So far, three studies using olfactory stimulants have been conducted. Thiele and Kobal (1984) and Pause, Sojka, and Ferstl (1997) found that the latencies of the CSERP decreased with increasing stimulus concentrations, whereas the amplitudes were unaffected by concentration changes. Pause et al. (1997) concluded that olfactory concentrations are encoded as qualitatively different stimuli and thus do not activate a stronger neuronal response. Those conclusions are supported by measurements of peripheral (Firestein and Werblin, 1989) and central (Cinelli, Hamilton, and Kauer, 1995) olfactory transduction, as well as by subjective rating data (O'Connell, Stevens, and Zogby, 1994). A more recent study (Tateyama et al., 1998) also found a strong effect of odor concentration on the latencies of the CSERP. As expected, those effects were strongest for the early components of the CSERP (e.g., the N1 component). However, they did find a weak effect of odor concentration on the amplitudes of the early CSERP components and a stronger effect on the late P3 component. An effect of stimulus intensity on the P3 component had previously been shown for auditory stimuli (Roth et al., 1980). That effect can be seen when very large concentration differences are used and can be explained as the processing of a higher subjective stimulus significance for very intense stimuli. A fourth very recent study examined CSERPs in response to different concentrations of isoamyl acetate (Covington et al., 1999). Even though that odor may also have trigeminal effects, the concentrations applied were below the threshold for irritation. Again, only the latencies (P2, N2) decreased with higher stimulus concentrations, and no tendency for the amplitudes to vary with odor concentration was found. Summarizing, all studies so far have underlined the importance of temporal characteristics in olfactory concentration encoding. Similarly, the most promising way to differentiate odor quality by means of CSERP research seems to be to analyze the latency differences within the N1 latency window. In a very recent study (Pause et al., 1999d), we found a very reliable effect of odor quality on the N1 latency that could be seen in each of 46 subjects. All of them responded earlier to the odor of phenylethyl alcohol (rose) than to the odor of isobutyraldehyde (rancid butter). As the odors were rated as being equally intense (Pause and Krauel, 2000), that effect was not due to concentration differences. However, as isobutyraldehyde may also have activated the trigeminal nerve, those latency differences may have resulted in part from activation of different transduction systems. The importance of temporal characteristics in odor quality coding is supported by studies indicating that identification of odorants in mixtures may depend on the speed of perception of the single substances (Laing et al., 1994). Recently, Lorig (1999) even speculated that the temporal coding of odors may have similarities to the temporal coding of speech.

The influence of selective attention on the olfactory N1 component has been examined by three studies. In a first study (Pause et al., 1997), two conditions were examined. While performing a special breathing technique, subjects had either to count the odors (active attention) or just relax (passive attention). Even though not statistically significant, there was a tendency for the N1 component to be larger in amplitude and to have a shorter latency when the subjects attended to the odors actively. In a second study (Krauel et al., 1998b), the difference between the attentional levels in two conditions was increased. In the active condition the subjects had to actively differentiate two odors, and in the passive condition the subjects had to count target words that served as an auditory distractor task. That study found a significant reduction in the N1 latency in the active attention condition. A third study should also be mentioned, one that did not explicitly address selective attention, but its effects could easily be interpreted as effects of attention. Pause et al. (1999b) found the N1 component to appear later when the subjects had to perform a special breathing technique, even though their attention was directed to the odors. Thus, carrying out two simultaneous tasks (counting odors and artificially breathing) seems to reduce the amount of selective attention that can be paid to the odors.

Another feature that seems to be unique to olfaction is that the N1 amplitude decreases even with relatively long inter-stimulus intervals (ISI). Effects of stimulus repetition on the N1 component in response to auditory stimuli can be observed only with ISIs shorter than about 10 sec. The amplitude of the olfactory N1, however, declines with ISIs of about 30 sec (Pause et al., 1996b). A full recovery of the N1 has been found for ISIs of 50 sec (Pause et al., 1997). Examining CSERPs in older subjects, Morgan et al. (1997) found a tendency for the N1 amplitude to show a recovery time of even 90 sec. For the auditory and visual modalities, it has been proposed that the decrease in N1 amplitude with stimulus repetition may be related to time uncertainty, neuronal refractoriness, or habituation (Näätänen, 1992). Time uncertainty refers to the phenomenon that it is more difficult to predict the time of stimulus onset with long ISIs than with short ISIs. As the effect of stimulus repetition on the olfactory N1 extends to very large ISIs, it is unlikely that time uncertainty can explain the amplitude reduction of the olfactory N1. However, further studies need to be carried out to investigate whether the olfactory effect is caused by refractory periods of the neuronal generators or by habituation. Habituation can occur only when some of the stimulus information is stored within short-term memory (Sokolov, 1977). If it could be proved that habituation is responsible for the amplitude decline (e.g., by simple habituation/dishabituation studies), then the recovery time for the olfactory N1 would indicate that the length of olfactory short-term memory exceeds that in other modalities.

The olfactory P2 reaches its maximum amplitude about 600 msec after odor presentation. That might reflect processing stages similar to those for the N1 component (Pause et al., 1996b; Tateyama et al., 1998) and thus probably is also involved in stimulus encoding (for other modalities, see Rockstroh et al., 1989). Under certain experimental conditions, a second negative component, the N2, can follow the N1 component. The N2 component can be divided into two subcomponents, the N2a and N2b (Figure 19.1). For the auditory modality, the N2 component has been analyzed in great detail (Näätänen, 1990). It has been shown that deviant, infrequent auditory stimuli presented within a train of homogeneous repetitive stimuli can elicit this second negativity. When the stimuli are actively attended to, a large negative wave, the N2b, is elicited, usually followed by a P3a. When the stimuli are not attended to, a small negative deflection, the N2a (also called the MMN, for "mismatch negativity"), appears in association with the deviant stimulus. The N2a precedes the N2b and can be extracted from the ERP by subtracting the standard-stimulus ERP from the deviant-stimulus ERP. As the N2a probably reflects mechanisms within short-term memory, it cannot be elicited by acoustic stimuli when the ISI exceeds 10 sec (Böttcher-Gandor and Ullsperger, 1992). To date, the N2b has not been examined using olfactory stimuli. Some studies, however, indicate that deviant odors that are processed on a pre-attentional level do evoke an N2a deflection (Krauel et al., 1999). The existence of such an automatic olfactory "mismatch detector" would indicate that olfactory information is automatically stored within a short-term register, even without the influence of attention. Furthermore, it has been shown that the olfactory N2a can be elicited with ISIs longer than 15 sec (Pause and Krauel, 1998), which indicates unusually long time frames for olfactory short-term memory.

For other modalities, the N2b has often been described as being associated with a P3a component, both of which indicate an attentional switch to a post-attentive evaluation of the stimulus (Näätänen, 1990). When an attentional switch occurs, a large late positivity sometimes dominates the ERP and has been called the P3 component (or P3b) (Figure 19.1). The P3 has a parietal dominance, and modality-independent generators are involved in its evocation. However, within CSERPs the P3 will also appear later (between 700 and 1,000 msec after stimulus onset) than in the visual or auditory ERP (between 300 and 500 msec after stimulus onset). The P3 has been described as varying with the stimulus probability and the task relevance in an independent manner whenever attention is directed toward the stimuli (Johnson, 1993). In fact, for the olfactory modality, dramatic effects of attention were observed for the late positivity. When attention was directed to another secondary task (Krauel et al., 1998b), the P3 component was almost absent. When the attention was mildly subtracted (Pause et al., 1999b) or not directed toward the odors (Pause et al., 1997), the P3 amplitude was reduced.

That the P3 amplitude is larger in response to infrequent stimuli has been interpreted in terms of a so-called context-updating model (Donchin and Coles, 1988). Donchin and Coles stated that an internal representation of the external environment needs to be updated whenever a change occurs within the external conditions. That process takes place whenever a change is not expected and may correlate with the subjective feeling of surprise (Donchin, 1981). That infrequently presented target odors elicit a P3 has been demonstrated by Pause et al. (1996b). Furthermore, Lorig et al. (1996) found the P3 in response to odors to be larger during exhalation than during inhalation. They attributed that effect to the evocation of surprise when odors are perceived during phases of exhalation.

Finally, the P3 amplitude varies considerably with the subjective stimulus significance. Thus, when odors have to be detected (Lorig et al., 1993; Pause et al., 1997; Krauel et al., 1998b) or to be counted (Pause et al., 1996b, 1999b), the P3 amplitude rises. The unique feature of odors whereby they also elicit a large P3 component when they are presented as frequent standard stimuli (Pause et al., 1996b) might be due to their inherent emotional significance (Pause et al., 1997). Odors mostly carry a specific hedonic value (Van Toller, 1988), and it has recently been shown that visual emotional stimuli elicit larger P3s than do neutral stimuli (Diedrich et al., 1997).

Summarizing, CSERP analysis divides odor processing in the central nervous system into three dimensions. With respect to the high time resolution of CSERPs, the perceptual speed of odor processing can be considered (latencies of the CSERP). The neuronal odor transmission can be further divided into the strength (amplitudes of the CSERP) and the distribution of the neuronal activity (topography of the CSERP). The three CSERP measures can in turn tell us about the perceptual dimensions of odor processing: olfactory sensitivity, odor evaluation, and olfactory learning. As indicated earlier, the temporal characteristics of the CSERPs seem to reflect quality and concentration encoding and can be taken as indicators of olfactory sensitivity. CSERP results also indicate that olfactory short-term memory may have longer time frames than short-term memory for visual or acoustic stimuli; therefore basic sensory olfactory learning seems to have modality-specific features. Moreover, findings reveal that whenever odors are attended to (actively or passively), they will be further processed, probably because of their emotional value.

2. The CSERP as an Indicator of Inter- and Intra-individual Differences in Odor Perception

In the following, all dimensions of odor perception will be treated as functions of the individual (Figure 19.2). This model will take into account that olfactory sensitivity, odor evaluation, and olfactory learning vary considerably between

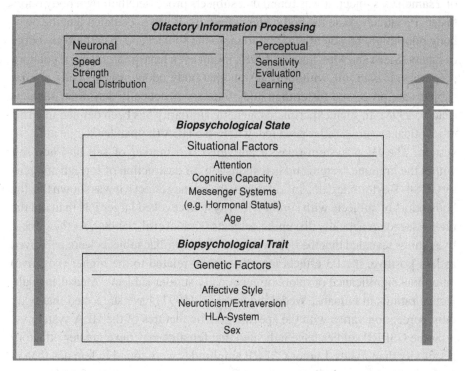

Figure 19.2. Olfactory information processing as a function of the individual. Olfactory information processing can be measured on a neuronal or perceptual level. CSERPs can give insight into the speed (latency), strength (amplitude), and local distribution (source analysis in multichannel EEG recording) of the neuronal activity related to olfactory stimulus processing. Perceptual measures can provide information about olfactory sensitivity, odor evaluation (intensity, quality, valence), and olfactory learning. Relating perceptual measures to neuronal measures can indicate the functionality of neuronal activity. In chapter 3 Olfactory information processing has been described as a function of the individual, and the characteristics of the individual are divided into biopsychological state and trait factors.

and within individuals. Furthermore, the characteristics of the individual will be separated in those that are stable over time (trait characteristics) and those that are variable (state characteristics).

2.1. Biopsychological Trait Markers Explain Differences in Odor Perception between Subjects

Even though CSERP studies using twins as subjects have not yet been carried out, some studies indicate that genetic factors contribute to the perceptual speed of processing of human body odors. In one study (Pause et al., 1999c), presentation of the subject's own body odor was alternated with presentation of the body odor of a same-sex subject. It was found that subjects processed their own body odors faster (P3 latency) than the body odors of others. It was speculated that the own-body odor might be potentially important to the subject and, like the body odors of others (Stern and McClintock, 1998), might even have pheromonal regulatory properties. It is in line with those results that body odors of genetically similar subjects are processed faster than body odors of genetically dissimilar subjects (Pause, 1998). In such experiments, genetic similarity has been defined in terms of identical or very similar gene loci within the HLA (human leukocyte antigen) system. The HLA system (class I antigens) is a marker of self and non-self within the immune system and is responsible for destruction of foreign antigens by CD8+ T-cytotoxic cells. In addition to the latency effects, it was shown that the body odors of subjects with similar HLA systems evoked larger P3s than did the body odors of genetically dissimilar subjects (Krauel et al., 1998a). As the subjective ratings revealed that the body odors of HLA-similar subjects were perceived as less positive, the P3 effects may have been related to the higher (negative) emotional significance of odors of genetically similar subjects. Measuring subjective ratings in humans, Wedekind and Füri (1997) have suggested that body odor perception varies with the specific genetic features of the HLA system.

Some CSERP studies have indicated that females may have greater olfactory sensitivity than males. Larger CSERP amplitudes were found in females than in males by Becker et al. (1993) and by Evans, Cui, and Starr (1995). Moreover, Morgan et al. (1997, 1999) found effects of age on CSERP amplitudes, which were strongest for older male subjects. However, none of those studies were experimentally controlled for the effects of task relevance. Given that odors are perceived mainly as emotional stimuli, sex differences in emotional stimulus processing (Grossman and Wood, 1993; Lang et al., 1993) may have contributed to the described effects.

Even though personality dimensions like extraversion or neuroticism might also contribute to olfactory perception (Pause, Ferstl, and Fehm-Wolfsdorf,

1998), CSERP studies of such factors have not yet been conducted. However, recent EEG studies (Davidson, 1998) have suggested that the affective style of a subject can explain differences in brain activity during perception of emotional stimuli. In accordance with that theory, we recently recorded CSERPs in clinically depressed subjects (Pause et al., 1999d, 2000) whose affective style can be described as a function of lowered activity within the approach/appetitive motivation system. By means of threshold tests we found that depressed subjects showed lower olfactory sensitivity than did healthy controls, and that effect was shown very clearly for the first time. The strength of the effect may have been caused by the selection of a homogeneous patient sample (only patients with diagnoses of major depression participated) and by the use of olfactory (phenylethyl alcohol and eugenol, according to Doty et al., 1978) rather than trigeminal or mixed substances. Moreover, CSERP analyses revealed that the early components (P2) were reduced in depressed subjects. However, the early components in response to visual slides were similar in healthy and depressed subjects. The P3 in response to odors and visual slides was also reduced in depressed patients. Thus, the reduction of the late P3 complex was not modality-specific and also has been reported by others using tones or slides as stimuli (e.g., Kayser et al., 2000). That study again points to the useful distinction between early (N1, P2) and late (P3) CSERP components. It was concluded that depressed subjects showed reduced modality-specific ability to encode olfactory information. That effect could be separated from a general deficit in selective attention that might have been responsible for the reduced P3 amplitudes. Follow-up studies that were conducted after the patients had recovered from the severe depressive phase indicated that their odor processing was still impaired, and that can be interpreted as a trait marker in depressed patients.

2.2. The Biopsychological State Explains Differences in Odor Perception within Subjects

Some CSERP studies have been conducted in order to explain differences in odor perception within a single subject. These differences are related to effects of age, menstrual-cycle phase, and olfactory learning. Four studies found significant effects of age on olfactory processing speed. Evans et al. (1995) examined subjects between 18 and 83 years and found the P2 latency to decline with age at a rate of 2.5 msec/year. A factorial design was used by Hummel et al. (1998), who compared three age groups (age ranges 15–34, 35–54, 55–74). They found the N1 latency to increase with increasing age. Morgan et al. (1997) and Covington et al. (1999) compared a group of young participants with a group of elderly participants and also found odors to be processed significantly faster in young subjects (latencies of N1, P2, N2). Moreover, three studies demonstrated that elderly participants also showed reduced CSERP amplitudes. Whereas Hummel et al. (1998) and Covington et al. (1999) found similar effects of age on amplitudes and latencies, Murphy et al. (1994) found the effects of age to be more pronounced for the amplitudes (N1, P2) than for the latencies. Measuring early olfactory stimulus processing, these results are in line with findings that olfactory sensitivity is reduced in older subjects (e.g., Doty et al., 1984). Recently, Morgan et al. (1999) focused on the olfactory P3 and found age effects for the amplitude as well as for the latency. They discussed the importance of separating a sensory impairment from a cognitive impairment in elderly subjects. Whereas the former can be addressed by investigating the N1 and P2 amplitudes, the latter can be evaluated by use of the P3 component.

In two studies, the influence of the menstrual-cycle phase on CSERPs was examined: In the first study (Pause et al., 1996a) it was found that a citrus odor was processed significantly faster (latencies of N1, P2, P3) during the ovulatory phase than during the follicular or luteal phase. Interestingly, that effect was found only after repeated stimulations, indicating that the plasticity of the olfactory system is higher during the ovulatory phase. Additionally, it was found that the P3 was larger during the ovulatory phase. That effect was interpreted as evidence for a higher significance of odorous information during the ovulatory phase. In the second study (Pause et al., 1999a), the body odors presented belonged to two groups of odor donors, one with HLA systems similar, and one dissimilar, to those of the subjects. It was confirmed that the largest P3s were obtained when the females smelled HLA-similar subjects; that effect was most pronounced when the subjects were examined during the ovulatory phase.

One of the most intriguing findings on olfactory plasticity in recent years has come from studies showing that subjects with specific anosmias can be olfactorily sensitized and thus can reach normal osmia for a specific odor (Wysocki, Dorries, and Beauchamp, 1989). Olfactory plasticity refers to the capability of the olfactory neurons for neurogenesis and reconnection of pathways in the olfactory system (Costanzo and Graziadei, 1987), as well as to the capacity of neurons within the main olfactory bulb to change their response characteristics through olfactory learning (Brennan and Keverne, 1997). In corresponding studies in humans, the odor most frequently investigated is that of androstenone. Up to one-third of the subjects examined were found to have a specific anosmia to that odor (Wysocki, Pierce, and Gilbert, 1991). Androstenone has been detected in human sweat and other body fluids and is thought to be a male pheromone (Gower and Ruparelia, 1993). However, it has never been shown that the odor of androstenone within the complex human body odor conveys any specific information to the perceiver. In a first CSERP study (Pause et al., 1999e), females were exposed to a foreign, male body odor and their own body odors. Whereas the potentials for an osmic control group became smaller in a second session conducted four weeks after the first session, a group of sensitized, initially anosmic females showed larger potentials exclusively for the male body odor during the second session, four weeks after exposure. It was concluded that the sensitivity to androstenone in females is associated with a stronger brain response to male body odor.

3. Conclusion

It has been shown how measurements of event-related brain activity can be used to better understand human odor perception. Especially the high time resolution of CSERPs has been addressed, which may serve as a good indicator of the temporal features in olfactory coding (Lorig, 1999). Olfactory perception has been shown to be a function of the individual. Each individual is made up of basic and variable psychological and biological features (Figure 19.2). Some of them have already been examined in CSERP research and can explain inter- and intra-individual differences in odor perception. However, further research is needed, for example, to examine the uniqueness of olfactory plasticity. So far, some intra-individual differences have indicated that olfactory performance can be improved by the actions of sexual hormones (Pause et al., 1996a, 1999a) and by repeated odor exposures (Möller, Pause, and Ferstl, 1999; Pause et al., 1999e). In addition, our first results (Pause and Krauel, 1998) indicate that olfactory learning can also occur unintentionally (olfactory priming), and that can be measured by CSERPs.

Acknowledgments

I would like to thank Roman Ferstl, Kerstin Krauel, Bernfried Sojka, Claudia Müller, and Ninja Raack for discussions and for their outstanding help in experimental work. Furthermore, I would like to thank two anonymous reviewers for their helpful suggestions on the manuscript. The preparation of the manuscript was in part supported by the German Research Council.

References

Becker E, Hummel T, Piel E, Pauli E, Kobal G, & Hautzinger M (1993). Olfactory Event-related Potentials in Psychosis-prone Subjects. *International Journal of Psychophysiology* 15:51–8.

Böttcher-Gandor C & Ullsperger P (1992). Mismatch Negativity in Event-related Potentials to Auditory Stimuli as a Function of Varying Interstimulus Interval. *Psychophysiology* 29:546–50.

Brennan P A & Keverne E B (1997). Neural Mechanisms of Mammalian Olfactory Learning. *Progress in Neurobiology* 51:457–81.

Cinelli A R, Hamilton K A, & Kauer J S (1995). Salamander Olfactory Bulb Neuronal Activity Observed by Video Rate, Voltage-sensitive Dye Imaging. III. Spatial and Temporal Properties of Responses Evoked by Odorant Stimuli. *Journal of Neurophysiology* 73:2053–71.

Costanzo R M & Graziadei P P C (1987). Development and Plasticity of the Olfactory System. In: *Neurobiology of Taste and Smell*, ed. T E Finger & W L Silver, pp. 233–50. New York: Wiley.

Covington J W, Geisler M W, Polich J, & Murphy C (1999). Normal Aging and Intensity Effects on the Olfactory Event-related Potential. *International Journal of Psychophysiology* 32:205–14.

Davidson R J (1998). Affective Style and Affective Disorders: Perspectives from Affective Neuroscience. *Cognition and Emotion* 12:307–30.

Diedrich O, Naumann E, Maier S, Becker G, & Bartussek D (1997). A Frontal Slow Wave in the ERP Associated with Emotional Slides. *Journal of Psychophysiology* 11:71–84.

Donchin E (1981). Surprise! . . . Surprise? *Psychophysiology* 18:493–513.

Donchin E & Coles M G H (1988). Is the P300 Component a Manifestation of Context Updating? *Behavioural and Brain Sciences* 11:357–74.

Doty R L, Brugger W E, Jurs P C, Orndorff M A, Snyder P J, & Lowry L D (1978). Intranasal Trigeminal Stimulation from Odorous Volatiles. Psychometric Responses from Anosmic and Normal Humans. *Physiology and Behavior* 20:175–85.

Doty R L, Shaman P, Appelbaum S L, Giberson R, Sikorsky L, & Rosenberg L (1984). Smell Identification Ability: Changes with Age. *Science* 226:1441–3.

Evans W J, Cui L, & Starr A (1995). Olfactory Event-related Potentials in Normal Human Subjects: Effects of Age and Gender. *Electroencephalography and Clinical Neurophysiology* 95:293–301.

Evans W J, Kobal G, Lorig T S, & Prah J D (1993). Suggestions for Collection and Reporting of Chemosensory (Olfactory) Event-related Potentials. *Chemical Senses* 18:751–6.

Firestein S & Werblin F (1989). Odor-induced Membrane Currents in Vertebrate Olfactory Receptor Neurons. *Science* 244:79–82.

Gower D B & Ruparelia B A (1993). Olfaction in Humans with Special Reference to Odorous 16-Androstenes: Their Occurrence, Perception and Possible Social, Psychological and Sexual Impact. *Journal of Endocrinology* 137:167–87.

Grossman M & Wood W (1993). Sex Differences in Intensity of Emotional Experience: A Social Role Interpretation. *Journal of Personality and Social Psychology* 65:1010–22.

Hummel T (2000). Assessment of Trigeminal Function. *International Journal of Psychophysiology* 36:147–55.

Hummel T, Barz S, Pauli E, & Kobal G (1998). Chemosensory Event-related Potentials Change with Age. *Electroencephalography and Clinical Neurophysiology* 108:208–17.

Johnson R (1993). On the Neural Generators of the P300 Component of the Event-related Potential. *Psychophysiology* 30:90–7.

Kayser J, Bruder G E, Tenke C E, Stewart J W, & Quitkin F M (2000). Event-related Potentials to Hemifield Presentations of Emotional Stimuli: Differences between Depressed Patients and Healthy Adults in P3 Amplitude and Asymmetry. *International Journal of Psychophysiology* 36:211–36.

Krauel K, Pause B M, Müller C, Sojka B, Müller-Ruchholtz W, & Ferstl R (1998a). Central Nervous Correlates of Chemical Communication in Humans. In: *Olfaction and Taste, XII*, ed. C Murphy, pp. 628–31. New York Academy of Sciences.

Krauel K, Pause B M, Sojka B, Schott P, & Ferstl R (1998b). Attentional Modulation of Central Odor Processing. *Chemical Senses* 23:423–32.

Krauel K, Schott P, Sojka B, Pause B M, & Ferstl R (1999). Is There a Mismatch Negativity Analogue in the Olfactory Event-related Potential? *Journal of Psychophysiology* 13:49–55.

Laing D G, Eddy A, Francis G W, & Stephens L (1994). Evidence for the Temporal Processing of Odor Mixtures in Humans. *Brain Research* 651:317–28.

Lang P J, Greenwald M K, Bradley M M, & Hamm A O (1993). Looking at Pictures: Affective, Facial, Visceral and Behavioral Reactions. *Psychophysiology* 30:261–73.

Lorig T S (1999). On the Similarity of Odor and Language Perception. *Neuroscience and Biobehavioral Reviews* 23:391–8.

Lorig T S, Matia D C, Peszka J J, & Bryant D N (1996). The Effects of Active and Passive Stimulation on Chemosensory Event-related Potentials. *International Journal of Psychophysiology* 23:199–205.

Lorig T S, Sapp A C, Campbell J, & Cain W S (1993). Event-related Potentials to Odor Stimuli. *Bulletin of the Psychonomic Society* 31:131–4.

Möller R, Pause B M, & Ferstl R (1999). Induzierbarkeit geruchlicher Sensitivität durch Duft-Exposition bei Personen mit spezifischer Anosmie. *Zeitschrift für Experimentelle Psychologie* 46:53–71.

Morgan C D, Covington J W, Geisler M W, Polich J, & Murphy C (1997). Olfactory Event-related Potentials: Older Males Demonstrate the Greatest Deficits. *Electroencephalography and Clinical Neurophysiology* 104:351–8.

Morgan C D, Geisler M W, Covington J W, Polich J, & Murphy C (1999). Olfactory P3 in Young and Older Adults. *Psychophysiology* 36:281–7.

Murphy C, Nordin S, De Wijk R A, Cain W S, & Polich J (1994). Olfactory-evoked Potentials: Assessment of Young and Elderly, and Comparison to Psychophysical Threshold. *Chemical Senses* 19:47–56.

Näätänen R (1990). The Role of Attention in Auditory Information Processing as Revealed by Event-related Potentials and Other Brain Measures of Cognitive Function. *Behavioral and Brain Sciences* 13:201–88.

Näätänen R (1992). *Attention and Brain Function*. Hillsdale, NJ: Lawrence Erlbaum.

Näätänen R & Picton T (1987). The N1 Wave of the Human Electric and Magnetic Response to Sound: A Review and an Analysis of the Component Structure. *Psychophysiology* 24:375–425.

O'Connell R J, Stevens D A, & Zogby L M (1994). Individual Differences in the Perceived Intensity and Quality of Specific Odors following Self- and Cross-adaptation. *Chemical Senses* 19:197–208.

Pause B M (1998). Differentiation of Body Odors by the Human Brain. *International Journal of Psychophysiology* 30:35–6.

Pause B M, Ferstl R, & Fehm-Wolfsdorf G (1998). Personality and Olfactory Sensitivity. *Journal of Research in Personality* 32:510–18.

Pause B M & Krauel K (1998). Smelling without Attending? *Aroma-Chology Review* 7:1, 4–6.

Pause B M & Krauel K (2000). Chemosensory Event-related Potentials (CSERP) as a Key to the Psychology of Odors. *International Journal of Psychophysiology* 36:105–22.

Pause B M, Krauel K, Eggert F, Müller C, Sojka B, Müller-Ruchholtz W, & Ferstl R (1999a). Perception of HLA-related Body Odors during the Course of the Menstrual Cycle. In: *Advances in Chemical Signals in Vertebrates*, ed. R E Johnston, D Müller-Schwarze, & P W Sorensen, pp. 201–7. New York: Kluwer.

Pause B M, Krauel K, Sojka B, & Ferstl R (1999b). Is Odor Processing Related to Oral Breathing? *International Journal of Psychophysiology* 32:251–60.

Pause B M, Krauel K, Sojka B, & Ferstl R (1999c). Body Odor Evoked Potentials: A New method to Study the Chemosensory Perception of self and Non-self in Humans. *Genetica* 104:285–94.

Pause B M, Miranda A, Nysterud M, & Ferstl R (2000). Geruchs- und emotionale Reizbewertung bei Patienten mit Major Depression. *Zeitschrift für Klinische Psychologie und Psychotherapie* 29:16–23.

Pause B M, Miranda A, Raack N, Sojka B, Göder R, Aldenhoff J B, & Ferstl R (1999d). Olfaction and Affective State: Is Odor Evaluation and Perception Altered in Patients with Major Depression? *Chemical Senses* 24:100.

Pause B M, Rogalski K P, Sojka B, & Ferstl R (1999e). Sensitivity to Androstenone in Female Subjects Is Associated with an Altered Brain Response to Male Body Odor. *Physiology and Behavior* 68:129–37.

Pause B M, Sojka B, & Ferstl R (1997). Central Processing of Odor Concentration Is a Temporal Phenomenon as Revealed by Chemosensory Event-related Potentials (CSERP). *Chemical Senses* 22:9–26.

Pause B M, Sojka B, Krauel K, Fehm-Wolfsdorf G, & Ferstl R (1996a). Olfactory Information Processing during the Course of Menstrual Cycle. *Biological Psychology* 44:31–54.

Pause B M, Sojka B, Krauel K, & Ferstl R (1996b). The Nature of the Late Positive Complex within the Olfactory Event-related Potential. *Psychophysiology* 33:376–84.

Rockstroh B, Elbert T, Canavan A, Lutzenberger W, & Birnbaumer N (1989). *Slow Cortical Potentials and Behaviour*. Baltimore: Urban & Schwarzenberg.

Roth W T, Doyle C M, Pfefferbaum A, & Kopell B S (1980). Effects of Stimulus Intensity on P300. *Progress in Brain Research* 54:296–300.

Sokolov E N (1977). Brain Functions: Neuronal Mechanisms of Learning and Memory. *Annual Review of Psychology* 28:85–112.

Stern K & McClintock M K (1998). Regulation of Ovulation by Human Pheromones. *Nature* 392:177–9.

Tateyama T, Hummel T, Roscher S, Post H, & Kobal G (1998). Relation of Olfactory Event-related Potentials to Changes in Stimulus Concentration. *Electroencephalography and Clinical Neurophysiology* 108:449–55.

Thiele V & Kobal G (1984). Vergleich der objektiven und subjektiven Methoden olfaktometrischer Bestimmungen – Beispiel H_2S. *Schriftenreihe der Landesanstalt für Immissionsschutz NW* 59:41–47.

Van Toller S (1988). Emotion and the Brain. In: *Perfumery*, ed. S Van Toller & G H Dodd, pp. 121–146. New York: Chapman & Hall.

Wedekind C & Füri S (1997). Body Odour Preferences in Men and Women: Do They Aim for Specific MHC Combinations or Simply Heterozygosity? *Proceedings of the Royal Society of London, Series B* 264:1471–9.

Wysocki C J, Dorries K, & Beauchamp G K (1989). Ability to Perceive Androstenone Can Be Aquired by Ostensibly Anosmic People. *Proceedings of the National Academy of Sciences USA* 86:7976–8.

Wysocki C J, Pierce J D, & Gilbert A N (1991). Geographic Cross-cultural and Individual Variation in Human Olfaction. In: *Smell and Taste in Health and Disease*, ed. T V Getchell, R L Doty, L M Bartoshuk, & J B Snow, pp. 287–314. New York: Raven Press.

20

Processing of Olfactory Affective Information: Contribution of Functional Imaging Studies

Robert J. Zatorre

A central concern in contemporary cognitive neuroscience is how stimulus information is represented in the human central nervous system. One approach to this issue is to examine the neural correlates of different types of stimuli in different modalities to determine the functional organization of distinct cortical and/or subcortical pathways associated with particular types of stimulus features. This approach has proved particularly powerful for understanding how visual processes are organized, for example. Much less is currently known about how the brain encodes olfactory sensory information. Despite progress in understanding the basic mechanisms of coding at the receptor level (Duchamp-Viret, Chaput, and Duchamp, 1999), cortical processing of odor quality remains largely unknown, especially in the human brain. One property of the chemical senses that differentiates them from other modalities is their strong hedonic association. Unlike visual and auditory stimuli, odors and tastes often result in strong affective reactions. That aspect of odor processing raises interesting questions about the neural substrates that account for the salient hedonic value of odors.

In the past decade, cognitive neuroscientists have benefited enormously from the development of functional neuroimaging, which allows noninvasive exploration of the brain. Those techniques measure changes in hemodynamic responses as functions of stimulus or task demands, thus permitting assessments of the neural activities associated with a wide variety of cognitive and perceptual processes. Neuroimaging complements more conventional approaches such as neurophysiological studies and lesioning studies of behavior, which continue to be of considerable importance in understanding brain–behavior relationships. This chapter offers a brief overview of recent studies in neuroimaging as applied to olfaction, and particularly as related to olfactory hedonics, followed by some recently published data from our laboratory examining specifically the processing of odor quality and intensity (Zatorre, Jones-Gotman, and Rouby, 2000).

Previous work with brain-lesioned patients has indicated that unilateral damage in the temporal lobe or in the orbitofrontal cortex (OFC), particularly in the right cerebral hemisphere, can result in significant disruptions in the higher-order processing of odors, including odor quality discrimination and memory tasks (Eskenazi et al., 1983; Zatorre and Jones-Gotman, 1991; Martinez et al., 1993; Jones-Gotman and Zatorre, 1993), whereas lower-level processes such as detection and intensity matching typically are unaffected (Eichenbaum et al., 1983; Jones-Gotman and Zatorre, 1988). Such dissociations suggest that the olfactory nervous system may be hierarchically organized, with the OFC responsible for more complex aspects of feature analysis and integration. Such a model is compatible with anatomical considerations, because neurons in the OFC receive inputs from the piriform cortex and from thalamic nuclei (Price et al., 1991), among other areas (Figure 20.1), presumably after having undergone several levels of prior processing. Also, electrophysiological data from tests with monkeys indicate greater degrees of selectivity in the profiles of neuronal responses proceeding up the olfactory neuraxis from the olfactory bulb to the piriform cortex to the OFC (Takagi, 1991). However, it remains unknown whether or not olfactory hedonic processing is associated with the same neural substrate as that associated with odor encoding, identification, and retention.

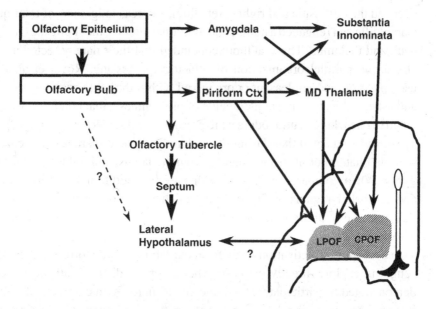

Figure 20.1. Diagram of known and putative primate olfactory pathways leading to OFCs: LPOF, lateroposterior orbital frontal; CPOF, centroposterior orbital frontal. (Adapted from Takagi, 1989.)

Several neuroimaging studies of human olfaction using positron-emission tomography (PET) and functional magnetic-resonance imaging (fMRI) have recently been carried out, and they have helped to clarify the precise functional neuroanatomy of olfactory processing (for a review, see Zatorre and Jones-Gotman, 2000). Among the more relevant findings are that the piriform cortex and the OFC usually show increased activity in response to presentation of odors, with an asymmetry favoring the right OFC (Zatorre et al., 1992; Small et al., 1997; Sobel et al., 1998; Savic et al., 2000). More recently, Royet et al. (1999) have also found an asymmetry favoring the right OFC, but it was present specifically in a condition in which familiarity judgments were elicited, suggesting that the right OFC may be involved in processing particular stimulus features. On the other hand, Zald and Pardo (1997) have shown that very unpleasant odors result in significant activation the left OFC, as well as limbic structures, notably the amygdala. That latter finding raises the question whether or not hedonic processes depend on a different hemispheric lateralization than other types of processes and suggests that the amygdala may be involved in evaluating the affective valence of an odor.

In what is perhaps a demonstration of scientific zeitgeist, three very recent papers, published within a few months of one another, have used PET to explore the neural substrates associated with judgments of odor quality, intensity, and pleasantness. Savic et al. (2000) showed that judgments of odor intensity engaged the left insula and right cerebellum, whereas judgments of odor quality (same/different) engaged a number of additional regions, including the OFC, caudate, and thalamus. Those authors concluded that their data reflected a hierarchical and parallel organization of olfactory processing. Royet et al. (2000) used pleasant and unpleasant odors mixed within the same scanning condition and asked subjects to make judgments of pleasantness; that condition was compared to one where neutral odors were presented and subjects made no judgment. The results indicated that during the hedonic judgment there was increased activity in bilateral orbitofrontal regions, hypothalamus, and right superior frontal gyrus. When compared with a no-odor baseline, additional activity in insular and cerebellar regions emerged. Many of those regions were similarly active in visual and auditory emotional tasks as well, suggesting some commonality across modalities.

In a recent PET activation study from our laboratory (Zatorre et al., 2000), we sought to explore directly the issue whether or not distinct pathways could be demonstrated to participate in hedonic versus non-hedonic aspects of olfactory information processing. We concentrated on two dimensions of odor perception, intensity and pleasantness, because they have been shown to be dissociable. In particular, Eichenbaum et al. (1983) found that the famous patient H.M., who has bilateral medial-temporal-lobe damage, was utterly unable to identify odors or assess their quality, but was nonetheless as accurate as control subjects in performing detection and intensity scaling tasks. That result indicates that odor intensity processing may depend on a neural substrate dissociable from that involved in other types of olfactory processing.

One practical problem in studying odor intensity and pleasantness is that these two dimensions typically are correlated at the behavioral level, with odors being judged less pleasant as they become more intense; however, it is possible to dissociate them psychophysically by a judicious choice of specific odorants and concentrations. Specifically, we used a set of odors of various concentrations previously selected and tested (see Rouby and Bensafi, Chapter 9, this volume) to be independent on the two dimensions in question (Figure 20.2). In that study,

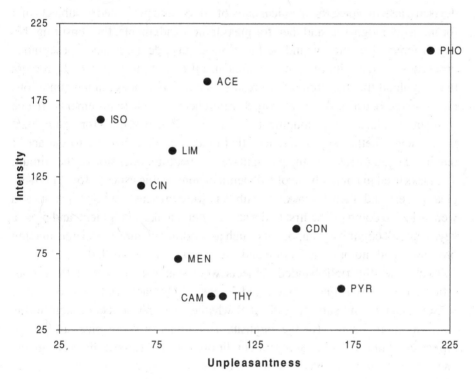

Figure 20.2. Scattergram of pleasantness and intensity responses obtained for 10 odorants. Each point corresponds to the ranking derived from pairwise comparisons of each odor with each other odor, presented once for pleasantness and once for intensity judgments. Odor concentrations were chosen to avoid a correlation between the two variables: PHO, thiophenol; ACE, acetophenone; ISO, isoamyl acetate; LIM, R-(+)-limonene; CIN, 1,8-cineole; CDN, cyclode-canone; MEN, (−)-menthol; CAM, (−)(+)-camphor; THY, thymol; PYR, pyri-dine. For the PET study, six odors were retained (excluding PHO, LIM, CAM, and MEN). See Zatorre et al. (2000) for further details and concentrations.

subjects performed pairwise comparison tests of certain concentrations of odors, and it was shown that judgments of relative intensity were uncorrelated with judgments of relative pleasantness. Thus, that procedure allowed the selection of a stimulus set that would permit us to examine odor pleasantness and intensity judgments separately. We hypothesized that different patterns of cerebral activity would be evoked when those two judgments were made.

The experimental design chosen was meant to control for two important issues that have complicated the interpretation of much prior research. One such issue is that it is often difficult to dissociate hedonic judgments from stimulus-specific effects. That is, if one compares, say, pleasant to neutral odors and differences are found, it is not clear whether the effects are due to differences in the stimuli themselves or to the intended factor of pleasantness. In order to avoid such confounding effects of the specific stimuli to be used for the two judgments, we devised a task in which the identical pairs of odors were presented to subjects first for intensity judgments and then for pleasantness judgments, thus ensuring that any differences elicited would be functions of task demands and not stimulus properties, as the stimuli were constant in all conditions. A second problem that can cloud the interpretation of results is that differences among conditions may emerge because of uncontrolled differences in task requirements. If one compares a pleasantness judgment task to a neutral baseline, for example, it is not clear whether any differences that emerge are attributable to the act of making the judgment or to the pleasant-versus-neutral dimension of the stimuli. The tasks used in our study involved identical nonspecific aspects, for in both the pleasantness and intensity tasks the subjects were presented with two items and were asked to compare the first to the second item, make a judgment, and press a key. Thus, all cognitive components, such as working memory, decision-making processes, and motor organization and execution, were controlled.

Twelve healthy right-handed subjects were scanned in each of three conditions, termed Baseline, Affect, and Intensity. On each trial, paired stimuli were presented, and subjects indicated whether the first or second item in the pair was more intense (in the Intensity condition) or more pleasant (in the Affect condition). In the Baseline condition, two odorless trials were given, and subjects simply inhaled and responded with a random keypress. Measures of cerebral blood flow were obtained according to standard techniques (Zatorre et al., 2000), and the resulting functional maps were co-registered with anatomical magnetic-resonance images after stereotaxic transformation. Data sets were averaged across subjects, and paired-image subtractions were performed to identify the presence and location of significantly different voxels, according to a *t*-statistic parametric map (Worsley et al., 1992).

Each of the two odor judgment conditions was compared to the baseline; in addition, the two odor conditions were compared directly to one another. Comparison to the common no-odor baseline revealed three main findings, each of which will be discussed in turn: First, the expected activity in the primary olfactory cortex, or piriform area, was *not* observed; second, and in accord with what we expected based on prior studies, activation was seen in the right OFC in both the Affect and Intensity conditions; third, the Affect condition, but not the Intensity condition, resulted in recruitment of the hypothalamus.

The absence of increased cerebral blood flow within the piriform region in either of the two odor conditions was somewhat surprising. There was no evidence of activity in that region, even when a more liberal statistical threshold was applied. In order to increase the power of the analysis, we pooled the data for the two odor presentation conditions together and compared them to the no-odor baseline to see if piriform activation could be detected, but once again no evidence of any significant change in that region was obtained. That negative finding cannot be dismissed simply as a false-negative response, as the analysis performed contained sufficient observations that piriform activity should have been detected had it been present (12 subjects scanned six times each, yielding 72 image volumes). Furthermore, previous studies in our laboratory (Zatorre et al., 1992; Small et al., 1997) had clearly shown piriform activity, even with fewer observations and using a less sensitive PET scanner than the one used in the current experiment. In addition, a similar lack of piriform activity was reported by Dade et al. (1998) in an olfactory working-memory task in which behavioral performance clearly indicated that the stimuli were being processed. Also, Zald and Pardo (1997) reported only weak or no activation of presumed piriform areas in their study with very noxious odors. In other studies (e.g., Royet et al., 2000; Savic et al., 2000), piriform activation was detected only weakly, by using region-of-interest analyses, or only in one hemisphere.

One possible explanation for the absence of piriform activity in our study is that sniffing odorless air can produce piriform activation (Sobel et al., 1998), and because subjects sniffed in both the baseline condition and the two odor conditions, it would have subtracted out. Although that possibility may account in part for the findings, it is insufficient to explain why piriform activity is observed in some studies but not others, all of which have used an odorless sniffing baseline condition. It is also possible, of course, that activity in the piriform cortex is present but is not detectable by PET methods, perhaps because it is of a very transient nature to which PET is not sensitive.

An alternative explanation is that the piriform cortex may respond to familiar odors, whereas the odors used in our study were chosen to be unfamiliar. That idea is consistent with the data of Dade et al. (1998), who noted piriform activation after odors had been familiarized, but not when they had first been presented. Regardless of the ultimate explanation for why piriform activity is so variably observed, the current data suggest a need to revise the traditional view of the piriform area as a simple sensory relay in a hierarchy. Instead, the findings suggest a more complex interaction among a network of regions including piriform and orbitofrontal cortices, with activity in orbitofrontal areas not necessarily dependent solely on input from piriform regions (Figure 20.1). In this regard, a recent anatomical study is of considerable relevance: Johnson et al. (2000) examined the connectivity of neurons in the rat piriform cortex and concluded that the piriform region "performs correlative functions analogous to those in association areas of neocortex rather than those typical of primary sensory areas with which it has been traditionally classed." Thus, the complex modulation of piriform areas seen across the various studies would be consistent with the idea that its function is much more complex than previously thought (see Ravel et al., Chapter 21, this volume).

The second result, which was in keeping with predictions, was the significant increase in cerebral blood flow in the right OFC in both the Affect and Intensity conditions (Figure 20.3). The positions of that OFC area were very similar across the two conditions and, more importantly, were remarkably consistent with that reported in several previous studies (see Zatorre and Jones-Gotman, 2000, for a meta-analysis of OFC activation sites). Such a result therefore confirms the greater importance of the right OFC for odor processing, as suggested by previous imaging studies (Zatorre et al., 1992; Small et al., 1997; Sobel et al., 1998; Savic et al., 2000) and lesioning studies of behavior (Zatorre and Jones-Gotman, 1991), and more generally by several studies implicating the right hemisphere in many, though by no means all, olfactory tasks (Abraham and Mathai, 1983; Eskenazi et al., 1983; Zatorre and Jones-Gotman, 1990; Carroll, Richardson, and Thompson, 1993; Martinez et al., 1993; Jones-Gotman and Zatorre, 1993). In terms of the main point of this study, however, the OFC does not appear to contribute differentially to affective processing, as opposed to non-affective processing, because the activations were essentially identical in the two conditions.

The third and most relevant finding in terms of the goals of the study was that a region of the hypothalamus was active in the Affect condition but not the Intensity condition. To verify that the hypothalamic activity constituted the main difference in brain activity associated with affective processing, we compared the two active conditions directly to one another. That subtraction yielded no differences in cerebral blood flow in the OFC areas, consistent with the fact that activation was present in both conditions in a similar location in the orbital region. There was, however, a region of significantly greater activity in the Affect condition

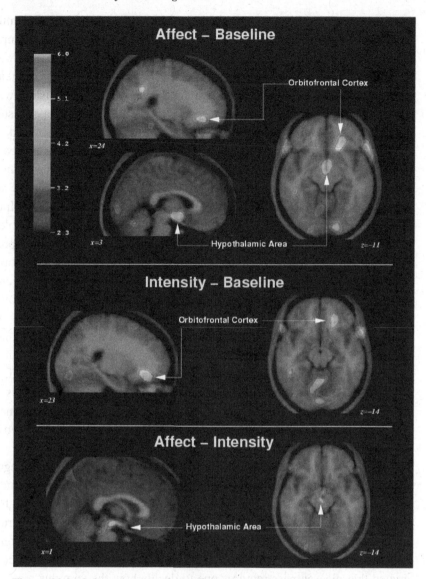

Figure 20.3. Selected areas of significant increases in cerebral blood flow in PET study. The average PET subtraction images are shown superimposed on averaged MRI scans. Changes in flow are shown as a *t*-statistic map, coded by the color scale shown at top left. Top: Comparison of Affect condition to Baseline condition. The two sagittal sections and the horizontal section illustrate the positions of activation sites in the right OFC and hypothalamic area. Middle: Comparison of Intensity and Baseline conditions. Sagittal and horizontal sections illustrate activation in right OFC. Bottom: Comparison of Affect and Intensity conditions. Sagittal and horizontal sections illustrate activation in hypothalamic area. (Adapted from Zatorre et al., 2000.)

compared with the Intensity condition in the vicinity of the hypothalamus, and in essentially the same location as that seen in the Affect–Baseline comparison (Figure 20.3).

Hypothalamic activity was also noted by Royet et al. (2000) in their odor pleasantness/unpleasantness judgment condition, which is consistent with our data; however, it is not clear if that effect was due to the affective state presumably elicited by the stimuli themselves or to the cognitive demands of making the pleasantness judgment. In our study, the hypothalamic activation appeared to be a function of task demands, rather than being stimulus-driven, for the stimuli were identical in the two conditions. Furthermore, because the hypothalamic activity was observed in both the Affect–Baseline subtraction and the Affect–Intensity comparison, it appears to be a robust finding that is specifically attributable to the condition of judging pleasantness/unpleasantness. One possible interpretation of that finding is that hypothalamic activity may be modulated by top-down mechanisms. Although there is scant converging evidence for such a mechanism from other studies, there is anatomical evidence (Takagi, 1991) that olfactory information travels through several parallel pathways, from olfactory bulb to OFC, and from olfactory bulb via the septal area, to the lateral hypothalamus (Figure 20.1). Although the evidence is unclear, there may also be connections between the hypothalamus and OFC areas (Tazawa, Onoda, and Takagi, 1983). Those pathways could provide the means for odor-processing neural systems in the OFC to interact with and thus influence the activity of hypothalamic neurons, a hypothesis that would be in keeping with the clear OFC activity observed during the odor judgment tasks.

The conventional view that the hypothalamus plays a crucial role in regulating the internal state may be relevant to our findings as well. If the affective judgment of relative pleasantness not only is carried out by analysis of stimulus features but also requires access to information about the internal state, then recruitment of the hypothalamus might be necessary. Whereas intensity can be determined solely on the basis of the physical characteristics of a stimulus, it could be argued that pleasantness requires one to decide how the stimulus may interact with one's internal environment (Might this odor indicate something good to eat? Might this odor indicate something that could make me sick?). Thus the stimulus features in themselves, which would be adequate for judging intensity, might be insufficient to make that safety judgment, and reference to neural systems involved in somatic and physiological status might be relevant. A similar function has previously been suggested for the OFC (Rolls, 1999), which undoubtedly plays a role in affective processes, for many neuroimaging studies have shown that it is active in response to various types of affectively valenced stimuli (Zald and Pardo, 1997; Blood et al., 1999; Francis et al., 1999). However, in our study the OFC was as active during judgments of intensity as it was during judgments of pleasantness. Clearly, much further experimental work will have to be done before these questions can be sorted out. The current state of our understanding of these complex systems has undoubtedly been enhanced by the contribution of neuroimaging studies, and one may be cautiously optimistic that future work will continue to yield additional insights.

Acknowledgments

I wish to thank my collaborators in this research, Marilyn Jones-Gotman and Catherine Rouby, and Drs. A. C. Evans and B. Pike of the Brain Imaging Center of the Montreal Neurological Institute. The work was supported by the Canadian Institutes of Health Research, and G.I.S. Sciences de la Cognition, France.

References

Abraham A & Mathai K V (1983). The Effect of Right Temporal Lobe Lesions on Matching of Smells. *Neuropsychologia* 21:277–81.

Blood A J, Zatorre R J, Bermudez P, & Evans A C (1999). Emotional Responses to Pleasant and Unpleasant Music Correlate with Activity in the Paralimbic Brain Regions. *Nature Neuroscience* 2:382–7.

Carroll B, Richardson J T, & Thompson P (1993). Olfactory Information Processing and Temporal Lobe Epilepsy. *Brain Cognition* 22:230–43.

Dade L A, Jones-Gotman M, Zatorre R J, & Evans A C (1998). Human Brain Function during Odor Encoding and Recognition. A PET Activation Study. *Annals of the New York Academy of Sciences* 855:572–4.

Duchamp-Viret P, Chaput M A, & Duchamp A (1999). Odor Response Properties of Rat Olfactory Receptor Neurons. *Science* 284:2171–4.

Eichenbaum H, Morton T H, Potter H, & Corkin S (1983). Selective Olfactory Deficits in Case H.M. *Brain* 106:459–72.

Eskenazi B, Cain W, Novelly R, & Friend K B (1983). Olfactory Functioning in Temporal Lobectomy Patients. *Neuropsychologia* 21:365–74.

Francis S, Rolls E T, Bowtell R, McGlone F, O'Doherty J, Browning A, Clare S, & Smith E (1999). The Representation of Pleasant Touch in the Brain and Its Relationship with Taste and Olfactory Area. *NeuroReport* 10:453–9.

Johnson D M G, Illig K R, Behan M, & Haberly L B (2000). New Features of Connectivity in Piriform Cortex Visualized by Intracellular Injection of Pyramidal Cells Suggest that "Primary" Olfactory Cortex Functions like "Association" Cortex in Other Sensory Systems. *Journal of Neuroscience* 20:6974–82.

Jones-Gotman M & Zatorre R J (1988). Olfactory Identification Deficits in Patients with Focal Cerebral Excision. *Neuropsychologia* 26:387–400.

Jones-Gotman M & Zatorre R J (1993). Odor Recognition Memory in Humans: Role of Right Temporal and Orbitofrontal Regions. *Brain and Cognition* 22:182–98.

Martinez B, Cain W, De Wijk R, Spencer D D, Novelly R A, & Sass K J (1993). Olfactory Functioning before and after Temporal Resection for Intractable Seizures. *Neuropsychology* 7:351–63.

Price J, Carmichael S, Carnes K, Clugnet M C, Kuroda M, & Ray J P (1991). Olfactory Input to the Prefrontal Cortex. In: *Olfaction: A Model System for Computational Neuroscience*, ed. J Davis & H Eichenbaum, pp. 101–20. Cambridge, MA: MIT Press.

Rolls E T (1999). *The Brain and Emotion*. Oxford University Press.

Royet J P, Koenig O, Gregoire M C, Cinotti L, Lavenne F, Le Bars D, Costes N, Vigouroux M, Farget V, Sicard G, Holley A, Manguiere F, Comar D, & Froment J C (1999). Functional Anatomy of Perceptual and Semantic Processing for Odors. *Journal of Cognitive Neuroscience* 11:94–109.

Royet J P, Zald D, Versace R, Costes N, Lavenne F, Koenig O, & Gervais R (2000). Emotional Responses to Pleasant and Unpleasant Olfactory, Visual and Auditory Stimuli: A Positron Emission Tomography Study. *Journal of Neuroscience* 20:7752–9.

Savic I, Gulyas B, Larsson M, & Roland P (2000). Olfactory Functions Are Mediated by Parallel and Hierarchical Processing. *Neuron* 26:735–45.

Small D M, Jones-Gotman M, Zatorre R J, Petrides M, & Evans A C (1997). Flavor Processing: More than the Sum of Its Parts. *NeuroReport* 8:3913–17.

Sobel N, Prabhakaran V, Desmond Glover G H, Goode R L, Sullivan E V, & Gabrieli J D (1998). Sniffing and Smelling: Separate Subsystems in the Human Olfactory Cortex. *Nature* 392:282–6.

Takagi S F (1989). *Human Olfaction*. University of Tokyo Press.

Takagi S F (1991). Olfactory Frontal Cortex and Multiple Olfactory Processing in Primates. In: *Cerebral Cortex*, vol. 9, ed. A Peters & E G Jones, pp. 133–52. New York: Plenum Press.

Tazawa Y, Onoda N, & Takagi S F (1983). Olfactory Pathway to the Lateral Hypothalamus in the Monkey. *Neuroscience Letters Suppl.* 13:S62.

Worsley K, Evans A, Marrett S, & Neelin P (1992). A Three-Dimensional Statistical Analysis for CBF Activation Studies in Human Brain. *Journal of Cerebral Blood Flow and Metabolism* 12:900–18.

Zald D & Pardo J (1997). Emotion, Olfaction, and the Human Amygdala: Amygdala Activation during Aversive Olfactory Stimulation. *Proceedings of the National Academy of Sciences USA* 94:4119–24.

Zatorre R J & Jones-Gotman M (1990). Right-Nostril Advantage for Discrimination of Odors. *Perception and Psychophysics* 47:526–31.

Zatorre R J & Jones-Gotman M (1991). Human Olfactory Discrimination after Unilateral Frontal or Temporal Lobectomy. *Brain* 114:71–84.

Zatorre R J & Jones-Gotman M (2000). Functional Imaging of the Chemical Senses. In: *Brain Mapping: The Applications*, ed. A Toga & J Mazziota, pp. 403–24. Orlando: Academic Press.

Zatorre R J, Jones-Gotman M, Evans A C, & Meyer E (1992). Functional Localization and Lateralization of Human Olfactory Cortex. *Nature* 360:339–40.

Zatorre R J, Jones-Gotman M, & Rouby C (2000). Neural Mechanisms Involved in Odor Pleasantness and Intensity Judgments. *NeuroReport* 11:2711–16.

21

Experience-induced Changes Reveal Functional Dissociation within Olfactory Pathways

Nadine Ravel, Anne-Marie Mouly, Pascal Chabaud,
and Rémi Gervais

Learning more about the neural basis of olfactory cognition should greatly improve our understanding of how different brain structures deal with information. Progress can be facilitated by simultaneous advances with animal models and human studies, for in the olfactory system, conservatism across mammalian species in the organization of olfactory pathways allows integration of data obtained in animals and humans. In each species, output neurons from the olfactory bulb (OB) monosynaptically reach the piriform cortex (PC), the peri-amygdaloid cortex, and the lateral entorhinal cortex (LEC). The LEC provides massive input to the hippocampus, and the PC sends information to the orbitofrontal neocortical area, both directly and after a relay in the dorsomedial thalamic nuclei (Haberly, 1998). As in other sensory systems, two strategies have been developed thus far in order to identify hierarchical organization: collecting information from one structure at a time, and looking at the system as a network of interconnected structures. The first strategy is typical for most animal studies and is implemented through single-cell recordings in anesthetized animals (Mori, Nagao, and Yoshiara, 1999) and active animals (Schoenbaum, Chiba, and Gallagher, 1999; Weibe and Staubli, 1999; Wood, Dudchenko, and Eichenbaum, 1999) or through surface EEG recordings in awake restrained animals (Freeman and Skarda, 1985). When using anesthetized animals, most studies have focused on the OB, and fewer on the PC. When using active rats, such studies have investigated hippocampal, amygdalar, and orbitofrontal electrophysiological characteristics. Those approaches provide data on cell reactivity with millisecond time resolutions, but give no information on how information is processed simultaneously at different levels. The hierarchical organization of central olfactory processing is beginning to be revealed by recent brain imaging studies in humans (Zatorre et al., 1992; Sobel et al., 1998, 2000; Royet et al., 1999, 2000; O'Doherty et al., 2000; Qureshy et al., 2000; Savic et al., 2000; Zatorre, Jones-Gotman, and Rouby, 2000; Zald and Pardo, 2000). Other chapters in this volume provide an overview of the main findings (see Phillips and Heining, Chapter 12, Zatorre, Chapter 20, and Rolls, Chapter 23). However, for at least two reasons, comparison between electrophysiological data and brain imaging data is difficult. First, the techniques have very different temporal and spatial resolutions. Second, whereas most electrophysiological data have been collected in anesthetized animals, brain imaging studies map alert brains involved in cognitive tasks. The purpose of some of our recent investigations has been to contribute to filling in the gap between the two approaches. Our experimental model relies on electrophysiological recordings from active rats. Ideally, one would like to determine electrophysiological activities simultaneously from the OB, several parts of the PC, the entorhinal cortex (EC), the amygdala, several sites of the hippocampal formation, and the orbitofrontal cortex. As a start, we focused our attention on three structures: the OB, PC, and EC. In addition, we were particularly interested in finding electrophysiological correlates of modulations of olfactory responses related to internal state, recent experience, and long-term memory. But before summarizing our findings, we shall briefly review what was available from recent studies regarding the physiology of the OB and the PC.

1. The OB and the PC Are More than Sensory Relays

Anatomical studies have clearly shown that information converging to the OB and PC does not originate exclusively from olfactory neuroreceptors. Indeed, both structures are densely interconnected with several other forebrain cortical areas. In addition, they receive dense innervations from ascending neuromodulatory systems such as the noradrenergic system, the serotoninergic system, and the basal forebrain cholinergic system (Gervais, Holley, and Keverne, 1988). Thus we can predict that in an active animal, responses elicited by odorant stimuli at the OB and PC levels are likely to be strongly modulated according to the animal's internal state and past experience.

At the OB level, we used three types of electrophysiogical studies to test that prediction. First, multi-unit mitral-cell activity was recorded from one site per bulb in response to repeated olfactory stimuli. Protocols were designed to investigate the differential rate of neuronal-response habituation according to the behavioral value of the stimulus. For example, food odor was presented every 3 min to food-deprived animals and satiated animals (Pager et al., 1972). Second, in awake restrained rabbits, an array of 64 electrodes recorded mass oscillatory activity (electroencephalogram, EEG) during odor presentation. In that case, the experimental paradigm included a simple associative-conditioning task in which presentation of odors was paired repeatedly over days with reinforcement (Freeman and Schneider, 1982). Both sets of data have been reviewed (Freeman and Skarda 1985; Gervais, 1993). The two key findings were that previous experience had a strong influence on populational OB responses, and that effect depended on the action of centrifugal control exerted at the level of the OB. However, in the two sets of experiments, the term "experience" referred to very different time scales. Indeed, the habituation paradigm tested for a simple form of short-term memory over minutes, whereas the associative-learning paradigm tested for changes in OB reactivity over days and weeks. A recent study in which single-cell mitral activity was recorded in rats engaged in a conditioning paradigm (Kay and Laurent, 1999) confirmed earlier findings (Pager, 1983). In fact, output cell activity was found to depend more strongly on an animal's experience and ongoing behavior than on the olfactory stimulation. On the whole, electrophysiological experiments show that in active animals, OB reactivity is under strong influences from more central structures.

But what might be the role of such central influence exerted at the input level? We hypothesized that it could be of importance for olfactory memory, and we designed a series of experiments to test that hypothesis. The experiments relied on examination of memory performances following transient perturbation of OB activity. In one experiment, transient blockade of cholinergic modulation of the OB (via intrabulbar injection of scopolamine) impaired short-term olfactory memory over a time scale of a few tens of seconds (Ravel, Elaagouby, and Gervais, 1994). In another experiment, transient inactivation of the OB activity (via intrabulbar lidocaine injection) just after each training session impaired retention of olfactory information over a time scale of a few days (Mouly et al., 1993). Both experiments are interpreted as evidence for OB integration into a functional network supporting both short-term and long-term memory: That is, cooperation between the OB and other structures such as the EC and hippocampus intervenes to support some forms of memory processes. That also suggests that neural computation performed at the OB level is doing more than "formatting" the message addressed by olfactory neuroreceptors before transmission to other areas.

At the PC level, some lines of evidence suggest that its role is not limited to pure sensory analysis, but also extends to olfactory learning. Electrophysiological studies in vitro (Jung, Larson, and Lynch, 1990; Kanter and Haberly, 1990) and in vivo (Stripling, Patneau, and Gramlich, 1988) have revealed that intrinsic PC connections can express long-term potentiation (LTP), and LTP in the PC has been shown to accompany behavioral learning (Roman, Staubli, and Lynch, 1987; Litaudon et al., 1997). Those experimental data are reinforced by theoretical and modeling approaches showing that the PC has several features of a network with content-addressable memory characteristics (Haberly, 1985; Wilson and Bower, 1988; Lynch and Granger, 1989). However, that general hypothesis needs definitive support from behavioral studies. In addition, the PC is commonly thought of as being functionally homogeneous. Some recent studies have suggested a functional dissociation along its anteroposterior axis (Litaudon and Cattarelli, 1996; Litaudon et al., 1997) although data have not been obtained from active animals. For that reason, in our experiments we obtained electrophysiological recordings from two sites in the PC cortex: one in the anterior part (aPC), and the other in the posterior part (pPC).

In summary, it is likely that experience modulates responsiveness at each level of olfactory processing, including the OB and PC. However, an overview of how and when changes occur at each of those levels in animals performing a given task is not yet available. Thus the purpose of our study is to provide new insights on how the OB-PC-EC complex works for information processing. Starting with the hypothesis that both recent and remote experiences modulate functioning at each of the three levels, we examined the functional consequences of experience on neural responsiveness in intact rats.

2. Local Field Potential Activity as a Measure of Functional Anatomy

One critical step in studying brain activities is the choice of the level of analysis. Electrophysiological investigations can rely on single-cell recordings, multi-unit recordings, local field potentials, or surface EEGs. At a microscopic level, single-cell recordings provide straightforward access to neural computation, but they are very difficult to obtain simultaneously from several structures. Furthermore, long-lasting effects of experience can be tested only with stable recordings over days and weeks. At a macroscopic level, surface EEGs provide valuable information on superficial brain structures such as the OB (Freeman, 1975), but are inappropriate for in vivo recordings of deep structures such as the PC and the EC. In a middle range between those two extremes, intracerebral local field potentials (LFPs) provide information on population activity from identified cell layers. In our approach, two types of LFPs were studied: electrically evoked LFPs and naturally occurring LFP oscillatory activities.

In a first approach, electrically evoked LFPs were recorded simultaneously in the PC and in the EC following stimulation of the OB. The characteristics of the evoked potentials (amplitude, latency) reflect the efficiency of synaptic functioning. Thus the characteristics of evoked field potentials (EFPs) were measured over days while animals learned an olfactory associative task, the implicit hypothesis being that changes in synaptic transmission could reflect some neural correlates of behavioral learning (Morris et al., 1986; Roman et al., 1987).

In a second approach, spontaneous and odor-evoked LFP activities were studied. In the olfactory system, it is well established that the activities of neuron populations are dominated by prominent oscillatory regimes with frequency ranges from 1 to 100 Hz, reflecting the observation that groups of neurons have a strong tendency to work in synchrony. One important reason for studying oscillatory activity is because of the current hypothesis suggesting that a widespread common oscillatory frequency could play a critical role in transient formation of neural assemblies. That refers to the so-called binding theory (Singer, 1999), one key idea being that the presence of a common oscillatory rhythm within a neuron population could favor synchronization of the individual discharges of neurons belonging to that population. That could be particularly important in the neural representation of odor, which is suspected to be widely distributed spatially within and across olfactory structures. Whereas strong evidence supporting that conception has been found in visual and somatosensory systems, such is not the case in the mammalian olfactory system. However, a detailed investigation of how oscillatory regimes arise during spontaneous activity and following odor presentation was found to be particularly revealing of how experience modulates the characteristics of large neuron assemblies (Freeman and Skarda, 1985). That impressive set of data focused mainly on OB activity, and very few studies have included data on oscillatory activities collected simultaneously from the OB, the PC, and the LEC (Boeijinga and Lopes da Silva, 1989; Kay and Freeman, 1998). In some recent experiments, we examined whether or not a common oscillatory regime was expressed simultaneously in the OB-PC-EC complex during odor processing. In addition, we tested whether or not odor-related oscillatory activities were modulated by internal states and conditioning.

3. Recent Findings

3.1. Effect of Associative Olfactory Learning on Synaptic Transmission

In this task, rats had to learn to discriminate between two olfactory cues in order to perform differential behavioral responses. For thirsty animals, one cue (S+) was constantly paired with a positive reward (a sucrose solution), and the other cue (S−) was paired with a negative reward (a quinine solution). Importantly, the olfactory cues were not real odors, but electrical impulses directly delivered at two different sites on the same OB (Mouly, Vigouroux, and Holley, 1985). The main advantage of using "electrical odors" is that the same bulbar electrode stimulated during the discrimination task can be reused following training to elicit EFPs in target structures. In a first series of experiments, rats learned the task within a few days. Then they were anesthetized, with a large fraction of the PC being exposed. A voltage-sensitive-dye method was used to map EFPs from 144 recording sites. One of the major findings was that learning was associated with an increase in the amplitude of the early EFP component in the posterior PC. That likely corresponded to improved synaptic efficiency at the level of afferent fibers to the PC following stimulation of bulbar electrodes involved in the acquisition of the task (Litaudon et al., 1997). Although that study suggested learning-induced plasticity restricted to the posterior part of the PC, it did not provide information on what happened in neighboring structures. In a recent study based on the same behavioral paradigm of learning "electrical odors," recordings of EFPs due to OB electrode stimulation were obtained simultaneously from the anterior piriform cortex (aPC), posterior piriform cortex (pPC), and lateral entorhinal cortex (LEC) (Figure 21.1).

Recordings were made on the day the animals mastered the learning task (at least 80% of correct choices for two successive daily training sessions) and following a 20-day period of training interruption. It is important to keep in mind that measures were not obtained during accomplishment of the task, but one hour later and several days later. Changes in signal amplitudes likely reflected

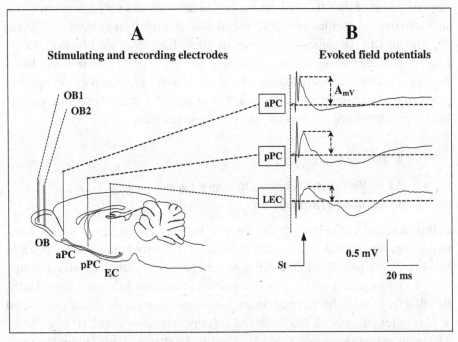

Figure 21.1. Schematic representation of the implanted electrodes and recorded signals. A: Two bipolar stimulating electrodes (OB1 and OB2) were implanted in the ventral mitral-cell layer of the OB. Three recording electrodes were implanted in the aPC, pPC, and LEC. B: Examples of the EFP signals induced at the three recording sites in response to stimulation of the OB. The amplitudes (A_{mV}) of the EFP main components are measured at each site, as indicated in the figure. St: stimulation artifact.

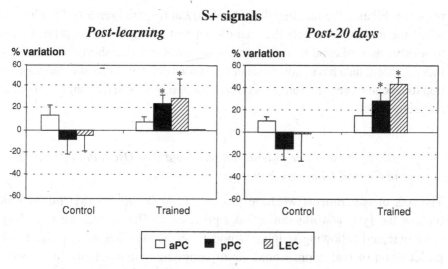

Figure 21.2. Mean (±SEM) variations in EFP amplitudes measured in control ($n = 7$) and trained ($n = 11$) animals at the four recording sites. In the absence of reinforcement, control animals received intrabulbar electrical pulses with the same characteristics (intensity, duration, number) as those received during training in the experimental group. Changes are expressed as percentage variations from the baseline signals collected before training. S+ signals: signals induced by stimulation of the OB electrode (which had been associated with delivery of sucrose in the learning task) after completion of learning (left) and after 20 days of training interruption (right). Note the absence of variation in the aPC following training and the significant increases in the pPC and the LEC (*$p < 0.05$: significant changes compared with the signals recorded before training).

long-lasting consequences of the learning process. That study confirmed that learning-induced changes in the PC developed specifically in its most posterior part. In addition, very similar increases in synaptic efficiency were revealed in the LEC. Changes observed at both levels can be considered as durable, because they developed just following learning completion and were still present or even amplified 20 days later (Figure 21.2) (Mouly et al., 2001).

Thus, to summarize, multiple-site recordings clearly showed functional dissociation between the aPC and the pPC and suggested that the pPC and the LEC have some common characteristics in the context of olfactory learning. Changes in synaptic efficiency observed at that level can be interpreted either as reflecting a specific substrate of memory traces or as reflecting more global changes. In the latter view, changes in electrical-signal amplitude observed following training could represent preferential gating of learned input. Such a gating mechanism could represent one aspect in the formation of long-term memory, which could result in rapid stimulus recognition and the appropriate behavioral response. Finally, the fact that the aPC electrical response remained unchanged following learning supports the idea of that part of the PC being related more to sensory analysis and to simple forms of nonassociative short-term memory. Indeed, recent data have shown that aPC neurons displayed pronounced habituation to repeated presentations of a given odor at intervals in the range of minutes (Wilson, 1998).

3.2. Effect of Odor Valence on Odor-induced Oscillatory Activities in Central Olfactory Pathways

The sets of experiments we have just examined do not provide information on how the system works during odor processing. That is because recordings were obtained following training. Another set of experiments examined how presentation of real odors would modify ongoing neural activity in the same olfactory structures. In freely moving rats, macroelectrodes (80 μm in diameter) simultaneously collected activities from the OB, aPC, pPC, and LEC. When the LFP signals were filtered between 1 and 300 Hz, each trace reflected the sum of extracellular current generated by cellular elements near the electrode tip. In each structure, the tip of the electrode was positioned near the layer of output cells.

As mentioned earlier, normal neuronal activity from olfactory structures developed into prominent oscillatory activity between 1 and 100 Hz. That was the case both in the absence and in the presence of odorant stimuli. Those oscillatory regimes belong to different frequency bands (theta, beta, gamma) nested together. In addition, most oscillatory regimes are expressed as short bursts of activity lasting for a few hundreds of milliseconds. In brief, oscillatory regimes are highly nonstationary and complex phenomena. In order to capture transient significant changes in signal power induced by odor presentation, signal analysis was based on time-frequency methods. The main method relied on fast-Fourier-transform sliding windows, which allowed the capture of changes in signal power with a 2–4-Hz frequency resolution and a 150–400-msec time resolution. With that technique, we had two sets of experiments aimed at identifying how the behavioral valences of different olfactory stimuli would modulate oscillatory regimes in neuron assemblies.

The first series of experiments tested the importance of an animal's motivation for food in its responses to food and non-food odors. The food odor consisted of the odor of the animal's usual diet received in its home cage for several days before the start of the experiment. The motivational state was controlled by recording an animal's responses following a period of free access to food (noted as "satiated condition" in the following) or after a 16-hour fast (noted as "food-deprived condition"). Another important aspect of the protocol was that no conditioned behavioral response was required. Indeed, odors were introduced into the recording cage from the top, and no other sensory cue announced odorant onset. Because rats had to perform no special behavioral response to the stimuli except normal exploratory sniffing, we refer to that situation as the "passive" one. Data analysis revealed several functional differences within central olfactory pathways. First, during spontaneous activity, power-spectrum analysis showed high degrees of similarity between the OB and the aPC, on one hand, and between the pPC and the LEC, on the other hand. The OB-aPC complex was dominated by high-frequency gamma (60–90 Hz) bursts of activity, and the pPC-EC complex was dominated by beta (15–40 Hz) activity (Chabaud et al., 2000). In addition, coherence analysis revealed that during spontaneous activity, the aPC was coupled more tightly to the OB than to the pPC (Chabaud et al., 1999). Thus the power and coherence analyses strongly suggested that the two parts of the PC had different functional characteristics. Another dissociation also appeared in response to food and non-food odors: A nutritional modulation of the food-odor response (difference in responding rate between satiated and food-deprived conditions) was observed at the level of the OB and the LEC. It was not expressed in the aPC and pPC. That suggests that the neuronal and neuroendocrine systems affected by changes in nutritional states exerted more pronounced effects on the OB and the LEC than on the PC. In addition, signal analysis revealed that the nutritional modulation was expressed in a beta-band oscillatory regime ranging from 15 to 30 Hz, with maximum effect near 20 Hz.

Repeated presentations of the food odor with a 1-min inter-stimulus interval showed that habituation was expressed differentially across the recorded structures, although the responses appeared as a general increase in activity in a wide 15–90-Hz frequency band. For example, over many repetitions, the rates of response declined in parallel in the OB and the LEC, remained stable in the aPC, and increased in the pPC. That was interpreted as evidence for differential expression of this simple form of short-term memory across the four recorded sites. In particular, the aPC and pPC seemed to react in very different manners, for reasons that remain to be investigated (Chabaud et al., 2000).

The data collected in the "passive" condition investigated, in an indirect manner, the effects of learning on neural responses. Indeed, the odor of the usual food did not have to be learned in a conditioning paradigm controlled experimentally, and odor presentation was not associated with a behavioral response indicating that the animal had recognized the stimulus. Thus, a second set of experiments was based on a more classic approach, referred to as the "active" condition: Rats had to learn that sampling odor A (S+) in an odor port should be followed by a "go" response, and sampling odor B (S−) should be followed by a "no-go" response. The "go" response consisted in rapidly crossing the entire length of the recording cage (60 cm) to obtain a food reward. For the "no-go" response the rats were to stay near the odor port. In that experiment, rats actively sampled the odor, and the exact time of odor onset was known. The rats had electrodes implanted in the OB, aPC, pPC, and LEC and were recorded at the very beginning of training ("beginners") and several days later once the responses had been learned ("experts"). Time-frequency analysis was performed with a 150-msec temporal resolution and a 4-Hz frequency resolution. The detailed results have been presented elsewhere (Ravel, Chabaud, and Gervais, 1999). The experiment led to several new observations. First, even for beginners, odor sampling (average duration 550 msec) was accompanied by clear-cut changes in ongoing oscillatory activities. For example, during the resting state, LFP recordings from the OB were dominated by high-frequency gamma bursts (60–90 Hz) occurring during each inspiration phase of the respiratory cycle. Such bursts have been extensively described (Freeman and Skarda, 1985) and are known to originate from the OB intrinsic circuitry. Concurrently, the activity in the beta band (15–30 Hz) was low. Active sampling of an odor was associated with a sharp decrease in rapid gamma bursts and with a significant increase in beta oscillation (Figure 21.3). The beta response in the OB corresponded, on average, to a twofold increase

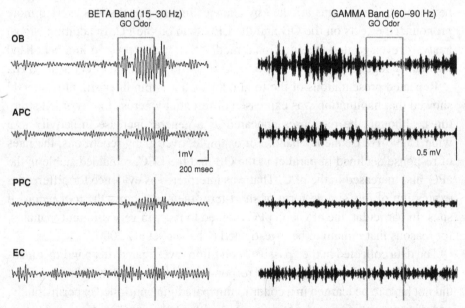

Figure 21.3. Individual example of LFP activities collected simultaneously from the OB, aPC, pPC, and LEC. The same signals are filtered in the beta band, on the left, and in the gamma band, on the right. In this example, the rat recorded in the "expert" condition sampled an odor learned as a "go" signal. Olfactory sampling, which lasted for 550 msec, was associated with obvious increases in amplitude in the beta-band oscillations for each recorded site, together with marked decreases in amplitude in the gamma-band oscillations for the OB and the aPC.

Table 21.1. *Differential effects of learning on odor-induced beta oscillations in olfactory pathways*[a]

Group	OB	aPC	pPC	LEC
Beginners	2.07 ± 0.11 (n = 41)	1.84 ± 0.1 (n = 42)	1.92 ± 0.11 (n = 37)	2.36 ± 0.87 (n = 40)
Experts	3.90 ± 0.26* (n = 177)	2.19 ± 0.08 (n = 157)	2.87 ± 0.12* (n = 177)	2.28 ± 0.07 (n = 153)

[a]The table expresses the average ratio (±SEM) of the signal amplitude during odor sampling to the signal amplitude within 500 msec before odor sampling. Data are from three rats for all odors used. The total numbers of trials are indicated in parentheses. For beginners, amplitudes varied on the order of twofold for each structure. For experts, that increased significantly for only the OB and the pPC. *$p < .001$ relative to beginners for the same structure.

in activity, which lasted for about 20% of the sampling duration, with maximum activity centered on 20 Hz. Moreover, similar responses in the beta band were observed in the aPC, pPC, and LEC, suggesting that active, brief-duration olfactory sampling transiently evokes oscillatory activity in neuron populations at a common frequency near 20 Hz within a large fraction of the central olfactory pathways. That response was found to be strongly modified following learning, and that effect was not due to changes in sampling strategy, because the durations of odor sampling, on the order of 400–600 msec, did not differ between the beginner and expert conditions. In contrast, for the experts, both the amplitude and duration of the odor-induced beta activity were enhanced.

Regarding amplitude, the average variation for experts during odor sampling reached fourfold relative to the baseline, compared with a twofold variation for beginners. In addition, that enhancement developed selectively in two structures only: the OB and pPC. Conversely, in the aPC and EC, response amplitudes remained at the same levels as in beginners (Table 21.1). Regarding the OB response duration, its values ranged from 40% to 60% of the sampling duration, instead of the 20% seen for beginners. As seen in Figure 21.3, the oscillatory beta-band response could be observed in experts on single trials. Finally, the depression in gamma 60–90-Hz activity within the OB was even more pronounced in experts than in beginners. Those data led to at least two conclusions: First, multi-site LFP recording revealed the existence of a common oscillatory regime near 20 Hz in central olfactory pathways during odor processing. In similarity to what has been found near 40 Hz in the visual system, the beta-band activity facilitated the emergence of synchronous neural assemblies in a widespread neuronal network. The extent of that synchronous oscillating network is still unknown, and we need to determine if it also spreads to other limbic structures and to neocortical areas. The cellular elements responsible for its generation also remain to be identified. Second, the effects observed following training suggest that learning improved the neural synchronization within each recorded structure.

4. Conclusion

The sets of recent data we have briefly described have revealed clear-cut functional dissociation in olfactory pathways and have led to more general considerations. Regarding functional dissociations, the most obvious one concerns the anteroposterior axis of the PC. The effects of the resting state, the rate of habituation, and the process of learning on the synaptic transmission and modulation of beta-band odor-induced activity revealed functional differences between the anterior and posterior parts of the PC. The current data suggests that the aPC is functionally more tightly linked to the OB, and the pPC is more tightly associated with the LEC.

Finally, the more we learn about the physiology of early olfactory areas studied in active animals, the more we realize the complexity of the computations performed at those levels. In a manner similar to what is observed in auditory pathways (Edeline, 1999), experience has profound effects on how neural responses to olfactory input are expressed. However, understanding their functional significance will require identifying how these changes affect or result from what happens in other structures. In other words, multi-site electrophysiological recordings in active animals can complement brain-imaging studies in humans. Looking more closely at network functioning, instead of merely structure, we predict that new data will reinforce our view that olfactory cognition cannot be considered as resulting exclusively from hippocampal or/and neocortical processing.

References

Boeijinga P H & Lopes da Silva F H (1989). Modulations of EEG Activity in the Entorhinal Cortex and Forebrain Olfactory Areas during Odour Sampling. *Brain Research* 478:257–68.

Chabaud P, Ravel N, Wilson D A, & Gervais R (1999). Functional Coupling in Rat Central Olfactory Pathways: A Coherence Analysis. *Neuroscience Letters* 273:1–4.

Chabaud P, Ravel N, Wilson D A, Mouly A M, Vigouroux M, Farget V, & Gervais R (2000). Exposure to Behaviourally Relevant Odour Reveals Differential Characteristics in Central Rat Olfactory Pathways as Studied through Oscillatory Activities. *Chemical Senses* 25:561–73.

Edeline J M (1999). Learning-induced Physiological Plasticity in the Thalamo-cortical Sensory Systems: A Critical Evaluation of Receptive Field Plasticity, Map Changes and Their Potential Mechanisms. *Progress in Neurobiology* 57:165–224.

Freeman W J (1975). *Mass Action in the Nervous System.* New York: Academic Press.

Freeman W J & Schneider W (1982). Changes in Spatial Patterns of Rabbit Olfactory EEG with Conditioning to Odors. *Psychophysiology* 19:44–56.

Freeman W J & Skarda C (1985). Spatial EEG Patterns, Nonlinear Dynamics and Perception: The Neo-Sherringhtonian View. *Brain Research Reviews* 10:147–75.

Gervais R (1993). Olfactory Processing Controlling Food and Fluid Intake. In: *Neurophysiology of Ingestion,* ed. D Booth, pp. 119–35. Oxford: Pergamon Press.

Gervais R, Holley A, & Keverne E B (1988). The Importance of Central Noradrenergic Influences on the Rat Olfactory Bulb in the Processing of Learned Olfactory Cues. *Chemical Senses* 13:3–12.

Haberly L B (1985). Neural Circuitry in Olfactory Cortex: Anatomy and Functional Implications. *Chemical Senses* 10:219–38.

Haberly L B (1998). Olfactory Cortex. In: *The Synaptic Organization of the Brain,* ed. G M Shepherd, pp. 377–416. Oxford University Press.

Jung M W, Larson J, & Lynch G (1990). Long-Term Potentiation of Monosynaptic EPSPs in Rat Piriform Cortex *in vitro. Synapse* 6:279–83.

Kanter E D & Haberly L B (1990). NMDA-dependent Induction of Long-Term Potentiation in Afferent and Association Fiber Systems of Piriform Cortex *in vitro. Brain Research* 525:175–9.

Kay L M & Freeman W J (1998). Bidirectional Processing in the Olfactory-Limbic Axis during Olfactory Behavior. *Behavioral Neuroscience* 112:541–53.

Kay L M & Laurent G (1999). Odor- and Context-dependent Modulation of Mitral Cell Activity in Behaving Rats. *Nature Neuroscience* 2:1003–9.

Litaudon P & Cattarelli M (1996). Olfactory Bulb Repetitive Stimulations Reveal Non-homogeneous Distribution of the Inhibitory Processes in the Rat Piriform Cortex. *European Journal of Neuroscience* 8:21–9.

Litaudon P, Mouly A M, Sullivan R, Gervais R, & Cattarelli M (1997). Learning-induced Changes in Piriform Cortex Evoked Activity Using Multisite Recordings with Voltage Sensitive Dye. *European Journal of Neuroscience* 9:1593–602.

Lynch G & Granger R (1989). Simulation and Analysis of a Simple Cortical Network. *Psychology: Learning and Motivation* 23:205–41.

Mori K, Nagao H, & Yoshiara Y (1999). The Olfactory Bulb: Coding and Processing of Odor Molecule Information. *Science* 286:711–15.

Morris R G M, Anderson E, Lynch G S, & Baudry M (1986). Selective Impairment of Learning and Blockade of Long-Term Potentiation by an *N*-methyl-D-aspartate Receptor Antagonist, AP5. *Nature* 319:774–6.

Mouly A M, Fort A, Ben-Boutayab N, & Gervais R (2001). Olfactory Learning Induces Differential Long-Lasting Changes in Rat Central Olfactory Pathways. *Neuroscience* 102:11–21.

Mouly A M, Kindermann U, Gervais R, & Holley A (1993). Evidence for the Involvement of the Olfactory Bulb in Consolidation Processes Associated with Long-Term Memory. *Behavioral Neuroscience* 107:451–7.

Mouly A M, Vigouroux M, & Holley A (1985). On the Ability of Rats to Discriminate between Microstimulations of the Olfactory Bulb in Different Locations. *Behavioral Brain Research* 17:47–58.

O'Doherty J, Rolls E, Francis S, Bowtell R, McGlone F, Kobal G, Renner B, & Ahne G (2000). Sensory-specific Satiety-related Olfactory Activation of the Human Orbitofrontal Cortex. *NeuroReport* 11:893–7.

Pager J (1983). Unit Responses Changing with Behavioral Outcome in the Olfactory Bulb of Unrestrained Rats. *Brain Research* 289:87–98.

Pager J, Giachetti I, Holley A, & Le Magnen J (1972). A Selective Control of Olfactory Bulb Electrical Activity in Relation to Food Deprivation and Satiety in Rats. *Physiology and Behavior* 9:573–9.

Qureshy A, Kawashima R, Imram M B, Sugiura M, Goto R, Okada K, Inoue K, Itoh M, Schormann T, Zilles K, & Fukuda H (2000). Functional Mapping of Human Brain in Olfactory Processing: A PET Study. *Journal of Neurophysiology* 84:1656–66.

Ravel N, Chabaud P, & Gervais R (1999). Learning Induces Specific Changes in Odor-evoked Oscillations in Beta and Gamma Bands in Rat Central Olfactory Pathways. Presented at a meeting of the Society for Neuroscience, Miami, 23–28 October (abstract).

Ravel N, Elaagouby A, & Gervais R (1994). Scopolamine Injection into the Olfactory Bulb Impairs Short-Term Olfactory Memory. *Behavioral Neuroscience* 108:317–24.

Roman F, Staubli U, & Lynch G (1987). Evidence for Synaptic Potentiation in a Cortical Network during Learning. *Brain Research* 418:221–6.

Royet J P, Koenig O, Gregoire M C, Cinotti L, Lavenne F, Le Bars D, Costes N, Vigouroux M, Farget V, Sicard G, Holley A, Mauguière F, Comar D, & Froment J C (1999). Functional Anatomy of Perceptual and Semantic Processing for Odors. *Journal of Cognitive Neuroscience* 11:94–109.

Royet J P, Zald D, Versace R, Costes N, Lavenne F, Koenig O, & Gervais R (2000). Emotional Responses to Pleasant and Unpleasant Olfactory, Visual, and Auditory Stimuli: A PET Study. *Journal of Neuroscience* 20:7752–9.

Savic I, Gulyas B, Larsson M, & Roland P (2000). Olfactory Functions Are Mediated by Parallel and Hierarchical Processing. *Neuron* 26:735–45.

Schoenbaum G, Chiba A, & Gallagher M (1999). Neural Encoding in Orbitofrontal Cortex and Basolateral Amydala during Olfactory Discrimination Learning. *Journal of Neuroscience* 19:1876–84.

Singer W (1999). Neural Synchrony: A Versatile Code for the Definition of Relations? *Neuron* 24:49–65.

Sobel N, Prabhakaran V, Desmond J E, Glover G H, Goodes R L, Sullivan E, & Gabrieli D E (1998). Sniffing and Smelling: Separate Subsystems in the Human Olfactory Cortex. *Nature* 392:282–6.

Sobel N, Prabhakaran V, Zhao Z, Desmond J, Glover G H, Sullivan E V, & Gabrieli J D E (2000). Time Course of Odorant-induced Activation in Human Primary Olfactory Cortex. *Journal of Neurophysiology* 83:537–51.

Stripling J S, Patneau D K, & Gramlich C A (1988). Selective Long-Term Potentiation in the Piriform Cortex. *Brain Research* 441:281–91.

Weibe S P & Staubli U (1999). Dynamic Filtering of Recognition Memory Codes in the Hippocampus. *Journal of Neuroscience* 19:10562–74.

Wilson D A (1998). Habituation of Odor Responses in the Rat Anterior Piriform Cortex. *Journal of Neurophysiology* 79:1425–40.

Wilson M A & Bower J M (1988). A Computer Simulation of Olfactory Cortex with Functional Implications for Storage and Retrieval of Olfactory Information. In: *Neural Information Processing Systems*, ed. D Anderson, pp. 114–26. New York: AIP Press.

Wood E R, Dudchenko P A, & Eichenbaum H (1999). The Global Record of Memory in Hippocampal Neuronal Activity. *Nature* 397:613–16.

Zald D H & Pardo J V (2000). Functional Neuroimaging of the Olfactory System in Humans. *International Journal of Psychophysiology* 36:165–81.

Zatore R J, Jones-Gotman M, Evans A, & Meyer E (1992). Functional Localization and Lateralization of Human Olfactory Cortex. *Nature* 360:339–40.

Zatore R, Jones-Gotman M, & Rouby C (2000). Neural Mechanisms Involved in Odor Pleasantness and Intensity Judgments. *NeuroReport* 11:2711–16.

22

Increased Taste Sensitivity by Familiarization to Novel Stimuli: Psychophysics, fMRI, and Electrophysiological Techniques Suggest Modulations at Peripheral and Central Levels

Annick Faurion, Barbara Cerf, Anne-Marie Pillias, and Nathalie Boireau

Several studies have shown that taste-aversion conditioning can modify the neural coding of taste in rodents. Chang and Scott (1984) reported that after aversive conditioning to saccharin, rats declined to drink saccharin solution, and, simultaneously, the neural code in the first relay, the nucleus of the solitary tract (NST), showed drastic changes compared with the neural code analyzed in unconditioned rats. Similarly, after aversive conditioning, c-*fos* staining showed changes in the locations of saccharin-responding neurons in the parabrachial nuclei (PBN) (Yamamoto, 1993) and in the NST (Houpt et al., 1994, 1996). Preference conditioning has been shown to produce changes in neural activation patterns in the NST (Giza et al., 1997).

In rodents, taste afferent pathways lead, on the one hand, to cortical taste areas through the NST, the PBN (the pontine taste relay), and thalamus and, on the other hand, to the amygdala, the lateral hypothalamus, and the bed nucleus of the stria terminalis (BST). In primates, the pontine taste relay is bypassed, and the NST projects directly to the parvicellular region of the thalamic ventroposteromedial nucleus (VPMpc). Efferent pathways from the amygdala, lateral hypothalamus, and BST have been traced down to the pons and the NST (Norgren, 1985). We know from a study by Mora, Rolls, and Burton (1976) that in primates, the lateral hypothalamus contains neurons responding to highly integrative information, such as the sight of a taste stimulus that a monkey likes. The rat lateral hypothalamus has been shown in operant-conditioning experiments to contain neurons that fire during electrical self-reinforcing self-stimulation.

The first part of the afferent pathway leads to cortical cognition-related projections, coding for quality and intensity of perception, whereas the second part, leading to the lateral hypothalamus and to the amygdala, is believed to be related to more subjective functions. The modifications in the neural coding of taste quality and taste intensity that can be produced by aversive conditioning may have their origins in feedback from the lateral hypothalamus to the NST and/or the pontine taste areas through efferent innervation.

Whatever the structures responsible for such modulations, it seems highly probable that modification of the neural code at the NST or PBN level should have consequences for taste coding at the thalamic and cortical levels because of the organization of successive projections (Figure 22.1). Hence, we suspected

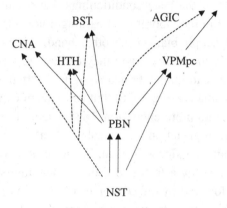

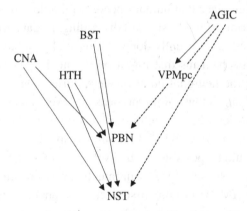

Figure 22.1. Schematic diagram of the central gustatory pathways in the rat. Top: Afferent projections to the nucleus of the solitary tract or nucleus tractus solitarius (NST), the pontine parabrachial nuclei (PBN), the parvicellular region of the thalamic ventroposteromedial nucleus (VPMpc), and the agranular insular cortex (AGIC), on the one hand, and to the hypothalamus (HTH), the central nucleus of the amygdala (CNA), and the bed nucleus of the stria terminalis (BST), on the other hand. Dashed projections from the NST probably convey visceral information; the function of those from the PBN is unknown. The dotted line from the NST to the thalamus indicates that the pontine taste relay is bypassed in the primate. Bottom: Efferent projections. Dashed arrows represent pathways not proven to arise from the AGIC. (Adapted from Norgren, 1985.)

that cortical neuronal activity would not be found to be invariant, but rather would depend on the subject's experience and on the nature of any stimulus presented. In other words, is the sensitivity to tastants simply fixed by genetically determined receptors, or can there be modulations of that sensitivity attributable to experience?

Extensive experiments in aversive conditioning cannot be carried out in humans, for obvious ethical reasons. But conditioned taste aversion can be considered as an innocuous example of the taste conditionings that occur in our daily lives and can be reproduced in the laboratory. Learning a novel taste, on the one hand, and extinction of neophobia, on the other hand, are two such situations that arise spontaneously in our lives, and they can be used as models for studying how "experience" can modulate neural signals. Often a novel food will initially produce neophobia (Rozin, 1976), but subsequent exposures to that food may turn neophobia into preference, or at least a neutral reaction. Learning is an essential function in gustation. As a matter of fact, no experiment can yield any relevant result in a single session. We have observed that supraliminal sensitivity can increase by a factor of 4 during learning studies, and detection thresholds can be lowered by a factor of 10 (A. Faurion, unpublished observations; Faurion, 1994). Our experience has shown that it is only after directed learning has taken place that humans become reliable subjects for evaluating taste intensity (Faurion, 1993). Instead of discarding the data accumulated during the learning period, as is usually done, in this study we have tried to understand the evolution mechanisms that might be revealed by such data. What actually happens, from the neuronal point of view, during such a learning period? Were there any changes that we might measure? Were any changes simply a matter of central plasticity in the central nervous system? Were there peripheral modifications involved as well? If there are modifications in taste coding with experience, it should be possible to observe correlates of such plasticity both with psychophysical sensitivity measurements and with functional magnetic-resonance imaging (fMRI) in human subjects. Peripheral modifications of sensitivity should be measurable using electrophysiological recordings of taste nerves in animals. Therefore, the studies we report here were aimed at detecting clues to changes in the neural code in cognition-related taste areas in humans, as well as changes at the chorda tympani level in hamsters, as the status of a stimulus changed from novel (and, later, preferred or neophobic) to familiar. The first experiment was designed to look for changes in sensitivity during familiarization using psychophysics. The second experiment was an fMRI study aimed at visualizing changes in the numbers of activated pixels in cortical taste areas during familiarization. The third experiment was aimed at detecting any modulation in signals from the peripheral taste receptors in hamsters during familiarization.

1. Increased Sensitivity during Familiarization to Novel Stimuli in Humans

Thirty healthy subjects, 20 to 24 years old, participated in the psychophysics experiment. They were not aware of the aim of the experiment and were not informed of their experimental performances until the end of the experiment. Five of them took part in the simultaneously ongoing neuroimaging study with fMRI. A group of 14 chemicals, not related to usual tastants, and chosen from a series of 182 chemicals previously tested in our laboratory, was submitted to preliminary testing by a different panel: Two stimuli, guanosine 5′-monophosphate (5′-GMP) and taurine, seemed to have a high probability of being judged as bad-tasting and inducing neophobia, so they were selected. Two other stimuli, D-threonine and glycyrrhizic acid, showed positive hedonic characteristics and were also selected. Those four stimuli were used for iso-intense concentration measurements. Subjects were submitted to a training period for 3 weeks with casual stimuli: quinine and sucrose. Then the familiarization experiment followed: at least 10 weeks with the four selected stimuli. Several parameters were recorded during the familiarization experiment. Iso-intense concentrations (concentrations evoking the same perceived intensity as the reference) were repetitively estimated for all four compounds during two 2-hour sessions per week. Estimates of the strength or concentrations of the solutions, here called "magnitude" estimations (at constant concentration), were repetitively recorded, and hedonic evaluations and semantic descriptions ("quantitative descriptive analysis") were given during one 90-min session per week. A food-intake questionnaire was filled out on each experimental day, and the hormonal status of female subjects was recorded. Subjects participated into three sessions per week, totaling 6 hours per week.

Stimuli were prepared every day with ultraviolet-sterilized water in sterile glassware. Tests for bacterial contamination were conducted randomly every day. The solution temperature was controlled at 0.1°C. For iso-intense concentration estimations, a forced-choice paired comparison of the stimulus with an NaCl reference solution (1.7 g/liter) was used. The subject was asked, Which is stronger? The stimulus concentration was varied according to the preceding response of the subject, following the staircase procedure of Dixon and Massey (1960), referred to as the up-and-down procedure. The dilution and distribution of solutions were computer-controlled in two home-built devices that could accommodate up to eight subjects; those devices diluted and distributed the solutions and controlled the opening and closing of the microvalve apertures. The concentration of the solution distributed to a subject depended on his/her response to the preceding paired presentation. Different up-and-down tests were intermingled so that the staircase procedure was not apparent to the subject. (The procedure was actually double-blind.) During each session, the up-and-down test was repeated four times for each stimulus, which meant that the test lasted about 30 min per stimulus (2 hours for all four stimuli), including about 30 paired presentations per stimulus. Results were calculated and printed automatically. The iso-intensity evaluations were averaged and compared from session to session for each subject individually. The effect of learning was determined by the evolution of the iso-intense concentration estimations using statistical evaluation (*t*-test). For magnitude estimation, hedonic evaluation, and quantitative descriptive analysis, three stimuli (L-valine, D-methyl mannopyranoside, naringin) were added to increase the variety of stimuli so that the subjects could not memorize their own evaluations from one session to the next. For magnitude estimation, quantitative descriptive analysis, and hedonic evaluation, solutions were also sterile and were prepared with the following: glycyrrhizic acid, 0.4 g/liter; 5'-GMP, 0.4 g/liter; D-threonine, 30 g/liter; taurine, 60 g/liter; D-methyl mannopyranoside, 20 g/liter; naringin, 0.15 g/liter; L-valine, 3 g/liter.

1.1. Hedonic Assessment for Each Stimulus

Preference evaluations (using a scale from −10 to +10) were positive at the first session for 24 or 25 stimuli out of 30, for glycyrrhizic acid (m = +4.6, s.d. 1.95), and for D-threonine (m = +2.5, s.d. 0.5). Evaluations were negative for 5'-GMP (m = −1.7, s.d. 0.80) and taurine (m = −1.6, s.d. 0.65), with a few subjects giving opposite hedonic evaluations.

1.2. Evolution of Subjects' Sensitivity during Familiarization

Figure 22.2A shows a typical familiarization curve for individual data. The estimations of iso-intense concentration diminished with successive sessions until stabilizing. Group data showed that for hedonic-positive glycyrrhizic acid and D-threonine, the iso-intense concentrations estimated with reference to NaCl (1.7 g/liter) diminished and then stabilized. For the hedonic-negative stimulus taurine, estimates of the iso-intense concentration diminished and stabilized as for glycyrrhizic acid and D-threonine. For the hedonic-negative stimulus 5'-GMP, estimates of the iso-intense concentration fluctuated but never diminished. In all cases, the group variance was much higher than the individual variances

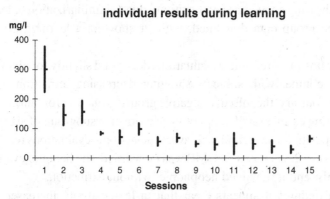

Figure 22.2A. Familiarization to a novel stimulus (glycyrrhizic acid), individual data (m ± s.e.m). Iso-intensity concentrations (ordinate) were estimated in 15 sessions (abscissa). The iso-intense concentration was determined by comparison with a solution of NaCl (1.7 g/liter) in a forced-choice paired comparison test, associated with the up-and-down presentation technique. Each point is the mean for four up-and-down tests, representing 30 min of repetitive testing for one stimulus (about 30 paired comparisons). A learning pattern is clearly observed: The decrease in iso-intense concentration indicates increased sensitivity.

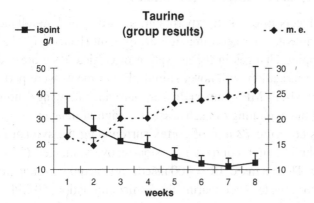

Figure 22.2B. Comparison of iso-intense concentrations and magnitude estimates. Data were averaged for the group of subjects. Iso-intensity concentrations (squares) diminished when magnitude estimates (diamonds) increased.

because of high and significant inter-individual differences in sensitivity. The estimations of magnitude and iso-intense concentration varied, as expected, as mirror images of each other for glycyrrhizic acid, D-threonine, and taurine (Figure 22.2B). Magnitude estimates showed that all three compounds seemed progressively stronger to subjects with increasing familiarization, and, simultaneously, subjects used lower and lower concentrations to match the reference intensity in iso-intense concentration estimations. The two techniques were in agreement in showing an increase in sensitivity due to familiarization or learning. In contrast, the group data fluctuated, without showing a learning effect, for 5'-GMP.

The initially positive preference evaluations decreased slightly for glycyrrhizic acid and D-threonine, with subjects showing decreasing attraction to those stimuli; on the contrary, the initially negative group evaluation for taurine tended to zero, indicating extinction of neophobia. Moreover, a subgroup of six subjects tended to give positive scores, and their averaged hedonic evaluations reached the positive value of +2. As far as 5'-GMP was concerned, the hedonic evaluations varied randomly, and it remained neophobic, without extinction.

From those findings it appears clear that taste sensitivity increased significantly with familiarization for three stimuli out of four: taurine, a hedonic-negative novel stimulus exhibiting extinction of neophobia; glycyrrhizic acid and D-threonine, two hedonic-positive stimuli. On the contrary, neither a reduction in neophobia nor any learning pattern was observed for 5'-GMP.

2. fMRI Discloses Plasticity in Cortical Areas during Familiarization to Novel Taste Stimuli

Among the 30 subjects in our first experiment, 5 participated simultaneously in the fMRI experiment. That neuroimaging experiment (Faurion et al., 1998) was aimed at mapping changes in the activation of cortical taste areas during the process of familiarization with novel stimuli. Experiments were performed using a 3-tesla whole-body MR scanner (Bruker, Germany) allowing echo-planar imaging (EPI). At the beginning of each session, a sagittal image was acquired to select 12 slices covering 72 mm of cortex surrounding the sylvian fissure. Twelve high-resolution, high-contrast inversion-recovery images (256×256 pixels, TR = 3 sec, TE = 8 msec, IR = 800 msec) were acquired for anatomical identification of the activated foci. During the 5-min stimulation, 50 EPI images per slice (64×64 pixels, FOV = 200 mm, thickness = 6 mm, TE = 40 msec, TR = 6 sec) were recorded.

Three stimuli from the psychophysics experiment were used: 5'-GMP, 0.4 g/liter (1-mM), glycyrrhizic acid, 0.4 g/liter (0.5-mM), D-threonine, 30 g/liter (250-mM). The first contact of the subjects with novel stimuli occurred on the first fMRI experimental day, prior to the psychophysical first experimental day. The second fMRI experiment was conducted after the first psychophysical experiment, by which time the subjects had experienced about 200 repetitive forced-choice pair tests in a 2-hour experimental session. The third fMRI experiment was conducted at the end of the familiarization (i.e., after 10 weeks of psychophysical experimentation).

Each subject sampled three stimuli and water, as a control, at each fMRI session. Liquids were manually pushed into the subject's mouth through microsyringes and silicon tubes (ID: 1.19 mm) at 50 microliters every 3 sec. Subjects could swallow freely, and no peculiar head-motion artifacts were recorded with such microquantities. The stimulation paradigm consisted in two ON (30-sec) and OFF (90-sec) periods after a first 60-sec OFF period of reference using water. For the wate-control experiment, the same ON and OFF periods were alternated, water being sent to the subject during ON and OFF periods, through two different syringes, just as tastant and water were sent during actual taste experiments. The perceived intensity was continuously measured throughout the fMRI experiment. Subjects used a linear potentiometer and matched the intensity of their perceptions by the distance between the thumb and forefinger according to the "finger-span" method (Berglund, Berglund, and Lindvall, 1978). The resulting plot of time versus perceived intensity was AD-converted and used for processing fMRI data by correlating the MR signal for each pixel with the time–intensity profile. The time–intensity profile obtained with the finger-span technique proved to be a powerful template to detect brain activation, particularly adapted for the case of the generally slowly rising taste responses, with no bias introduced relative to finger movement (Cerf et al., 1996a,b; Van de Moortele et al., 1997).

Images were processed using custom IDL software (Le Bihan et al., 1993) (Interactive Data Language, Research System Inc., Boulder, CO). Activations were calculated for each pixel based on the correlation coefficient between the MR signal and the perception profile recorded during the experiment. Pixels with $r > 0.4$ and belonging to clusters of at least two pixels were selected. Clusters with $r > 0.4$ and $p < .001$ (Student t-test) were finally considered as activated. After pixels had been selected, the regions in which they were found were identified with Duvernoy and Talairach atlases; the Voxtool 3D software was used for localization of sulci at the surface of the brain.

Taste activations (Figure 22.3) were found bilaterally, surrounding and buried in the sylvian fissure (Cerf et al., 1996a,b, 1997, 1998, 1999; Cerf, 1998; Faurion et al., 1998, 1999; Cerf-Ducastel et al., 2001). The upper part of the insula, the frontal operculum, the feet of pre-central and post-central gyri (rolandic operculum), and the temporal operculum were usually activated – hereafter collectively referred to as the peri-insular area. The locations of the detected clusters of activated pixels were in agreement with electrophysiological studies performed in monkeys (Patton and Ruch, 1946; Burton and Benjamin, 1971; reviewed by Ogawa, 1994; Norgren, 1995) and clinical observations gathered in humans (reviewed by Norgren, 1990). Positron-emission tomography (Kinomura et al., 1994) has revealed taste-related activations in the insula and temporal lobe,

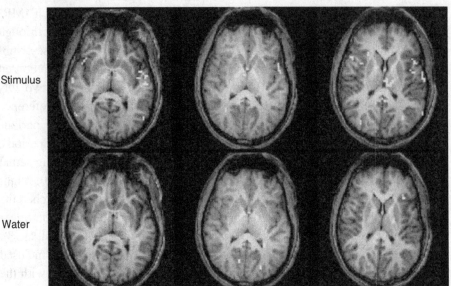

Figure 22.3. Examples of activation pixels observed for taste stimulation by FMRI: horizontal anatomical sections presented according to radiological orientation (the subject's left side is at the right in the figure). Top: Taste stimulation. Bottom: Control presentation of water. Activation was found in and around the insular cortex: anterior insula, frontal operculum, superior temporal gyrus (opercular part), bordering the sylvian fissure and the inferior parts of pre- and post-central gyri. Other activation foci were observed in the temporal lobe, frontal lobe, medio dorsal thalamus, anterior cingulate gyrus, angular gyrus, and supramarginal gyrus.

as further confirmed by Small et al. (1999). Magnetoencephalographic studies (Kobayakawa et al., 1996, 1999; Murayama et al., 1996) have shown that some dipoles located in the insula exhibit the shortest latencies. That finding supports the hypothesis that regions of the insula or peri-insular area should receive the primary sensory projection (i.e., a direct projection from the thalamus). More recently, Barry et al. (2000) found insular activation with fMRI using "electric taste" as a stimulus. In addition, we have recorded activations in the anterior cingulate gyrus, the centromedial thalamus, and other areas related to emotional or cognitive processes.

We observed that the percentages of activated pixels in the peri-insular area increased or decreased with familiarization and hence looked for psychophysical correlates of those modifications. We categorized experiments (one experiment being one subject stimulated with one stimulus) according to the evolution of the hedonic assessments. We found (Figure 22.4) that an experiment that at first yielded a low hedonic evaluation that increased with familiarization also showed

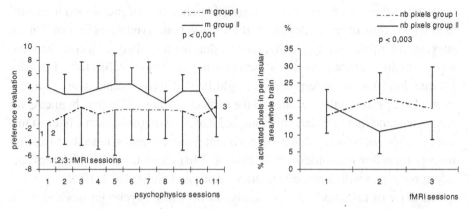

Figure 22.4. Effects of familiarization on the evolution of hedonic assessments and the evolution of the percentages of activation pixels in the peri-insular region. Left: Preference data for 5 subjects and 3 stimuli (from experiment I). Experiments are grouped according to increasing (group I) or decreasing (group II) hedonic assessments during the familiarization experiment. The difference between the two groups of experiments was highly significant ($p < .001$, χ^2). Right: fMRI data for the same two groups (groups I and II). The evolutions of the percentages of activation pixels in the peri-insular areas were significantly different for the two groups ($p < .003$, χ^2). High hedonic values corresponded to high percentages of activation pixels in the peri-insular region, which decreased in parallel with hedonic assessments during the familiarization. Low hedonic values corresponded to low percentages of pixels in the peri-insular region, increasing in parallel with the hedonic assessments during familiarization.

a low percentage of activated pixels in the peri-insular region, which increased with familiarization. Conversely, experiments yielding higher hedonic evaluations that diminished with familiarization also showed higher percentages of activated pixels in the peri-insular area, which diminished with familiarization. The differences in activated pixels occurred between the first and second fMRI sessions, separated by only one psychophysical session, and were highly significant ($p < .003$, χ^2), and so were the differences between corresponding hedonic evaluations ($p < .001$, χ^2). Subjects needed only one session for learning before showing significant differences in fMRI results.

In conclusion, in the peri-insular area, the percentages of activated pixels varied with familiarization, depending on the evolution of hedonic assessments during familiarization. If the dislike for the stimulus disappeared, the number of pixels increased. When the positive hedonic score decreased, a phenomenon we interpreted in terms of loss of interest in the repeated stimulus, the number of activated pixels also diminished. Simultaneously, it should be remembered that the perceived intensity was statistically increased for all subjects when the stimulus became more familiar, whether it be preferred or recovering from neophobia.

Whether these findings relate to a possible encoding of preference intermingled with coding of taste intensity in the areas we observed is not known, but we suggest, for further study, the hypothesis that the so-called subjective hedonic aspect might contribute to and structure putatively more "objective" intensity information. Another interpretation might be that those quantitative modifications in the pixel numbers related to the subject's "interest" in the chemical. We can understand that familiarization might interfere with the subject's interest in a tastant. It should be noted that the experimental procedures included no ingestion (except for a few microliters in the case of fMRI experiments), which prevents any comparison with a reward situation (Rolls and Rolls, 1977, 1982).

Plasticity of taste coding was clearly shown using psychophysics and brain imaging cooperatively. We saw that sensory information could actually be modified by the previous experience of the subject, and we saw a correlate of the subject's experience of modulating both the perceived intensity and the neural code reflected in the relative numbers of activated pixels in the cortical peri-insular area observed with fMRI. Those findings correlate with earlier results showing modification of neural taste coding with aversive or preference conditioning at the hindbrain level. Indeed, the cortical taste message does not seem to be invariant, but rather subject to modulation in relation with the subject's experience. As has been well documented in the earlier literature, the effects seen in both of the first two synaptic relays have consequences for coding in the higher forebrain structures. Apart from aversive conditioning, which is seldom encountered, many events in everyday life do act as mild conditionings that can eventually modulate and shape our taste information.

3. Peripheral Electrophysiology in the Hamster: Modulation of Sensitivity with Familiarization at the Receptor Level

After our demonstration of central plasticity, the question naturally arose whether the peripheral chemoreceptor equipment is invariant and genetically determined or possibly is adaptable to environmental situations. An approach to that question is possible because of the availability of animal models. Our earlier recordings from the chorda tympani nerve using a series of organic stimuli involved animals that had been familiarized to the stimuli before our experiments, for better reproducibility of responses (Faurion and Courchay, 1990; Faurion, 1993). This third experiment (Boireau, Pillias, and Faurion, 1999, in press) was designed to detect any differences in the sensitivity of taste chemoreceptions at the peripheral level. We examined the reproducibility of chorda tympani responses over time, looking for eventual modulation of peripheral sensitivity following familiarization to novel stimuli. Several groups of hamsters were pre-exposed to various novel tastants, and their responses were compared with those of control, hamsters without pre-exposure.

Pre-exposure consisted in overnight two-bottle choice tests in the home cage lasting at least two weeks (23 nights ± 7.5). One group of hamsters was exposed to dulcin (5-mM), a sucrose-like stimulus (Dul group), one group was exposed to potassium glutamate (50-mM) (KGlu group), and one group was exposed to 5′-GMP (1-mM) (5′-GMP group). After the familiarization period, animals were anesthetized, and the right chorda tympani (CT) nerve was monitored without being cut, to prevent an eventual effect of deafferentation on the sensitivity of the peripheral receptor system (Oakley, Jones, and Hosley, 1979; Oakley, Chu, and Jones, 1981).

Seventeen stimuli were chosen to be used in evaluating any eventual differences in sensitivity to chemicals among the various familiarized and control groups. The stimuli were presented six times. The chemicals were dissolved in ultraviolet-sterilized tapwater and frozen at −24°C. Prior to the experiment, samples were equilibrated at controlled room temperature to avoid thermal stimulation of the tongue. A continuous water flow, at a rate of 40 ml/min, was applied to the animal's tongue, and stimuli were applied for 12 sec at intervals of 75 sec. The 17 stimuli were randomly ordered, and a different random order was used for each of the six presentations, and 102 responses were recorded in less than 2.5 hours. For each hamster and each stimulus, the percentage increase in response amplitude was calculated by comparison to the first stimulation with that stimulus. The individual percentage increases were averaged in each group (m ± SD) for each stimulus and checked for significance with a paired Student *t*-test for each stimulus in each group and with ANOVA (Fischer LSD) among the five hamster groups.

Significant increases in CT response amplitudes between the first and the sixth presentations were observed for 15 of 17 stimuli in the control group, with a mean percentage increase varying from 16% ± 24% for D-threonine to 109% ± 65% for KGlu (50-mM).

For animals familiarized to 5′-GMP, no increases in CT response amplitudes to 5′-GMP were observed with repetition of stimulation (0% ± 22%), as compared with groups pre-exposed to other stimuli, and with the control group ($p < .05$; Fischer LSD). CT response amplitudes to KGlu were not increased in the group familiarized to KGlu (50-mM) nor in the group familiarized to 5′-GMP. The effect of familiarization to 5′-GMP generalized to KGlu. In the case of familiarization to dulcin, a different effect was observed. The CT response amplitudes to the conditioning stimulus dulcin increased as much in the Dul group as in the control group during acute recording. However, the two-week familiarization to dulcin had effects on the amplitudes of responses recorded in the first series of stimulations, for they were significantly higher than in the control group ($p < .05$).

From this experiment, it is clear that familiarization to novel stimuli induces modifications in quantitative peripheral responses. Not only plasticity in brain areas but also peripheral adaptation may be responsible for the objective increases in sensitivity measured in human subjects in our first experiment in psychophysics. Such observations may reflect enhancement of specific gene expressions in taste cells. In the liver, metabolites like glucose can influence the regulation of specific gene expressions by receptors, an effect that is stimulus- and concentration-dependent (Foufelle, Girard, and Ferre, 1998). The findings in our study suggest a possible induction of supplementary synthesis of taste receptor protein. Organic tastants are detected by G-protein-coupled receptors on the apical surface of a taste cell (Striem et al., 1989; Bernhardt et al., 1996; Kinnamon and Margolskee, 1996; Uchida and Sato, 1997; Herness and Gilbertson, 1999). Either supplementary receptors might be induced or the transductional coupling might be enhanced.

Immediate early genes such as c-*fos*, C-*jun*, and so forth, could be induced during the repetitive stimulation and during familiarization. Known mechanisms coupling neuronal activity and intracellular biochemical processes leading to gene expression include Ca^{2+} influx (Finkbeiner and Greenberg, 1998), neurotropic factors, and membrane-depolarizing agents. All these conditions coalesce during the taste-cell response, along with the presence of growth factors (Nosrat et al., 1996, 1997; Oakley et al., 1998; Cooper and Oakley, 1998). Greenberg, Thompson, and Sheng (1992) showed that different pathways activate distinct subsets of immediate early genes, and proteins encoded by immediate early genes may subserve different functions – among them, long-term adaptive responses. We suggest that the observed modulations of the increases in CT response amplitudes might involve such immediate early genes in long-term adaptive transcriptional responses.

4. Conclusion

Plasticity in cortical taste areas might be responsible for increases in perceived intensity, which can be measured in human subjects during learning. There is a hint that this increase in sensitivity might interfere with preference for the stimulus or some relevant aspect of the stimulus. Our results are in accordance with the idea that the taste neuronal code may be modifiable by learning, as suggested for the hindbrain levels by experiments on aversive and preference conditionings by the groups of Scott, Houpt, and Yamamoto. Moreover, we suggest that the plasticity might extend to every level of the central nervous system, in accordance with the study of Montag-Sallaz et al. (1999) on rodents, using novel stimuli. Moreover, our study showed increases in CT response amplitudes to novel stimuli with repetitive stimulation, effects that could be reduced or suppressed after a familiarization period. Not only brain plasticity but also supplementary receptors or facilitation of transduction coupling, induced by stimulation, may modify individual sensitivity after familiarization to a novel stimulus. It is clear that the multiple small conditionings of everyday life are potential modulating agents for taste sensitivity, either through plastic neural coding or through peripheral modulation of receptor sensitivity, or both.

Acknowledgments

We gratefully acknowledge financial support from the *Aliment Demain* program (Ministry of Agriculture), in cooperation with Danone and Orsan, for our experiment 1 and from GIS (CNRS-CEA) for our experiment 2.

References

Barry M A, Gatenby J C, Zeigler J D, & Gore J C (2000). Cortical Activity Evoked by Focal Electric-Taste Stimuli. Presented at the ISOT/ECRO Congress, Brighton, 20–24 July (abstract).

Berglund U, Berglund B, & Lindvall T (1978). Separate and Joint Scaling of Perceived Odor Intensity of *n*-Butanol and Hydrogen Sulfide. *Perception and Psychophysics* 23:313–20.

Bernhardt S J, Naim M, Zehavi U, & Lindemann B (1996). Changes in IP3 and Cytosolic Ca^{2+} in Response to Sugars and Non-Sugar Sweeteners in Transduction of Sweet Taste in the Rat. *Journal of Physiology* 490:325–36.

Boireau N, Pillias A M, & Faurion A (1999). Modulation of Chorda Tympani Sensitivity. *Chemical Senses* 24:53 (abstract).

Boireau N, Pillias A M, & Faurion A (in press). Modulation of Chorda Tympani Sensitivity with Repeated Exposure to Novel Tastants.

Burton H & Benjamin R M (1971). Central Projections of the Gustatory System. In: *Handbook of Sensory Physiology, Chemical Senses 2, Taste*, ed. L M Beidler, pp. 148–63. Berlin: Springer-Verlag.

Cerf B (1998). Exploration par IRM fonctionnelle des aires corticales impliquées dans la perception gustative chez l'homme, Thèse, Paris VII, Décembre 1998.

Cerf B, Faurion A, MacLeod P, Van de Moortele P F, & Le Bihan D (1996a). Functional MRI Study of Human Gustatory Cortex. *NeuroImage* 3:342.

Cerf B, MacLeod P, Van de Moortele P F, Le Bihan D, & Faurion A (1997). Functional MRI Study of Human Gustatory Cortex: An Insular Secondary Projection Related to Handedness. *NeuroImage* 5:200.

Cerf B, Van de Moortele P F, Giacomini E, MacLeod P, Faurion A, & Le Bihan D (1996b). Correlation of Perception to Temporal Variations of fMRI Signal: A Taste Study. Presented at the Fourth Scientific Meeting of the International Society for Magnetic Resonance in Medicine (ISMRM), New York.

Cerf B, Van de Moortele P F, Le Bihan D, & Faurion A (1999). Cortical Activation Related to Both Gustatory and Lingual Somatic Stimulation in the Human: A fMRI Study. *Chemical Senses* 24:58.

Cerf B, Van de Moortele P F, Pillias A M, MacLeod P, Le Bihan D, & Faurion A (1998). Plasticity of Cortical Taste Responses in the Human: A Joint Psychophysical and fMRI Study. *NeuroImage* 7:442.

Cerf-Ducastel B, Van de Moortele P F, MacLeod P, Le Bihan D, & Faurion A (2001). Interaction of Gustatory and Lingual Somatosensory Perceptions at the Cortical Level in the Human: A Functional Magnetic Resonance Imaging Study. *Chemical Senses* 26:371–83.

Chang F C T & Scott T R (1984). Conditioned Taste Aversions Modify Neural Responses in the Rat Nucleus Tractus Solitarius. *Journal Neuroscience* 4:1850–62.

Cooper D & Oakley B (1998). Functional Redundancy and Gustatory Development in BDNF Null Mutant Mice. *Developmental Brain Research* 105:79–84.

Dixon W J & Massey F (1960). Introduction to Statistical Analysis. In: *Sensitivity Experiments*, pp. 377–94. New York: McGraw-Hill.

Faurion A (1993). The Physiology of Sweet Taste and Molecular Receptors. In: *Sweet-Taste Chemoreception*, ed. M Mathlouthi, J Kanters, & G G Birch, pp. 291–317. London: Elsevier.

Faurion A (1994). Structure and Dimension of the Taste Sensory Space: Central and Peripheral Data. In: *Olfaction and Taste XI*, ed. K Kurihara, N Suzuki, & H Ogawa, pp. 301–4. Tokyo: Springer-Verlag.

Faurion A, Cerf B, Le Bihan D, & Pillias A M (1998). fMRI Study of Taste Cortical Areas in Humans (Activations Observed in Relation to the Hedonic and Semantic Status of the Stimulus). *Annals of the New York Academy of Sciences* 585:535–45.

Faurion A, Cerf B, Van de Moortele P F, Lobel E, MacLeod P, & Le Bihan D (1999). Human Taste Cortical Areas Studied with fMRI: Evidence of Functional Lateralization Related to Handedness. *Neuroscience Letters* 277:189–92.

Faurion A & Courchay C (1990). Taste as a Highly Discriminative System: A Hamster Intrapapillar Single Unit Study with 18 Compounds. *Brain Research* 512:317–32.

Finkbeiner S & Greenberg M E (1998). Ca^{2+} Channel-regulated Neuronal Gene Expression. *Journal of Neurobiology* 37:171–89.

Foufelle F, Girard J, & Ferre P (1998). Glucose Regulation of Gene Expression. *Current Opinion in Clinical Nutrition and Metabolic Care* 1:323–8.

Giza B K, Ackroff K, McCaughey S A, Sclafani A, & Scott T R (1997). Preference Conditioning Alters Taste Responses in the Nucleus of the Solitary Tract of the Rat. *American Journal of Physiology* 273:1230–40.

Greenberg M E, Thompson M A, & Sheng M (1992). Calcium Regulation of Immediate Early Gene Transcription. *Journal of Physiology* 86:99–108.

Herness M S & Gilbertson T A (1999). Cellular Mechanisms of Taste Transduction. *Annual Reviews of Physiology* 61:873–900.

Houpt T A, Philopena J M, Joh T H, & Smith G P (1996). C-*fos* Induction in the Rat Nucleus of the Solitary Tract by Intraoral Quinine Infusion Depends on Prior Contingent Pairing of Quinine and Lithium Chloride. *Physiology and Behavior* 60:1535–41.

Houpt T A, Philopena J M, Wessel T C, Joh T H, & Smith G P (1994). Increased c-*fos* Expression in Nucleus of the Solitary Tract Correlated with Conditioned Taste Aversion to Sucrose in Rats. *Neuroscience Letters* 172:1–5.

Kinnamon S C & Margolskee R F (1996). Mechanisms of Taste Transduction. *Current Opinion in Neurobiology* 6:506–13.

Kinomura S, Kawashima R, Yamada K, Ono S, & Itoh M (1994). Functional Anatomy of Taste Perception in the Human Brain Studied with Positron Emission Tomography. *Brain Research* 659:263–6.

Kobayakawa T, Endo H, Ayabe-Kanamura S, Kumagai T, Yamaguchi Y, Kikuchi Y, Takeda T, Saito S, & Ogawa H (1996). The Primary Gustatory Area in Human Cerebral Cortex Studied by Magnetoencephalography. *Neuroscience Letters* 212:155–8.

Kobayakawa T, Ogawa H, Kaneda H, Ayabe-Kanamura S, Endo H, & Saito S (1999). Spatio temporal Analysis of Cortical Activity Evoked by Gustatory Stimulation in Humans. *Chemical Senses* 24:201–10.

Le Bihan D, Jezzard P, Turner R, Cuenod C A, Pannier L, & Prinster A (1993). Practical Problems and Limitations in Using Z-maps for Processing of Brain Function MR Images. In: *Abstracts of the 12th Annual Meeting of the Society of Magnetic Resonance in Medicine*, p. 11. New York: SMRM.

Montag-Sallaz M, Welzl H, Kuhl D, Montag D, & Schachner M (1999). Novelty-induced Increased Expression of Immediate-Early Genes c-*fos* and *arg* 3.1 in the Mouse Brain. *Journal of Neurobiology* 38:234–46.

Mora F, Rolls E T, & Burton M J (1976). Modulation during Learning of the Responses of Neurones in the Lateral Hypothalamus to the Sight of Food. *Experimental Neurology* 53:508–19.

Murayama N, Nakasato N, Hatanaka K, Fujita S, Igasaki T, Kanno A, & Yoshimoto T (1996). Gustatory Evoked Magnetic Fields in Humans. *Neuroscience Letters* 210:121–3.

Norgren R (1985). The Sense of Taste and the Study of Ingestion. In: *Taste, Olfaction and the Central Nervous System*, vol. 10, ed. D W Pfaff, pp. 233–49. New York: Rockefeller University Press.

Norgren R (1990). Gustatory System. In: *The Human Nervous System*, vol. 25, ed. G Paxinos, pp. 845–61. San Diego: Academic Press.

Norgren R (1995). Gustatory System. In: *The Rat Nervous System*, vol. 29, ed. G Paxinos, pp. 751–71. San Diego: Academic Press.

Nosrat C A, Blomlöf J, Elshamy W M, Ernfors P, & Olson L (1997). Lingual Deficits in BDNF and NT3 Mutant Mice Leading to Gustatory and Somatosensory Disturbances, Respectively. *Development* 124:1333–42.

Nosrat C A, Ebendal T, & Olson L (1996). Differential Expression of Brain-derived Neurotrophic Factor and Neurotrophin 3MRA in Lingual Papillae and Taste Buds Indicates Roles in Gustatory and Somatosensory Innervation. *Journal of Comparative Neurology* 376:587–602.

Oakley B, Brandemihl A, Cooper D, Lau D, Lawton A, & Zhang C (1998). The Morphogenesis of Mouse Vallate Gustatory Epithelium and Taste Buds Requires BDNF-dependent Taste Development. *Brain Research* 105:85–96.

Oakley B, Chu J S, & Jones L B (1981). Axonal Transport Maintains Taste Responses. *Brain Research* 221:289–98.

Oakley B, Jones L B, & Hosley M A (1979). Decline of IXth Nerve Taste Responses following Nerve Transection. *Chemical Senses and Flavour* 4:287–99.

Ogawa H (1994). Gustatory Cortex of Primates: Anatomy and Physiology. *Neuroscience Research* 20:1–13.

Patton H D & Ruch T C (1946). The Relation of the Deep Opercular Cortex to Taste. *Federation Proceedings* 5:89–90.

Rolls E T & Rolls B J (1977). Activity of Neurons in Sensory, Hypothalamic and Motor Areas during Feeding in the Monkey. In: *Food Intake and Chemical Senses*, ed. Y Katsuki et al., pp. 525–49. University of Tokyo Press.

Rolls E T & Rolls B J (1982). Brain Mechanisms Involved in Feeding. In: *Psychobiology of Human Food Selection*, ed. L Barker, pp. 33–62. Westport, CT: AVI Publishing.

Rozin P (1976). The Selection of Food by Rats, Humans and Other Animals. In: *Advances in the Study of Behavior*, ed. J S Rosenblatt et al., pp. 21–76. New York: Academic Press.

Small D M, Zald D H, Jones-Gotman M, Zatorre R J, Pardo J V, Frey S, & Petrides M (1999). Human Cortical Gustatory Areas: A Review of Functional Neuroimaging Data. *NeuroReport* 10:7–14.

Striem B J, Pace U, Zehavi U, Naim M, & Lancet D (1989). Sweet Tastants Stimulate Adenylate Cyclase Coupled to GTP-Binding. *Biochemical Journal* 260:121–6.

Uchida Y & Sato T (1997). Changes in Outward K^+ Currents in Response to Two Types of Sweeteners in Sweet Taste Transduction of Gerbil Taste Cells. *Chemical Senses* 22:163–9.

Van de Moortele PF, Cerf B, Lobel E, Paradis AL, Faurion A, & Le Bihan D (1997). Latencies in fMRI Time-Series: Effect of Slice Acquisition Order and Perception. NMR. *Biomedicine* 10:230–6.

Yamamoto T (1993). Neural Mechanisms of Taste Aversion Learning. *Neuroscience Research* 16:181–5.

23

The Cortical Representation of Taste and Smell

Edmund T. Rolls

The aims of this chapter are to describe the rules that appear to govern the cortical processing of taste and smell, how taste and smell inputs combine with each other to form flavor, how visual and oral somatosensory inputs also converge with taste and smell, and how hunger affects the representations in different cortical areas. Particular attention is paid to investigations in a non-human primate, the macaque, because they have provided fundamental evidence relevant to understanding information processing in the same areas in humans, and to neuroimaging studies in humans to complement the primate studies. A broad perspective on the brain processing involved in emotion and motivation is provided elsewhere (Rolls, 1999a).

1. Taste Pathways in Primates

A diagram of the taste and related pathways in primates is shown in Figure 23.1. Of particular interest is that in primates there is a direct projection from the rostral part of the nucleus of the solitary tract (NTS) to the taste thalamus and thus to the primary taste cortex in the frontal operculum and adjoining insula (Figure 23.2), with no pontine taste area and associated subcortical projections as in rodents (Norgren, 1984; Pritchard et al., 1986). The emphasis on cortical processing of taste in primates appears to be related to the extensive development of the cerebral cortex in primates and to the advantage of using similar cortical analyses of inputs from every sensory modality before the analyzed representations from all modalities are brought together in multimodal regions, as will be documented here.

A secondary cortical taste area in primates was discovered by Rolls, Yaxley, and Sienkiewicz (1990) in the caudolateral orbitofrontal cortex, extending several millimeters in front of the primary taste cortex (Figure 23.2). Injections of horseradish peroxidase in that region produced labeled cell bodies in the primary taste cortex (Baylis, Rolls, and Baylis, 1994).

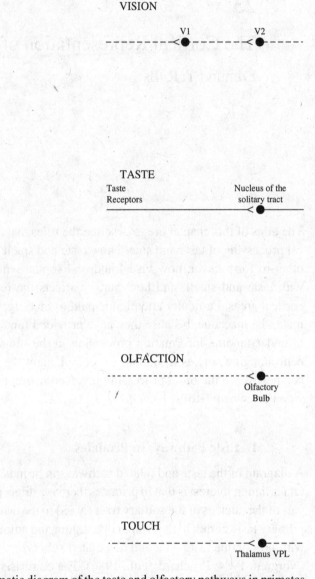

Figure 23.1. Schematic diagram of the taste and olfactory pathways in primates showing how they converge with each other and with visual pathways. The gate functions refer to the finding that the responses of taste neurons in the orbitofrontal cortex and the lateral hypothalamus are modulated by hunger. VPMpc, parvicellular region of the ventraoposteromedial thalamic nucleus; V1, V2, V4, visual cortical areas.

2. Taste Processing in the NTS and the Taste Thalamus

Taste neurons have been found, and their responses analyzed, in the rostral part of the NTS in macaque monkeys (Scott et al., 1986a). It was found that different

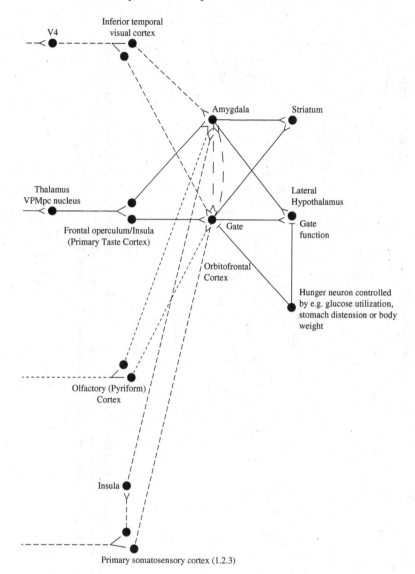

Figure 23.1. (*cont.*)

neurons responded best to various stimuli – orally administered glucose, NaCl, HCl (sour), and quinine HCl (bitter) – but the tuning of the neurons was in most cases broad, in that, for example, 84% of the neurons had at least some response to three or four of these four prototypical taste stimuli. Pritchard, Hamilton, and Norgren (1989) were able to confirm that the parvicellular division of the ventroposteromedial (VPMpc) nucleus of the thalamus is the relay nucleus for thalamic taste in primates: In macaque monkeys, single neurons in the VPMpc responded to taste stimuli.

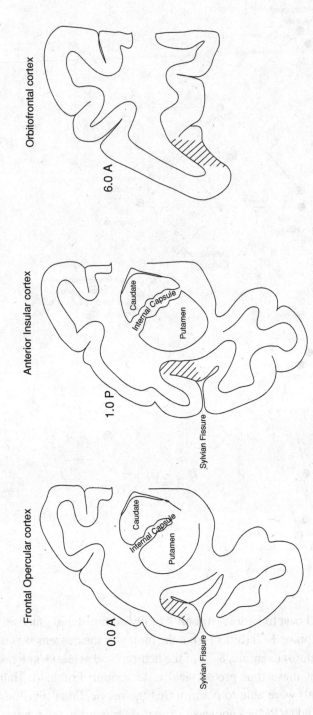

Figure 23.2. Coronal sections to show the locations of the primary taste cortices in the macaque in the frontal operculum and rostral insula and the location of the secondary taste cortex in the caudolateral orbitofrontal cortex. The coordinates are in millimeters anterior (A) or posterior (P) to the sphenoid (Aggleton and Passingham, 1981).

3. Gustatory Responses in the Primary Taste Cortex

Taste neurons in the frontal opercular cortex and rostral insular cortex in the cynomolgus macaque monkey, *Macaca fascicularis* (Figure 23.2) were found to be more specifically tuned to the prototypical stimuli (glucose, NaCl, HCl, and quinine HCl) than were neurons recorded in the same monkeys in the NTS (Scott et al., 1986a,b; Yaxley, Rolls, and Sienkiewicz, 1990). With the use of glucose and NaCl, a corresponding area has been demonstrated by fMRI investigations in humans (e.g., Francis et al., 1999; O'Doherty et al., 2001).

4. Gustatory Responses in the Secondary Cortical Taste Area: Caudolateral Orbitofrontal Cortex

In a study of the role of the orbitofrontal cortex in learning, Thorpe, Rolls, and Maddison (1983) found a small proportion of neurons (7.9%) showing gustatory responses in the main part of the orbitofrontal cortex. In recordings made from 3,120 single neurons, Rolls et al. (1990) showed that to be a secondary cortical taste area in the dysgranular field of the orbitofrontal cortex (Ofdg) (Figure 23.2), situated anterior to the primary taste cortical area. Quite sharp tuning to several taste stimuli (glucose, NaCl, HCl, quinine HCl, water, blackcurrant juice) was evident in many cases, in that some of the neurons responded primarily to one tastant. That tuning was much finer than that seen for neurons in the NTS in the monkey, and finer than that for neurons in the primary frontal opercular and the insular taste cortices. With the use of glucose and NaCl, a corresponding area in the medial orbitofrontal cortex close to the subgenual cingulate area has been demonstrated by fMRI investigations in humans (e.g., Francis et al., 1999; O'Doherty et al., 2001).

In the secondary taste cortex and also in the primary taste cortex, populations of neurons were found that responded to the taste stimuli monosodium L-glutamate (MSG), glutamic acid, and inosine 5'-monophosphate (IMP), which are tastants that produce the protein or "umami" taste (Baylis and Rolls, 1991; Rolls et al., 1996a; Rolls, 2000a). Across those populations of neurons, the responsiveness to glutamate was poorly correlated with the responsiveness to NaCl, so the representation of glutamate was clearly different from that of NaCl. Further, the representation of glutamate was shown to be approximately as different from the representation of each of the other four tastants as they were from each other, as shown by multidimensional scaling and cluster analysis. It was concluded that in primate taste cortical areas, the taste of glutamate, which produces the umami taste in humans, is approximately as well represented as are the tastes produced by glucose (sweet), NaCl (salty), HCl (sour), and quinine HCl (bitter). Thus the neurophysiological evidence in primates indicates that there is a representation

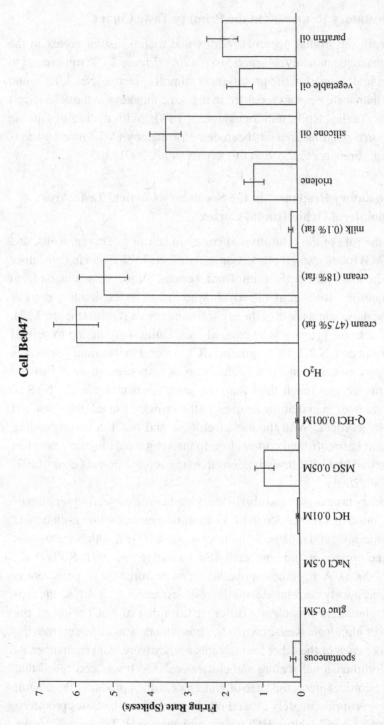

Figure 23.3. Responses of a primate neuron in the orbitofrontal cortex to the texture of fat in the mouth. The cell (Be047) increased its firing rate to cream (double and single cream), and responded to texture rather than to the chemical structure of the fat, in that it also responded to 0.5 ml of silicone oil [Si(CH$_3$)$_2$O$_n$] or paraffin oil (hydrocarbon). The cell also had a taste input, in that it had a consistent but small response to umami taste (monosodium glutamate, MSG); gluc, glucose; NaCl, salt; HCl, sour; Q-HCl, quinine, bitter. The spontaneous firing rate of the cell is also shown (data of H. Critchley, E. T. Rolls, A. D. Browning, and I. Hernadi).

of umami flavor in the cortical areas that can be distinguished from those for the prototypical tastants sweet, salt, bitter, and sour, and in that respect umami can be considered as a fifth prototypical taste. That representation probably is important for the taste produced by proteins, and to complement those findings, recently evidence has started to accumulate that there may be taste receptors on the tongue specialized for umami taste (Chaudhari et al., 1996; Chaudhari and Roper, 1998).

5. The Mouth Feel of Fat

Texture monitored in the mouth is an important indicator of whether or not fat is present in a food, which is important not only as a high-value energy source but also as a potential source of essential fatty acids. In the orbitofrontal cortex, Rolls et al. (1999) have found a population of neurons that respond when fat is in the mouth. An example of such a neuron is shown in Figure 23.3, illustrating that information about fat as well as information about taste can converge onto a given neuron in that region. It was shown that such a neuron could respond to taste, because its firing rates were significantly different for the tastants sweet, salt, bitter, and sour. However, its response to fat in the mouth was larger. The fat-related responses of these neurons are produced at least in part by the texture of the food, rather than by chemical receptors sensitive to certain chemicals, in that such neurons typically respond not only to foods such as cream and milk containing fat but also to paraffin oil (which is a pure hydrocarbon) and to silicone oil $[Si(CH_3)_2O_n]$. However, some of the fat-related neurons do have convergent inputs from the chemical senses, in that in addition to taste inputs, some of these neurons respond to the odor associated with a fat, such as the odor of cream (Rolls et al., 1999). Feeding to satiety with fat (e.g., cream) decreases the responses of these neurons to zero for the food eaten to satiety, but if the neuron receives a taste input from, for example, glucose taste, that is not decreased by feeding to satiety with cream. Thus there is a representation of the macronutrient fat in that brain area, and the activation produced by fat is reduced by eating fat to satiety.

6. Effects of Hunger and Satiety on Taste Processing at Different Stages of the Taste Pathway and Their Relevance to the Control of Eating and Sensory-specific Satiety

In order to analyze the neural control of feeding, we have been recording the ac-tivities of single neurons during feeding in brain regions implicated in feeding in the monkey (Rolls, 1994, 1997, 1999a, 2000c). It has been found that populations of neurons in the lateral hypothalamus and adjoining substantia innominata in the monkey respond to the sight and/or taste of food (Rolls, Burton, and Mora, 1976). Part of the evidence that these neurons are involved in the control of responses to food is that they respond to food only when the monkey is hungry (Burton,

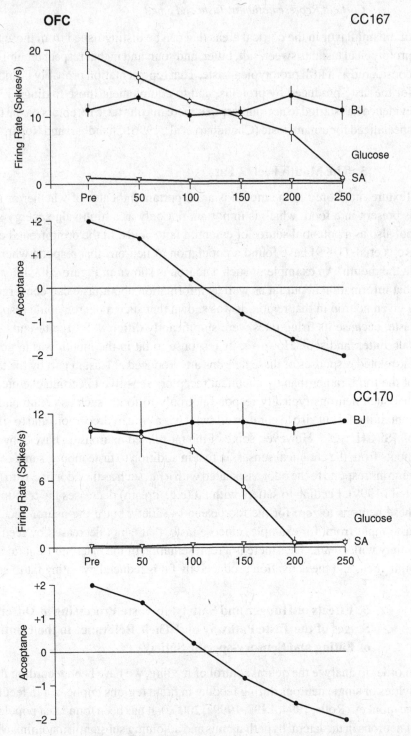

Figure 23.4. Effect of feeding to satiety with glucose solution on the responses of two neurons in the secondary taste cortex to the tastes of glucose and of blackcurrant juice (BJ). The spontaneous firing rate is also indicated (SA).

Rolls, and Mora, 1976; Rolls et al., 1986). Indeed, modulation of the reward or incentive value of a motivationally relevant sensory stimulus, such as the taste of food, by an animal's motivational state (e.g., hunger) is one important way in which motivational behavior is controlled (Rolls, 1975, 1999a). The subjective correlate of this modulation is that food tastes pleasant when one is hungry, but tastes hedonically neutral when it has been eaten to satiety.

These findings raise the question of the stage in sensory processing at which satiety modulates responsiveness. We have found that this modulation of taste-evoked signals by motivation is not a property found in the early stages of the primate gustatory system. The responsiveness of taste neurons in the NTS (Yaxley et al., 1985) and in the primary taste cortex – frontal opercular (Rolls et al., 1988) and insular (Yaxley, Rolls, and Sienkiewicz, 1988) – is not attenuated by feeding to satiety. In contrast, in the secondary taste cortex, in the caudolateral part of the orbitofrontal cortex, it was shown that the responses of neurons to the taste of glucose decreased to zero when the monkey ate it to satiety, during the course of which its behavior turned from avid acceptance to active rejection (Rolls, Sienkiewicz, and Yaxley, 1989). That modulation of the responsiveness of the gustatory responses of the orbitofrontal cortex neurons by satiety could not have been due to peripheral adaptation in the gustatory system or to altered efficacy of gustatory stimulation after satiety was reached, because modulation of neuronal responsiveness by satiety was not seen at the earlier stages of the gustatory system, including the NTS, the frontal opercular taste cortex, and the insular taste cortex. We also found evidence that gustatory processing involved in thirst became interfaced to motivation in the caudolateral orbitofrontal cortex taste projection area, in that neuronal responses there to water were decreased to zero when water was drunk to satiety (Rolls et al., 1989). In the secondary taste cortex, it was also found that decreases in the responsiveness of neurons were relatively specific to the food with which the monkey had been fed to satiety. For example, in seven experiments in which monkeys were fed glucose solution, neuronal responsiveness to the taste of glucose decreased, but not that to the taste of blackcurrant juice (Figure 23.4). Conversely, in two experiments in which

Figure 23.4 (*cont.*). Below the neuronal response data for each experiment, the behavioral measure of the acceptance or rejection of the solution on a scale from +2 to −2 is shown. The solution used to feed to satiety was 20% glucose. The monkey was fed 50 ml of the solution at each stage of the experiment, as indicated along the abscissa, until he was satiated, as shown by whether he accepted or rejected the solution. "Pre" indicates the firing rate of the neuron before the satiety experiment started. The values shown are the mean firing rate and its s.e. (Adapted from Rolls et al., 1989.)

monkeys were fed to satiety with fruit juice, the responses of the neurons decreased to fruit juice, but not to glucose (Rolls et al., 1989).

Such evidence shows that the reduced acceptance of food that occurs when food is eaten to satiety and the reduction in the pleasantness of its taste (Cabanac, 1971; Rolls and Rolls, 1977, 1982; Rolls et al., 1981a,b; Rolls, Rowe, and Rolls, 1982; Rolls, Rolls, and Rowe, 1983c) are not produced by reductions in the responses of neurons in the NTS or frontal opercular or insular gustatory cortices to gustatory stimuli. Indeed, after eating to satiety, humans reported that the taste of the food on which they had become satiated was almost as intense as when they had been hungry, though much less pleasant (Rolls et al., 1983c). That comparison is consistent with the possibility that activity in the frontal opercular and insular taste cortices, as well as the NTS, does not reflect the pleasantness of the taste of a food, but rather its sensory qualities independently of motivational state. On the other hand, the responses of the neurons in the caudolateral orbitofrontal cortex taste area and in the lateral hypothalamus (Rolls et al., 1986) are modulated by satiety, and presumably it is in areas such as those that neuronal activity may be related to whether or not a food tastes pleasant, and to whether or not the food should be eaten (Scott, Yan, and Rolls, 1995; Critchley and Rolls, 1996b; Rolls, 1996, 1999a, 2000b,c).

7. Representation of Flavor: Convergence of Olfactory and Taste Inputs

At some stage in taste processing, it is likely that taste representations are brought together with inputs from different modalities, such as olfactory inputs, to form a representation of flavor. Takagi and his colleagues (Tanabe et al., 1975a,b) have found an olfactory area in the medial orbitofrontal cortex. In a mid-mediolateral part of the caudal orbitofrontal cortex there is an area investigated by Thorpe et al. (1983) in which they found many neurons with visual responses and some with gustatory responses. We therefore sought to investigate whether or not there were neurons in the secondary taste cortex and the adjoining, more medial orbitofrontal cortex that would respond to stimuli in other modalities, including the olfactory and visual modalities, and whether or not single neurons in that cortical region would in some cases respond to stimuli from more than one modality. We found (Rolls and Baylis, 1994) that in the orbitofrontal cortex taste areas, of 112 single neurons that responded to any of those modalities, many were unimodal (taste 34%, olfactory 13%, visual 21%), but they were found in close proximity to one another. Some single neurons showed convergence, responding, for example, to taste and visual inputs (13%), taste and olfactory inputs (13%), and olfactory and visual inputs (5%). Some of those multimodal single neurons had corresponding sensitivities in the two modalities, in that they responded best to sweet tastes (e.g., 1-м glucose), and in a visual-discrimination task they responded more to the visual stimulus that signified sweet fruit juice than to one that signified saline, and in an olfactory-discrimination task they responded more to fruit odor. The different types of neurons (unimodal in different modalities, and multimodal) frequently were found close to one another in tracks made into that region, consistent with the hypothesis that the multimodal representations are actually being formed from unimodal inputs to that region.

It thus appears to be in these orbitofrontal cortex areas that flavor representations are built, where "flavor" is taken to mean a representation that is evoked best by a combination of gustatory and olfactory inputs. This orbitofrontal region does appear to be an important region for convergence, for there are only low proportions of bimodal taste and olfactory neurons in the primary taste cortex (Rolls and Baylis, 1994).

8. Rules Underlying the Formation of Olfactory Representations in the Primate Cortex

Critchley and Rolls (1996a) showed that 35% of olfactory neurons in the orbitofrontal cortex categorized odors in an olfactory–taste discrimination task on the basis of the taste–reward associations of the odorants. Rolls et al. (1996b) found that 68% of the odor-responsive neurons in the orbitofrontal cortex modified their responses in some way following changes in the taste–reward associations of the odorants during olfactory–taste-discrimination reversals. (In an olfactory-discrimination experiment, if the monkey made a lick response when one odor, the S+, was delivered, he obtained a drop of glucose reward; if the monkey incorrectly made a lick response to another odor, the S−, he obtained a drop of aversive saline. At some time during the experiment, the contingency between the odor and the taste was reversed. The monkey relearned the discrimination, showing reversal. It will be of interest to investigate in which parts of the olfactory system the neurons show reversal, for where they do, it can be concluded that the neuronal response to the odor depends on the taste with which it is associated.) Full reversal of neuronal responses was seen in 25% of the neurons analyzed. (In full reversal, the odor to which the neuron responded was reversed when the taste with which it was associated was reversed.) Extinction of the differential neuronal responses after task reversal was seen in 43% of the neurons. (Those neurons simply stopped discriminating between the two odors after the reversal.) Those findings demonstrate directly a coding principle in primate olfaction whereby the responses of some orbitofrontal cortex olfactory neurons are modified by and depend upon the taste with which the odor is associated. It was of interest, however, that those modifications were less extensive, and much slower, than the modifications found for orbitofrontal visual neurons during visual–taste reversal (Rolls et al., 1996b). That relative inflexibility of olfactory responses is consistent with the need for some stability in odor–taste associations to fa- • cilitate the formation and perception of flavors. In addition, some orbitofrontal cortex olfactory neurons did not code in relation to the taste with which the odor was associated (Critchley and Rolls, 1996a), so there is also a taste-independent representation of odor in this region.

9. Olfactory and Visual Sensory-specific Satiety and Representations in the Primate Orbitofrontal Cortex

It has also been possible to investigate whether or not the olfactory representation in the orbitofrontal cortex is affected by hunger. In satiety experiments, Critchley and Rolls (1996b) showed that the responses of some olfactory neurons to a food odor were decreased when the monkey was fed to satiety with a food (e.g., fruit juice) having that odor. In particular, seven of nine olfactory neurons that were responsive to the odors of foods, such as blackcurrant juice, were found to decrease their responses to the odor of the satiating food. Typically the decrease was at least partly specific to the odor of the food that had been eaten to satiety, potentially providing part of the basis for sensory-specific satiety. It was also found for eight of nine neurons that had selective responses to the sight of food that they demonstrated sensory-specific reductions in their visual responses to foods following satiation. Those findings show that the olfactory and visual representations of food, as well as the taste representation of food, in the primate orbitofrontal cortex are modulated by hunger. Usually a component related to sensory-specific satiety can be demonstrated.

Those findings link at least part of the processing of olfactory and visual information in this brain region to the control of feeding-related behavior. This is further evidence that part of the olfactory representation in this region is related to the hedonic value of the olfactory stimulus, and in particular that at this level of the olfactory system in primates, the pleasure elicited by the food odor is at least part of what is represented.

As a result of the neurophysiological and behavioral observations showing the specificity of satiety in the monkey, experiments were performed to determine whether or not satiety was specific to foods eaten by humans. It was found that the pleasantness of the taste of food eaten to satiety decreased more than that for foods that had not been eaten (Rolls et al., 1981a). One consequence of this is that if one food is eaten to satiety, appetite reduction for other foods is often incomplete, and that will lead to increased eating when a variety of foods is offered (Rolls et al., 1981a,b; Rolls, Van Duijenvoorde, and Rolls, 1984). Because sensory factors, such as similarities in color, shape, flavor, and texture, usually are more important than metabolic factors, such as protein, carbohydrate, and fat content, in influencing how foods interact in this type of satiety, it has been termed "sensory-specific satiety" (Rolls and Rolls, 1977, 1982; Rolls et al., 1981a,b, 1982; Rolls, 1990). It should be noted that this effect is distinct from alliesthesia, in that alliesthesia is a change in the pleasantness of sensory inputs produced by internal signals (such as glucose in the gut) (Cabanac and Duclaux, 1970; Cabanac, 1971; Cabanac and Fantino, 1977), whereas sensory-specific satiety is a change in the pleasantness of sensory inputs that is accounted for, at least in part, by the external sensory stimulation received (such as the taste of a particular food), in that, as shown earlier, it is at least partly specific to the external sensory stimulation received.

The parallel between these studies of eating by humans and studies of the neurophysiology of hypothalamic and orbitofrontal cortex neurons in the monkey has been extended by observations that in humans, sensory-specific satiety occurs for the sight of food as well as for the taste of food (Rolls et al., 1982). Further, to complement the finding that in the hypothalamus there are neurons that respond differently to food and to water (Rolls, 1999a), and that satiety with water can decrease the responsiveness of hypothalamic neurons that respond to water, it has been shown that in humans, motivation-specific satiety can also be detected. For example, satiety with water decreases the pleasantness of the sight and taste of water, but not of food (Rolls et al., 1983c).

To investigate whether or not the sensory-specific reduction in the responsiveness of orbitofrontal olfactory neurons might be related to a sensory-specific reduction in the pleasure produced by the odor of a food when it has been eaten to satiety, Rolls and Rolls (1997) measured humans' responses to the smell of a food that was eaten to satiety. It was found that the pleasantness of the odor of a food (and also, but much less significantly, the perceived intensity of its odor) was decreased when the subjects ate it to satiety (Figure 23.5). It was also found that pleasantness ratings for the smells of other foods (i.e., foods not eaten in the meal) showed much smaller decreases. That finding has clear implications for the control of food intake, for ways to make foods presented in a meal appealing, and for odor pleasantness ratings following meals. In an investigation of the mechanisms of this odor-specific sensory-specific satiety, Rolls and Rolls (1997) had humans chew a food, without swallowing, for approximately as long as the food normally would be in the mouth during eating. They demonstrated sensory-specific satiety with that procedure, showing that the sensory-specific satiety did not depend on food reaching the stomach. Thus at least part of the mechanism is likely to involve a change in processing in the olfactory pathways.

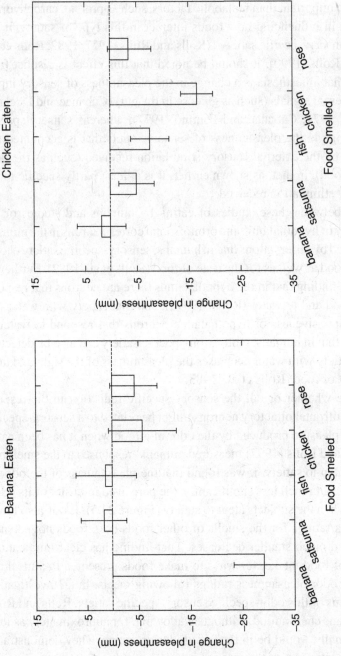

Figure 23.5. Olfactory sensory-specific satiety in humans. The pleasantness of the smell of a food became less when the humans ate that food (banana or chicken) to satiety. A similar reduction was not found for other foods not eaten in the meal. The changes in pleasantness were measured on a 100-mm visual-analogue rating scale. The number of subjects was 12, and the results (as shown by the interaction term in a two-way within-subjects ANOVA) were very significant ($p < .001$). (Adapted from Rolls and Rolls, 1997.)

It is not yet known at which of the earliest stages of olfactory processing this modulation occurs. It is unlikely to be in the receptors, because the change in pleasantness found was much more significant than the change in intensity (Rolls and Rolls, 1997).

The pattern of increased consumption when a variety of foods is available, as a result of the operation of sensory-specific satiety, may have been advantageous during evolution to ensure that different foods with a variety of important nutrients would be consumed, but for humans today, when a wide variety of foods is readily available, it may be a factor that can lead to overeating and obesity. In a test of that in the rat, it was found that variety itself could lead to obesity (Rolls, Van Duijenvoorde, and Rowe, 1983b; Rolls and Hetherington, 1989).

10. Responses of Orbitofrontal Cortex Taste and Olfactory Neurons to the Sight and Texture of Food

Many of the neurons that respond to visual stimuli in this region also show olfactory or taste responses (Rolls and Baylis, 1994), reverse rapidly in visual-discrimination reversal (Rolls et al., 1996b), and respond to the sight of food only if hunger is present (Critchley and Rolls, 1996b). That part of the orbitofrontal cortex thus seems to implement a mechanism that can flexibly alter responses to visual stimuli depending on the reinforcement (e.g., the taste) associated with a visual stimulus (Thorpe et al., 1983; Rolls, 1996). That facilitates prediction of the taste that will be associated with ingestion of what is seen, thus aiding in visual selection of foods (Rolls 1993, 1994, 1999a, 2000c).

The orbitofrontal cortex in primates is also important as an area of convergence for somatosensory inputs, such as those related to the perceived textures of fatty foods in the mouth. We have shown, for example, in recent recordings that single neurons influenced by taste in that region can in some cases have their responses modulated by the texture of the food. That was shown in experiments in which the texture of food was manipulated by addition of methylcellulose or gelatin, or by puréeing a semisolid food (E. T. Rolls and H. Critchley, unpublished data; Rolls, 1997, 1999a).

11. Taste and Olfactory Responses of Primate Amygdala Neurons

The orbitofrontal cortex has rich interconnections with the amygdala, and some amygdala neurons also show taste and olfactory responses. Sanghera, Rolls, and Roper-Hall (1979) described taste responses in some amygdala neurons in macaques, and Wilson and Rolls (1993, in press) found more. Nishijo, Ono, and Nishino (1988a,b) found ingestion-related amygdala neurons, at least some of which probably were taste-related. Scott et al. (1993) analyzed the responses of 35 taste neurons in the macaque amygdala. Although individual neurons were quite broadly tuned to different tastes, the population as a whole clearly discriminated among three tastes: sweet, salt (NaCl), and umami (monosodium glutamate, protein taste) (Rolls, 1997; Rolls et al., 1998). To test whether or not the representation of taste in the primate amygdala may encode its reward value, Yan and Scott (1996) fed monkeys to satiety to determine if decreasing the reward value of the taste to zero in that way would decrease the responses of primate amygdala sweet-taste neurons to zero, as had been found for orbitofrontal cortex neurons (Rolls et al., 1989). Yan and Scott found that only partial reductions of the responses of amygdala neurons were produced by feeding to satiety (an average of 62%). The implication is that amygdala neurons in primates can respond differently to rewarding and punishing stimuli, but they react less extensively than do orbitofrontal cortex neurons to the reward value when it is changed rapidly, as during feeding to satiety, or, as described earlier for visual neurons, in visual-discrimination reversal. Olfactory responses have also been found in the primate amygdala (Sanghera et al., 1979), and the anterior cortical nucleus of the amygdala and the periamygdaloid cortex receive direct connections from the olfactory bulb (Carmichael, Clugnet, and Price, 1994).

12. Imaging Studies in Humans

In humans, it has been shown in neuroimaging studies with functional magnetic-resonance imaging (fMRI) that taste activates an area of the anterior insula, which probably is the primary taste cortex, and part of the orbitofrontal cortex, which probably is the secondary taste cortex (Francis et al., 1999). The orbitofrontal cortex taste area is distinct from areas activated by odors and by pleasant touch (Francis et al., 1999). It has been shown that within individual subjects separate areas of the orbitofrontal cortex are activated by sweet (pleasant) and salt (unpleasant) tastes (O'Doherty et al., 2001). Francis et al. (1999) also found activation of the human amygdala by the taste of glucose. Extending that study, O'Doherty et al. (2001) showed that the human amygdala was as much activated by the affectively pleasant taste of glucose as by the affectively negative taste of NaCl and thus provided evidence that the human amygdala is not especially involved in processing aversive stimuli as compared with rewarding stimuli.

It is of interest that in humans there is an area of the far anterior insula that is activated by olfactory stimuli (Francis et al., 1999; O'Doherty et al., 2000). It is not clear whether or not that area is separate from the part of the insula activated by taste. In macaques, the primary taste cortex (in the anterior insula and adjoining frontal operculum) does not appear to be strongly activated by olfactory stimuli (though further studies on this are in progress), and the human anterior insular olfactory area may thus correspond to what in macaques is the caudal transitional area of the orbitofrontal cortex, where it adjoins the insula, area Ofdg, where part of the secondary taste cortex is located (Baylis et al., 1994). In humans, there is strong and consistent activation of the right orbitofrontal cortex by olfactory stimuli (Zatorre et al., 1992; Francis et al., 1999). In an investigation of where the pleasantness of olfactory stimuli might be represented in humans, O'Doherty et al. (2000) showed that activation of an area of the orbitofrontal cortex to banana odor was decreased (relative to a control vanilla odor) after bananas were eaten to satiety. Thus activity in a part of the human orbitofrontal cortex olfactory area is related to sensory-specific satiety.

13. Conclusions

The primate orbitofrontal cortex is an important site for convergence of representations of the taste, smell, sight, and mouth feel of food, and that convergence allows the sensory properties of each food to be represented and defined in detail. The primate orbitofrontal cortex is also the region where a short-term, sensory-specific control of appetite and eating is implemented, in the form of sensory-specific satiety. Moreover, it is likely that visceral and other satiety-related signals reach the orbitofrontal cortex and there modulate the representation of food, resulting in an output that reflects the reward (or appetitive) value of each food. Part of the evidence that the reward value (and the pleasantness of food in humans) is represented in the orbitofrontal cortex is that macaques will work to obtain electrical stimulation of this brain region if they are hungry, but much less so if they are satiated (Rolls, 1999a). Further, monkeys and humans with damage to this brain region show altered, often less selective, food preferences (e.g., Baylis and Gaffan, 1991; Rolls, 1999b). The orbitofrontal cortex contains not only representations of taste and olfactory stimuli but also representations of other types of rewarding and punishing stimuli, including pleasant touch, and all these inputs, together with the functions of the orbitofrontal cortex in stimulus–reward and stimulus–punishment learning, provide a basis for understanding its functions in emotional and motivational behavior (Rolls, 1999a, 2000a–d).

Acknowledgments

Some of the experiments described here were conducted in association with Drs. A. Browning, H. Critchley, T. R. Scott, Z. J. Sienkiewicz, E. A. Wakeman, L. L. Wiggins (L. L. Baylis), and S. Yaxley, and their collaboration is sincerely acknowledged. The research in our laboratory was supported by the Medical Research Council (PG8513790).

References

Aggleton J P & Passingham R E (1981). Syndrome Produced by Lesions of the Amygdala in Monkeys (*Macaca mulatta*). *Journal of Comparative Physiology and Psychology* 95:961–77.

Baylis L L & Gaffan D (1991). Amygdalectomy and Ventromedial Prefrontal Ablation Produce Similar Deficits in Food Choice and in Simple Object Discrimination Learning for an Unseen Reward. *Experimental Brain Research* 86:617–22.

Baylis L L & Rolls E T (1991). Responses of Neurons in the Primate Taste Cortex to Glutamate. *Physiology and Behavior* 49:973–9.

Baylis L L, Rolls E T, & Baylis G C (1994). Afferent Connections of the Orbitofrontal Cortex Taste Area of the Primate. *Neuroscience* 64:801–12.

Burton M J, Rolls E T, & Mora F (1976). Effects of Hunger on the Responses of Neurones in the Lateral Hypothalamus to the Sight and Taste of Food. *Experimental Neurology* 51:668–77.

Cabanac M (1971). Biological Role of Pleasure. *Science* 173:1103–7.

Cabanac M & Duclaux R (1970). Specificity of Internal Signals in Producing Satiety for Taste Stimuli. *Nature* 227:966–7.

Cabanac M & Fantino M (1977). Origin of Olfacto-gustatory Alliesthesia: Intestinal Sensitivity to Carbohydrate Concentration? *Physiology and Behavior* 10:1039–45.

Carmichael S T, Clugnet M C, & Price J L (1994). Central Olfactory Connections in the Macaque Monkey. *Journal of Comparative Neurology* 346:403–34.

Chaudhari N & Roper S D (1998). Molecular and Physiological Evidence for Glutamate (Umami) Taste Transduction via a G-Protein-coupled Receptor. *Annals of the New York Academy of Sciences* 855:398–406.

Chaudhari N, Yang H, Lamp C, Delay E, Cartford C, Than T, & Roper S (1996). The Taste of Monosodium Glutamate: Membrane Receptors in Taste Buds. *Journal of Neuroscience* 16:3817–26.

Critchley H D & Rolls E T (1996a). Olfactory Neuronal Responses in the Primate Orbitofrontal Cortex: Analysis in an Olfactory Discrimination Task. *Journal of Neurophysiology* 75:1659–72.

Critchley H D & Rolls E T (1996b). Hunger and Satiety Modify the Responses of Olfactory and Visual Neurons in the Primate Orbitofrontal Cortex. *Journal of Neurophysiology* 75:1673–86.

Francis S, Rolls E T, Bowtell R, McGlone F, O'Doherty J, Browning A, Clare S, & Smith E (1999). The Representation of Pleasant Touch in the Brain and Its Relationship with Taste and Olfactory Areas. *NeuroReport* 10:453–9.

Nishijo H, Ono T, & Nishino H (1988a). Single Neuron Responses in Amygdala of Alert Monkey during Complex Sensory Stimulation with Affective Significance. *Journal of Neuroscience* 8:3570–83.

Nishijo H, Ono T, & Nishino H (1988b). Topographic Distribution of Modality-specific Amygdalar Neurons in Alert Monkey. *Journal of Neuroscience* 8:3556–69.

Norgren R (1984). Central Neural Mechanisms of Taste. In: *Handbook of Physiology – The Nervous System III, Sensory Processes 1*, ed. I Darien-Smith, pp. 1087–128. Washington, DC: American Physiological Society.

O'Doherty J, Rolls E T, Francis S, Bowtell R, McGlone F, Kobal G, Renner B & Ahne G (2000). Sensory-specific Satiety Related Olfactory Activation of the Human Orbitofrontal Cortex. *NeuroReport* 11:893–7.

O'Doherty J, Rolls E T, Francis S, McGlone F, & Bowtell R (2001). The Representation of Pleasant and Aversive Taste in the Human Brain. *Journal of Neurophysiology* 85:1315–21.

Pritchard T C, Hamilton R B, Morse J R, & Norgren R (1986). Projections of Thalamic Gustatory and Lingual Areas in the Monkey, *Macaca fascicularis*. *Journal of Comparative Neurology* 244:213–28.

Pritchard T C, Hamilton R B, & Norgren R (1989). Neural Coding of Gustatory Information in the Thalamus of *Macaca mulatta*. *Journal of Neurophysiology* 61:1–14.

Rolls B J (1990). The Role of Sensory-specific Satiety in Food Intake and Food Selection. In: *Taste, Experience, and Feeding*, ed. E D Capaldi & T L Powley, pp. 197–209. Washington, DC: American Psychological Association.

Rolls B J & Hetherington M (1989). The Role of Variety in Eating and Body Weight Regulation. In: *Handbook of the Psychophysiology of Human Eating*, ed. R Shepherd, pp. 57–84. Chichester: Wiley.

Rolls B J, Rolls E T, Rowe E A, & Sweeney K (1981a). Sensory Specific Satiety in Man. *Physiology and Behavior* 27:137–42.

Rolls B J, Rowe E A, & Rolls E T (1982). How Sensory Properties of Foods Affect Human Feeding Behavior. *Physiology and Behavior* 29:409–17.

Rolls B J, Rowe E A, Rolls E T, Kingston B, & Megson A (1981b). Variety in a Meal Enhances Food Intake in Man. *Physiology and Behavior* 26:215–21.

Rolls B J, Van Duijenvoorde P M, & Rolls E T (1984). Pleasantness Changes and Food Intake in a Varied Four Course Meal. *Appetite* 5:337–48.

Rolls B J, Van Duijenvoorde P M, & Rowe E A (1983b). Variety in the Diet Enhances Intake in a Meal and Contributes to the Development of Obesity in the Rat. *Physiology and Behavior* 31:21–7.

Rolls E T (1975). *The Brain and Reward*. Oxford: Pergamon.

Rolls E T (1993). The Neural Control of Feeding in Primates. In: *Neurophysiology of Ingestion*, ed. D A Booth, pp. 137–69. Oxford: Pergamon.

Rolls E T (1994). Neural Processing Related to Feeding in Primates. In: *Appetite: Neural and Behavioural Bases*, ed. C R Legg & D A Booth, pp. 11–53. Oxford University Press.

Rolls E T (1996). The Orbitofrontal Cortex. *Philosophical Transactions of the Royal Society B* 351:1433–44.

Rolls E T (1997). Taste and Olfactory Processing in the Brain and Its Relation to the Control of Eating. *Critical Reviews in Neurobiology* 11:263–87.

Rolls E T (1999a). *The Brain and Emotion*. Oxford University Press.

Rolls E T (1999b). The Functions of the Orbitofrontal Cortex. *Neurocase* 5:301–12.

Rolls E T (2000a). The Representation of Umami Taste in the Taste Cortex. *Journal of Nutrition* 130:S960–5.

Rolls E T (2000b). The Orbitofrontal Cortex and Reward. *Cerebral Cortex* 10:284–94.

Rolls E T (2000c). Taste, Olfactory, Visual and Somatosensory Representations of the Sensory Properties of Foods in the Brain, and Their Relation to the Control of

Food Intake. In: *Neural and Metabolic Control of Macronutrient Intake*, ed. H R Berthoud & R J Seeley, pp. 247–62. Boca Raton: CRC Press.

Rolls E T (2000d). Neurophysiology and Functions of the Primate Amygdala, and the Neural Basis of Emotion. In: *The Amygdala: A Functional Analysis*, ed. J P Aggleton, pp. 447–78. Oxford University Press.

Rolls E T & Baylis L L (1994). Gustatory, Olfactory and Visual Convergence within the Primate Orbitofrontal Cortex. *Journal of Neuroscience* 14:5437–52.

Rolls E T, Burton M J, & Mora F (1976). Hypothalamic Neuronal Responses Associated with the Sight of Food. *Brain Research* 111:53–66.

Rolls E T, Critchley H D, Browning A, & Hernadi I (1998). The Neurophysiology of Taste and Olfaction in Primates, and Umami Flavor. *Annals of the New York Academy of Sciences* 855:426–37.

Rolls E T, Critchley H D, Browning A S, Hernadi A, & Lenard L (1999). Responses to the Sensory Properties of fat of Neurons in the Primate Orbitofrontal Cortex. *Journal of Neuroscience* 19:1532–40.

Rolls E T, Critchley H, Mason R, & Wakeman E A (1996b). Orbitofrontal Cortex Neurons: Role in Olfactory and Visual Association Learning. *Journal of Neurophysiology* 75:1970–81.

Rolls E T, Critchley H, Wakeman E A, & Mason R (1996a). Responses of Neurons in the Primate Taste Cortex to the Glutamate Ion and to Inosine 5′-monophosphate. *Physiology and Behavior* 59:991–1000.

Rolls E T, Murzi E, Yaxley S, Thorpe S J, & Simpson S J (1986). Sensory-specific Satiety: Food-specific Reduction in Responsiveness of Ventral Forebrain Neurons after Feeding in the Monkey. *Brain Research* 368:79–86.

Rolls E T & Rolls B J (1977). Activity of Neurons in Sensory, Hypothalamic and Motor Areas During Feeding in the Monkey. In: *Food Intake and Chemical Senses*, ed. Y Katsuki, M Sato, S Takagi, & Y Oomura, pp. 525–49. University of Tokyo Press.

Rolls E T & Rolls B J (1982). Brain Mechanisms Involved in Feeding. In: *Psychobiology of Human Food Selection*, ed. L M Barker, pp. 33–62. Westport, CT: AVI Publishing.

Rolls E T & Rolls J H (1997). Olfactory Sensory-specific Satiety in Humans. *Physiology and Behavior* 61:461–73.

Rolls E T, Rolls B J, & Rowe E A (1983c). Sensory-specific and Motivation-specific Satiety for the Sight and Taste of Food and Water in Man. *Physiology and Behavior* 30:185–92.

Rolls E T, Scott T R, Sienkiewicz Z J, & Yaxley S (1988). The Responsiveness of Neurones in the Frontal Opercular Gustatory Cortex of the Macaque Monkey is Independent of Hunger. *Journal of Physiology* 397:1–12.

Rolls E T, Sienkiewicz Z J, & Yaxley S (1989). Hunger Modulates the Responses to Gustatory Stimuli of Single Neurons in the Orbitofrontal Cortex. *European Journal of Neuroscience* 1:53–60.

Rolls E T, Yaxley S, & Sienkiewicz Z J (1990). Gustatory Responses of Single Neurons in the Orbitofrontal Cortex of the Macaque Monkey. *Journal of Neurophysiology* 64:1055–66.

Sanghera M K, Rolls E T, & Roper-Hall A (1979). Visual Responses of Neurons in the Dorsolateral Amygdala of the Alert Monkey. *Experimental Neurology* 63:610–26.

Scott T R, Karadi Z, Oomura Y, Nishino H, Plata-Salaman C R, Lenard L, Giza B K, & Aou S (1993). Gustatory Neural Coding in the Amygdala of the Alert Macaque Monkey. *Journal of Neurophysiology* 69:1810–20.

Scott T R, Yan J, & Rolls E T (1995). Brain Mechanisms of Satiety and Taste in Macaques. *Neurobiology* 3:281–92.

Scott T R, Yaxley S, Sienkiewicz Z J, & Rolls E T (1986a). Taste Responses in the Nucleus Tractus Solitarius of the Behaving Monkey. *Journal of Neurophysiology* 55:182–200.

Scott T R, Yaxley S, Sienkiewicz Z J, & Rolls E T (1986b). Gustatory Responses in the Frontal Opercular Cortex of the Alert Cynomolgus Monkey. *Journal of Neurophysiology* 56:876–90.

Tanabe T, Iino M, & Takagi S F (1975a). Discrimination of Odors in Olfactory Bulb, Pyriform-Amygdaloid Areas, and Orbitofrontal Cortex of the Monkey. *Journal of Neurophysiology* 38:1284–96.

Tanabe T, Yarita H, Iino M, Ooshima Y, & Takagi S F (1975b). An Olfactory Projection Area in Orbitofrontal Cortex of the Monkey. *Journal of Neurophysiology* 38:1269–83.

Thorpe S J, Rolls E T, & Maddison S (1983). Neuronal Activity in the Orbitofrontal Cortex of the Behaving Monkey. *Experimental Brain Research* 49:93–115.

Wilson F A W & Rolls E T (1993). The Effects of Stimulus Novelty and Familiarity on Neuronal Activity in the Amygdala of Monkeys Performing Recognition Memory Tasks. *Experimental Brain Research* 93:367–82.

Wilson F A W & Rolls E T (in press). The Primate Amygdala and Reinforcement: A Dissociation between Rule-based and Associatively-mediated Memory Revealed in Amygdala Neuronal Activity.

Yan J & Scott T R (1996). The Effect of Satiety on Responses of Gustatory Neurons in the Amygdala of Alert Cynomolgus Macaques. *Brain Research* 740:193–200.

Yaxley S, Rolls E T, & Sienkiewicz Z J (1988). The Responsiveness of Neurones in the Insular Gustatory Cortex of the Macaque Monkey Is Independent of Hunger. *Physiology and Behavior* 42:223–9.

Yaxley S, Rolls E T, & Sienkiewicz Z J (1990). Gustatory Responses of Single Neurons in the Insula of the Macaque Monkey. *Journal of Neurophysiology* 63:689–700.

Yaxley S, Rolls E T, Sienkiewicz Z J, & Scott T R (1985). Satiety Does Not Affect Gustatory Activity in the Nucleus of the Solitary Tract of the Alert Monkey. *Brain Research* 347:85–93.

Zatorre R J, Jones-Gotman M, Evans A C, & Meyer E (1992). Functional Localization and Lateralisation of the Human Olfactory Cortex. *Nature* 360:339–40.

Section Six
Individual Variations

The final four chapters of this volume probe the theme of inter-individual variations in perceptual and cognitive performances in the chemical senses, a topic often encountered briefly in earlier chapters. Several chapters have described how group performances depend on previous exposure to odorants (e.g., Chapters 3, 8–10, and 21) and to tastants (Chapters 22 and 23). All those chapters have outlined emotional and memory processes evoked by odors and tastes that are at the core of an individual's functioning.

The chapters in this section more specifically address issues of chemosensory variability linked with individual constitution and with the interactions between individuals and given environments. Katharine Fast and colleagues (Chapter 24) offer a survey of the relationship between taste ability and an individual's genetic makeup. They highlight the phenomenon of taste blindness to bitter tastants as a window to individual variability in taste function. From that starting point they examine how it is possible to develop standard tools for measurements of taste intensity and hedonicity despite the considerable variability of individuals, ranging from nontaster to supertaster status. They examine the links between individual psychophysical data and the anatomical variations of the tongue, indicating that supertasters have higher fungiform papilla counts and rate stimuli as more intense. Similar structure–function correlates can explain sex differences.

In Chapter 25, Robyn Hudson and Hans Distel develop the argument that we will not be able to properly account for olfactory function without paying closer attention to the role of the conditions that have molded it in the individual-specific environment. The multidimensionality of the odor environment makes it difficult for the olfactory system to cope with the features of it that become psychologically salient. Two evolved properties of olfactory cognition deal with that difficulty, detecting a wide spectrum of molecules and developing selective responsiveness as a function of individual experience. To substantiate those points, they compare ratings of intensity, familiarity, and pleasantness evoked by a set of odorants in subjects from contrasting cultural backgrounds. Their studies lead to the conclusion that both the qualitative and intensity ratings of odors are fine-tuned by prior exposure.

In Chapter 26, Benoist Schaal and colleagues follow a similar line in showing how olfaction develops from the onset of its functioning. Even at the fetal and neonatal stages, sensory and affective responses to odor stimuli clearly are influenced by previous encounters with odorants, and the adaptability and functional malleability of olfaction can have consequences in terms of the diversity of odor qualities that can influence behavior. It is also pointed out that, at least during early development, preferential responses to odors may be evolutionarily or developmentally constrained: Some odor substrates appear to facilitate learning more readily than others, and more easily in some contexts than in others. For example, milk odors are highly attractive to newborns even when they have never been exposed to them (i.e., formula-fed since birth). Thus prior exposure may not always predict increased attractiveness. Olfactory predispositions whose development precedes postnatal experience most probably combine with learned abilities to determine stimulus salience in young organisms.

Finally, in Chapter 27, Thomas Hummel and colleagues analyze how another variable, age, contributes to the dynamic changes in chemosensory abilities. They develop multilevel measurements to assess (1) which performance (sensitivity, discrimination, identification) is most affected by age in cross-sectional samples ranging from young adulthood to old age, (2) which nasal subsystem is most affected, and (3) which central locus is concerned. Age-related changes in olfactory function are noted at all functional levels that are considered. But those effects are heterogeneous across tasks, the decreases being more pronounced for sensitivity than for higher-level functions. Across the different subsystems and periods of development, trigeminal responses are more affected by age in early adulthood, whereas olfactory responses are more reduced in old age. Finally, at the central level, the volume and distribution of brain regions activated by odor stimuli are reduced and different in the oldest subjects as compared with the youngest subjects.

Taken together, these chapters provide encouragement to our efforts to improve the existing methods and tools to assess sensory responses and use them with subjects who bring various prior experiences from various developmental periods in order to better understand how, why, and when individuals differ in odor and taste perceptions and related hedonic experience and knowledge.

24

New Psychophysical Insights in Evaluating Genetic Variation in Taste

Katharine Fast, Valerie B. Duffy, and Linda M. Bartoshuk

How awful is "awful"? And is my "awful" the same as yours? The answers to these questions require comparing sensory or hedonic experiences across individuals, one of the most difficult tasks for psychophysicists. This chapter aims to trace the evolution of sensory scaling techniques intended to provide such comparisons, focusing on taste and using the discovery of taste blindness as a starting line. The past 70 years have been exciting times in psychophysics and have witnessed the development of methods useful for quantifying not only the oral impact of a stimulus but also its appeal.

For several generations, psychophysicists have been concerned about our ability to scale sensory experiences. A 1,000-Hz, 98-decibel blast is a 1,000-Hz, 98-decibel blast, but we recognize that it may sound far more intense to the department chair's grandson than to the department chair herself. We recognize this because we accept that a certain auditory deficit may accompany the blooming of wisdom, but how do we go about quantifying perceived sound intensity so that we can compare the experiences of the young and old directly? Our scale may start in silence, but if we have assimilated the idea that a given sound will be of different perceived intensities to different people, where do we anchor our scale besides the bottom? The perceived strength of a cleanser's odor works the same way: The same concentration of scent is added to each bottle at the factory, but the aroma may strike some as overpowering, while being barely detectable to others. The sweetness of lemonade is similar: Sips from the same glass will not be equally sweet to all sharing. The sense of taste, though, is singular among its sensory siblings in that it arrives with a built-in anatomical tool with which we can check our scales for taste. We can look backward at our scaling procedures and see them in a new – and brighter – light.

1. Taste Blindness

The phenomenon of taste blindness was discovered accidentally (Fox, 1931): While Fox was transferring a quantity of phenylthiocarbamide (PTC) to a jar, some became airborne, and a colleague remarked on its bitterness, whereas Fox tasted nothing. Thus began a series of experiments to determine the scope of taste blindness. Fox and Blakeslee, a geneticist, took PTC to the 1931 meeting of the American Association for the Advancement of Science and gathered taste responses from more than 2,500 attendees (Blakeslee and Fox, 1932). To most (called "tasters"), Fox's PTC crystals were bitter, but to 28% (called "nontasters") the crystals had essentially no taste. When the findings were published in *The Journal of Heredity*, the journal included a piece of paper impregnated with PTC crystals; that PTC paper was the ancestor of the "Prop" (6-n-propylthiouracil) paper we use today (see Figure 24.4). Prop, chemically similar to PTC, is odorless, unlike PTC, and is less toxic as well.

Family studies (e.g, Blakeslee, 1932) had suggested that taste blindness was a simple Mendelian recessive trait. However, reliance on thresholds in gathering information about the range of genetic sensitivity to PTC missed the breadth of the issue. Whereas thresholds were helpful in separating nontasters from tasters, they failed to predict, never mind capture, the range of suprathreshold experience (e.g., Bartoshuk, 2000).

Fernberger tried tackling the question of suprathreshold experience in early 1932. He presented his subjects with PTC crystals. "They were told to swallow the substance . . . then report the experience in terms of one of the five following categories: tasteless, slightly bitter, bitter, very bitter, and extremely bitter. The category of 'tasteless' more or less defines itself; 'extremely bitter,' the other limiting category, was defined in terms of raw quinine . . ." (Fernberger, 1932, p. 323). We can agree on the importance of a zero ("tasteless"), but did Fernberger's "extremely bitter" (equated to the bitterness of raw quinine) mark a reasonable top for his scale? The answer is no. We now know that raw quinine, the taste at the top of Fernberger's scale, is more bitter to those who can taste PTC than to those who cannot. As will be discussed later, scaling developments in the wake of Fernberger's experiment gave us the tools to reveal that.

1.1. Stevens and Fechner

Stevens revolutionized psychophysics. In a paper published to mark the 100th anniversary of Fechner's groundbreaking *Elemente der Psychophysik* (Fechner, 1860), Stevens identified an error in the great scientist's work (Stevens, 1961). Fechner had used the JND (the amount a stimulus must be changed to produce a "just noticeable difference") as the unit in intensity scales. Stevens undid Fechner by noting that if the JND did indeed mark a unit of sensation, a scale based on

JNDs would have to have ratio properties. But Fechner's scale did not. For example, a tone 20 JNDs above threshold is much more than twice as loud as a tone 10 JNDs above threshold (Stevens, 1951).

Stevens devised a method for ratio scaling called "magnitude estimation" (e.g., Stevens, 1969). In magnitude estimation's earliest days, experimenters presented subjects with the first stimulus in a series and assigned a number designating its intensity. Subjects were instructed to rank all following stimuli on a ratio scale in terms of that first sensation. An experimenter might first present a subject with a 1.0-M salt solution, declare it a 20, and explain that if the next stimulus tasted twice as strong, it should be rated a 40, and if it tasted half as strong, it should be rated 10, and so on. Later, subjects were allowed to rate without anchors: Subjects assigned numbers to stimuli based on their perceived intensities, without regard to an experimenter-assigned number designation.

Though, across subjects, the numbers assigned to stimuli using magnitude estimation can be all over the map – subject Wilma might assign 1-M sucrose a 10, but subject Fred might call it 1,000 – they do reveal the perceived ratios among stimuli for each subject. Normalization can help make these ratios clearer by bringing the numbers assigned by a subject pool into line. We do this by assigning the data from each subject a factor by which we multiply each rating: That maintains the ratio properties of the magnitude-estimate data from each subject, as every rating is multiplied by the same factor. We might assign a given stimulus a certain number designation or use the average of ratings for a group of stimuli to assign the factor. In the preceding example, we might assign 1-M sucrose a 100. To normalize Wilma's sucrose rating, we divide 100 by 10; to do the same for Fred, we divide 100 by 1,000. Wilma's factor is then 10, and Fred's is 0.1. The sucrose solutions Wilma called 20, 25, and 15 are now 200, 250, and 150; if Fred earlier assigned those same solutions 2,000, 2,500, and 1,500, his numbers are now 200, 250, and 150. Although normalization helps us to better see the sizes of ratios within subjects' data, it cannot reveal differences among subjects. Wilma's rating of the 1-M sucrose solution, whether her raw 10 or her normalized 100, does not tell us whether or not her 10 is as strong as Fred's 1,000 (or his 100, or his 75 or 7,500). But what if we could find a stimulus that would be equal to all? Then we could normalize magnitude-estimate data to that standard.

The trick is to find such a standard, one independent of the stimuli of interest. In early studies, when there was little reason to assume that salt intensity and bitterness intensity were at all linked, NaCl seemed such a standard (Blakeslee and Salmon, 1935). For example, subjects could scale the saltiness of NaCl and the bitterness of Prop using magnitude estimation. If we use NaCl as our standard, we have to assume that NaCl intensity is unrelated to bitterness intensity; then, on average, each of the Prop groups will perceive NaCl as equally intense. Once we normalized our data to NaCl, we were able to identify a group of tasters (we call them supertasters) to whom Prop was the most bitter (Bartoshuk, 1991). We found that a variety of substances tasted most intense to supertasters; for example, in one study (Bartoshuk, 1993), quinine was about 1.4 times as bitter to supertasters as to nontasters. We can do the same thing by using another sense – say hearing – instead of another taste. This is magnitude matching (e.g., Marks et al., 1988). We assume that if hearing and taste are truly unrelated, and if we have a large group scaling both taste and sound, the average hearing will be the same across any subgroups we identify. At first, NaCl and sound appeared to be about equally good as standards (more about this later).

1.2. Context Effects and Prop Studies

Psychophysical responses are not absolute; the context of a stimulus affects the responses to it. See Riskey, Parducci, and Beauchamp (1979), Rankin and Marks (1991), Schifferstein (1994), and Lawless, Horne, and Spiers (2000) for examples concerning taste. Gescheider (1988) reviewed the matter and noted that "context effects in psychophysical scaling ... can be found with the use of every psychophysical-scaling procedure. Investigators respond to this situation in two distinct ways. On the one hand, the sensory scientist in pursuit of pure sensory functions has attempted to eliminate context from the measurement situation. ... In contrast to the sensory scientist, the cognitive scientist welcomes context effects in scaling because they supply a wealth of information about judgment processes. ... The stimulus contexts that are of great interest to the sensory scientist are not those that affect judgment but those that affect sensation" (p. 130).

In that latter category, Marks (1992) has demonstrated an assimilation context effect that operates across modalities; this has particular relevance to Prop studies. Marks found that experiencing strong stimuli in one modality can cause a stimulus in a second modality to be perceived as stronger than normal. Thus in a magnitude-matching study with a sound standard, if a supertaster tastes Prop and then hears a sound, the sound will be perceived as louder than it is perceived by a nontaster. When such data are normalized to sound, taste differences between the supertaster and the nontaster will seem smaller than they really are. Recognizing the impact of this context effect has changed the way we run experiments: Prop comes last in our taste studies now, and we always begin with our standard (tones). This methodological advance has revealed that the differences across nontasters, medium tasters, and supertasters are larger than we originally thought. For example, a recent study using sound as a standard showed that quinine was about twice as bitter to supertasters as to nontasters (Ko et al., 2000).

1.3. Prop and Tongue Anatomy

Miller and Reedy developed a simple procedure to permit the counting of fungi-form papillae on the tongue, as well as the taste buds on those papillae (Miller and Reedy, 1990). The surfaces of fungiform papillae do not absorb dyes, but the taste pores (conduits to the taste buds) do. This means that a quick swab of dye (even food coloring – we use blue, for contrast) will render the un-stained fungiform papillae relatively easy to count on their suddenly darkened surroundings. Fungiform papillae density can be assessed with a flashlight and magnifying glass. Christine Bechtel, a 13-year-old working on a science project, used a paste-on notebook-paper reinforcer, with its inner circle aligned with the tongue tip and midline (Figure 24.3), to provide a standard template for counting papillae; we have now adopted that procedure for our own studies. Miller and Reedy demonstrated that Prop medium tasters have a greater density of fungi-form papillae on their tongues than nontasters (Miller and Reedy, 1990), and later work demonstrated that supertasters have the most fungiform papillae of all (Bartoshuk, Duffy, and Miller, 1994). Counting taste pores is much more difficult, because of the high magnification needed, but the numbers show the same disparities: Supertasters have the most, and nontasters the fewest, taste pores. Perceived taste intensities are proportional to the density of fungiform papillae/taste pores. Anatomical measurements have supported the early obser-vations that women were more sensitive to the bitterness of PTC/Prop; in our most recent data, over 17% of the women had more taste pores than any of the men (Prutkin et al., 2000).

That fungiform papillae density can be correlated with taste intensity is not the end of this measure's usefulness. It provides insight into the genetic variations in perceived intensities of other oral stimuli as well. Only 25% of fungiform papillae innervation is for taste (carried by the chorda tympani nerve); the rest is trigem-inal innervation coding for pain and touch sensations (Whitehead, Beeman, and Kinsella, 1985; Silver and Finger, 1991). Supertasters perceive the most burn from oral irritants like capsaicin (chili peppers) and ethanol, just as they perceive the most tactile sensation (e.g., oiliness, thickness) from fats in foods (Duffy et al., 1996; Tepper and Nurse, 1997; Bartoshuk et al., 1999; Prutkin et al., 1999).

1.4. The Labeled Magnitude Scale of Green and Colleagues

Around the same time as Marks's and Miller's work, Green and his colleagues devised their Labeled Magnitude Scale (LMS) (Green, Shaffer, and Gilmore, 1993; Green et al., 1996). Their motivation was to create a scale with the ra-tio properties of magnitude estimation that would also incorporate some indi-cation of absolute intensity through the use of adjectives. Some years earlier,

Borg (1982) had explored the placement of common intensity adjectives (e.g., "weak," "moderate," "strong," "extremely strong," "maximal") on a scale to see if he could give that scale ratio properties. That resulted in roughly logarithmic spacings. He believed that correct spacings of the adjectives would have the potential to produce a universal sensory ruler that could be used in any modality to compare sensory intensities across individuals. That belief was based on the assumption that the range from zero to maximal perceived intensity would be the same for different individuals and in different modalities (Teghtsoonian, 1971, 1973). Green et al. (1993) had doubts that "zero to maximal" (Green substituted "strongest imaginable" for Borg's "maximal") would have the same meaning in different modalities, and they set out to create a scale limited to oral sensations. They took a very important step beyond Borg by empirically determining the locations of the adjectives. Subjects were asked to give "magnitude estimates of adjectives within the context of numerous recalled, 'real-life' experiences." The experiences were all oral sensations (e.g., "the bitterness of celery," "the burn of cinnamon gum"). The distances between the adjectives were similar to but not identical with those of Borg. In particular, the distance between "very strong" and "strongest imaginable" (Borg's "maximal") was larger on Green's LMS. Of special importance, Green's LMS produced functions like those produced by magnitude estimation (Green et al., 1993, 1996).

1.5. A New Label for the Top of Green's LMS

As Green's LMS is portable and some subjects and patients find its use easier than magnitude matching, we were eager to put it to use in Prop studies. Because we knew that oral sensations were not equivalent for nontasters, medium tasters, and supertasters, we knew that the scale as originally designed could not allow valid comparisons across the three groups. So we instructed our subjects to consider the top of the scale to be the "strongest imaginable sensation of any kind." If everyone had experienced strong sensations (e.g., sight, smell, sound, touch, pain, thermal sensations) that were not related to their taste experiences, then their "strongest imaginable sensation of any kind" should provide anchors for the scale that would be equivalent, on average, for nontasters, medium tasters, and supertasters.

We used Prop to compare Green's LMS and magnitude matching (Bartoshuk et al., 2000). Our 100 subjects rated the bitterness of Prop using magnitude estimation with a sound standard (magnitude matching) in one session, and Green's LMS with our label at the top in another session (order counterbalanced). As noted earlier, the validity of magnitude matching for comparison across groups rests on the assumption that there is no systematic relationship between hearing and taste, that the perceived intensities of sounds will be, on average, the same for nontasters, medium tasters, and supertasters. On the other hand, the validity of Green's LMS (with our label at the top) rests on the assumption that "strongest imaginable sensation of any kind" establishes that the domain to which the adjectives apply comprises all sensations. If there is no systematic relationship between this anchor and taste, then the anchor will reflect, on average, the same perceived intensity for nontasters, medium tasters, and supertasters. Although these two methods rely on very different assumptions, they produce very similar pictures of the differences across nontasters, medium tasters, and supertasters. That supports the conclusion that the assumptions underlying both methods are correct; both methods appear to provide valid comparisons of sensory experiences across nontasters, medium tasters, and supertasters.

1.6. Tongue Anatomy and Scaling

Figure 24.1 shows an example of the correlation between perceived taste intensity and density of fungiform papillae using sucrose. The left panel shows the results of a study using Green's LMS with our label at the top. Perceived sweetness correlates with density of fungiform papillae. The right panel shows a study using a nine-point scale. Once we have established that a relationship exists between perceived taste intensity and fungiform papillae density, we can compare scales with regard to their ability to show that relation. Scales that make valid comparisons across groups of subjects show the relation; scales invalid for such comparisons do not. We could, of course, use taste pore density to test these scales as well, but the measurement of taste pore density is far more labor-intensive and requires special equipment. Measurement of fungiform papillae density is much easier and is adequate for scale evaluation.

The nine-point scale in Figure 24.1 fails on two fronts. As is evident, supertasters tend to rate the sweetness of 1-M sucrose as more intense than "very strong": This is a ceiling effect. Also, the absolute taste intensity reflected by "very strong" is greatest for supertasters, less for medium tasters, and least for nontasters.

1.7. Large Mice and Small Elephants: The Relativity of Intensity Adjectives

As the foregoing discussion shows, intensity adjectives do not reflect the same absolute intensities for all individuals under all conditions. Stevens made that point more than 40 years ago: "Mice may be called large or small, and so may

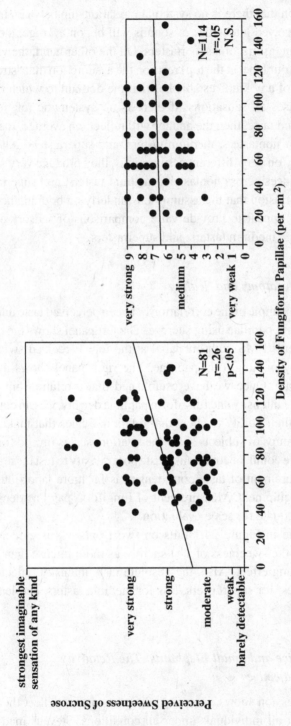

Figure 24.1. Perceived sweetness of 1-M sucrose versus density of fungiform papillae. For the graph on the left, perceived sweetness was measured with the Green scale (portions of the data are from Prutkin, 1997). For the graph on the right, perceived sweetness was measured with the nine-point scale (e.g., Kveton and Bartoshuk, 1994).

elephants, and it is quite understandable when someone says it was a large mouse that ran up the trunk of the small elephant" (Stevens, 1958, p. 633). How could the obvious relativity of intensity adjectives have been overlooked for so long even as they were being used to label intensity scales? The answer is, of course, that intensity adjectives refer to a specific domain (e.g., the domain of mice or the domain of elephants). In experiments, the domains may be specified explicitly by the experimenter or implicitly by the setting (subjects in a taste experiment expect to rate tastes). So in specific experiments, adjectives may have *seemed* to be conveying roughly the same intensities for all subjects. But we now know that that was not true, at least for taste.

We can illustrate the consequences of incorrectly assuming that the adjectives mean the same to all by returning to Fernberger. We noted earlier that magnitude matching with a tone standard shows that quinine is about twice as bitter to supertasters as it is to nontasters. Fernberger's scale started at "tasteless" and went up to "extremely bitter" (the taste of raw quinine). We see, now, that Fernberger was not using the same scale for all subjects. "Tasteless" is tasteless for everyone, but the taste of raw quinine is not the same for everyone. Let us consider the consequence of that for an experiment that Fernberger might have conducted. We now know that saltiness does not vary as much across Prop groups as does quinine, which means that the ratio of quinine to salt is largest for supertasters and smallest for nontasters. Because Fernberger forced the ratings for quinine to be equal for all subjects, had he asked his subjects to rate salt, he would have seen a bizarre result: Nontasters would have rated salt as saltier than would supertasters. Figure 24.2 illustrates this. We call this a reversal artifact.

Adjective-labeled scales – be they Fernberger's, the nine-point, Likert, or the visual analogue used with sensations or feelings – assume that the adjective labels hold the same meaning for all subjects. Whenever that assumption is false, comparisons across groups (or individuals) either can fail to show differences that really exist or can produce reversal artifacts, as illustrated in Figure 24.2. Many variables can contribute to systematic differences in the ways groups use adjectives (e.g., sex, clinical status, age, ethnic status), and that is a serious issue for any field that makes use of these scales.

Green's LMS as originally labeled is subject to the same problem when used for Prop studies (but not necessarily with other taste studies), because we know that the "strongest imaginable oral sensations" are not equivalent for nontasters, medium tasters, and supertasters. Green's LMS does have a virtue not shared by any of the other scales, however, and it is this: The distances between the intensity adjectives were determined empirically. Although it has not yet been proved, we suspect that those relative distances are invariant. One can even imagine a family of Green LMSs of varying sizes in which the relative distances among adjectives are always roughly the same. Assume for a moment that the strongest tea flavor

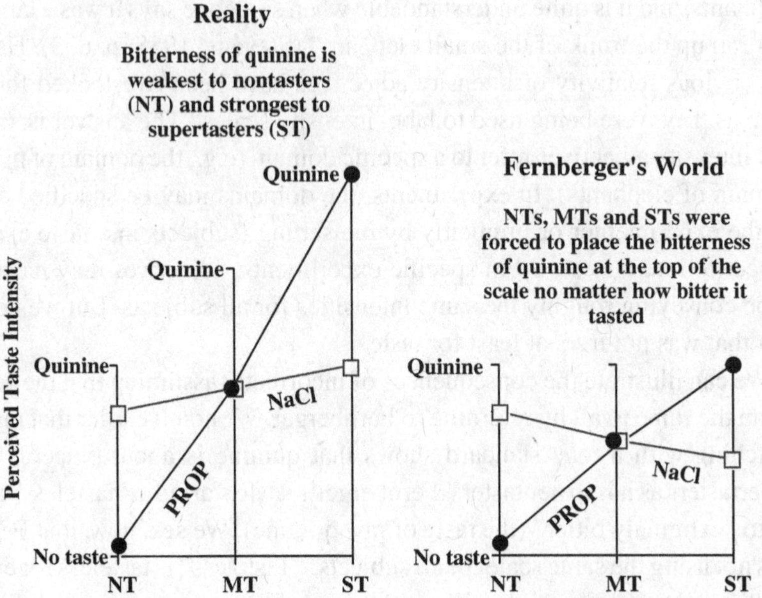

Reality

**Bitterness of quinine is
weakest to nontasters
(NT) and strongest to
supertasters (ST)**

Fernberger's World

**NTs, MTs and STs were
forced to place the bitterness
of quinine at the top of the
scale no matter how bitter it
tasted**

Figure 24.2. Left: The bitterness of Prop and quinine and the saltiness of NaCl as shown by magnitude matching and the Green scale (ratio scales). Note that quinine, the stimulus that Fernberger used to define the top of his scale, is not equally bitter to all. Nontasters (NT) perceived the least bitterness, supertasters (ST) the most, and medium tasters (MT) an intermediate degree of bitterness. The saltiness of NaCl was also associated with the ability to taste Prop, but the effect was smaller than for quinine. Right: The distortions induced by Fernberger's mistake. If all subjects were forced to place the bitterness of quinine at the top of this scale, the medium tasters and supertasters would have to compress their ratings for NaCl proportionately. That would cause the reversal artifact: The saltiness of NaCl would appear most intense to nontasters, and least intense to supertasters.

is less intense than the strongest odor, which in turn is less intense than the strongest pain. A Green LMS for tea would be smaller than that for odor, which in turn would be smaller than that for pain. But for all of the scales, a "very strong" sensation (tea, odor, pain) would be halfway between no sensation and the "strongest imaginable sensation" in that domain. We also suspect that the closest we will get to a universal sensory ruler may be the one we created earlier, labeled "strongest imaginable sensation of any kind." Yet we note that should we find a good candidate for designation as an experience that we can assume to be equally intense for all (e.g., Borg, 1990, proposed maximal perceived exertion), such a standard would make a universal sensory ruler possible.

Incidentally, Green's LMS can be used for magnitude matching with its original label ("strongest imaginable oral sensation") at the top. Including a standard permits the responses to be normalized to the standard. For example, including tones would permit taste data to be normalized to the tones, allowing comparisons across nontasters, medium tasters, and supertasters.

2. Valid Taste Comparisons across Groups: What Have We Learned?

Data collected from control subjects have allowed us to assess the effects of various abnormalities on taste and oral somatosensation. When a patient's perceived intensity ratings for taste or tactile stimuli are well below those we would expect for someone with a similar array of fungiform papillae, the presence

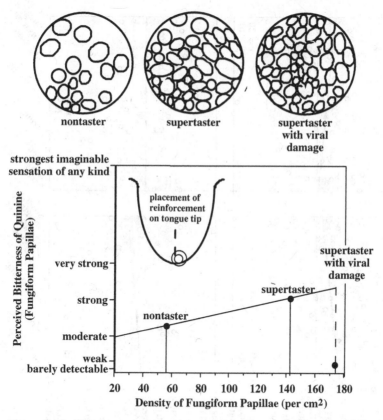

Figure 24.3. The three circles at the top of the figure are sketches of the fungiform papillae in a circular area (6 mm in diameter) inside a paper reinforcement placed at the tip of the tongue next to the midline (see sketch of tongue on graph). The graph shows the regression line for the perceived bitterness (Green scale) of quinine hydrochloride (0.001 M) swabbed onto fungiform papillae on the anterior tongue. The filled circles indicate the bitterness ratings by the subjects whose fungiform papillae densities are illustrated above.

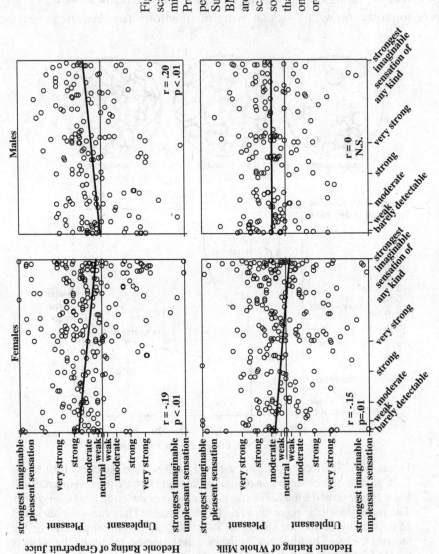

Figure 24.4. Hedonic ratings (Green scale) for grapefruit juice and whole milk plotted against the bitterness of Prop paper (3-cm circles of filter paper impregnated with 1.6 mg Prop). Subjects were 20–53 years old, with BMIs < 35. Correlation coefficients are shown on each panel. The hedonic scale was derived from the Green sensory scale as described in the text. Note that the labels "barely detectable" were omitted because of lack of space on the ordinate.

of pathologic conditions often can explain the dissociation. The integrity of fungiform papillae appears to be controlled by the trigeminal nerve in humans. We know that in extreme cases, such as when the chorda tympani nerve is cut during surgery, fungiform papillae remain on the tongue even though their service as a first stop for taste is meaningless (T. Janjua and S. R. Schwartz, unpublished data). Thus damage to the sense of taste produced by head trauma, ear infections, and upper respiratory tract infections (Solomon et al., 1991; Bartoshuk et al., 1996) and cyclical changes in taste due to hormonal changes (e.g., Glanville and Kaplan, 1965; Prutkin et al., 2000), which occur in the absence of damage to the trigeminal nerve, affect taste, but not the density of fungiform papillae.

Figure 24.3 shows the correlation between bitterness and fungiform papillae density on the anterior tongue for normal subjects. Three circles are pictured; they show the fungiform papillae in a 6-mm circle near the tip of the tongue (the diameter of the hole in a paper reinforcement). The left-hand circle is from a normal nontaster tongue, and the middle circle is from a normal supertaster tongue. The right-hand circle is from the tongue of a supertaster who has suffered viral damage; her tongue shows fungiform papillae density analogous to that of the other supertaster, though her bitterness perception is less than the nontaster's. Taste pathologic states and hormonal variation will alter taste experience, but not the density of fungiform papillae. The effects of such taste alterations on the density of taste pores are still unknown.

2.1. Valid Hedonic Scaling

Can Green's LMS be used to scale hedonic measures? A valid scale for measuring the hedonic properties of food would have great benefit, as what we eat has significant impact on our health. Popeye's preference for spinach may have reduced his risk for cancer, whereas Wimpy's love of burgers may have added to his girth and placed him on the road to cardiovascular disease.

Given the lessons of sensory scaling, we might pursue magnitude matching as a means of gathering hedonic data by looking for a standard experience, sensory or hedonic, against which to measure subjects' responses. We could also try the Green LMS, with the assumption that the "strongest imaginable pleasant" and "strongest imaginable unpleasant" experiences will have some universal meaning. Figure 24.4 shows data collected in such fashion (Duffy et al., 1999). Two mirror-image Green LMSs were presented to participants: "strongest imaginable dislike" lay at the left end of the line, "strongest imaginable like" lay on the right, and "neutral" marked where the zeros aligned. Because the liking and disliking of any food are subject to many factors – past experience, culture, geography – it makes sense to limit our search for links between sensory intensity and preference to homogeneous groups. The subjects in Figure 24.4 all had body mass indexes (BMIs) below 35; as BMI is a measure of obesity, that level excluded the morbidly obese, who might have exhibited distinct food preferences. The subjects were between the ages of 20 and 53; that excluded postmenopausal women. Data from men and women are shown separately, as we know that sex is a variable in perceived taste intensity. Despite the remaining diversity, there is some order in the data. For example, female supertasters in our sample liked grapefruit juice less than female nontasters, confirming the observations of Drewnowski, Henderson, and Shore (1997), and male supertasters liked it more than male nontasters. Female supertasters also liked whole milk less than female nontasters, confirming the data of Duffy and Bartoshuk (1996); no effect was seen for males.

The category scales so commonly used in sensory studies are also being used as hedonic scales. Is the Green LMS superior to them in this context as well? We suspect so, given that many of the logical problems inherent in category scales also apply to their hedonic incarnations (e.g., lack of ratio properties, ceiling effects, reversal artifacts).

3. Perspectives

Taste offers a unique tool for evaluation of psychophysical scales. Counts of fungiform papillae and perceived taste intensities are linked, and we can evaluate the power of a psychophysical scale to capture differences across subjects by how well that scale captures this linkage. Hedonic scales pose additional challenges, but the prize may be worth the effort. Valid measures of the intensity of affective experiences may reveal new phenomena as well as challenge long-held views. The impact of what we have learned about taste may have implications that will reach far beyond this one sensory modality. Now that we have identified the serious consequences of the current use of adjective-labeled scales in taste, we must be prepared to look for similar examples in other fields.

Acknowledgments

We thank the National Institutes of Health, grant DC00283, The Donaghue Women's Health Investigator Program at Yale, and the U.S. Department of Agriculture, grant 9603745 NRICG, for financial support for these studies.

References

Bartoshuk L M (1991). Sweetness: History, Preference, and Genetic Variability. *Food Technology* 45:108–13.

Bartoshuk L M (1993). Genetic and Pathological Taste Variation: What Can We Learn from Animal Models and Human Disease? In: *The Molecular Basis of Smell and Taste Transduction*, ed. D Chadwick, J Marsh, & J Goode, pp. 251–67. New York: Wiley.

Bartoshuk L M (2000). Comparing Sensory Experiences across Individuals: Recent Psychophysical Advances Illuminate Genetic Variation in Taste Perception. *Chemical Senses* 25:447–60.

Bartoshuk L M, Cunningham K E, Dabrila G M, Duffy V B, Etter L, Fast K R, Lucchina L A, Prutkin J M, & Synder D J (1999). From Sweets to Hot Peppers: Genetic Variation in Taste, Oral Pain, and Oral Touch. In: *Tastes and Aromas. The Chemical Senses in Science and Industry*, ed. G A Bell & A J Watson, pp. 12–22. Sydney: University of New South Wales Press.

Bartoshuk L M, Duffy V B, & Miller I J (1994). PTC/PROP Tasting: Anatomy, Psychophysics, and Sex Effects. *Physiology and Behavior* 56:1165–71.

Bartoshuk L M, Duffy V B, Reed D, & Williams A (1996). Supertasting, Earaches, and Head Injury: Genetics and Pathology Alter Our Taste Worlds. *Neuroscience and Biobehavioral Reviews* 20:79–87.

Bartoshuk L M, Green B G, Synder D J, Lucchina L A, Hoffmann H J, & Weiffenbach J M (2000). Valid Across-Group Comparisons: Supertasters Perceive the Most Intense Taste Sensations by Magnitude Matching or the LMS Scale. *Chemical Senses* 25:639.

Blakeslee A F (1932). Genetics of Sensory Thresholds: Taste for Phenylthiocarbamide. *Proceedings of the National Academy of Sciences USA* 18:120–30.

Blakeslee A F & Fox A L (1932). Our Different Taste Worlds. *Journal of Heredity* 23:97–107.

Blakeslee A F & Salmon T N (1935). Genetics of Sensory Thresholds: Individual Taste Reactions for Different Substances. *Proceedings of the National Academy of Sciences USA* 21:84–90.

Borg G (1982). A Category Scale with Ratio Properties for Intermodal and Interindividual Comparisons. In: *Psychophysical Judgment and the Process of Perception*, ed. H G Geissler & P Petxold, pp. 25–34. Berlin: VEB Deutscher Verlag der Wissenschaften.

Borg G (1990). Psychophysical Scaling with Applications in Physical Work and the Perception of Exertion. *Scandinavian Journal of Work, Environment and Health* (*Suppl 1*) 16:55–8.

Drewnowski A, Henderson S A, & Shore A B (1997). Taste Responses to Naringin, a Flavonoid, and the Acceptance of Grapefruit Juice Are Related to Genetic Sensitivity to 6-*n*-propylthiouracil. *American Journal of Clinical Nutrition* 66:391–7.

Duffy V B & Bartoshuk L M (1996). Genetic Taste Perception and Food Preferences. *Food Quality and Preference* 7:309 (abstract).

Duffy V B, Bartoshuk L M, Lucchina L A, Snyder D J, & Tym A (1996). Supertasters of PROP (6-*n*-propylthiouracil) Rate the Highest Creaminess to High-Fat Milk Products. *Chemical Senses* 21:598 (abstract).

Duffy V B, Fast K, Cohen Z, Chodos E, & Bartoshuk L M (1999). Genetic Taste Status Associates with Fat Food Acceptance and Body Mass Index in Adults. *Chemical Senses* 24:545–6 (abstract).

Fechner G T (1860). *Elemente der Psychophysik*. Leipzig: Breitkopf & Härterl.

Fernberger S W (1932). A Preliminary Study of Taste Deficiency. *American Journal of Psychology* 44:322–6.

Fox A L (1931). Six in Ten "Tasteblind" to Bitter Chemical. *Science News Letter* 9:249.

Gescheider G A (1988). Psychophysical Scaling. *Annual Review of Psychology* 39:169–200.

Glanville E V & Kaplan A R (1965). Taste Perception and the Menstrual Cycle. *Nature* 206:930–1.

Green B G, Dalton P, Cowart B, Rankin K, & Higgins J (1996). Evaluating the Labeled Magnitude Scale for Measuring Sensations of Taste and Smell. *Chemical Senses* 21:323–34.

Green B G, Shaffer G S, & Gilmore M M (1993). A Semantically-labeled Magnitude Scale of Oral Sensation with Apparent Ratio Properties. *Chemical Senses* 18:683–702.

Ko C W, Hoffmann H J, Lucchina L A, Snyder D J, Weiffenbach J M, & Bartoshuk L M (2000). Differential Perceptions of Intensity for the Four Basic Taste Qualities in PROP Supertasters versus Nontasters. *Chemical Senses* 25:639–40.

Kveton J F & Bartoshuk L M (1994). The Effect of Unilateral Chorda Tympani Damage on Taste. *Laryngoscope* 104:25–9.

Lawless H T, Horne J, & Spiers W (2000). Contrast and Range Effects for Category, Magnitude and Labeled Magnitude Scales in Judgements of Sweetness Intensity. *Chemical Senses* 25:85–92.

Marks L E (1992). The Slippery Context Effect in Psychophysics: Intensive, Extensive, and Qualitative Continua. *Perception and Psychophysics* 51:187–98.

Marks L E, Stevens J C, Bartoshuk L M, Gent J G, Rifkin B, & Stone V K (1988). Magnitude Matching: The Measurement of Taste and Smell. *Chemical Senses* 13:63–87.

Miller I J & Reedy F E (1990). Variations in Human Taste Bud Density and Taste Intensity Perception. *Physiology and Behavior* 47:1213–19.

Prutkin J (1997). PROP Tasting and Chemesthesis. Unpublished senior essay at Yale University.

Prutkin J, Duffy V B, Etter L, Fast K, Gardner E, Lucchina L A, Snyder D J, Tie K, Weiffenbach J, & Bartoshuk L M (2000). Genetic Variation and Inferences about Perceived Taste Intensity in Mice and Men. *Physiology and Behavior* 69:161–73.

Prutkin J M, Fast K, Lucchina L A, & Bartoshuk L M (1999). Prop (6-*n*-propylthiouracil) Genetics and Trigeminal Innervation of Fungiform Papillae. *Chemical Senses* 24:243 (abstract).

Rankin K M & Marks L E (1991). Differential Context Effects in Taste Perception. *Chemical Senses* 16:617–29.

Riskey D R, Parducci A, & Beauchamp G K (1979). Effects of Context in Judgements of Sweetness and Pleasantness. *Perception and Psychophysics* 26:171–6.

Schifferstein H N J (1994). Sweetness Suppression in Fructose/Citric Acid Mixtures: A Study of Contextual Effects. *Perception and Psychophysics* 56:227–37.

Silver W L & Finger T E (1991). The Trigeminal System. In: *Smell and Taste in Health and Disease*, ed. T V Getchell, R L Doty, L M Bartoshuk, & J B Snow, pp. 97–108. New York: Raven Press.

Solomon G M, Catalanotto F, Scott A, & Bartoshuk L M (1991). Patterns of Taste Loss in Clinic Patients with Histories of Head Trauma, Nasal Symptoms, or Upper Respiratory Infection. *Yale Journal of Biology and Medicine* 64:280 (abstract).

Stevens S S (1951). Mathematics, Measurement, and Psychophysics. In: *Handbook of Experimental Psychology*, ed. S S Stevens, pp. 1–49. New York: Wiley.

Stevens S S (1958). Adaptation-Level vs. the Relativity of Judgment. *American Journal of Psychology* 4:633–46.

Stevens S S (1961). To Honor Fechner and Repeal His Law. *Science* 133:80–6.

Stevens S S (1969). Sensory Scales of Taste Intensity. *Perception and Psychophysics* 6:302–8.

Teghtsoonian R (1971). On the Exponents in Stevens' Law and the Constant in Ekman's Law. *Psychological Review* 78:71–80.

Teghtsoonian R (1973). Range Effects in Psychophysical Scaling and a Revision of Stevens' Law. *American Journal of Psychology* 86:3–27.

Tepper B J & Nurse R J (1997). Fat Perception Is Related to PROP Taster Status. *Physiology and Behavior* 61:949–54.

Whitehead M C, Beeman C S, & Kinsella B A (1985). Distribution of Taste and General Sensory Nerve Endings in Fungiform Papillae of the Hamster. *American Journal of Anatomy* 173:185–201.

25

The Individuality of Odor Perception

Robyn Hudson and Hans Distel

1. Odors Are Cognitive Constructs

The central argument of this chapter is that odors are the products of plastic nervous systems and that to adequately understand how olfactory stimuli influence physiology and behavior it is necessary to understand how experience shapes the way odorants are received and processed. Fundamental to the argument is the conceptual distinction between odorants and odors (Hudson, 1999). Odorants are molecules, entities of the external world objectively definable in terms of physicochemical characteristics and capable of being interpreted by particular nervous systems to yield the perceptions we call odors. Odors are the subjective products – constructs, if you like – of individual nervous systems and thus potentially are open to the many modulating influences of what might broadly be thought of as "mind." They are among the phenomena that drive behavioral, physiological, and psychological functioning and thus are phenomena that we need to understand.

Following from that, we shall not be able to provide a proper account of olfactory function without considering the role of individual experience in shaping it. Indeed, increasing numbers of studies are suggesting that learning significantly influences aspects of olfactory function as diverse as the evocation of memories and associations, hedonic judgment, the early acquisition of preferences, and the ability to perceive and discriminate odors (Engen, 1991), as reviewed by Ayabe-Kanamura et al. (1998); see also the chapters in Sections 2 and 3 of this volume.

This experience-dependent, subjective aspect of odor perception helps explain one of the central puzzles of olfaction: the inability to provide an objective classification of the odor world, as reviewed by Hudson (1999). Despite persistent and imaginative attempts to identify the true relationships that determine the way we perceive and respond to odors, it has thus far not proved possible to order the odor world according to a limited number of objective, physicochemical parameters, nor to identify populations of receptor cells consistent with any such order. Recent breakthroughs in identifying ligands for specific olfactory receptors (Krautwurst, Yau, and Reed, 1998; Zhao et al., 1998) and identifying zones of the sensory epithelium that express particular receptor populations (Ressler, Sullivan, and Buck, 1993; Strotmann, Konzelmann, and Breer, 1996), as well as demonstration that these project to form epitope maps in the olfactory bulb (Ressler, Sullivan, and Buck, 1994; Vassar, Ngai, and Axel, 1994; Mori, 1995), are undoubtedly of major importance in advancing our understanding of odorant reception. And yet, given that most receptor neurons appear capable of responding to a variety of odorants (Sicard and Holley, 1984; Sicard, 1986; Sato et al., 1994; Bozza and Kauer, 1998; Duchamp-Viret, Chaput, and Duchamp, 1999; Sicard, Chapter 18, this volume), that most natural odorants consist of many, even hundreds, of volatiles (Laing et al., 1989; Maarse, 1991), and that lesioning of large and arbitrarily selected areas of the olfactory bulb has shown few measurable effects on behavioral responses to odor stimuli (Hudson and Distel, 1987; Lu and Slotnick, 1998), it is difficult to see how those advances alone can account for odor coding. Adding to this complexity are reports of extensive polymorphism in human olfactory receptors (Lancet et al., 1993; Glusman et al., 1996).

2. An Evolutionary Perspective

The failure of olfactory research to provide a satisfactory objective classification of odor quality or an account of odor coding at the primary receptor level is understandable when we consider the task with which the olfactory system is confronted. Under natural conditions the olfactory system faces a particular problem not encountered in other sensory modalities, namely, the large numbers and inherent unpredictability of the potentially relevant stimuli. The problem lies not only in the wide range of potential molecular effects but also in their seemingly endless possibilities for combining. That makes it extremely difficult to try to reduce the chemical world to a few primary features and to map them onto the receptor surface. In an evolutionary sense, that has also made it difficult for the system to anticipate which features of the chemical environment are going to be of particular relevance and to construct neural filters accordingly. In only a few specific contexts – for example, the development of pheromonal signals involved in reproduction (Vandenberg, 1983; Hudson and Distel, 1995) – are odorants reliable predictors of events and resources that will promote survival and reproductive fitness.

Olfactory systems appear to have evolved several means for coping with the problem of such vast numbers of stimuli. One has been to develop receptor surfaces capable of responding to a wide range of molecular features. That is reflected by the large number of receptor proteins – perhaps as many as 1,000 in mammals (Buck and Axel, 1991) – now believed to exist and by the presence of some broadly tuned receptor neurons capable of responding in distinctive ways to a spectrum of odorants (Sicard and Holley, 1984; Sicard, 1986; Duchamp-Viret et al., 1999; Sicard, Chapter 18, this volume).

A second means for coping is by learning. Learning permits flexibility and selectivity, enabling individuals to acquire or enhance their neural representations of the most appropriate information in a particular environment (Hudson et al., 1998). Thus, early in the chain of perception, within the olfactory bulb, learning processes appear to play an important part in selecting, defining, amplifying, and storing patterns of activation that are of particular significance and in helping to distinguish those patterns from the continually shifting patterns of background "noise" impinging on receptors, as reviewed by Brennan and Keverne (1997).

A third means for coping with the wide range of stimuli is to alter the receptive properties of the system according to local conditions and individual experience so as to enhance sensitivity at the receptor surface to those molecules that are of potential or actual relevance. Again, there is a growing body of evidence supporting the existence of such phenomena (Wang, Wysocki, and Gold, 1993; Nevitt et al., 1994; Semke Distel, and Hudson, 1995), as reviewed by Hudson and Distel (1998).

3. Empirical Evidence

We shall now discuss three examples from our recent work in human olfaction illustrating how individual experience can influence various aspects of olfactory function. Consistent with our belief in the importance of using natural stimuli and natural routes of stimulus delivery, those studies had two methodological features in common: Subjects were exposed to multimolecular, complex odorants drawn from everyday life (cf. Laing et al., 1989; Maarse, 1991), and the odorants were presented in squeeze bottles, with the subjects being encouraged to use their own natural sniffing techniques when sampling them (cf. Laing, 1982, 1983, 1985).

3.1. Cross-Cultural Comparisons

Obviously in humans it is difficult to define or control ecologically relevant olfactory experience – an essential requirement for investigating the relationship between naturally occurring individual learning and olfactory function. A possible solution is to make use of presumed geographical differences in olfactory experiences and compare responses to everyday odors across cultures. Because the findings from our first application of that strategy have been published elsewhere (Ayabe-Kanamura et al., 1998; Distel et al., 1999), we report them only briefly here.

We compared the responses of 40 Japanese subjects and 44 age-matched German subjects to 18 everyday odorants (Ayabe-Kanamura et al., 1998). Subjects were asked to rate the stimuli on intensity, familiarity, pleasantness, and edibility, to describe associations elicited by them, and, if possible, to name them. One-third of the odorants were presumed to be familiar to the Japanese, one-third to the Germans, and one-third to both populations. Better performances by the Japanese in providing appropriate descriptors for "Japanese" odorants, and by the Germans for "European" odorants, supported our pre-selection of the stimuli as culture-typical. Particularly clear differences between the two populations were found in pleasantness ratings. In general, a positive relationship was found between the pleasantness rating and judgment of a stimulus as edible, suggesting that culture-specific experiences – particularly of foods – significantly influence odor perception. More surprising, there were significant differences between the two populations in regard to intensity ratings for some odorants. Those differences did not seem to be artifacts of the test situation and raised the possibility that experience might influence such basic aspects of odor perception as subjective perception of stimulus strength.

That was examined by extending the original Japanese-German sample to include a group of 39 age-matched Mexican subjects tested using the same substances and in the same way (Distel et al., 1999). Despite large individual differences in the responses to a given odorant among the subjects in the three populations, significant positive correlations were found between rating scores for three measures: intensity, pleasantness, and familiarity. Most notable were the consistent positive correlations between perceived intensity and the ratings of familiarity and pleasantness. That suggested that the subjective perceptions of the intensities of odors may depend not only on stimulus concentration but also on aspects of an individual's experiences of odorants in the context of everyday life.

It would seem to make sense that odors that have acquired meaning for an individual should be perceived and attended to more readily than stimuli of little relevance, and that should result in stronger subjective impressions of their strengths. However, that is not the way we are accustomed to thinking about stimuli in classic psychophysical research, where it has long been assumed that there are close and invariant relationships between our subjective perceptions and the objectively measurable physical properties of the external world.

3.2. Intra-cultural Comparisons

Given the clear results (and, in the case of perceived intensity, the surprising results) of our cross-cultural comparisons, we were interested in examining the generality of those findings by seeing if they could be replicated in culturally more homogeneous populations. Because the more surprising and potentially most controversial finding in our studies concerned the relationship between experience or knowledge of an odor and its perceived intensity, that aspect has been particularly emphasized.

Influence of prior knowledge. A first, preliminary study involved testing the responses of a northern European, mainly Scandinavian, population of nine women and four men (mean age 26.5 ± 2.3 years) to 30 everyday odorants, some of which are listed in Figure 25.1. The method of stimulus presentation was

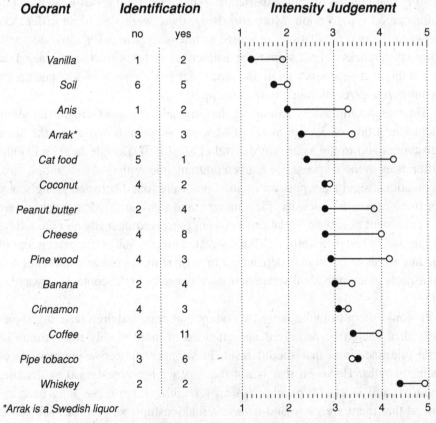

Figure 25.1. Median rating scores for intensity judgments when subjects could identify the odorant (open circles) or could not (filled circles).

the same as in our cross-cultural study. In a first trial, subjects sampled each sniff bottle and responded "certainly know," "certainly don't know," or "uncertain." Of the 30 odorants, 20 were then selected as being well known to some subjects and unknown to others. In a second trial on the same day, subjects were presented with 4 of the odorants, 2 they had claimed to recognize and 2 they did not know, and, in addition, the odor of coffee, which all subjects claimed to know. An attempt was made to select odorants in such a way as to have equal numbers of subjects knowing and not knowing a given odor, in order to control for possible intrinsic effects of odorants on perception.

Subjects first rated the intensities of the five odors on a 5-point scale that included intermediate values (e.g., 4.6). In a second round they were asked to rate the odors for pleasantness and familiarity on similar 5-point scales, to indicate whether or not the odors represented edible substances and whether or not they would be willing to eat those substances, and finally to name each odorant or provide an association.

The responses to the odorants were divided into two groups, on the basis of the apparent certainty of subjects' odor knowledge inferred from accurate naming or provision of an appropriate association. Because several of the knowledge ratings from the first trial with 20 odorants appeared inaccurate, the final groups were not equal in number, and 6 odorants had to be dropped from the final analysis. So 14 odorants were judged by subjects with different levels of odor knowledge – on average, by four subjects knowing the odor and by three not knowing it. Median scores for the two groups were calculated for each substance.

Subjects who knew an odor gave higher intensity ratings, on average, than did those not knowing it (3.8 vs. 2.9) (Figure 25.1), as well as higher pleasantness ratings (3.5 vs. 2.7) and higher familiarity ratings (3.7 vs. 3.0). Those differences were significant when compared across all odor pairs combined (Wilcoxon signed-ranks test, $p < .006$, $p < .02$, $p < .01$, respectively).

The findings basically confirmed those in our cross-cultural study. We found the same pattern of positive association between subjects' familiarity with or knowledge of an odor and their judgments of its intensity and pleasantness. Thus, although it was based on a very small sample, the study suggested that such relationships may exist even within culturally more homogeneous groups and do not depend on the marked differences in experience of particular odorants presumed to exist in different cultures – differences that would have been further enhanced by our selection of culture-typical stimuli in the cross-cultural study. The similarity between the findings in the Scandinavian study and the cross-cultural study also reduces the probability that the differences between national groups found in the cross-cultural study were genetically based.

Prior knowledge versus context. Encouraged by such results, we conducted a larger study (Distel and Hudson, 2001) along similar lines with German subjects, but with one modification. In addition to assessing the relationships between subjects' prior knowledge of a set of everyday odors and their ratings of intensity, pleasantness, and familiarity, we evaluated the association between the subjects' judgments of the goodness of fit between (1) their perceptions of the odors and (2) information about the odor sources provided by the experimenter.

The responses of 76 medical students (mean age 24.2 ± 5.2 years) of both sexes (47% women) to the 24 everyday odorants listed in Figure 25.3 were tested under two conditions: (A) without information about the odor source; (B) together with the veridical name of the odorant. In both conditions we asked the subjects to rate the intensity, pleasantness, and familiarity of the odors on a scale from 1 to 10. In condition A we also asked them to identify the source, whereas in condition B we asked them to rate how well the odor corresponded to the name provided (on a scale from 1 to 5). Subjects were randomly assigned to one of the two groups, with 38 subjects in each. Each group rated half of the odorants using condition A, and the other half using condition B, so that each odorant was rated the same number of times under the two conditions.

No significant differences were found between men and women, and so their scores have been combined. Again, the findings have basically confirmed those of our earlier studies. As summarized in Figure 25.2, for the three parameters

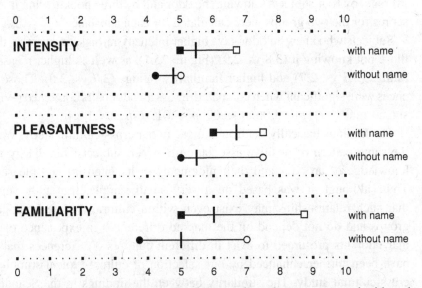

Figure 25.2. Median rating scores for everyday odorants presented with and without veridical names: filled circles, odor not identified; open circles, odor identified; filled squares, name inappropriate for odor; open squares, name appropriate for odor.

of intensity, pleasantness, and familiarity, knowledge of the odor, either because of a subject's ability to identify the source or because the veridical name was provided by the experimenter, was associated with higher ratings for all three parameters. Knowledge of the veridical name was associated with increases of 0.7, 1.3, and 1.0 points for intensity, pleasantness, and familiarity, respectively, and the effect was significant (Wilcoxon signed-ranks test, $p < .025$ across all odorants combined). If subjects were able to identify the odor sources unaided (37% of presentations under condition A), the increase in rating scores (compared with inability to identify) was significant for intensity ($p < .001$) and was even more pronounced for pleasantness and familiarity (both $p < .0001$).

However, when subjects believed that the names provided did not fit the odors (59% of presentations under condition B), they were judged to be less intense, less pleasant, and less familiar than when names and odors were perceived to coincide (Figure 25.2; 5 vs. 6.75, 6 vs. 7.5, and 5 vs. 8.75 points, respectively; Wilcoxon signed-ranks test, $p < .0001$).

Figure 25.3 shows ratings of intensity for all of the odorants to illustrate in more detail the influence of subjects' prior knowledge on their judgments of intensity. The pattern in the left panel (condition A) confirms the positive association between subjects' knowledge of an odor (correct identification of its source) and their perceptions of its intensity, and the right panel (condition B) shows a clear association between the perceived intensity of an odor and subjects' judgments of the goodness of fit between the veridical name provided and their own perceptions of the odor.

In addition to confirming our earlier findings, the study just described suggests that odor perception can be strongly influenced by the goodness of fit between subjects' expectations or imagination of how a particular odorant should smell and their external contextual information (in this case, provision of the veridical name). If the fit between internal expectations and external information is good, ratings of intensity, pleasantness, and familiarity will be higher. If the fit is poor, the resulting dissonance will lead to lower ratings for those parameters.

But how can we account for such variability in responding? There surely are as many answers to that as there are individual lives and olfactory experiences, but there is one general factor that may well be important: Most of the stimuli used were foods or beverages and thus would normally have been experienced in the context of ingestion. In our study, the absence of taste input and of the aromas released by chewing and mixing with saliva – that is, the sensory complex of flavor – may have contributed to the dissonance between the subjects' expectations generated by provision of the veridical name and their perceptions arising from smell alone.

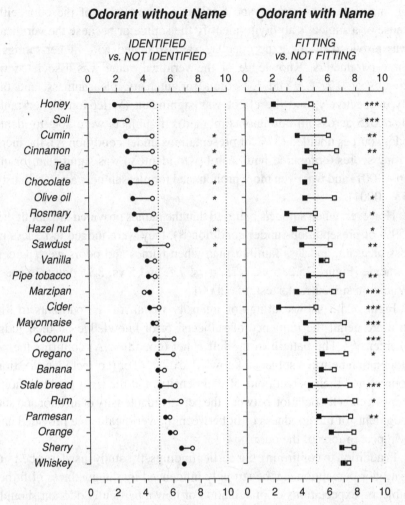

Figure 25.3. Median scores for intensity judgments for everyday odorants presented with and without veridical names: filled circles, odor not identified; open circles, odor identified; filled squares, name inappropriate for odor; open squares, name appropriate for odor; *$p < .05$; **$p < .01$; ***$p < .001$; Wilcoxon-Mann-Whitney test for independent samples.

And what might these results tell us about the nature of odor representation in the brain? It has often been debated whether or not we can truly remember or imagine odors, and thus to what extent we possess accurate, detailed, and behaviorally relevant odor images (Crowder and Schab, 1995). The findings here suggest that we do indeed possess such internal representations and that they are of sufficient definition and force to shape our perceptions of odors, depending on the goodness of fit among our images, the actual olfactory stimuli, and the information available from our immediate environment.

4. Implications for Future Research

At least three broad implications emerge from the studies described here.

4.1. Importance of Experience

The findings from all four of our studies point to the importance of individual experience or knowledge in shaping responses to odors and thus in influencing odor-guided behavior. In doing so, the findings underscore the individuality of olfactory perception. Therefore we should not expect to be able to identify finely graded psychophysical universals – at least not with respect to behaviorally relevant, everyday odorants – and in assessing olfactory function we will need to take into account the individual amounts and types of experience with the particular test stimuli used.

That raises the question to what extent the ages of subjects and the particular contexts in which their olfactory learning occurred can influence the intensity and longevity of memory for odors. For example, the widespread phenomenon of odors evoking vivid memories from early childhood (Engen, 1991), combined with reports of early olfactory imprinting in animals, as reviewed by Hudson (1993), would suggest that early experience may be particularly salient. Certainly there is now ample evidence that human newborns possess a good sense of smell and rapidly learn olfactory features of their environment from birth and probably even in utero (Schaal et al., Chapter 26, this volume).

4.2. Influence of Context

It follows from all the foregoing evidence and discussion that odors are experienced and learned in individual contexts that are culturally specific. Not only can a particular odor signify or predict a particular context or event (a bottom-up process, if you will), but also a particular context can activate expectations of the odors to be encountered (a top-down process), and depending on the goodness of fit between those two processes, one's behavioral responses and decisions are made either for good or ill. That is particularly clear in the context of eating, the basic biological function in which the chemical senses still play the critical and dominant role: We order from a menu and then judge the dish or the wine, to a considerable extent, on the basis of the degree of congruence between our experience-based expectations and our immediate chemosensory perceptions. Of course, there are some lucky surprises, occasional positive mismatches between expectation and actuality, but as many newlyweds and market researchers know, those tend to be the exceptions.

4.3. Explanatory Limits of Peripheral Coding

As we review our findings and those from other studies, we see that perhaps their most far-reaching consequences, and thus the ones that have to be thought through most carefully, are the implications for our efforts to understand what we generally refer to as olfactory coding. Initially based on chemistry-inspired psychophysics, and more recently advanced by spectacular achievements in molecular biology, olfactory research has long been guided by the reductionist dream of being able to identify neural and perceptual universals and to understand the whole as the sum of its parts.

Without doubt, reductionist approaches have provided and will continue to provide essential information and perhaps breakthroughs of the kind mentioned in Section 2 of this chapter. At the same time, for reasons outlined in that same section and supported by many examples in this volume (e.g., Sicard, Chapter 18), we need to recognize the limitations imposed on reductionist levels of explanation by the increasingly well-documented operation of cognitive processes. Recognition of that necessity should not lead to annoyance or despair. Rather, it should increase our appreciation of a sensory system that allows such individual, ecologically appropriate fine tuning to an infinitely complex chemical world.

Acknowledgments

The data for the Scandinavian study were collected during a practical class at the 1998 International Course in Sensory Ecology, Lund University, Sweden. We thank the director of the course, Bill Hansson, for making this possible.

References

Ayabe-Kanamura S, Schicker I, Laska M, Hudson R, Distel H, Kobayakawa T, & Saito S (1998). Differences in the Perception of Everyday Odors – A Japanese-German Cross-Cultural Study. *Chemical Senses* 23:31–8.

Bozza T C & Kauer J S (1998). Odorant Response Properties of Convergent Olfactory Receptor Neurons. *Journal of Neurophysiology* 63:175–87.

Brennan P A & Keverne E B (1997). Neural Mechanisms of Mammalian Olfactory Learning. *Progress in Neurobiology* 51:457–81.

Buck L B & Axel R (1991). A Novel Multigene Family May Encode Odorant Receptors: A Molecular Basis for Odor Recognition. *Cell* 65:175–87.

Crowder R G & Schab F R (1995). Imagery for Odors. In: *Memory for Odors,* ed. R G Crowder & F R Schab, pp. 93–107. Hillsdale, NJ: Lawrence Erlbaum.

Distel H, Ayabe-Kanamura S, Schicker I, Martínez-Gómez M, Kobayakawa T, Saito S, & Hudson R (1999). Perception of Everyday Odors – Correlation between Intensity, Familiarity and Strength of Hedonic Judgement. *Chemical Senses* 24:191–9.

Distel H & Hudson R (2001). Odor Judgements Are Influenced by the Subject's Knowledge of the Odor Source. *Chemical Senses* 26:247–52.

Duchamp-Viret P, Chaput M A, & Duchamp A (1999). Odor Response Properties of Rat Olfactory Receptor Neurons. *Science* 284:2171–4.

Engen T (1991). *Odor Sensation and Memory.* New York: Praeger.

Glusman G, Clifton S, Roe B, & Lancet D (1996). Sequence Analysis in the Olfactory Receptor Gene Cluster on Human Chromosome 17: Recombinatorial Events Affecting Receptor Diversity. *Genomics* 37:147–60.

Hudson R (1993). Olfactory Imprinting. *Current Opinion in Neurobiology* 3:548–52.

Hudson R (1999). From Molecule to Mind: The Role of Experience in Shaping Olfactory Function. *Journal of Comparative Physiology* 185:297–304.

Hudson R & Distel H (1987). Regional Autonomy in the Peripheral Processing of Odor Signals in Newborn Rabbits. *Brain Research* 421:85–94.

Hudson R & Distel H (1995). On the Nature and Action of the Rabbit Nipple-Search Pheromone: A Review. In: *Chemical Signals in Vertebrates. VII. Advances in the Biosciences,* ed. R Apfelbach, D Müller-Schwarze, K Reuter, & E Weiler, pp. 223–32. Oxford: Pergamon.

Hudson R & Distel H (1998). Induced Peripheral Sensitivity in the Developing Vertebrate Olfactory System. *Annals of the New York Academy of Sciences* 855:109–15.

Hudson R, Bacheralier J, Doupe A J, Fanselow M S, Kuhl P K, Menzel R, Morris M G, Rudy J W, & Squire L R (1998). What Does Behavior Tell Us about the Relationship between Development and Learning? In: *Mechanistic Relationships between Development and Learning. Dahlem Workshop Report,* ed. T J Carew, R Menzel, & C J Shatz, pp. 75–92. Chichester: Wiley.

Krautwurst D, Yau K W, & Reed R R (1998). Identification of Ligands for Olfactory Receptors by Functional Expression of a Receptor Library. *Cell* 95:917–26.

Laing D G (1982). Characterisation of Human Behaviour during Odour Perception. *Perception* 11:221–30.

Laing D G (1983). Natural Sniffing Gives Optimum Perception for Humans. *Perception* 12:99–117.

Laing D G (1985). Optimum Perception of Odor Intensity by Humans. *Physiology and Behavior* 34:569–74.

Laing D G, Cain W S, McBride R I, & Ache B W (ed.) (1989). *Perception of Complex Smells and Tastes.* New York: Academic Press.

Lancet D, Gross-Isseroff R, Margalit T, Seidemann E, & Ben-Arie N (1993). Olfaction: From Signal Transduction and Termination to Human Genome Mapping. *Chemical Senses* 18:217–25.

Lu X C & Slotnick B (1998). Olfaction in Rats with Extensive Lesions of the Olfactory Bulbs: Implications for Odor Coding. *Neuroscience* 84:849–66.

Maarse H (1991). *Volatile Compounds in Foods and Beverages.* New York: Marcel Dekker.

Mori K (1995). Relation of Chemical Structure to Specificity of Response in Olfactory Glomeruli. *Current Opinion in Neurobiology* 5:467–74.

Nevitt G A, Dittman A D, Quinn T P, & Moody W J (1994). Evidence for a Peripheral Olfactory Memory in Imprinted Salmon. *Proceedings of the National Academy of Sciences USA* 91:4288–92.

Ressler K J, Sullivan S L, & Buck L B (1993). A Zonal Organization of Odorant Receptor Gene Expression in the Olfactory Epithelium. *Cell* 73:597–609.

Ressler K J, Sullivan S L, & Buck L B (1994). Information Coding in the Olfactory System: Evidence for a Stereotyped and Highly Organized Epitope Map in the Olfactory Bulb. *Cell* 79:1245–55.

Sato T, Hirono J, Tonoike M, & Takebayashi M (1994). Tuning Specificities to Aliphatic Odorants in Mouse Olfactory Receptor Neurons and Their Local Distribution. *Journal of Neurophysiology* 72:2980–9.

Semke E, Distel H, & Hudson R (1995). Specific Enhancement of Olfactory Receptor Sensitivity Associated with Foetal Learning of Food Odours in the Rabbit. *Naturwissenschaften* 82:148–9.

Sicard G (1986). Electrophysiological Recordings from Olfactory Receptor Cells in Adult Mice. *Brain Research* 397:405–8.

Sicard G & Holley A (1984). Receptor Cell Responses to Odorants: Similarities and Differences among Odors. *Brain Research* 292:283–96.

Strotmann J, Konzelmann S, & Breer H (1996). Laminar Segregation of Odorant Receptor Expression in the Olfactory Epithelium. *Cell and Tissue Research* 284:347–54.

Vandenberg J G (1983). *Pheromones and Reproduction in Mammals*. New York: Academic Press.

Vassar R, Ngai J, & Axel R (1994). Spatial Segregation of Odorant Receptor Expression in the Mammalian Olfactory Epithelium. *Cell* 74:309–18.

Wang H W, Wysocki C J, & Gold G H (1993). Induction of Olfactory Receptor Sensitivity in Mice. *Science* 260:998–1000.

Zhao H, Ivic L, Otaki J M, Hashimoto M, Mikoshiba K, & Firestein S (1998). Functional Expression of a Mammalian Odorant Receptor. *Science* 279:237–42.

26

Olfactory Cognition at the Start of Life: The Perinatal Shaping of Selective Odor Responsiveness

Benoist Schaal, Robert Soussignan, and Luc Marlier

In the days following birth, human newborns are swept into a whirlwind of stimuli that are both complex and unpredictable in terms of space, time, and multimodal contingencies. In their very first excursion into the postnatal world, newborns can respond to only a limited array of those stimuli, but their brains possess effective perceptive and integrative capacities that soon will allow them to organize differential perceptions and directional actions in what might otherwise be a "blooming, buzzing confusion." Our current knowledge of the initial states of perceptual and cognitive functions and their development has been derived almost exclusively from studies of unimodal and cross-modal visual and auditory perceptions (e.g., Gottlieb and Krasnegor, 1985; Lewkowicz and Lickliter, 1994; Kellman and Arterberry, 1998; Slater, 1998; Rochat, 1999). However, evidence that nonvisual, nonaural cues are involved in the partitioning of the neonatal *Umwelt* has begun to accumulate, both for animals and for humans. Studies have demonstrated, for example, that some newborn animals rely so profoundly and pervasively on olfactory cues that "we cannot appreciate their behavior without understanding the roles of olfaction" (Alberts, 1981, p. 352). Compared with the findings in animal studies, the evidence for olfactory modulation of behavior in our own species remains narrow and shallow, although the body of data is steadily growing. Accordingly, our current knowledge of human cognitive development does not encompass the psychological facets of the chemical senses, a situation that may distort our understanding of the world and mind of the infant (Turkewitz, 1979).

This chapter deals with the initial state of olfactory functioning in the human infant and its changes during early development. We shall first survey some recent research on sensory and hedonic discrimination of odors at the very beginning of cognition. Then we shall address such issues as whether or not olfactory competence is predisposed to process certain kinds of stimuli during early development. By which experience-dependent or experience-independent mechanisms is olfactory cognition shaped? How variable are individuals' trajectories of olfactory development as functions of the early odor environment?

1. Discriminative, "Categorical," and Hedonic Processing of Odors by Newborns

Two basic processes have been investigated in attempts to describe infant perception: discrimination and categorization. Success in discrimination is generally considered to indicate an ability to process a given stimulus in a given context. The most straightforward measure of early discrimination ability relies on expression of differential responses to distinct odors presented in succession. The frequency with which subjects will respond and the magnitude and latency of their motor or autonomic reactions clearly will vary for different odor stimuli (e.g., Engen, Lipsitt, and Kaye, 1963; Self, Horowitz, and Paden, 1972; Yasumatsu et al., 1994; Soussignan et al., 1997).

Simultaneous presentation of odor stimuli can provide even stronger evidence for discrimination. Thus, paired odor-choice paradigms (Macfarlane, 1975) have been used in a number of studies of infants' ability to *actively* show odor discrimination, recognition, and preference. Some of those studies will be summarized later.

Infants readily detect and differentiate odors, but little is known about the "dimensions" they use in discriminative processing. It is generally assumed that the differential stimulative power of odorants results both from sensory attributes (quality, intensity, irritation) and from psychological attributes (hedonic value, familiarity). There is empirical evidence that newborns treat some of those dimensions discriminatively, such as quality (Engen and Lipsitt, 1965), intensity (Rovee, 1969), and familiarity (Schleidt and Genzel, 1990). Most such studies have relied on the notion that hedonic responses are basic in early development (Schneirla, 1965). Hence, they have made use of the dichotomous nature of neonatal behavior, that is, use of the two response tendencies termed, in various theoretical contexts, positive versus negative, approach versus withdrawal, attraction versus aversion, orientation versus defense, or low versus high arousal.

The use of behavioral markers of opposite response tendencies allowed some interesting insights into the early ability to differentiate odor stimuli. Early experiments suggested that "naive" newborns (i.e., tested right after birth, with minimal postnatal chemosensory experience) showed different facial expressions when exposed to odorants that had opposite hedonic valences. Steiner (1979) took advantage of that observation to try to assess empirically the hedonic responses of newborns exposed to odorants categorized a priori, by adults, as hedonically positive or negative: The photographs of infants' facial expressions elicited by putatively pleasant odors (e.g., banana, vanilla, milk) were rated by a panel, blind to the stimuli presented, as indicating attraction or indifference. In contrast, the infants' facial responses to unpleasant odors (having fishy or rotten qualities) were consistently rated as indicating rejection. On the basis of the precocity indicated by those differential responses, as well as similar data from an anencephalic newborn, Steiner concluded that "their innate, possibly inherited character is demonstrated" (Steiner, 1979, p. 278). He speculated on the existence of a "hedonical monitor" that would release the facial responses in a stereotyped, reflexive way. However, it was not clear whether infants were responding to stimulus differences in intensity, quality, or familiarity. In addition, the newborns' facial responses to the odors were judged from single photographs of each expression taken at some undefined time after stimulus administration and assessed by a rather coarse judgment method. Despite those concerns, Steiner's observations remain important in that they raised the issue of whether newborns respond selectively to specific odors or to categories of odors when smelling stimuli for the first time.

In a reinvestigation of neonatal ability to discriminate hedonically between odors, we attempted to control for several of the shortcomings in the Steiner study (Soussignan et al., 1997): Three-day-old infants were presented with odorants from their perinatal environment or with pure, reagent-grade odorants. All biological odorants were unfamiliar and were either of human origin (amniotic fluid, breast milk, but not from the infant's own mother) or of commercial origin (cow's-milk-based formula and protein hydrolysate formula, but not the brand usually consumed by the infant); pure vanilla and butyric acid were diluted to match, in intensity and trigeminal potency, each of the biological odorants. Subjects were thus presented with 12 odor stimuli (plus a wet control) while their respiratory rates and facial responses were recorded. Changes in respiratory rates (relative to control) demonstrated that all odors had been detected. Facial behaviors (in terms of proportion of responding infants, and numbers and durations of facial actions) also differentiated between the odor stimuli and the control. However, a fine-grained quantification of facial movements, using the infant version of Ekman and Friesen's Facial Action Coding System (Oster and Rosenstein, 1993), did not confirm that odorants classified by adults as pleasant (vanillin) and unpleasant (butyric acid) triggered facial expressions reflecting attraction and aversion, respectively. In fact, the topographies and combinations of facial actions were extremely variable across subjects (Figure 26.1). Thus the olfacto-facial responses were far from having the stereotyped character of a reflex. Nevertheless, butyric acid was significantly more effective in evoking facial markers of disgust than was vanillin, whereas vanillin did not trigger more smiling or mouthing movements than did butyric acid. Thus, reliable facial

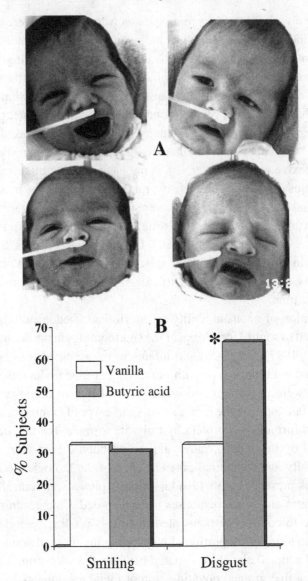

Figure 26.1. A: Samples of the facial expressions of newborns in response to odors: (upper left) formula milk (age 82 hours, latency from stimulus onset 5.12 sec); (upper right) butyric acid (age 115 hours, latency 2.88 sec, concentration 0.0078%); (lower left) butyric acid (age 124 hours, latency 2.76 sec, concentration 0.125%); (lower right) vanillin (age 68 hours, latency 3.56 sec, concentration 0.0078%). B: Proportions of infants displaying smiling (FACS action unit 12) and disgust (FACS action units 9–10) to successive presentations of the odors of vanillin and butyric acid (rated by an adult panel as pleasant and unpleasant, respectively); *$p < .05$. (Adapted from Soussignan et al., 1997.)

responses of human newborns seemed to be more easily elicited by an unpleasant (to adults) stimulus than by an iso-intense pleasant stimulus. Among the biological odorants, only the odor of unfamiliar amniotic fluid increased the numbers of negative facial actions, as compared with the control stimulus. The odor of the unfamiliar formula did not provoke increased expressions of disgust, although it was rated slightly unpleasant to very unpleasant by adults. Thus infants' facial responses may, by far, not correspond to adults' hedonic ratings of odorants.

Two points from our study pertaining to the early development of olfactory cognition should be highlighted. First, positive and negative behavioral tendencies, as demonstrated by the infants' facial responses to odors, seem to be relatively independent. Thus, pleasantness and unpleasantness may constitute independent perceptual dimensions, and the opposite repertoires of responses may emerge at different rates during development. It has been suggested that a general positive emotional responsiveness develops earlier than negative responsiveness (Schneirla, 1965; Izard, 1977). That rule, however, may not apply equally to all modalities: Although the data on neonatal olfaction are limited, they suggest that olfacto-facial systems for responding may be better integrated on the negative pole than on the positive pole of hedonic space.

Second, in our experiment described earlier (Soussignan et al., 1997), the autonomic measure revealed an influence of odor familiarity on responsiveness: Whereas bottle-fed infants responded with their highest increases in respiratory rates to the unfamiliar formula milk, breast-fed infants showed the highest increases in their respiratory activity to unfamiliar breast milk. Such differential responsiveness can be interpreted as reflecting abilities (1) to extract information from complex, unfamiliar chemical stimuli, (2) to compare variations of such information to previously stored information, and (3) to respond discriminatively to a stimulus that fits (or does not fit) to an already encoded odor image.

Deriving a single psychological category from different versions of a stimulus has been termed, in the realm of vision, "extraction of invariants" (Antell, Caron, and Myers, 1985) or "perceptual equivalence categorization" (Bornstein, 1984). With the use of a habituation paradigm, similar abilities for invariance extraction from olfactory stimuli have been shown in newborns by Engen and Lipsitt (1965). They found that 3-day-old newborns were able to discriminate the odor of a mixture and the odors of its components. Following a habituation of respiratory disruption produced by an anise/asafetida mixture, dishabituation was obtained with asafetida alone, but not with the anise component. Judgments of similarity by adults indicated that the odor of the mixture was dominated by the anise note, suggesting that newborns had difficulty in discriminating between pure anise and the mixture. When the procedure was repeated with a more

balanced odorant mixture of isoamyl acetate and heptanal, both odorants efficiently dishabituated the respiratory disruption; in addition, the dishabituation
efficacy of each compound was proportional to its dissimilarity from the mixture
(as judged by adults). That approach not only confirmed that the newborns were
able to make fine qualitative discriminations but also suggested that they may
have perceived the qualitative distance between those odors much as did adults.

Neonatal ability to extract invariants from salient odor stimuli was further assessed by using the phenomenon of negative alliesthesia. "Alliesthesia" refers to a
shift in one's hedonic evaluation of food that has been eaten to satiation (Cabanac
and Duclaux, 1973). It is called "negative" ("positive") when the change in one's
internal state is a decrease (increase) in the perceived pleasantness of a given
stimulus. That model would predict that motivation-dependent fluctuations in
hedonic responsiveness would pertain only to the odor of a familiar food or to
odors resembling that of the familiar food. Accordingly, we set out to determine
if infants would show hedonic discrimination for milk odors as a function of their
prandial state (Soussignan, Schaal, and Marlier, 1999). Taking advantage of the
qualitative variety and stability of the brands of milk formulas, we tested only
bottle-fed newborns. Their facial responses were recorded as they were exposed
to five odorants: familiar (regular formula); unfamiliar but qualitatively similar
(unfamiliar regular formula); unfamiliar and dissimilar to the familiar formula
(protein hydrolysate formula and vanillin); neutral control (the similarities had
been assessed by an adult panel). Measurements of facial movements revealed
that, after feeding, infants displayed more hedonically negative expressions to
their familiar formula and, to a lesser extent, to the unfamiliar but qualitatively
similar formula than they had before feeding (Figure 26.2). In contrast, their
prandial state had no effect on hedonic facial expressions when responding to
the odors of the protein hydrolysate formula and the vanillin. In brief, newborns
showed post-ingestive increases in aversive facial expressions to odorants that
shared quality or intensity features with their habitual food, but not to odorants
with unfamilar features (vanillin, protein hydrolysate formula).

Those data can be interpreted as evidence that newborns are able to generalize
the sensory features of a familiar odorant to a similar, but qualitatively different,
odorant they have never encountered. Interestingly, such transfers were seen
across substantial variations in stimulation contexts and behavioral states. Infants
were able to encode qualitative features of their milk perceived retronasally while
being fed and to transfer their responses into a testing situation in which the
stimulus was presented orthonasally, quite independent of oral chemoreception.
In addition, they were asleep during the olfactory tests, but were awake during
the acquisition phase (i.e., the feeding episodes), which suggests that low-level
cognitive processes were involved in the sensory encoding.

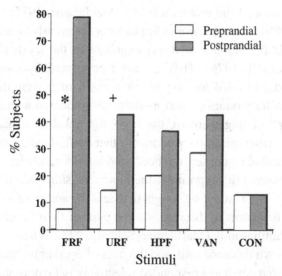

Figure 26.2. Proportions of newborns ($n = 22$) displaying aversive facial responses to the odors of familiar and unfamiliar regular formulas (FRF and URF, respectively), an unfamiliar protein hydrolysate formula (HPF), vanillin (VAN), and a control stimulus (CON, water) as functions of the prandial state; $*p < .01$; $^+p = .07$. (Adapted from Soussignan et al., 1999.)

In sum, human infants, from the very beginning of extrauterine life, are competent to extract qualitative information from environmental odors and to translate that into hedonic responsiveness. To a great extent, the hedonic direction of an infant's response to a given odor is dependent on previous exposure. Some of the mechanisms mediating such experiential effects will be surveyed next.

2. Shaping of Olfactory Preferences in Early Ontogeny

Mammalian newborns have a general ability to associate distinct odors with different contexts and later to exhibit differential behavior patterns to those odors. Early acquisition of olfactory competence thus is characterized by high degrees of plasticity and specificity of odor–response associations; for reviews, see Alberts (1987), Johanson and Terry (1986), Wilson and Sullivan (1994), and Hudson, Schaal, and Bilko (1999). Human newborns are equally adept olfactory learners, though much less studied. They not only can detect and discriminate odorants but also will behave in ways indicating that they can use olfactory "representations" derived from previous experience.

Some of the processes by which odor-mediated preferences are acquired by human newborns have been uncovered in social contexts. The first empirical evidence that responses to social odors could be highly selective came from choice tests using the odor of the mother's breast (Macfarlane, 1975; Russell, 1976; Schaal et al., 1980). The response to breast odor apparently is related to both postnatal age and the amount of direct exposure to the mother's body (Macfarlane, 1975; Russell, 1976). That has been confirmed by comparing 2-week-old breast- and bottle-fed infants: Whereas the former were differentially attracted to the axillary odors of their mothers (as compared with axillary odors from unfamiliar lactating women), the latter did not differentiate their mothers' axillary odors from the axillary odors of other bottle-feeding mothers. The absence of a differential response from bottle-fed infants was believed to be related to their less extensive prior exposure to their mothers' skin odors (Cernoch and Porter, 1985). In a further study, we sought to assess the ability of 4-day-old breast- and bottle-fed newborns to discriminate the odor of their familiar milk and the odor of an unfamiliar milk (Marlier and Schaal, 1997). When presented with the odors of their own mothers' milk and the odor of another mother's milk simultaneously, breast-fed newborns responded selectively in favor of their own milk by longer-duration head turns and appetitive mouthing. In contrast, bottle-fed newborns did not exhibit clear differential responses when presented with their familiar formula and an unfamiliar, qualitatively different formula. The amount of exposure to milk odor may not necessarily be causal to the differential effects of breast feeding and bottle feeding, because factors such as the daily chemical environment obviously can have an impact, in at least two respects: (1) breast- and bottle-fed infants are exposed to distinct arrays of milk-borne substances, possibly affecting olfactory performance differentially (Marlier, Schaal, and Soussignan, 1998b); (2) they are exposed to different chemosensory contexts: somewhat variable for breast-fed infants versus monotonous for bottle-fed infants (Jiang et al., 1998). Although, it has not yet been shown that early variability in food-related odors will later lead to discriminative keenness in human infants, empirical evidence for a functional consequences from early food variability (i.e., reduced neophobia toward unfamiliar foods) has been obtained in young rats (Capretta, Petersik, and Steward, 1975).

The effectiveness of early odor learning has been further substantiated by experiments that have examined to what extent artificial odorants can become preferred in different reinforcing contexts, such as mother–infant contact outside the feeding situation (Schaal, 1986), breast feeding per se (Schleidt and Genzel, 1990), and tactile stimulation (Sullivan et al., 1991b). In Schleidt and Genzel's study, mothers perfumed their breasts for 2 weeks after birth. At the ages of 1 and 2 weeks, their infants were given a choice test between the familiar perfume and a novel odor. At both times, the infants oriented longer to their mothers' perfume than to the control. After the test at 2 weeks, the breast-odor association was discontinued for some of the infants, while their peers remainded exposed

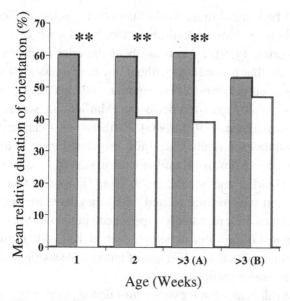

Figure 26.3. Mean relative durations of head (nose) orientations toward the odor of a perfume associated with the mother's breast (solid bars) and toward a novel control odor (empty bars) by newborns aged 1 week ($n = 17$), 2 weeks ($n = 18$), and more than 3 weeks (subgroup A, $n = 6$; subgroup B, $n = 10$). Infants were exposed to the perfume in conjunction with breast feeding during weeks 1 and 2; odorization of the mother's breast was continued after week 2 for subgroup A, but was discontinued for subgroup B; choice tests lasted 4 minutes; **$p < .01$. (Adapted from Schleidt and Genzel, 1990.)

to it. In a third paired-choice test, at 3–6 weeks, the preference for the mothers' perfume had disappeared in the former subgroup of infants, but it was maintained in the latter subgroup (Figure 26.3). Thus, an artificial odor can acquire a positive value after repeated association with the mother's breast or with suckling; but its positive value apparently can rapidly reverse to neutrality when it is no longer reinforced. In the study by Sullivan et al. (1991b), 1-day-old newborns were assigned to four treatment conditions: odor paired with massage of the newborn, odor and massage applied successively, odor only, and massage only – each for 10 periods of only 30 sec each. On the following day, general activity and head-turning responses were recorded in the presence of the odor. Only those infants who had been given odor and massage concomitantly exhibited "general arousal" and positive head turns toward the odorant. Thus the variability in hedonic valence for such artificial odor stimuli revealed by the use of various reinforcements shows the considerable plasticity of neonatal odor preferences.

However, there is some empirical evidence to suggest that the early malleability of odor hedonics may be developmentally or evolutionarily constrained:

Certain odors seem to be learned more easily than others. Some studies have suggested that there is something special about the breast odor of lactating women (Makin and Porter, 1989), as well as about the odor of human milk itself. Colostrum or milk that is collected without any contamination by areolar skin secretions can elicit positive head orientation and appetitive mouthing responses in 2- and 4-day-old breast-fed newborns (Marlier and Schaal, 1997; Marlier, Schaal, and Soussignan, 1997, 1998a). Furthermore, 4-day-olds who had been fed milk formula exclusively since birth responded more strongly to the odor of breast milk taken from an unfamiliar woman than to the odor of their own familiar formula (Marlier and Schaal, 1999). Thus, it appears that infants have a stronger attraction toward milk-related odors they have never directly experienced than toward an odor repeatedly experienced in the feeding context. In other words, the amount of previous exposure, even in the highly reinforcing contexts of comfort contact with the mother and of intake and satiation, does not explain all cases of positive responses to odors.

Several mechanisms might account for such odor-elicited positive orientations at first whiff that are not mitigated by postnatal feeding experience. It cannot be excluded that to bottle-fed infants, the unfamiliar odor of human milk might be reminiscent of their own mothers' body odor. It is also possible that infants may react to an intensity contrast, with breast or milk odors being weaker than formula odors (Jiang et al., 1998). Recent data, however, indicate that after the two stimuli were matched in intensity, the greater arousing power of human milk remained unchanged (Marlier and Schaal, 1999). Alternatively, the power of human milk odor may derive from (1) an inborn attraction to odors bearing special psychobiological properties, (2) rapid postnatal odor learning, (3) prenatal odor learning, or (4) the combined actions of those mechanisms. Those hypothetical possibilities will be briefly surveyed:

First, the potential action of special substances with pheromone-like properties, contained in mammary secretions and minimally dependent on prior exposure, has been mentioned for the human species (Russell, 1976), but thus far never seriously investigated. However, studies of other mammals have offered suggestive evidence for the existence of such pheromonal cues present around the nipple and/or in milk (e.g., Hudson and Distel, 1995; Coureaud et al., 1999, 2001).

Second, states of calm arousal can promote very rapid engagement of odor learning in both animal and human newborns (Wilson and Sullivan, 1994; Sullivan et al., 1991b). Thus the salience of maternal odors might become rapidly established through internal and external events related to the birth process. For example, uterine contractions induce a generalized arousal state that leads to immediate olfactory learning in rat pups (Ronca, Abel, and Alberts, 1996). In the first hours after birth, human infants are in a prolonged state of calm arousal

(McLaughlin, O'Connor, and Denni, 1981), during which they demonstrate coordinated motor activity (head orientation, rooting movements, and sucking). That state of relative "sensory receptiveness" after birth correlates with neurochemical states (e.g., high concentrations of norepinephrine) (Lagerkrantz, 1996) that favor rapid odor learning (Sullivan, McGaugh, and Leon, 1991a).

Third, neonatal attraction to the odor of human milk might be traceable back to the olfactory experience of the fetus. That possibility was examined in a recent series of experiments that consisted in exposing newborns simultaneously to salient prenatal and postnatal odorants. When facing a choice between the odor of their mothers' colostrum and the odor of their own amniotic fluids, 2-day-old breast-fed newborns did not orient toward either stimulus (Marlier et al., 1997, 1998a). That equal treatment for the odors of amniotic fluid and colostrum waned, however, at 4 days of age, when the infants were more (i.e., longer) attracted by their mothers' breast milk than by amniotic fluid (Figure 26.4). That developmental shift in relative orientation to a prenatal odor and a postnatal odor was interpreted as reflecting (1) chemosensory continuity of the substrates that contact the chemosensors of the fetus/newborn before and after birth (the odors of both being influenced by the mother's diet) (Schaal, Orgeur, and Rognon, 1995b) and (2) the possibility that with lactogenesis, around postpartum day 3, the odor of the lacteal secretions departs from the odor of colostrum and amniotic fluid and thus becomes increasingly distinguishable.

3. Odor Learning in the Fetus, Newborn, and Infant

Obviously the hypothesis of transnatal continuity in chemosensory substrates depends on the functioning of odor perception and learning in the fetus and on the use of that sensory information by the newborn. We therefore sought to examine the responsiveness to odors encountered in utero by exposing newborns at day 3 to samples of their own amniotic fluid (Schaal, Marlier, and Soussignan, 1995a) and to unfamiliar amniotic fluid (Schaal, Marlier, and Soussignan, 1998). Both breast- and bottle-fed groups responded selectively in favor of the odor of their own familiar amniotic fluid. Findings in a subgroup of cesarean-born, bottle-fed infants strengthened the case for prenatal acquisition of amniotic odor, because they had their breathing pathways cleared of amniotic remnants immediately after birth, were thoroughly washed, and were not reexposed to any amniotic-like cues that might have been transferred into their mothers' milk. Thus, those newborns seemed to have the ability to retrieve the unique amniotic odor signature acquired 3 days earlier in the aquatic conditions of the womb.

To further test the ability of fetuses to learn an isolated odor note, we compared the olfactory responses of infants born to mothers who had consumed anise

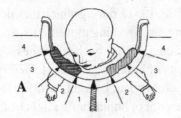

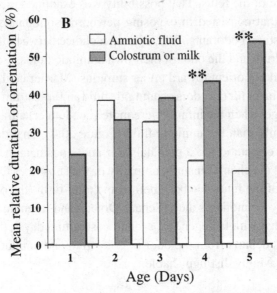

Figure 26.4. A: Device used for simultaneous, symmetrical presentation of paired odor pads (hatched zones). B: Mean relative durations of head orientations for five groups of breast-fed infants of different ages exposed to choice tests contrasting the odor of their own amniotic fluid with the odor of their mother's lacteal secretion of the day: group 1 ($n = 9$, mean age 9.7 hours); group 2 ($n = 20$, age 33.6 hours); group 3 ($n = 12$, age 55.4 hours); group 4 ($n = 16$, age 85.1 hours); group 5 ($n = 9$, age 114.2 hours). The choice tests lasted 2 minutes; **$p < .01$ (Adapted from Marlier et al., 1997.)

during pregnancy and infants born to mothers who had never ingested anise during pregnancy (Schaal, Marlier, and Soussignan, 2000). Both groups of infants were tested at 3 hours and at 4 days of age for behavioral markers of attraction (positive head-turning, appetitive mouthing responses) and markers of aversion (negative facial actions) when presented with pure, diluted anise. Infants born to anise-consuming mothers showed stable preference for anise at both times. In contrast, the newborns from non-anise-consuming mothers displayed aversion responses at the 3-hour testing, and neutral responses at the 4-day testing. That finding clearly demonstrates that pregnant women influence, partly through diet, the olfactory preferences of their offspring. Furthermore, infants are able to extract an olfactory component from the highly noisy chemical environment of the amniotic fluid.

The odor knowledge of fetuses at birth is further enriched by their ability as infants to learn rapidly from early interactions with their odor world. Without going into the question of which subtype of memory is involved, we can say that several studies have shown that olfactory memory is well developed in the newborn. The first evidence came from habituation experiments indicating that newborns (aged 2–3 days) showed short-term retention of odor information (i.e., over several minutes within a single test session) (Engen et al., 1963; Engen and Lipsitt, 1965). Much longer retention capacity of the immature brain has been reported with conditioning or exposure paradigms. In an experiment cited earlier (Sullivan et al., 1991b), the pairing of citrus odor with massage for infants aged 4–16 hours resulted, the next day, in differential arousal and preferential head orientation toward the citrus odor, as compared with a novel floral odor; see Hermans and Baeyens (Chapter 8, this volume) for a similar effect in adults. That approximately 24-hour recall ability was corroborated in a study that consisted in "merely" odorizing the newborns' crib atmosphere over a period beginning, on average, 12 hours after birth and lasting for 23 hours (Balogh and Porter, 1986). The next day (on average, 42 minutes after the exposure had ended), subjects' head-turning was examined in a test comparing the familiar odor and a novel odor. Infants oriented longer to the familiar stimulus, but that effect reached significance only for females. Using the same familiarization procedure (beginning 20 hours after birth and lasting for 22 hours) and the same testing paradigm, Davis and Porter (1991) found that even after a 2-week delay, without reexposure to the familiarization odor, the infants spent longer periods oriented toward the familiar odor than toward a novel odor. In that case, no sex difference was noted.

To sum up, these data indicate that from the earliest stages of ontogeny, the human mind can acquire and store olfactory information and demonstrate selective recall not only of perceptions imprinted during the postnatal period but also of those acquired during the prenatal period. Thus far, the postnatal stability of prenatal odor learning has been shown to last for at least 3 to 4 days, and neonatal odor memory can still be shown after a 2-week delay.

One may wonder if there are sensitive periods for odor learning. Although that has not yet been investigated in ad hoc experiments in humans, animal data clearly suggest developmental windows more favorable for olfactory acquisition. For example, the one-trial odor-learning ability of newborn rabbits follows a steep decline over postnatal days 1–5 (Kindermann, Hudson, and Distel, 1994). Likewise, 3-month-old rats will respond preferentially to an odor presented during the first week (but not later) in association with tactile stimulation

(Leon et al., 1987). More recently, rodents have been shown to be highly sensitive to odor exposure following eye-opening, around postnatal day 10 (Voznessenskaya, Feoktistova, and Wysocki, 1999). Thus the first postnatal days may correspond to a state of neural plasticity, correlated with specific endocrine levels and neurochemical states. Particular reinforcing events, such as mother-initiated contacts, arousal, suckling, and gastrointestinal stimulation, may act synergistically to facilitate odor learning in mammalian neonates. Whether or not such facilitating circumstances for odor learning also apply to the human newborn needs further investigation.

Finally, whereas it is known that odor memories are especially resistant to decay over time in adult subjects (Herz and Engen, 1996), the longevity of infantile odor and flavor associations has not been studied systematically over the life span. Although there is some clear evidence of very long-term retention of early chemosensory experience in mice and dogs (e.g., Mainardi, Marsan, and Pasquali, 1965; Hepper, 1994), the human data are thus far only suggestive of similar processes in the functional context of ingestion. For example, in a retrospective study of self-reported food aversions among adults, the origins of many aversive responses could be traced back to events that had occurred before the age of 5 years (Garb and Stunkart, 1974). A stronger indication of the long-term consequences of early chemosensory experience was provided by a study (Mennella and Beauchamp, 1996) on infants' acceptance of formula milks – especially protein hydrolysates that often are reported to have an aversive smell (to adults). Only those infants who had been fed such formulas during their first 2 months were willing to ingest them at age 7 months; infants who had never tasted such formulas prior to the ingestion test at age 7 months refused them. Finally, a quasi-experimental survey has contributed to our speculations about long-term impact of early flavor exposure. Haller et al. (1999) took advantage of the fact that before 1992, when the European Community policy changed, many brands of formula milk had been vanilla-flavored. When testing preferences for a regular ketchup versus vanillin-added ketchup in persons aged 12 to 59 years (mean 28.8 years), they noted that 60% of subjects who had been bottle-fed before 1992 preferred the vanilla-added ketchup, but the reverse was noted for breast-fed subjects who had not been exposed to vanilla in the nursing context (71% preferring regular ketchup).

4. Conclusions and Prospects

The results discussed here paint a picture of olfaction as a sophisticated means for information extraction and perceptual elaboration operating from the earliest stages of cognitive development. The best-established fact of early olfactory cognition is that infantile responsiveness to odors is strongly affected by experience. Both discriminative and hedonic responses to odors are clearly influenced by previous encounters with odorants, and such adaptability has far-reaching consequences in terms of the diversity of stimuli that can influence behavior, as well as the multiplicity of contexts (with or without reinforcers) that can promote learning. The high plasticity of odor learning presumably has evolved to ensure optimal adaptive responses in the unpredictible postnatal environment. The immediate registration of novel odors by newborns in social and ingestive situations probably helps them to rapidly deploy such learning for selective social interaction and energy conservation responses, leading to normal species-specific development.

Our interaction with environmental odors does not begin at birth. Newborns emerge from the amniotic niche with their responses biased in favor of cues that they encountered therein. That selective responsiveness supports the notion of a multilevel continuity in the sensory processing of odor stimuli before and after birth: (1) Peripheral chemoreceptive mechanisms capture odorants in either aqueous and aerial media. (2) The fetal brain is mature enough to encode complex mixtures as well as isolated olfactory cues. (3) Odor experience in the amnion induces differential attention and behavior in the newborn. The fetal odor memory is durable enough to help activate adaptive responses (e.g., orientation, searching, soothing, mouthing) to prenatal odors in the postnatal environment (e.g., amniotic fluid spread on the mother's body by the newly born infant) or to odors encountered in both environments (e.g., aromas present in utero and in colostrum). The progressive disappearance of odor cues from the original (prenatal) acquisition context leads to a declining probability of reactivation of memory traces. However, at the same time, memory stores are further enriched and elaborated through encounters with new odors in new contexts provided by the current developmental niche. Olfaction thus serves to track continuity against a background of continual change that constitutes the normal course of development.

The exact delineations of such acquisition and memory processes, as well as their resistance to decay over time, are directions for future research. The data discussed here suggest that both associative and nonassociative mechanisms are involved and that early memory can remain accessible to later activation, rather than being superseded or simply erased by later learning. Research on very young animals, and also humans, should prove especially useful in further delineating the basis of early olfactory plasticity, such as the time scale (sensitive periods) and neural mechanisms involved, the development of perceptual processes, short- and long-term memory, and cognitive top-down mechanisms (e.g., priming, olfactory lexicon).

The early integrative functions in olfaction most probably favor the development of expectancies and categorizations that engage more general cognitive competence. Whether or not newborns can use olfactory cues to build cross-modal relationships has not yet been addressed, but older infants are able to associate and recall the pairing of an odor with an object. For example, 4-month-olds who had been exposed to an object A paired with cherry odor and to an odor-free object B looked longer toward object A (as compared with the scent-less object B) when the cherry odor was subsequently diffused in the test room (Fernandez and Bahrick, 1994). Similarly, in an everyday setting, previous exposure to vanilla (via mother's milk or the home environment) was correlated positively with the patterns of exploration of a vanilla-scented toy by 6–13-month-old infants (Mennella and Beauchamp, 1998). Reardon and Bushnell (1988) have speculated that chemosensory–visual pairings might be easier for young infants to learn than somatosensory–visual or visual–visual relationships, because the hedonic values of chemical signals are more salient and hence more easily monitored as cues to the reinforcing event. That proposal marks chemosensation, by virtue of its tight connections with affective processes, as an important starting point for understanding the organization of cognitive development (e.g., Van Toller and Kendal-Reed, 1995). Thus, one solution to Bruner's developmental riddle (If we must already know something in order to learn anything, how do we get started learning at all?) may well be to learn about odors, flavors, and tastes!

Acknowledgments

The reseach reported in this chapter was partially funded by the Direction Générale de l'Alimentation (programme "Aliment Demain"), Ministères de la Recherche et de la Technologie, et de l'Agriculture. Dr. Tao Jiang is acknowledged for her help.

References

Alberts J R (1981). Ontogeny of Olfaction: Reciprocal Roles of Sensation and Behavior in the Development of Perception. In: *Development of Perception: Psychobiological Perspectives*, vol. 1, ed. R N Aslin, J R Alberts, & M R Petersen, pp. 322–52. New York: Academic Press.

Alberts J R (1987). Early Learning and Ontogenetic Adaptation. In: *Perinatal Development. A Psychobiological Perspective*, ed. N E Krasnegor, E M Blass, M A Hofer, & W P Smotherman, pp. 11–37. Orlando: Academic Press.

Antell S E, Caron A J, & Myers R S (1985). Perception of Relational Invariants by Newborns. *Developmental Psychology* 21:942–8.

Balogh R D & Porter R H (1986). Olfactory Preferences Resulting from Mere Exposure in Human Neonates. *Infant Behavior and Development* 9:395–401.

Bornstein M H (1984). A Descriptive Taxonomy of Psychological Categories Used by Infants. In: *Origins of Cognitive Skills*, ed. C Sophian, pp. 313–38. Hillsdale, NJ: Lawrence Erlbaum.

Cabanac M & Duclaux R (1973). Alliesthésie olfactive et prise alimentaire chez l'homme. *Journal de Physiologie* 66:113–35.

Capretta P J, Petersik J T, & Steward D J (1975). Acceptance of Novel Flavours Is Increased after Early Exposure of Diverse Taste. *Nature* 254:689–91.

Cernoch J M & Porter R H (1985). Recognition of Maternal Axillary Odors by Infants. *Child Development* 56:1593–8.

Coureaud G, Schaal B, Langlois D, & Perrier G (1999). The Mammary Pheromone of the Rabbit: A Milk Odor Fraction That Equals Odor Cues from Lactating Females' Abdomen and from Milk. Presented at the sixteenth annual meeting of the International Society for Chemical Ecology, Marseille, France.

Coureaud G, Schaal B, Langlois D, & Perrier G (2001). Orientation of Newborn Rabbits to Odours Emitted by Lactating Females: Relative Effectiveness of Surface and Milk Cues. *Animal Behavior* 61:153–62.

Davis L B & Porter R H (1991). Persistent Effects of Early Odor Exposure on Human Neonates. *Chemical Senses* 16:169–74.

Engen T & Lipsitt L P (1965). Decrement and Recovery of Responses to Olfactory Stimuli in the Human Neonate. *Journal of Comparative and Physiological Psychology* 59:312–16.

Engen T, Lipsitt L P, & Kaye H (1963). Olfactory Response and Adaptation in the Human Neonate. *Journal of Comparative and Physiological Psychology* 56:73–7.

Fernandez M & Bahrick L E (1994). Infants' Sensitivity to Arbitrary Object–Odor Pairings. *Infant Behavior and Development* 17:471–4.

Garb J L & Stunkart A J (1974). Taste Aversion in Man. *American Journal of Psychiatry* 131:1204–7.

Gottlieb G G & Krasnegor N A (eds.) (1985). *Measurement of Audition and Vision in the First Year of Postnatal Life*. Norwood, NJ: Ablex.

Haller R, Rummel C, Henneberg S, Pollmer U, & Köster E P (1999). The Influence of Early Experience with Vanillin on Food Preference Later in Life. *Chemical Senses* 24:465–7.

Hepper P G (1994). Long Term Retention of Kinship Recognition Established during Infancy in the Domestic Dog. *Behavioural Processes* 33:3–14.

Herz R & Engen T (1996). Odor Memory: Review and Analysis. *Psychonomic Bulletin Review* 3:300–13.

Hudson R & Distel H (1995). On the Nature and Action of the Rabbit Nipple-Search Pheromone: A Review. In: *Chemical Signals in Vertebrates VII*, ed. R Apfelbach, D Müller-Schwarze, K Reuter, & E Weiler, pp. 223–32. London: Elsevier.

Hudson R, Schaal B, & Bilko A (1999). Transmission of Olfactory Information from Mother to Young in the European Rabbit. In: *Mammalian Social Learning: Comparative and Ecological Perspectives*, ed. H O Box & K R Gibson, pp. 141–57. Cambridge University Press.

Izard C E (1977). *Human Emotions*. New York: Plenum Press.

Jiang T, Schaal B, Marlier L, & Soussignan R (1998). The Food-related Odor Environment of French Newborns: Human and Formula Milk Odors Compared by Adult Nose. *Chemical Senses* 24:80–1.

Johanson I B & Terry L M (1986). Learning in Infancy. A Mechanism for Behavioral Change during Development. In: *Handbook of Behavioral Neurobiology. Vol. 9. Developmental Psychobiology and Behavioral Ecology*, ed. E M Blass, pp. 245–81, New York: Plenum Press.

Kellman P J & Arterberry M E (1998). *The Cradle of Knowledge*. Cambridge, MA: MIT Press.

Kindermann U, Hudson R, & Distel H (1994). Learning of Suckling Odors by Newborn Rabbits Declines with Age and Suckling Experience. *Developmental Psychobiology* 27:111–22.

Lagerkrantz H (1996). Stress, Arousal, and Gene Activation at Birth. *News in Physiological Sciences* 11:214–8.

Leon M, Coopersmith R, Lee S, Sullivan R M, Wilson D A, & Woo C C (1987). Neural and Behavioral Plasticity Induced by Early Olfactory Learning. In: *Perinatal Development. A Psychobiological Perspective*, ed. N E Krasnegor, E M Blass, M A Hofer, & W P Smotherman, pp. 145–67. Orlando: Academic Press.

Lewkowicz D J & Lickliter R (eds.) (1994). *The Development of Intersensory Perception: Comparative Perspectives*. Hillsdale, NJ: Lawrence Erlbaum.

Macfarlane A J (1975). Olfaction in the Development of Social Preferences in the Human Neonate. *Ciba Foundation Symposia* 33:103–17.

McLaughlin F J, O'Connor S, & Denni R (1981). Infant State and Behavior during the First Postpartum Hour. *Psychological Record* 31:455–8.

Mainardi D, Marsan M, & Pasquali A (1965). Causation of Sexual Preferences of House Mouse. The Behaviour of Mice Reared by Parents Whose Odour Was Artificially Altered. *Atti della Societa Italiana de Scienze Naturali* 104:325–38.

Makin J W & Porter R H (1989). Attractiveness of Lactating Females' Breast Odors to Neonates. *Child Development* 60:803–10.

Marlier L & Schaal B (1997). Familiarité et discrimination olfactive chez le nouveau-né: influence différentielle du mode d'alimentation? *Enfance* 1:47–61.

Marlier L & Schaal B (1999). Feeding-Experience Dependent and Feeding-Experience Independent Odour Preference Responses in Human Infants: In Search of an Inborn Attractive Factor in Human Milk. Presented at the first European Symposium on Olfaction and Cognition, Lyon, France, June 10–12.

Marlier L, Schaal B, & Soussignan R (1997). Orientation Responses to Biological Odours in the Human Newborn. Initial Pattern and Postnatal Plasticity. *Comptes Rendus de l'Académie des Sciences (Paris) Life Sciences* 320:999–1005.

Marlier L, Schaal B, & Soussignan R (1998a). Neonatal Responsiveness to the Odor of Amniotic and Lacteal Fluids: A Test of Perinatal Chemosensory Continuity. *Child Development* 69:611–23.

Marlier L, Schaal B, & Soussignan R (1998b). Bottle-fed Neonates Prefer an Odor Experienced in Utero to an Odor Experienced Postnatally in the Feeding Context. *Developmental Psychobiology* 33:133–45.

Mennella J A & Beauchamp G K (1996). Developmental Changes in the Acceptance of Protein Hydrolysate Formula. *Developmental and Behavioral Pediatrics* 17:386–91.

Mennella J A & Beauchamp G K (1998). Infants' Exploration of Scented Toys: Effects of Prior Experiences. *Chemical Senses* 23:11–17.

Oster H & Rosenstein D (1993). Baby FACS: Analyzing Facial Movements in Infants. Unpublished manuscript.

Reardon P & Bushnell E W (1988). Infants' Sensitivity to Arbitrary Pairings of Color and Taste. *Infant Behavior and Development* 11:245–50.

Rochat P (ed.) (1999). *Early Social Development*. Hillsdale, NJ: Lawrence Erlbaum.

Ronca A E, Abel R A, & Alberts J A (1996). Perinatal Stimulation and Adaptation of the Neonate. *Acta Paediatrica, Supplement* 416:8–15.

Rovee C K (1969). Psychophysical Scaling of Olfactory Response to the Aliphatic Alcohols in Human Neonates. *Journal of Experimental Child Psychology* 7:245–54.

Russell M J (1976). Human Olfactory Communication. *Nature* 260:520–2.

Schaal B (1986). Presumed Olfactory Exchanges between Mother and Neonate in Humans. In: *Ethology and Psychology*, ed. J Le Camus & J Cosnier, pp. 101–10. Toulouse: Privat-IEC.

Schaal B, Marlier L, & Soussignan R (1995a). Responsiveness to the Odour of Amniotic Fluid in the Human Neonate. *Biology of the Neonate* 67:397–406.

Schaal B, Marlier L, & Soussignan R (1998). Olfactory Function in the Human Fetus: Evidence from Selective Neonatal Responsiveness to the Odor of Amniotic Fluid. *Behavioral Neuroscience* 112:1438–49.

Schaal B, Marlier L, & Soussignan R (2000). Human Fetuses Learn Odors from Their Pregnant Mother's Diet. *Chemical Senses* 25:729–37.

Schaal B, Montagner A, Hertling E, Bolzoni D, Moyse A, & Quichon R (1980). Les stimulations olfactives dans les relation entre l'enfant et la mère. *Reproduction, Nutrition, Développement* 20:843–58.

Schaal B, Orgeur P, & Rognon C (1995b). Odor Sensing in the Human Fetus: Anatomical, Functional and Chemo-ecological Bases. In: *Prenatal Development. A Psychobiological Perspective*, ed. J P Lecanuet, N A Krasnegor, W A Fifer, & W P Smotherman, pp. 205–37. Hillsdale, NJ: Lawrence Erlbaum.

Schleidt M & Genzel C (1990). The Significance of Mother's Perfume for Infants in the First Weeks of Their Life. *Ethology and Sociobiology* 11:145–54.

Schneirla T C (1965). Aspects of Stimulation and Organization in Approach/Withdrawal Processes underlying Vertebrate Behavioral Development. In: *Advances in the Study of Behavior*, vol. 1, ed. D S Lehrman, R A Hinde, & E Shaw, pp. 1–74. New York: Academic Press.

Self P A, Horowitz F D, & Paden L Y (1972). Olfaction in Newborn Infants. *Developmental Psychology* 7:349–63.

Slater A (ed.) (1998). *Perceptual Development. Visual, Auditory and Speech Perception in Infancy*. Exeter: Psychology Press.

Soussignan R, Schaal B, & Marlier L (1999). Olfactory Alliesthesia in Human Neonates: Prandial State and Stimulus Familiarity Modulate Facial and Autonomic Responses to Milk Odors. *Developmental Psychobiology* 35:3–14.

Soussignan R, Schaal B, Marlier L, & Jiang T (1997). Facial and Autonomic Responses to Biological and Artificial Olfactory Stimuli in Human Neonates: Re-examining Early Hedonic Discrimination of Odors. *Physiology and Behavior* 62:745–58.

Steiner J E (1979). Human Facial Expressions in Response to Taste and Smell Stimulations. In: *Advances in Child Development*, vol. 13, ed. L P Lipsitt & H W Reese, pp. 257–95. New York: Academic Press.

Sullivan R M, McGaugh J L, & Leon M (1991a). Norepinephrine-induced Plasticity and One-Trial Olfactory Learning in Neonatal Rats. *Brain Research, Developmental Brain Research* 60:219–28.

Sullivan R M, Taborsky-Barba S, Mendoza R, Itano A, Leon M, Cotman C W, Payne T, & Lott I (1991b). Olfactory Classical Conditioning in Neonates. *Pediatrics* 87:511–17.

Turkewitz G (1979). The Study of Infancy. *Canadian Journal of Psychology* 33:408–12.

Van Toller S & Kendal-Reed M (1995). A Possible Protocognitive Role for Odors in Human Infant Development. *Brain and Cognition* 29:275–93.

Voznessenskaya V, Feoktistova N, & Wysocki C (1999). Is There a Time during
    Neonatal Development for Maximal Imprinting of Odor? In: *Advances in
    Chemical Signals in Vertebrates*, ed. R E Johnston, D Müller-Schwarze, & P W
    Sorensen, pp. 617–22. New York: Plenum Press.
Wilson D A & Sullivan R M (1994). Neurobiology of Associative Learning in the
    Neonate: Early Olfactory Learning. *Behavioral and Neural Biology* 61:1–18.
Yasumatsu K, Uchida S, Sugano H, & Suzuki T (1994). The Effect of the Odour of
    Mother's Milk and Orange on the Spectral Power of EEG in Infants. *Journal of
    UOEH* 16:71–83.

27

Age-related Changes in Chemosensory Functions

Thomas Hummel, Stefan Heilmann, and Claire Murphy

Early in this century it was shown that aging is accompanied by a decrease in intranasal chemosensory sensitivity to camphor (Vaschide, 1904). Numerous studies have confirmed such findings for various odorants (Venstrom and Amoore, 1968; Schiffman, Moss, and Erickson, 1976; Stevens and Cain, 1987). Others have reported decreased ability to identify odorants with increasing age (Doty et al., 1984; Wood and Harkins, 1987; Cain and Gent, 1991), as well as greater tendencies for olfactory adaptation and slower recovery of threshold sensitivity (Stevens et al., 1989). In contrast, few investigators have reported stable olfactory function over a life span (e.g., Rovee, Cohen, and Shlapack, 1975).

1. What Is the Anatomical Substrate for Age-related Loss of Chemosensory Function?

The olfactory system is part of a living organism that undergoes significant changes related to aging. Both peripheral sensory and central processing units are affected in that process. In the following we shall try to summarize some of those effects concerning the aging olfactory epithelium, as well as changes in the central nervous system that may contribute to the deterioration of the aging sense of smell.

1.1. Central Nervous System and Olfactory Bulb

Olfactory receptor neurons (ORNs) may be subject to alterations similar to those seen in neurons in the aging brain. Alterations in the aging central nervous system (CNS) include, for example, deficits in the regulation of intracellular calcium levels, increased leakage of synaptic transmitters, and changes in neuronal arborization (Smith, 1988). Calcium potentials in neurons from older individuals also exhibit decreased amplitude and prolonged excitation (Verkhratsky and

Toescu, 1998). In addition, there is a decrease in glucose metabolism that is associated with changes in transmitter release and an increase in free-radical production, as reviewed by Busciglio et al. (1998), and aging mitochondria are thought to be potential sources of oxidative lesions (Ames, Shigenaga, and Hagen, 1995). As endogenously produced oxidants add to cellular aging, susceptibility to toxic stress is increased (Verkhratsky and Toescu, 1998). Higher centers of the olfactory pathway are reported to undergo histological changes, as described in the mouse (Machado-Salas, Scheibel, and Scheibel, 1977). There is also evidence suggesting that there is a loss of plasticity of the aging CNS (Mori, 1993; Verkhratsky and Toescu, 1998) that might affect the perception of odors.

In 1941, Smith reported atrophy of the olfactory bulb (OB) with increasing age (Smith, 1941). He believed the OB to contain approximately 10,000 glomeruli, their number decreasing with age, leading to a reduced overall size of the OB. Atrophy of the OB and olfactory tract seems to be part of the aging process of the olfactory system (Bhatnagar et al., 1987b; Jones and Rog, 1998). Meisami et al. (1998) reported that the numbers of both glomeruli and mitral cells (MCs) in the human OB decline with age. These findings were consistent with those of Bhatnagar et al. (1987b), who reported the MCs declined in number and size with aging, and the thickness of the glomerular layer decreased. In contrast, the total numbers and densities of granule cells in the OB increase in aging rats (Kaplan, 1985). It was shown by Hinds and McNelly (1981) that the MC volume in the OB declines in aged animals, with a delay of 3.5 months following loss of ORNs, suggesting that such changes occur as a consequence of the loss of ORNs in the mucosa. In the study of Hinds and McNelly (1981), the number of MCs remained constant, which was not consistent with the findings in a different strain of rats (Kaplan, 1985) or findings in humans (Bhatnagar et al., 1987b; Meisami et al., 1998). Changes also occur in the number of glomerular dendrites: Their number peaks at the beginning of the last third of the average life span in experimental animals.

Gap-43 is a protein expressed mostly by immature ORNs; it is presumed to be a marker for differentiating cells. Its expression declines significantly in the OB in older rats (Casoli et al., 1996), although MCs maintain the ability to express Gap-43 (Weiler and Farbman, 1997); the reduced Gap-43 level may eventually lead to a decrease in neuronal plasticity.

Declining sensory functions in audition, balance, and olfaction have been explained by bone apposition at places where the respective nerves pass through the skull. Krmpotic-Nemancic (1969) described a narrowing of the holes of the cribriform plate and the bony passage of the acoustic and vestibular nerves with increasing age. Thus, in some individuals, functional loss may simply be due to nerve compression (Kalmey, Thewissen, and Dluzen, 1998).

1.2. Olfactory Epithelium

A patchy appearance of the olfactory epithelium (OE) has frequently been reported (von Brunn, 1892; Morrison and Costanzo, 1990). Early investigators assumed that replacement of OE by respiratory mucosa might be due to inflammatory processes (Kolmer, 1927). The fetal OE is described as thick and highly cellular, covering the upper part of the nasal cavity in a continuous fashion. In contrast, the adult OE is thinner than the respiratory mucosa and has lost that structure. It appears to be degenerated, with an inconsistent distribution surrounded by respiratory tissue (Naessen, 1971; Paik et al., 1992). According to Weiler and Farbman (1997), the higher number of proliferating cells in younger animals (rat) is due to the growth-associated increase in the area of the OE. In rats, the number of ORNs increases considerably between months 3 and 24, and it decreases after month 29 (Hinds and McNelly, 1981). The adult epithelium shows changes especially in the zonal distribution of the supporting cell and sensory cell nuclei. Supporting cells were shown to accumulate pigment granules with age (Naessen, 1971). Intraepithelial capillaries are not present in adult OE (Naessen, 1971; Nakashima, Kimmelman, and Snow, 1984, 1985).

Continuous neural cell death in the OE is a phenomenon well described in rats (Carr and Farbman, 1993). It occurs during the entire life of the individual among immature olfactory neurons; it is not limited to receptors reaching the end of their life spans. Both environmental and genetic factors may influence the regeneration of the OE. The rostral part of the OE shows more signs of deterioration than the caudal part (Farbman, 1990; Loo et al., 1996). It is not clear whether infection, air flow, or loss of regeneration potential is responsible for those changes; however, the pattern suggests environmental influences, as deterioration is found predominantly in the anterior part of the nasal cavity (Loo et al., 1996).

In contrast to fetal and neonatal OE, tissue obtained from adult humans has been reported to contain dystrophic neurites. Dystrophic neurites show neurofilament expression, which is not observed in normal ORNs. Those dystrophic neurites appeared to originate from morphologically altered ORNs in deeper layers of the OE (Trojanowski et al., 1991). The changes were present even in one young adult (age 16 years). Thus, if degeneration was the reason for that observations, then those alterations may begin in childhood.

The histological changes in ORNs mentioned earlier are accompanied by molecular changes within the cells. Specifically, it has been reported that intracellular localization of neuron-specific tubulin in immature ORNs differs between young and old animals, suggesting differences in the regeneration of ORNs at different ages (Lee and Pixley, 1994). Decreased adrenergic innervation of the

OE (in mice) may also contribute to changes in olfactory sensitivity (Chen et al., 1993). In addition, reduced P-450 expression in the aging human nasal mucosa may contribute to olfactory loss. This may be the basis for facilitated access of harmful substances to ORNs, which in turn may produce a higher degree of damage because of reduced detoxification abilities (Getchell et al., 1993).

ORNs from older humans, however, retain their ability to respond to odorants. They even seem to be capable of responding to more substances than ORNs from younger subjects (Rawson et al., 1998). This may represent a compensatory mechanism at a peripheral level for the overall loss of ORNs.

2. Age-related Loss of Olfactory Function Documented with Different Approaches

2.1. Psychophysical Studies

As indicated earlier, numerous investigations (e.g., Venstrom and Amoore, 1968; Schiffman, 1979; Doty et al., 1984; Murphy et al., 1985; Wysocki and Gilbert, 1989) have indicated age-related decreases in olfactory function. This has been shown for odor thresholds [e.g., coffee, citral (Megighian, 1958), *n*-butanol (Kimbrell and Furchtgott, 1963), domestic gas (Chalke and Dewhurst, 1957), commercial food odors (Schiffman et al., 1976), musk odor (Schiffman and Pasternak, 1979), 10 normal fatty acids from formic to caproic (Venstrom and Amoore, 1968), dibutylamine, spray thinner, isoamyl alcohol, acrylic acid methyl ester, hydrogen sulfide (Bahnmüller, 1984), phenol (Fordyce, 1961; Joyner, 1963), *d*-limonene, isoamyl butyrate, benzaldehyde (Stevens and Cain, 1987), 1-butanol, isoamyl butyrate, phenyl ethylmethyl ethylcarbinol, and pyridine (Cain and Gent, 1991)], suprathreshold odor intensity perception (e.g., Stevens and Cain, 1987; Wysocki and Gilbert, 1989), odor discrimination (e.g., Schiffman and Pasternak, 1979), odor identification (e.g., Schiffman et al., 1976; Schemper, Voss, and Cain, 1981; Doty et al., 1984; Stevens and Cain, 1987; Wood and Harkins, 1987; Cain, 1989; Wysocki and Gilbert, 1989; Cain, Reid, and Stevens, 1990; Ship and Weiffenbach, 1993), and odor memory (e.g., Nordin and Murphy, 1998; Lehrner, Gluck, and Laska, 1999; Lehrner and Walla, Chapter 17, this volume).

Few studies have directly compared multiple olfactory functions in different tests. One of our investigations (Hummel et al., 1998) focusing on olfactory event-related potentials (discussed later) found only an age-related decrease in the subjects' ability to discriminate odors; similar changes were present, but non-significant, for odor thresholds (odorants: phenylethyl alcohol, pyridine; method of ascending limits) and odor identification (identification of eight odorants). Whereas the psychophysical methods used in that study may have been somewhat inadequate (cf. Stevens and Dardawala, 1993; Doty et al., 1995), a much more sophisticated technique used by Cain et al. (1995) showed age-related reductions in thresholds that actually preceded reductions in ability to identify odors. Similar findings emerged when recalculating data from a study (Hummel et al., 1997) in which olfactory testing was performed using "Sniffin' Sticks" (Kobal et al., 1996). That procedure combines tests for odor discrimination (16 triplets of odorants, forced choice), odor identification (16 odorants, multiple forced choice), and assessment of *n*-butanol odor thresholds (single staircase, seven reversals, triple forced choice). A total of 104 caucasian subjects participated (52 male and 52 female); for details, see Hummel et al. (1997). For statistical analyses, 12 groups ($n = 8$–11; age 18–84 years) were defined based on the subjects' ages. After normalization of the data with regard to average results in the youngest subjects (age 18–22 years), results were investigated using analysis of variance for repeated measures with the factors "age group" and "test." For all measurements, performances decreased with increasing age (Figure 27.1) [$F(184/2) = 12.00$; $p < .001$; power $= 0.995$]; the decreases were most pronounced in subjects older than 65 years. However, they did not appear to be uniform for all olfactory tests, but were found to be relatively more pronounced for odor thresholds [interaction "test" × "age group": $F(184/22) = 2.61$; $p < .001$;

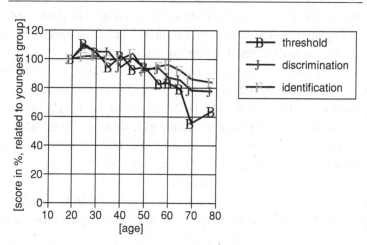

Age related changes for odor identification, odor discrimination, and odor thresholds

Figure 27.1. Mean scores for odor identification, odor discrimination, and odor thresholds ($n = 8$–11) in relation to the subjects' ages. Mean scores were normalized to the average score for the youngest group tested. A significant effect of the factor "age" was observed. Decreases in sensitivity were found to be most pronounced for threshold measurements in the oldest subjects investigated.

power = 0.999]. Thus, assuming that threshold measurements may reflect the function of the peripheral olfactory system to a larger degree than other olfactory tests (e.g., Jones-Gotman and Zatorre, 1988; Hornung et al., 1998; Moberg et al., 1999), those findings might indicate that age-related changes in olfactory function are largely due to damage of the OE (cf. Nakashima et al., 1984; Rosli, Breckenridge, and Smith, 1999). They also appear to indicate that changes in olfactory threshold may not immediately affect other, presumably more complex olfactory functions like odor discrimination and odor identification. In other words, it may be that the subjects' ability to discriminate odorants is partly independent of olfactory thresholds, especially in view of the data from Rawson et al. (1997), who reported an age-related, broader "tuning" of ORNs. In fact, Cain and co-workers recently presented evidence that odor discrimination appears to be independent of absolute sensitivity (Cain et al., in press; cf. Hummel et al., 1998).

2.2. Studies Using Chemosensory Event-related Potentials

Chemosensory event-related potentials (ERPs) have become important tools not only in olfactory research (Livermore, Hummel, and Kobal, 1992; Lorig et al., 1996) but also for diagnosis of olfactory disorders (Kobal and Hummel, 1994, 1998). Chemosensory ERPs reflect the intensity of a stimulus (Kobal and Hummel, 1988) and can discriminate between the excitation of trigeminal nerves and olfactory nerves (Hummel and Kobal, 1992).

Age-related decreases in chemosensory function are also indicated by ERP measures. Specifically, the ERP's P2 amplitude and the composite N1P2 amplitudes decreased with increasing age, but a prolongation of N1 peak latencies was observed. Using amyl acetate, Evans, Cui, and Starr (1995) reported decreases in peak-to-peak N1P2 amplitudes in relation to the subjects' ages. They also reported a significant life-span prolongation of the P2 latency, accompanied by a trend for mean N1 latencies to increase with age. In addition, Hawkes and co-workers (personal communication) found both an age-related decrease in ERP amplitudes and an increase of latencies for the stimulant H_2S ($n = 60$). As that has been confirmed by numerous other studies (e.g., Murphy et al., 1994; Hummel et al., 1998; Covington et al., 1999), it can be assumed that especially the earlier ERP components (e.g., N1) reliably reflect loss of olfactory function.

Olfactory ERP data indicate that changes in the processing of olfactory information may appear relatively early in life. Specifically, data from Murphy et al. (1998) and Hummel et al. (1998) suggest that ERP amplitudes already exhibit decreases in middle age (Figure 27.2). In elderly individuals, that may, at least

Age related changes of olfactory and trigeminal ERP amplitudes N1P2

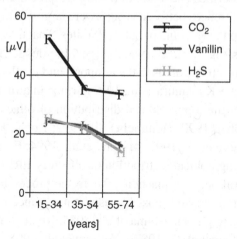

Figure 27.2. Mean ERP amplitudes N1P2 at recording position Cz in response to stimulation with CO_2, vanillin, and H_2S for three different age groups of subjects (data from Hummel et al., 1998). The age-related decrease in N1P2 amplitudes was different for the trigeminal stimulant (CO_2) compared with the two olfactory stimulants. Specifically, for CO_2 the decrease was found to be most pronounced between the youngest group and the middle-aged group. In contrast, responses to olfactory stimulation decreased between the middle-aged group and the oldest group.

in part, be due to increased adaptation to repetitive olfactory stimuli (Stevens et al., 1989; Morgan et al., 1997).

In a series of studies, Murphy and co-workers investigated age-related changes in ERPs with a special focus on the late component of the potential, the P3 component (Murphy et al., 1996). That component has been demonstrated to be related to endogenous processes (Pause et al., 1996), such as whether or not a stimulus contains new and relevant information compared with previous stimuli (e.g., Sutton, Braren, and Zubin, 1965). Findings have indicated that the age-related decrease in the P3 amplitude is related to overall reduction of neuropsychological performance (Geisler et al., 1999; Morgan et al., 1999). Thus, the olfactory P3 may serve as a reliable indicator of attentional resources related to olfactory cognitive processing (Murphy et al., 1998). However, the question remains in what specific way the changes indicated by the olfactory P3 differ from age-related changes seen in the visual and auditory systems (Ford and Pfefferbaum, 1985; Knight, 1987; Tachibana, Toda, and Sugita, 1992).

Murphy et al. (1994) investigated chemosensory ERPs to amyl acetate in the elderly (mean age 66 years, $n = 7$) and in young adults (mean age 27 years,

$n = 7$). They found differences in amplitudes at central and parietal recording positions, but not at the frontal site Fz. That agrees with data provided by
Hummel et al. (1998), who described a leveling of the topographical differences
between ERP amplitudes: Differences between groups were largest at the central
site Cz, and were smaller at frontal (Fz) and parietal (Pz) sites. That may indicate
changes in the cortical processing of olfactory sensations. In younger subjects,
olfactory stimulation seems to activate areas in the temporal and insular cortices
(Kettenmann et al., 1996; Ayabe-Kanamura et al., 1997), which in turn produces
the specific surface pattern of the topographical distribution of olfactory ERP
amplitudes (Hummel and Kobal, 1992; Hummel et al., 1992; Kobal, Hummel,
and Van Toller, 1992; Livermore et al., 1992; Murphy et al., 1994; Evans et al.,
1995). The change in that topographical distribution of olfactory ERPs in older
subjects may be due to central nervous plasticity. There the processing of olfactory information may be reorganized, possibly as a consequence of loss of
neurons both in the periphery (e.g., Nakashima et al., 1985; Rosli et al., 1999)
and at central sites (e.g., Bhatnagar et al., 1987a; Yousem et al., 1998).

Electrophysiological studies have also indicated differences in age-related
changes in responses to trigeminal or olfactory stimuli. Specifically, responses
to trigeminal and olfactory stimulation may follow different courses across a life
span (Figure 27.2). However, more research is needed to investigate the behavioral correlates of such changes in trigeminal responsiveness, as discussed later.

2.3. Imaging Studies

To our knowledge there has been only one study that has investigated age effects
on olfaction using functional magnetic-resonance imaging (Yousem et al., 1999).
The 10 participating subjects underwent olfactory testing using the University
of Pennsylvania Smell Identification Test (UPSIT) (Doty, 1995). Five subjects
(3 women, 2 men; age 62–80 years) were compared to five younger subjects
(3 women, 2 men; age 18–27 years). The older group had an UPSIT score of 33.0;
the younger group had a mean UPSIT score of 38.0. Each task paradigm consisted of alternating rest–stimulus cycles (30 sec each) over 6 minutes. Eugenol,
phenylethyl alcohol, and hydrogen sulfide were used for olfactory stimulation;
they were delivered using a continuous-flow olfactometer (OM4-B; Burghart,
Wedel, Germany). Stimuli were delivered intranasally to both sides (4 liters/min;
stimulus duration 1 sec) every 4 sec during the 30-sec "on period." During the
30-sec "off period" the patient received room air; for details, see Yousem et al.
(1999).

On group maps, the older subjects showed the largest number of activated
voxels in the right perisylvian region, followed by the right superior frontal

region. Single voxels of activation were noted in the right inferior frontal and left superior frontal zones. Most individual maps demonstrated activation in the right perisylvian, right and left superior frontal, and right and left inferior frontal regions. In general, right-side activation was greater than left-side activation. The younger subjects showed greater volumes of activation. The sites of activation identified in the older group were also activated to the greatest degree in the younger group, namely, the right perisylvian, right superior frontal, right inferior frontal, and left superior frontal regions. In addition, the group maps for the younger subjects showed activation in the left superior frontal and left perisylvian zones and both cingulate regions. All younger subjects showed activation in the right inferior frontal and left perisylvian regions.

Thus the major finding of that study was that there was a decreased volume of activation after the age of 60. In general, the same areas that were found to be activated in older subjects were also found to be activated in younger subjects. However, with regard to group maps, younger subjects exhibited activation in the left and right cingulate regions, and they exhibited left-side activity in the superior frontal and perisylvian regions. That was not seen in group maps for older subjects. Those differences in activated brain areas may, at least in part, explain the age-related changes observed in the topographical distribution of olfactory ERPs discussed earlier.

3. Trigeminal Sensitivity Also Appears to Decrease in an Age-related Manner

Compared with the data for the olfactory system, considerably fewer data are available regarding age-related changes in trigeminal chemoreception. Elevated thresholds to trigeminal stimuli (e.g., menthol) were reported in elderly subjects (Minz, 1968; Murphy, 1983); in addition, a steeper slope of the intensity function was found for younger adults. Stevens, Plantinga, and Cain (1982) found an age-related decrease in the perceived intensity of CO_2 (a trigeminal stimulus with minimal or no odor qualities), and Stevens and Cain (1986) reported a strong age-related elevation of the threshold for transitory apnea in response to CO_2. As mentioned earlier, that age-related loss of trigeminal function is also seen in ERP studies (Hummel et al., 1998). Interestingly, the cross-sectional results were also confirmed, at least in part, by data obtained from a single subject, where trigeminal ERPs were recorded over a period of 10 years. There an intra-individual decrease in response amplitudes to CO_2 stimuli occurred between the ages of 28 and 38 years.

Similar age-related changes have been reported for other nociceptive, pain-related processes. Specifically, responsiveness of Adelta-fibers to nociceptive heat stimuli appears to decrease in relation to age, but C-fiber function seems to be largely unaffected (Harkins, Price, and Martelli, 1986; Chakour et al., 1996; Harkins et al., 1996). Those functional observations have been confirmed on a histological level, where the number of myelinated fibers (A-fibers) appears to decrease with increasing age (Ochoa and Mair, 1969; Kenshalo, 1986). Translating that into terms of intranasal trigeminal function of chemosensors, it should result in decreased stinging sensations (Adelta-fibers), and the perception of burning sensations (C-fibers) should remain mostly unchanged. Those observations correlate closely with changes found for ERPs to intranasal trigeminal stimuli that mostly were due to the activation of Adelta-fibers (Harkins, Price, and Katz, 1983; Hummel et al., 1994). Overall, the indication is that the trigeminal chemoreceptive system exhibits an age-related functional decrease, aspects of which appear to be similar to those of the olfactory system.

4. Perspectives

The aging process in the olfactory system cannot be reduced to a single phenomenon. It involves all levels, from anatomical to molecular changes. It seems to be a process influenced by genetic and environmental factors, involving the olfactory system itself and the higher centers of information processing. Age-related changes in olfactory function are observed at all levels of information processing. However, they do not appear to be the inevitable fate of each individual (Rowe and Kahn, 1987; Almkvist, Berglund, and Nordin, 1992). Thus, especially when considering potential therapeutic interventions, further studies are needed to determine in more detail which parts of the olfactory system are most relevant to the decreases in olfactory function and to find out what can be done to prevent or delay the age-related changes.

Acknowledgments

Part of the work reported here was based on research performed in collaboration with G. Kobal, S. Barz, B. Sekinger, E. Pauli, S. R. Wolf (all at Universität Erlangen-Nürnberg, Germany), D. M. Yousem, J. A. Maldjian, D. C. Alsop, R. J. Geckle, and R. L. Doty (all at the University of Pennsylvania, USA). It was supported by the Marohn-Stiftung, Germany, and by grant P01 DC 00161 from the National Institute on Deafness and Other Communication Disorders, USA. The help of Dragoco and Harmann & Reimer (both Holzminden, Germany) and Givaudan (Hannover, Germany) in some of the studies is also appreciated. In addition, we are indebted to Dr. W. S. Cain for helpful comments on this manuscript.

References

Almkvist O, Berglund B, & Nordin S (1992). Odor Detectability in Successfully Aged Elderly and Young Adults. *Reports from the Department of Psychology, Stockholm University* 744:1–12.

Ames B N, Shigenaga M K, & Hagen T M (1995). Mitochondrial Decay in Aging. *Biochimica Biophysica Acta* 1271:165–70.

Ayabe-Kanamura S, Endo H, Kobayakawa T, Takeda T, & Saito S (1997). Measurement of Olfactory Evoked Magnetic Fields by a 64-Channel Whole-Head SQUID System. *Chemical Senses* 22:214–15.

Bahnmüller H (1984). Olfaktometrie von Dibutylamin, Acrylsäuremethylester, Isoamylalkohol und eines Spritzverdünners für Autolacke – Ergebnisse eines VDI-Ringvergleichs. *Staub-Reinhaltung der Luft* 44:352–8.

Bhatnagar K P, Kennedy R C, Baron G, & Greenberg R A (1987a). Number of Mitral Cells and the Bulb Volume in Aging Human Olfactory Bulb: A Quantitative Morphological Study. *Anatomical Record* 218:73–87.

Bhatnagar K P, Kennedy R C, Baron G, & Greenberg R A (1987b). Number of Mitral Cells and the Bulb Volume in the Aging Human Olfactory Bulb: A Quantitative Morphological Study. *Anatomical Record* 218:73–87.

Busciglio J, Andersen J K, Schipper H M, Gilad G M, McCarty R, Marzatico F, & Toussaint O (1998). Stress, Aging, and Neurodegenerative Disorders. Molecular Mechanisms. *Annals of the New York Academy of Sciences* 851:429–43.

Cain W S (1989). Testing Olfaction in a Clinical Setting. *Ear, Nose and Throat Journal* 68:321–8.

Cain W S, De Wijk R A, Nordin S, Nordin M, & Murphy C (in press). Odor Discrimination in Aging: Autonomy of Quality. *Psychology and Aging*.

Cain W S & Gent J F (1991). Olfactory Sensitivity: Reliability, Generality, and Association with Aging. *Journal of Experimental Psychology* 17:382–91.

Cain W S & Stevens J C (1989). Uniformity of Olfactory Loss in Aging. *Annals of the New York Academy of Sciences* 561:29–38.

Cain W S, Stevens J C, Nickou C M, Giles A, Johnston I, & Garcia-Medina M R (1995). Life-Span Development of Odor Identification, Learning, and Olfactory Sensitivity. *Perception* 24:1457–72.

Carr V M & Farbman A I (1993). The Dynamics of Cell Death in the Olfactory Epithelium. *Experimental Neurology* 124:308–14.

Casoli T, Spagna C, Fattoretti P, Gesuita R, & Bertoni-Freddari C (1996). Neuronal Plasticity in Aging: A Quantitative Immunohistochemical Study of GAP-43 Distribution in Discrete Regions of the Rat Brain. *Brain Research* 714:111–17.

Chakour M C, Gibson S J, Bradbeer M, & Helme R D (1996). The Effect of Age on Adelta- and C-fibre Thermal Pain Perception. *Pain* 64:143–52.

Chalke H D & Dewhurst J R (1957). Coal Gas Poisoning: Loss of Sense of Smell as a Possible Contributory Factor with Old People. *British Medical Journal* 2:1915–17.

Chen Y, Getchell T V, Sparks D L, & Getchell M L (1993). Patterns of Adrenergic and Peptidergic Innervation in Human Olfactory Mucosa: Age-related Trends. *Journal of Comparative Neurology* 334:104–16.

Covington J W, Geisler M W, Polich J, & Murphy C (1999). Normal Aging and Odor Intensity Effects on the Olfactory Event-related Potential. *International Journal of Psychophysiology* 32:205–14.

Doty R L (1995). *The Smell Identification Test™ Administration Manual*, 3rd ed. Haddon Heights, NJ: Sensonics.

Doty R L, McKeown D A, Lee W W, & Shaman P (1995). A Study of the Test–Retest Reliability of Ten Olfactory Tests. *Chemical Senses* 20:645–56.

Doty R L, Shaman P, Applebaum S L, Giberson R, Sikorski L, & Rosenberg L (1984). Smell Identification Ability: Changes with Age. *Science* 226:1441–3.

Evans W J, Cui L, & Starr A (1995). Olfactory Event-related Potentials in Normal Human Subjects: Effects of Age and Gender. *Electroencephalography and Clinical Neurophysiology* 95:293–301.

Farbman A I (1990). Olfactory Neurogenesis: Genetic or Environmental Controls? *Trends in Neuroscience* 13:362–5.

Ford J M & Pfefferbaum A (1985). Age-related Changes in Event-related Potentials. *Advances in Psychophysiology* 1:301–39.

Fordyce I D (1961). Olfaction Tests. *British Journal of Industrial Medicine* 18:213–15.

Geisler M W, Morgan C D, Covington J W, & Murphy C (1999). Neuropsychological Performance and Cognitive Olfactory Event-related Brain Potentials in Young and Elderly Adults. *Journal of Clinical and Experimental Neuropsychology* 21:108–26.

Getchell M L, Chen Y, Ding X, Sparks D L, & Getchell T V (1993). Immunohistochemical Localization of a Cytochrome P-450 Isozyme in Human Nasal Mucosa: Age-related Trends. *Annals of Otology Rhinology and Laryngology* 102:368–74.

Harkins S W, Davis M D, Bush F M, & Kasberger J (1996). Suppression of First Pain and Slow Temporal Summation of Second Pain in Relation to Age. *Journal of Gerontology* 51A:M260–5.

Harkins S W, Price D D, & Katz M A (1983). Are Cerebral Evoked Potentials Reliable Indices of First or Second Pain? In: *Advances in Pain Research and Therapy*, vol. 5, ed. J J Bonica, U Lindblom, & A Iggo, pp. 185–91. New York: Raven Press.

Harkins S W, Price D D, & Martelli M (1986). Effects of Age on Pain Perception: Thermonociception. *Journal of Gerontology* 41:58–63.

Hinds J W & McNelly N A (1981). Aging in the Rat Olfactory System: Correlation of Changes in the Olfactory Epithelium and Olfactory Bulb. *Journal of Comparative Neurology* 203:441–53.

Hornung D E, Kurtz D B, Bradshaw C B, Seipel D M, Kent P F, Blair D C, & Emko P (1998). The Olfactory Loss That Accompanies an HIV Infection. *Physiology and Behavior* 15:549–56.

Hummel T, Barz S, Pauli E, & Kobal G (1998). Chemosensory Event-related Potentials Change as a Function of Age. *Electroencephalography and Clinical Neurophysiology* 108:208–17.

Hummel T, Gruber M, Pauli E, & Kobal G (1994). Event-related Potentials in Response to Repetitive Painful Stimulation. *Electroencephalography and Clinical Neurophysiology* 92:426–32.

Hummel T & Kobal G (1992). Differences in Human Evoked Potentials Related to Olfactory or Trigeminal Chemosensory Activation. *Electroencephalography and Clinical Neurophysiology* 84:84–9.

Hummel T, Livermore A, Hummel C, & Kobal G (1992). Chemosensory Event-related Potentials: Relation to Olfactory and Painful Sensations Elicited by Nicotine. *Electroencephalography and Clinical Neurophysiology* 84:192–5.

Hummel T, Sekinger B, Wolf S R, Pauli E, & Kobal G (1997). "Sniffin' Sticks": Olfactory Performance Assessed by the Combined Testing of Odor Identification, Odor Discrimination, and Olfactory Thresholds. *Chemical Senses* 22:39–52.

Jones N & Rog D (1998). Olfaction: A Review. *Journal of Laryngology and Otology* 112:11–24.

Jones-Gotman M & Zatorre R J (1988). Olfactory Identification Deficits in Patients with Focal Cerebral Excision. *Neuropsychologia* 26:387–400.

Joyner R E (1963). Olfactory Acuity in an Industrial Population. *Journal of Occupational Medicine* 5:37–42.

Kalmey J K, Thewissen J G, & Dluzen D E (1998). Age-related Size Reduction of Foramina in the Cribriform Plate. *Anatomical Record* 251:326–9.

Kenshalo D R (1986). Somesthetic Sensitivity in Young and Elderly Humans. *Journal of Gerontology* 41:732–42.

Kettenmann B, Jousmäki V, Portin K, Salmelin R, Kobal G, & Hari R (1996). Odorants Activate the Human Superior Temporal Sulcus. *Neuroscience Letters* 203:143–5.

Kimbrell G M & Furchtgott E (1963). The Effect of Aging on Olfactory Threshold. *Journal of Gerontology* 18:364–5.

Knight R T (1987). Aging Decreases Auditory Event-related Potentials to Unexpected Stimuli in Humans. *Neurobiology of Aging* 8:109–13.

Kobal G & Hummel C (1988). Cerebral Chemosensory Evoked Potentials Elicited by Chemical Stimulation of the Human Olfactory and Respiratory Nasal Mucosa. *Electroencephalography and Clinical Neurophysiology* 71:241–50.

Kobal G & Hummel T (1994). Olfactory (Chemosensory) Event-related Potentials. *Toxicology and Industrial Health* 10:587–96.

Kobal G & Hummel T (1998). Olfactory and Intranasal Trigeminal Event-related Potentials in Anosmic Patients. *Laryngoscope* 108:1033–5.

Kobal G, Hummel T, Sekinger B, Barz S, Roscher S, & Wolf S R (1996). "Sniffin' Sticks": Screening of Olfactory Performance. *Rhinology* 34:222–6.

Kobal G, Hummel T, & Van Toller S (1992). Differences in Chemosensory Evoked Potentials to Olfactory and Somatosensory Chemical Stimuli Presented to Left and Right Nostrils. *Chemical Senses* 17:233–44.

Kolmer W (1924). Über die Regio Olfactoria des Menschen. *Monatsschrift für Ohrenheilkunde* 58:626.

Krmpotic-Nemanic J (1969). Presbycusis, Presbystasis and Presbyosmia as Consequences of the Analogous Biological Process. *Acta Oto-laryngologica (Stockholm)* 67:217–23.

Lee V M & Pixley S K (1994). Age and Differentiation-related Differences in Neuron-specific Tubulin Immunostaining of Olfactory Sensory Neurons. *Brain Research. Developmental Brain Research* 83:209–15.

Lehrner J P, Gluck J, & Laska M (1999). Odor Identification, Consistency of Label Use, Olfactory Threshold and Their Relationships to Odor Memory over the Human Lifespan. *Chemical Senses* 24:337–46.

Livermore A, Hummel T, & Kobal G (1992). Chemosensory Event-related Potentials in the Investigation of Interactions between the Olfactory and the Somatosensory (Trigeminal) Systems. *Electroencephalography and Clinical Neurophysiology* 83:201–10.

Loo A T, Youngentob S L, Kent P F, & Schwob J E (1996). The Aging Olfactory Epithelium: Neurogenesis, Response to Damage, and Odorant-induced Activity. *International Journal of Developmental Neuroscience* 14:881–900.

Lorig T S, Matia D C, Pezka J J, & Bryant D N (1996). The Effects of Active and Passive Stimulation on Chemosensory Event-related Potentials. *International Journal of Psychophysiology* 23:199–205.

Machado-Salas J, Scheibel M E, & Scheibel A B (1977), Morphologic Changes in the Hypothalamus of the Old Mouse. *Experimental Neurology* 57:102–11.

Megighian D (1958). Variazioni della soglia olfattiva nell'eta senile. *Minerva Otolaringologica* 9:331–7.

Meisami E, Mikhail L, Baim D, & Bhatnagar K P (1998). Human Olfactory Bulb: Aging of Glomeruli and Mitral Cells and a Search for the Accessory Olfactory Bulb. *Annals of the New York Academy of Sciences* 855:708–15.

Minz A I (1968). Condition of the Nervous System in Old Men. *Zeitschrift für Altersforschung* 21:271–7.

Moberg P J, Agrin R, Gur R E, Gur R C, Turetsky B I, & Doty R L (1999). Olfactory Dysfunction in Schizophrenia: A Qualitative and Quantitative Review. *Neuropsychopharmacology* 21:325–40.

Morgan C D, Covington J W, Geisler M W, Polich J, & Murphy C (1997). Olfactory Event-related Potentials: Older Males Demonstrate the Greatest Deficits. *Electroencephalography and Clinical Neurophysiology* 104:351–8.

Morgan C D, Geisler M W, Covington J W, Polich J, & Murphy C (1999). Olfactory P3 in Young and Older Adults. *Psychophysiology* 36:281–7.

Mori N (1993). Toward Understanding of the Molecular Basis of Loss of Neuronal Plasticity in Ageing. *Age and Ageing* 22:S5–18.

Morrison E E & Costanzo R M (1990). Morphology of the Human Olfactory Epithelium. *Journal of Comparative Neurology* 297:1–13.

Murphy C (1983). Age-related Effects on the Threshold, Psychophysical Function, and Pleasantness of Menthol. *Journal of Gerontology* 38:217–22.

Murphy C, Morgan C D, Covington J, Geisler M W, & Polich J M (1996). Late (Cognitive) Components of the Olfactory Event-related Potential Are More Sensitive to Aging Than Early Components. *Chemical Senses* 21:648–9.

Murphy C, Nordin S, de Wijk R A, Cain W S, & Polich J (1994). Olfactory-evoked Potentials: Assessment of Young and Elderly, and Comparison to Psychophysical Threshold. *Chemical Senses* 19:47–56.

Murphy C, Nunez K, Whithee J, & Jalowayski A A (1985). The Effects of Age, Nasal Airway Resistance and Nasal Cytology on Olfactory Threshold for Butanol. *Chemical Senses* 10:418.

Murphy C, Wetter S, Morgan C D, Ellison D W, & Geisler M W (1998). Age Effects on Central Nervous System Activity Reflected in the Olfactory Event-related Potential. Evidence for Decline in Middle Age. *Annals of the New York Academy of Sciences* 855:598–607.

Naessen R (1971). An Inquiry on the Morphological Characteristics and Possible Changes with Age in the Olfactory Region of Man. *Acta Oto-laryngologica (Stockholm)* 71:49–62.

Nakashima T, Kimmelman, C P, & Snow J B, Jr (1984). Structure of Human Fetal and Adult Olfactory Neuroepithelium. *Archives of Otolaryngology* 110:641–6.

Nakashima T, Kimmelman C P, & Snow J B, Jr (1985). Immunohistopathology of Human Olfactory Epithelium, Nerve and Bulb. *Laryngoscope* 95:391–6.

Nordin S & Murphy C (1998). Odor Memory in Normal Aging and Alzheimer's Disease. *Annals of the New York Academy of Science* 855:686–93.

Ochoa J & Mair W G P (1969). The Normal Sural Nerve in Man. I. Ultrastructure and Numbers of Fibres and Cells. *Acta Neuropathologica (Berlin)* 13:197–216.

Paik S I, Lehman M N, Seiden A M, Duncan H J, & Smith D V (1992). Human Olfactory Biopsy: The Influence of Age and Receptor Distribution. *Archives of Otolaryngology* 118:731–8.

Pause B M, Sojka B, Krauel K, & Ferstl R (1996). The Nature of the Late Positive Complex within the Olfactory Event-related Potential. *Psychophysiology* 33:168–72.

Rawson N E, Gomez G, Cowart B, Brand J G, Lowry L D, Pribitkin E A, & Restrepo D (1997). Selectivity and Response Characteristics of Human Olfactory Neurons. *Journal of Neurophysiology* 77:1606–13.

Rawson N E, Gomez G, Cowart B, & Restrepo D (1998). The Use of Olfactory Receptor Neurons (ORNs) from Biopsies to Study Changes in Aging and Neurodegenerative Diseases. *Annals of the New York Academy of Sciences* 855:701–7.

Rosli Y, Breckenridge L J, & Smith R A (1999). An Ultrastructural Study of Age-related Changes in Mouse Olfactory Epithelium. *Journal of Electron Microscopy (Tokyo)* 48:77–84.

Rovee C K, Cohen R Y, & Shlapack W (1975). Life-Span Stability in Olfactory Sensitivity. *Developments in Psychology* 11:311–18.

Rowe J W & Kahn R L (1987). Human Aging: Usual and Successful. *Science* 237:143–9.

Schemper T, Voss S, & Cain W S (1981). Odor Identification in Young and Elderly Persons: Sensory and Cognitive Limitations. *Journal of Gerontology* 36:446–52.

Schiffman S (1979). Changes in Taste and Smell with Age: Psychophysical Aspects. In: *Sensory Systems and Communication in the Elderly*, ed. J M Ordy & K Brizzee, pp. 227–46. New York: Raven Press.

Schiffman S, Moss J, & Erickson R P (1976). Thresholds of Food Odors in the Elderly. *Experimental Aging Research* 2:389–98.

Schiffman S & Pasternak M (1979). Decreased Discrimination of Food Odors in the Elderly. *Journal of Gerontology* 34:73–9.

Ship J A & Weiffenbach J M (1993). Age, Gender, Medical Treatment, and Medication Effects on Smell Identification. *Journal of Gerontology* 48: M26–32.

Smith C G (1941). Incidence of Atrophy of the Olfactory Nerves in Man. *Archives of Otolaryngology* 34:533–9.

Smith D O (1988). Cellular and Molecular Correlates of Aging in the Nervous System. *Experimental Gerontology* 23:399–412.

Stevens J C & Cain W S (1986). Aging and the Perception of Nasal Irritation. *Physiology and Behavior* 37:323–8.

Stevens J C & Cain W S (1987). Old-Age Deficits in the Sense of Smell as Gauged by Thresholds, Magnitude Matching, and Odor Identification. *Psychology and Aging* 2:36–42.

Stevens J C, Cain W S, Schiet F T, & Oatley M W (1989). Olfactory Adaptation and Recovery in Old Age. *Perception* 18:265–76.

Stevens J C. & Dardawala A D (1993) Variability of Olfactory Threshold and Its Role in Assessment of Aging. *Perceptual Psychophysiology* 54:296–302.

Stevens J C, Plantinga A, & Cain W S (1982). Reduction of Odor and Nasal Pungency Associated with Aging. *Neurobiology of Aging* 3:125–32.

Sutton S, Braren M, & Zubin J (1965). Evoked-Potential Correlates of Stimulus Uncertainty. *Science* 150:1187–8.

Tachibana H, Toda K, & Sugita M (1992). Age-related Changes in Attended and Unattended P3 Latency in Normal Subjects. *International Journal of Neuroscience* 66:277–84.

Trojanowski J Q, Newman P D, Hill W D, & Lee V M (1991). Human Olfactory Epithelium in Normal Aging, Alzheimer's Disease, and Other Neurodegenerative Disorders. *Journal of Comparative Neurology* 310:365–76.

Vaschide N (1904). L'état de la sensibilité olfactive dans la vieillesse. *Bulletin de Laryngolagie, Otologie et Rhinologie* 7:323–33.

Venstrom D & Amoore J E (1968). Olfactory Threshold in Relation to Age, Sex, or Smoking. *Journal of Food Science* 33:264–5.

Verkhratsky A & Toescu E C (1998). Calcium and Neuronal Ageing. *Trends in Neuroscience* 21:2–7.

Weiler E & Farbman A I (1997). Proliferation in the Rat Olfactory Epithelium: Age-dependent Changes. *Journal of Neuroscience* 17:3610–22.

Wood J B & Harkins S W (1987). Effects of Age, Stimulus Selection, and Retrieval Environment on Odor Identification. *Journal of Gerontology* 42:584–8.

Wysocki C J & Gilbert A N (1989). National Geographic Smell Survey: Effects of Age Are Heterogeneous. In: *Nutrition and the Chemical Senses in Aging: Recent Advances and Current Research Needs*, ed. C. Murphy, W S Cain, & D M Hegsted, pp. 12–28. Washington, DC: National Geographic.

Yousem D M, Geckle R J, Bilker W B, & Doty R L (1998). Olfactory Bulb and Tract and Temporal Lobe Volumes. Normative Data Across Decades. *Annals of the New York Academy of Sciences* 855:546–55.

Yousem D M, Maldjian J A, Hummel T, Alsop D C, Geckle R J, Kraut M A, & Doty R L (1999). The Effect of Age on Odor-stimulated Functional MR Imaging. *American Journal of Neuroradiology* 20:600–8.

Index

of degree N and with coefficients in K, and a power series

$$h(X) = 1 + c_1 X + c_2 X^2 + \dots$$

with coefficients in K, satisfying:

i) $f(X) = g(X)h(X)$,

ii) $|b_N| = \max |b_n|$, i.e., $\|g(X)\|_1 = |b_N|$,

iii) $\lim_{n \to \infty} c_n = 0$, so that $h(x)$ converges for $x \in \hat{\mathcal{O}}$ (i.e., $h(X) \in A_1$),

iv) $|c_n| < 1$ for all $n \geq 1$, i.e., $\|h(X) - 1\|_1 < 1$, and

v) $\|f(X) - g(X)\|_1 < 1$.

In particular, $h(X)$ has no zeros in $\hat{\mathcal{O}}$.

This clearly related to Strassman's Theorem. Since $h(X)$ has no zeros in $\hat{\mathcal{O}}$, it is clear that the zeros of $f(X)$ in $\hat{\mathcal{O}}$ are exactly the same as the zeros of $g(X)$. Since $g(X)$ is a polynomial of degree N, there are at most N of these, and we get Strassman's Theorem. If we move to $\mathbb{C}_p$, we can say more: since $\mathbb{C}_p$ is algebraically closed, we get that, counting multiplicities, $g(X)$ has exactly N zeros in $\mathbb{C}_p$, and the condition on its coefficients means that all of them are in $\hat{\mathcal{O}}$ (see the problem below). So we know that, counting multiplicities, $f(X)$ has exactly N zeros in $\hat{\mathcal{O}}$, which gives a stronger version of Strassman's Theorem.

Problem 319 Suppose the polynomial $g(X) = b_0 + b_1 X + \dots + b_N X^N$ satisfies the condition in the theorem: $|b_N| = \max |b_n|$. Show that if $g(\alpha) = 0$, then $|\alpha| \leq 1$.

Proving the Weierstrass Preparation Theorem will take a while and will require some effort. We will do it by means of a series of lemmas of various kinds. To begin to set everything up, recall that A_1 is the ring of power series that converge in $\hat{\mathcal{O}}$; in other words, a power series

$$f(X) = \sum a_n X^n = a_0 + a_1 X + a_2 X^2 + \dots$$

belongs to A_1 if and only if $a_n \to 0$ as $n \to \infty$. We already know that defining

$$\|f(X)\| = \|f(X)\|_1 = \max_n |a_n|$$

gives a norm on A_1 (we will drop the subscript 1, since this is the only norm we'll be working with in this proof). The first step of the proof is to prove that A_1, with this norm, has very nice properties.

Lemma 7.2.7 A_1 is complete with respect to the norm $\| \ \|$.

PROOF: We have to show that a Cauchy sequence in A_1 (with respect to the norm $\| \ \|$) converges. So consider a sequence of power series

$$f_i(X) = a_{i0} + a_{i1}X + a_{i2}X^2 + a_{i3}X^3 + \cdots$$

Saying this sequence is Cauchy amounts to saying that for each $\varepsilon > 0$ there exists an M such that we have $\|f_i(X) - f_j(X)\| < \varepsilon$ whenever $i, j > M$.

Translating that inequality, we get that

$$\max_n |a_{in} - a_{jn}| < \varepsilon, \qquad \text{whenever } i, j > M,$$

which certainly implies that

$$|a_{in} - a_{jn}| < \varepsilon \qquad \text{for each } n, \text{ whenever } i, j > M.$$

In other words, each of the sequences $(a_{in})_i$ is Cauchy. Since K is complete, that means they are all convergent.

So, for each n, let

$$a_n = \lim_{i \to \infty} a_{in},$$

and consider the series

$$g(X) = a_0 + a_1 X + a_2 X^2 + a_3 X^3 + \cdots$$

We obviously want to say that it is the limit of the sequence of series. To see why, we need two things: first, we need to estimate

$$\|f_i(X) - g(X)\| = \max_n |a_{in} - a_n|$$

and show that it goes to zero; next, we need to show that $g(X)$ is actually in A_1.

The first part is easy: we know that if $i, j > M$ we have $|a_{in} - a_{jn}| < \varepsilon$ for every n. Letting $j \to \infty$, it follows that if $i > M$ we have $|a_{in} - a_n| \leq \varepsilon$ for all n, which means that if $i > M$ we have

$$\|f_i(X) - g(X)\| = \max_n |a_{in} - a_n| \leq \varepsilon,$$

so that $f_i(X) \to g(X)$ with respect to $\| \ \|$.

For the second part, we use what we have just proved. For $i > M$, we know that $|a_{in} - a_n| < \varepsilon$ for every n. Now, since $f_i(X) \in A_1$, we know that $|a_{in}| \to 0$ as $n \to \infty$, i.e., that for each i there exists an M_i such that $|a_{in}| < \varepsilon$ for all $n > M_i$. Choose any $i > M$. Then if n is greater than the corresponding M_i, we have

$$|a_n| \leq |a_{in} - a_n| + |a_{in}| < \varepsilon + \varepsilon.$$

It follows that $a_n \to 0$ as $n \to \infty$.

Thus, $f_i(X) \to g(X)$ and $g(X) \in A_1$; since this works for any Cauchy sequence of power series, it shows that A_1 is complete. $\square$